Noncompact Lie Groups and Some of Their Applications

NATO ASI Series

Advanced Science Institutes Series

A Series presenting the results of activities sponsored by the NATO Science Committee, which aims at the dissemination of advanced scientific and technological knowledge, with a view to strengthening links between scientific communities.

The Series is published by an international board of publishers in conjunction with the NATO Scientific Affairs Division

A Life Sciences **B Physics**	Plenum Publishing Corporation London and New York
C Mathematical and Physical Sciences **D Behavioural and Social Sciences** **E Applied Sciences**	Kluwer Academic Publishers Dordrecht, Boston and London
F Computer and Systems Sciences **G Ecological Sciences** **H Cell Biology** **I Global Environmental Change**	Springer-Verlag Berlin, Heidelberg, New York, London, Paris and Tokyo

NATO-PCO-DATA BASE

The electronic index to the NATO ASI Series provides full bibliographical references (with keywords and/or abstracts) to more than 30000 contributions from international scientists published in all sections of the NATO ASI Series.
Access to the NATO-PCO-DATA BASE is possible in two ways:

– via online FILE 128 (NATO-PCO-DATA BASE) hosted by ESRIN, Via Galileo Galilei, I-00044 Frascati, Italy.

– via CD-ROM "NATO-PCO-DATA BASE" with user-friendly retrieval software in English, French and German (© WTV GmbH and DATAWARE Technologies Inc. 1989).

The CD-ROM can be ordered through any member of the Board of Publishers or through NATO-PCO, Overijse, Belgium.

Series C: Mathematical and Physical Sciences - Vol. 429

Noncompact Lie Groups and Some of Their Applications

edited by

Elizabeth A. Tanner

and

Raj Wilson
Division of Mathematics,
University of Texas,
San Antonio, Texas, U.S.A.

Springer Science+Business Media, B.V.

Proceedings of the NATO Advanced Research Workshop on
Noncompact Lie Groups and Their Physical Applications
San Antonio, Texas, U.S.A.
January 4–8, 1993

A C.I.P. Catalogue record for this book is available from the Library of Congress.

ISBN 978-94-010-4470-7 ISBN 978-94-011-1078-5 (eBook)
DOI 10.1007/978-94-011-1078-5

Printed on acid-free paper

---

in honour of
Asim Orhan Barut

CONTENTS

Preface xi

1. E. A. Tanner and R. Wilson 1
Noncompact Lie groups, their algebras and some of their applications.

Lie Groups and Lie Algebras

2. S. Helgason 55
Harish-Chandra's **c**-*function. A mathematical jewel.*

3. E. Van Den Ban, M. Flensted-Jensen and H. Schlichtkrull 69
Basic harmonic analysis on pseudo-Riemannian symmetric spaces.

4. A. O. Barut 103
The extensions of space-time. Physics in the 8-dimensional homogeneous space $\mathcal{D} = SU(2,2)/K$.

5. P. Budinich 123
Ordinary - and momentum - space conformal compactifications: Some possible observable consequences.

6. J. Hilgert 141
Radon transform on halfplanes via group theory.

7. B. Speh 147
Analytic torsion and automorphic forms.

8. A. Figà-Talamanca 157
Diffusion on compact ultrametric spaces.

9. J.-P. Antoine 169
Generalized square integrability and coherent states.

10. P. Winternitz, M. A. del Olmo and M. A. Rodríguez 181
Maximal abelian subgroups of $SU(p,q)$ *and integrable Hamiltonian systems.*

11. A. Inomata and G. Junker 199
Path integrals and Lie groups.

12. T. Hirai 225
Representations of diffeomorphism groups and the infinite symmetric group.

13. M. Anoussis 239
Characters of Lie groups.

14. W. Schmid and K. Vilonen 243
Weyl group actions on Lagrangian cycles and Rossmann's formula.

15. E. Angelopoulos 251
Taylor formula, tensor products, and unitarizability.

16. G. W. Mackey 265
A connection between Lie algebra roots and weights and the Fock space construction.

17. D. J. Rowe 285
Applications of $Sp(3, \mathbf{R})$ in nuclear physics.

18. W. H. Klink 301
Nilpotent groups and anharmonic oscillators.

19. C. H. Conley 315
Extensions of the mass 0 helicity 0 representation of the Poincare group.

20. W. F. Heidenreich 325
Invariant causal propagators in conformal space.

21. R. F. Streater 333
Gauge groups, anomalies and non-abelian cohomology.

22. R. V. Moody and J. Patera 341
The E_8 family of quasicrystals.

23. R. O. Wells, Jr. and X. Zhou 349
Wavelet interpolation and approximate solutions of elliptic partial differential equations.

Lie Superalgebras and Lie Supergroups

24. V. Hussin and L. M. Nieto 367
From super Lie algebras to supergroups: Matrix realizations and the factorisation problem.

25. J. Mickelsson 373
Current algebras as Hilbert space operator cocycles.

26. J. Niederle 391
Non-linear realization technique - The most convenient way of deriving $N = 1$ supergravity.

27. L. O'Raifeartaigh 405
Toda systems as constrained linear systems.

Quantum Groups

28. A. Guichardet 413
On the definitions of the quantum group $\mathcal{U}_h(sl(2,k))$ *and the restricted dual of* $\mathcal{U}_h(sl(n,k))$.

29. C. Frønsdal 423
Universal T-*matrix for twisted quantum* $gl(N)$.

30. W. Pusz and S. L. Woronowicz 453
Unitary representations of quantum Lorentz group.

31. R. Giachetti 473
Contraction of quantum groups and lattice physics.

32. L. C. Biedenharn and M. Tarlini 487
A quantum Poincaré group and the Dirac-Coulomb problem.

PREFACE

During the past two decades representations of noncompact Lie groups and Lie algebras have been studied extensively, and their application to other branches of mathematics and to physical sciences has increased enormously. Several theorems which were proved in the abstract now carry definite mathematical and physical significance. Several physical observations which were not understood before are now explained in terms of models based on new group-theoretical structures such as dynamical groups and Lie supergroups. The workshop was designed to bring together those mathematicians and mathematical physicists who are actively working in this broad spectrum of research and to provide them with the opportunity to present their recent results and to discuss the challenges facing them in the many problems that remain. The objective of the workshop was indeed well achieved.

This book contains 31 lectures presented by invited participants attending the NATO Advanced Research Workshop held in San Antonio, Texas, during the week of January 3-8, 1993. The introductory article by the editors provides a brief review of the concepts underlying these lectures (cited by *author* [*]) and mentions some of their applications. The articles in the book are grouped under the following general headings: Lie groups and Lie algebras, Lie superalgebras and Lie supergroups, and Quantum groups, and are arranged in the order in which they are cited in the introductory article.

We are very thankful to Dr. Luigi Sertorio, the Programme Director of the NATO Advanced Research Workshop, for sponsoring the meeting and for his invaluable advice at each stage of its organization. The workshop was also partially supported by The University of Texas at San Antonio, Technical Concepts Corporation, and the St. Anthony Hotel. We should like to thank, respectively, Dr. Shair Ahmad, Mr. Charles Becker and Mrs. Michele Lock for their efforts in arranging financial support. We are also grateful to Drs. Asim Barut, Alain Guichardet, and Alessandro Figà-Talamanca for serving on the International Advisory Committee and giving us valuable advice. The St. Anthony Hotel, host for the worhshop, provided the participants with an outstanding academic environment accompanied by a very warm hospitality. We would like to thank each and every staff member of the hotel for their help during this workshop.

Finally, especially, we thank all the contributors and participants for making the workshop a lively and fruitful meeting and the book a very useful contribution to the field of noncompact Lie groups and Lie algebras.

We are very proud to publish this book in honour of Professor Asim Barut who, in fact, first recognized the need for this meeting.

Elizabeth Tanner
Raj Wilson
San Antonio, 25 December 1993

Noncompact Lie Groups, their Algebras and some of their Applications

Elizabeth A. Tanner and Raj Wilson

Abstract

In this introductory article we review briefly the underlying concepts of noncompact Lie groups and Lie algebras, Lie supergroups and Lie superalgebras, and of quantum groups. We also discuss some of their representations and applications.

1 Introduction

It is reasonably modest to say that for the last three decades the theory of Lie groups and Lie algebras as developed earlier by Lie, Killing and Elie Cartan has undergone considerable maturation, infiltrating many branches of mathematics and other sciences. In §2 we give an outline of the important basic results in Lie groups and Lie algebras and recapitulate some of their important applications.

Furthermore, starting in the late 1960's interest in the theory of graded Lie algebras, now more commonly called *Lie superalgebras*, grew rapidly among mathematicians and physicists. In §3 we discuss the basic concepts of these algebras and their respective *Lie supergroups*, then sketch in particular their application to gauge theories.

More recently, some very specific Hopf algebras, now referred to as *quantum groups*, emerged in the study of the quantum inverse scattering transform method. In §4 we define these algebras, discuss their salient properties and outline some of the recent results.

In this article the papers contained in this volume are cited as *author*[*]. We have attempted to provide a review of the underlying theory and some of the definitions for fundamental and relevant terms pertinent to the papers which follow. We have also provided an extensive list of references to supplement this introduction and, of course, further references may be found as provided by each author.

E. A. Tanner and R. Wilson (eds.), Noncompact Lie Groups and Some of Their Applications, 1–54.

2 Lie groups and Lie algebras

2.1 Introduction

In 1859 Arthur Cayley announced that "*.....descriptive geometry is all geometry*". Pursuing this notion Felix Klein after 13 years developed what is now called *the Erlangen Program*[58], in which he employed group-theoretic concepts to unify different kinds of geometries. In particular, he showed that one could obtain affine, Euclidean and non-Euclidean geometries from the group of projective (*descriptive*) transformations (see [2]). (For physical implications of the Erlangen Program, see [6d]). While Klein studied *discontinuous transformation groups*, Marius Sophus Lie studied the theory of *continuous transformation groups* [26], both of them working from a geometric standpoint. In 1854 Georg Friedrich Bernhard Riemann published his idea of a metric geometry, which later came to be known as *Riemannian geometry.* Generally lacking congruence transformations (isometries) this geometry fell outside the Erlangen Program. The Riemannian geometry and the Erlangen Program were later reconciled by Elie Cartan [17b,18] using the notion of *connection.*

Marius Sophus Lie was also motivated to study the relationship between continuous transformation groups (later known as *Lie groups*) and systems of differential equations whose solutions are left invariant under those groups. The local action of a Lie group can be represented by the action of linear differential operators which are infintesimal operators forming an *infintesimal Lie group* later known as the *Lie algebra* of the corresponding Lie group. It was observed that the properties of Lie groups are remarkably reflected by the properties of Lie algebras. Since physical problems are so often structured geometrically or as a system of differential equations, the concept of Lie groups and Lie algebras as mathematical tools play crucial roles in obtaining physical solutions.

We define a Lie group in the following way [46b,c]. A real (complex) Lie group is a group G which is also a real (complex) analytic manifold such that the mapping $(\sigma,\tau) \to \sigma\tau^{-1}$ of the product manifold $G \times G$ into G is real (complex) analytic. Let H be a subgroup of G. Then, by the works of Kuranishi and Yamabe, H is a *connected Lie subgroup* if and only if H is arcwise connected. Let $T_e(G)$ denote the tangent space of G at the identity element $e \in G$. Let $\rho \in G$. For a given tangent vector $\mathfrak{X} \in T_e(G)$ there exists a unique left invariant analytic vector field $\widetilde{\mathfrak{X}}$ on G such that $\widetilde{\mathfrak{X}}_e = \mathfrak{X}$ and is defined by

$$\widetilde{\mathfrak{X}} f(\rho) = \left\{ \frac{d}{dt} f(\rho(\gamma(t))) \right\}_{t=0} \quad \forall \rho \in G, \tag{1}$$

where $f \in C^\infty(G)$ and $\gamma(t) = \exp(tx)$ is a curve in G. Let $\mathfrak{L}(G)$ be the set of all left invariant vector fields on G. If $\widetilde{\mathfrak{X}}, \widetilde{\mathfrak{Y}} \in \mathfrak{L}(G)$ then $\left[\widetilde{\mathfrak{X}}, \widetilde{\mathfrak{Y}}\right] = (\widetilde{\mathfrak{X}}\widetilde{\mathfrak{Y}} - \widetilde{\mathfrak{Y}}\widetilde{\mathfrak{X}}) \in \mathfrak{L}(G)$. The mapping $\mathfrak{L}(G) \to T_e(G)$ is a linear isomorphism, and is defined by $\widetilde{\mathfrak{X}} \mapsto \widetilde{\mathfrak{X}}_e = \mathfrak{X}$.

Therefore $\left[\tilde{\mathfrak{X}},\tilde{\mathfrak{Y}}\right] \mapsto \left[\tilde{\mathfrak{X}},\tilde{\mathfrak{Y}}\right]_e$, which is denoted by $[\mathfrak{X},\mathfrak{Y}] \in T_e(G)$. The vector space $T_e(G)$ with the composition rule $(\mathfrak{X},\mathfrak{Y}) \to [\mathfrak{X},\mathfrak{Y}]$ forms an algebra $\mathfrak{g}$ over $\mathbf{R}$ called the *Lie algebra of the Lie group G*.

In general [82a,b], a Lie algebra over a field F of characteristic zero is a vector space $\mathfrak{a}$ over F equipped with a bilinear composition $(\mathfrak{X},\mathfrak{Y}) \to [\mathfrak{X},\mathfrak{Y}]$ $\forall\mathfrak{X},\mathfrak{Y} \in \mathfrak{a}$ such that

1. $0 = [\mathfrak{X},\mathfrak{X}]$ $\forall \mathfrak{X} \in \mathfrak{a}$.
2. $0 = [\mathfrak{X},[\mathfrak{Y},\mathfrak{Z}]] + [\mathfrak{Y},[\mathfrak{Z},\mathfrak{X}]] + [\mathfrak{Z},[\mathfrak{X},\mathfrak{Y}]]$ $\forall \mathfrak{X},\mathfrak{Y},\mathfrak{Z} \in \mathfrak{a}$. *(Jacobi identity)*

A Lie algebra $\mathfrak{g}$ over F is *semisimple* if its radical is zero (i.e. $\mathfrak{g}$ has no abelian ideals other than zero) and is *simple* if it is not abelian and its only ideals are zero and $\mathfrak{g}$. Every semisimple Lie algebra can be decomposed into a direct sum of simple Lie algebras. A Lie algebra $\mathfrak{g}$ is said to be *reductive* if the radical of $\mathfrak{g}$ coincides with the center of $\mathfrak{g}$. Whenever the Lie algebra $\mathfrak{g}$ of a connected Lie group G is semisimple, simple, solvable, nilpotent or abelian, then G also will be, respectively.

For every $\mathfrak{X} \in \mathfrak{g}$ the linear transformation $\mathfrak{g} \to \mathfrak{g}$ defined by $\mathfrak{Y} \mapsto [\mathfrak{X},\mathfrak{Y}]$ $\forall \mathfrak{Y} \in \mathfrak{g}$ is denoted by $\mathrm{ad}\mathfrak{X}$. The *Killing form* is given by $B(\mathfrak{X},\mathfrak{Y}) = \mathrm{Tr}(\mathrm{ad}\mathfrak{X}\cdot\mathrm{ad}\mathfrak{Y})$. By Weyl's theorem [19,92] a semisimple Lie algebra is called a *compact Lie algebra* if its Killing form on the algebra is negative definite or, equivalently, if there exists a positive definite quadratic form $(\cdot,\cdot)$ satisfying the condition

$$([\mathfrak{X},\mathfrak{Y}],\mathfrak{Z}) + (\mathfrak{Y},[\mathfrak{X},\mathfrak{Z}]) = 0. \tag{2}$$

All remaining Lie algebras are said to be *noncompact.* In a complex Lie algebra any invariant quadratic form is indefinite, thus all complex Lie algebras are noncompact.

A subalgebra $\mathfrak{h}$ of a Lie algebra $\mathfrak{g}$ is a *Cartan subalgebra* of $\mathfrak{g}$ if and only if $\mathfrak{h}$ is a maximal abelian subalgebra of $\mathfrak{g}$ such that for every $H \in \mathfrak{h}$ the endomorphism $\mathrm{ad}H$ of $\mathfrak{g}$ is semisimple (i.e. $\mathrm{ad}H$ is diagonalizable). The dimension of $\mathfrak{h}$ is called the *rank* of $\mathfrak{g}$.

In 1889 Killing [56] succeeded in classifying the simple Lie algebras over $\mathbf{C}$ into four *classical algebras* $(A_\ell(\ell \geq 1), B_\ell(\ell \geq 2), C_\ell(\ell \geq 3), D_\ell(\ell \geq 4))$ and five *exceptional algebras* $(E_\ell, \ell = 6,7,8; F_4, G_2)$, where ℓ is the rank of the algebra. This classification was made rigorous by Cartan [17a] and in 1914 he also gave a complete solution to the classification problem for Lie algebras over $\mathbf{R}$; later simplified by Gantmacher [32]. The classification of Lie algebras over an arbitrary field F has been developed by Jacobson [52]. Since compact semisimple Lie algebras over $\mathbf{R}$ are in one-to-one correspondence with their complexifications, the classification of compact simple Lie algebras over $\mathbf{R}$ reduces to that of simple Lie algebras over $\mathbf{C}$.

If a connected Lie group G has a compact Lie algebra over $\mathbf{R}$, then G is compact. Corresponding to the classical complex [compact real] simple Lie algebras $A_\ell, B_\ell, C_\ell, D_\ell$, the classical complex [compact] connected simple Lie groups [70] are $Sl(\ell+1,\mathbf{C})$, $SO(2\ell+1,\mathbf{C})$, $Sp(\ell,\mathbf{C})$, $SO(2\ell,\mathbf{C})$ [$SU(\ell+1)$, $SO(2\ell+1)$, $Sp(\ell)$, $SO(2\ell)$]. Exceptional complex (compact) simple Lie groups are defined similarly by their corresponding algebras.

2.2 Representations

Let G be a noncompact (locally compact) Lie group. Let V be a complex locally convex Hausdorff topological vector space [41,60,90], $\dim V \neq 0$. A *representation* of G (or a *topological G-module* or simply a *G-module*) is a pair (ρ, V), where $\rho \in \mathrm{Hom}(G, \mathrm{Aut}(V))$. The dimension of V is called the *dimension (degree)* of the representation $(\rho, \mathcal{V})$. The elements of $\rho(G)$ are called the *operators of the representation*, and the space V is called the *representation space*. $(\rho, \mathcal{V})$ is said to be *finitely generated* if there exists a finite set $V_f \subseteq V$ such that the span of the vectors $\rho(g)(v_f)$ $\forall v_f \in V_f$ is dense in V for every $g \in G$. A representation is called *finite-dimensional (infinite-dimensional)* if the representation space is finite-dimensional (infinite-dimensional) [15,67]. By a *continuous representation* of G we mean a representation $(\rho, \mathcal{H})$ of G, where

1. $\mathcal{H}$ is a separable complex Hilbert space, and
2. the mapping $G \times \mathcal{H} \to \mathcal{H}$ defined by $(g, v) \mapsto \rho(g)v$ $\forall g \in G, v \in \mathcal{H}$ is continuous.

For what follows, $(\rho, \mathcal{H})$ denotes a continuous representation. Let $\mathcal{H}'$ be a closed subspace of $\mathcal{H}$. A representation $(\rho, \mathcal{H}')$ is called a *subrepresentation* of $(\rho, \mathcal{H})$ provided $\mathcal{H}'$ is invariant under all operators $\rho(g) \in \rho(G) \subseteq \mathrm{Aut}(\mathcal{H})$. The representation $(\rho, \mathcal{H})$ is said to be *irreducible* if there does not exist any subrepresentations other than $(\rho, \{0\})$ and $(\rho, \mathcal{H})$, and a*unitary representation* if the operators in $\rho(G)$ are all unitary. By Schur's lemma an irreducible finite-dimensional representation of an abelian group is one-dimensional. Representations for which every invariant subspace admits an invariant complement are called *completely reducible.* A finite-dimensional unitary representation of any group is completely reducible. Finite-dimensional unitary representations of a finite group were studied by Frobenius and Schur during the early 20th centuary. In 1925 Hermann Weyl studied the finite-dimensional unitary representations of compact Lie groups. The theory of infinite-dimensional unitary representations was initiated by Eugene Wigner [94] in his work on the inhomogeneous Lorentz group, which was motivated by problems in quantum mechanics.

Let $(\rho_1, \mathcal{H}_1)$ and $(\rho_2, \mathcal{H}_2)$ be two representations. We say that they are *unitarily equivalent* if there exists a $\phi \in \mathrm{Iso}(\mathcal{H}_1, \mathcal{H}_2)$ satisfying

1. $\rho_2(g) \cdot \phi = \phi \cdot \rho_1(g)\ \forall g \in G$, and
2. the *isometry condition* $\|\phi(v)\| = \|v\|\ \forall v \in \mathcal{H}_1$.

If the isometry condition is not satisfied then the two representations are said to be *equivalent.* The set of all unitary irreducible representations of G is partitioned by unitary equivalence and the set of eqivalence classes is called the *unitary dual* of G, denoted by $\widehat{G}$.

By an *extension* of the representation $(\rho_1, \mathcal{H}_1)$ by the representation $(\rho_2, \mathcal{H}_2)$ we shall mean [10] a representation $(\rho, \mathcal{H} = \mathcal{H}_1 \oplus \mathcal{H}_2)$ and an exact sequence

$$\{0\} \longrightarrow \mathcal{H}_1 \xrightarrow{\phi_1} \mathcal{H} \xrightarrow{\phi_2} \mathcal{H}_2 \longrightarrow \{0\}$$

such that

1. $\rho(g)(v_1 \oplus v_2) = (\rho_1(g)(v_1) + A(g)(v_2)) \oplus \rho_2(g)(v_2)$ where $v_i \in \mathcal{H}_i$, $i = 1, 2$, and $A(g) : \mathcal{H}_2 \to \mathcal{H}_1$ is a bounded linear operator, and
2. ϕ_1 is an isometrey onto the subspace $\mathcal{H}_1 \oplus \{0\}$ and the quotient mapping $\mathcal{H}/\mathcal{H}_1 \to \mathcal{H}_2$ is an isometry.

As $\rho(g) \in \mathrm{Aut}(\mathcal{H})$, we have

$$\rho_1(g)A(g') + A(g)\rho_2(g') = A(gg')\ \forall g, g' \in G. \tag{3}$$

If $A(g) = \mathbf{0}\, \forall g \in G$, the extension is said to be *trivial.* A representation $(\rho, \mathcal{H})$ is called an *indecomposable representation* if

1. $(\rho, \mathcal{H})$ is not a trivial extension, and
2. there exists an ascending chain of closed invariant subspaces

$$\{0\} \subseteq \mathcal{H}_1 \subseteq \mathcal{H}_2 \subseteq \cdots \subseteq \mathcal{H}_n = \mathcal{H}$$

   such that the representations on the factor space $\mathcal{H}_m/\mathcal{H}_{m-1}$ are all irreducible.

By an *intertwining operator* for the representations $(\rho_1, \mathcal{H}_1)$ and $(\rho_2, \mathcal{H}_2)$ we mean a bounded linear operator $B : \mathcal{H}_1 \to \mathcal{H}_2$ such that $B\rho_1(g) = \rho_2(g)B\, \forall g \in G$. The set of all such intertwining operators forms a vector space, denoted by $\mathcal{R}((\rho_1, \mathcal{H}_1), (\rho_2, \mathcal{H}_2))$. Two representations are said to be *boundedly equivalent* if there exists an intertwining operator for them. As defined by Mackey [65f], two unitary representations $(\rho_1, \mathcal{H}_1)$, $(\rho_2, \mathcal{H}_2)$ are called *disjoint* if and only if no subrepresentation of $(\rho_1, \mathcal{H}_1)$ is equivalent to any subrepresentation of $(\rho_2, \mathcal{H}_2)$; and are called *quasi-equivalent* if no nonzero subrepresentation of$(\rho_1, \mathcal{H}_1)$ is disjoint from $(\rho_2, \mathcal{H}_2)$ and no nonzero subrepresentation of $(\rho_2, \mathcal{H}_2)$ is disjoint from $(\rho_1, \mathcal{H}_1)$. Thus, two inequivalent irreducible representations are always disjoint (see also [23d],[27]). Two equivalent representations are necessarily quasi-equivalent, and two irreducible quasi-equivalent representations are necessarily equivalent.

A representation $(\rho, \mathcal{H})$ of G is said to be the *direct sum* of representations $(\rho_i, \mathcal{H}_i)$ $i = 1, 2, \cdots$ of G if

1. all $\mathcal{H}_i$ are invariant subspaces of $\mathcal{H}$ such that

$$\mathcal{H} = \sum_i \oplus \mathcal{H}_i, \text{ and} \tag{4}$$

2. each $(\rho_i, \mathcal{H}_i)$ is a subrepresentation of $(\rho, \mathcal{H})$.

One may then write

$$(\rho, \mathcal{H}) = \sum_i \oplus (\rho_i, \mathcal{H}_i). \tag{5}$$

Where each subrepresentation $(\rho_i, \mathcal{H}_i)$ is irreducible, $(\rho, \mathcal{H})$ is *primary* if and only if $(\rho_i, \mathcal{H}_i)$ is equivalent to $(\rho_j, \mathcal{H}_j)$ for all i and j; and is *multiplicity free* if and only if the equivalence of $(\rho_i, \mathcal{H}_i)$ and $(\rho_j, \mathcal{H}_j)$ implies $i = j$.

Let G be a Lie group. Let $C_c(G)$ denote the space of all continuous complex -valued functions on G with compact support. As defined in [23b], the linear mapping of $C_c(G)$ into $\mathbf{R}$ given by $\mu_\ell : f \mapsto \int f \, d_\ell g$ is a *measure on* G, where $d_\ell g$ is a left invariant positive n-form on G. μ_ℓ is positive and is *left invariant*, that is, $\mu_\ell(f \circ L_x) = \mu_\ell(f) \; \forall x \in G$, $f \in C_c(G)$ and $L_x : g \mapsto xg$ is the left translation. Similarly one defines the right invariant positive measure $\mu_r : f \mapsto \int f \, d_r g$ on G, where $d_r g$ is a right invariant positive n-form on G. A Lie group G is called *unimodular* if the left invariant measure μ_ℓ is also right invariant, that is, if and only if $|\det \mathrm{Ad}(x)| = 1 \; \forall x \in G$. Semisimple Lie groups, connected nilpotent groups, and Lie groups G for which $\mathrm{Ad}(G)$ is compact are unimodular.

Let $\mathscr{B}$ be a Borel space with a positive measure μ. Let $\beta \in \mathscr{B}$. A μ - *measurable field of Hilbert spaces* is a pair $(\{\mathcal{H}(\beta)\}, \mathscr{V}) = \mathscr{E}$, where $\{\mathcal{H}(\beta)\}$ is a family of Hilbert spaces *indexed by* $\mathscr{B}$, and $\mathscr{V}$ is a set of *measurable vector fields* of $\mathscr{E}$, satisfying the conditions:

1. $\mathscr{V}$ is a subspace of the *continuous product* of all elements of $\{\mathcal{H}(\beta)\}$.

2. There exists a sequence $\{v_n\}$ in $\mathscr{V}$ such that for every $\beta \in \mathscr{B}$, $\{v_n(\beta)\}$ form a *total* sequence in $\mathcal{H}(\beta)$. By *total sequence* we mean that $\beta = 0$ is the only element in $\mathscr{B}$ for which $v_i(\beta) = \mathbf{0} \in \mathcal{H}(\beta) \; \forall i = 1, 2, \cdots$.

3. For every $v \in \mathscr{V}$, the function $\mathscr{B} \to \mathbf{C}$ defined by $\beta \mapsto ||v(\beta)||$ is μ-measurable.

4. Let u be an arbitrary vector field associated with $\mathscr{B}$. If for every vector field $v \in \mathscr{V}$, the function $\mathscr{B} \to \mathbf{C}$ defined by $\beta \mapsto (u(\beta), v(\beta))$ is μ-measurable, then $u \in \mathscr{V}$.

A vector field v is said to be *square-integrable* if

1. $v \in \mathscr{V}$, and
2. $\int_{\mathscr{B}} \|v(\beta)\|^2 \, d\mu(\beta) < \infty$.

If v_1 and v_2 are square-integrable, then $v_1 + v_2$ and λv , $\lambda \in \mathbf{C}$ are square-integrable, and the function $\beta \mapsto (v_1(\beta), v_2(\beta))$ is integrable. Define an inner product

$$(v_1, v_2) = \int_{\mathscr{B}} (v_1(\beta), v_2(\beta)) \, d\mu(\beta). \tag{6}$$

With respect to this inner product the square-integrable vector fields constitute a Hilbert space

$$\mathscr{H} = \int_{\mathscr{B}}^{\oplus} \mathcal{H}(\beta) \, d\mu(\beta) \tag{7}$$

called the *direct integral of* $\mathcal{H}(\beta)$. Also, an element $v \in \mathscr{H}$ is written as

$$v = \int_{\mathscr{B}}^{\oplus} v(\beta) \, d\mu(\beta). \tag{8}$$

If μ is a *standard measure* as defined in [23d], then $\mathscr{H}$ is separable. Let, for each $\beta \in \mathscr{B}$, $(\rho(\beta), \mathcal{H}(\beta))$ be a representation of G in $\mathcal{H}(\beta)$. The mapping $\beta \mapsto (\rho(\beta), \mathcal{H}(\beta))$, called *a field of representations of* G, is said to be *measurable* if, for each $g \in G$, the *field of operators* $\beta \to \rho(\beta)(g)$ is measurable. If the field of representations of G is measurable then one can construct, for every $g \in G$, a continuous unitary operator

$$\rho(g) = \int_{\mathscr{B}}^{\oplus} \rho(\beta)(g) \, d\mu(\beta) \tag{9}$$

on the Hilbert space $\mathscr{H}$. The mapping $g \mapsto \rho(g)$ is a unitary representation of G in $\mathscr{H}$, denoted by $(\rho, \mathscr{H})$. This unitary representation is called the *direct integral* of the family of unitary representations $\{(\rho(\beta), \mathcal{H}(\beta))\}$ and is denoted by

$$(\rho, \mathscr{H}) = \int_{\mathscr{B}}^{\oplus} (\rho(\beta), \mathcal{H}(\beta)) \, d\mu(\beta). \tag{10}$$

Let $(\rho, \mathcal{H})$ be decomposed as in (5). Then one may classify $(\rho, \mathcal{H})$ as one of three types of representations as a consequence of the classification of von Neumann algebras [23d,65f]:

1. A representation $(\rho, \mathcal{H})$ is of *type* I if it is quasi-equivalent to a multiplicity-free representation.

2. A representation $(\rho, \mathcal{H})$ is of *type* II if it is quasi-equivalent to a finite representation and if any representation quasi-equivalent to $(\rho, \mathcal{H})$ is of the form of a direct sum of two representations.

3. A representation $(\rho, \mathcal{H})$ is of *type* III if quasi-equivalence of two such representations implies equivalence.

Every representation quasi-equivalent to a representation of type I (II or III) is of type I (II or III). Every subrepresentation of a representation of type I (II or III) is of type I (II or III). Every direct sum $(\rho, \mathcal{H})$ of representations $(\rho_i, \mathcal{H}_i)$ of type I (II or III) is of type I (II or III). It is known that the representations of the compact groups, the locally compact commutative groups, the connected semisimple Lie groups, the connected nilpotent Lie groups and the connected real or complex linear algebraic groups are only of type I. That a connected semisimple Lie group is of type I is proved by Harish-Chandra [45a]. The regular representation of a discrete group - under certain restrictions - is of type II. By a *(left) regular representation* we mean the unitary representation $(\rho, \mathcal{H} = \mathcal{L}^2(X, \mu))$ satisfying

$$(\rho(g)v)(x) = v(g^{-1}x),$$

where X is a locally compact Hausdorff space, μ is a G-invariant Radon measure on X, $v \in \mathcal{L}^2(X,\mu)$, $x \in X$. A *(right) regular representation* is defined similarly. By definition a linear functional Φ defined on $C_c(X)$, the linear space of all real-valued continuous functions f on X having compact support, is a *Radon measure* if and only if the set $\{\Phi(h) : |h| \leq |f|, h \in C_c(X)\}$ is bounded for an arbitrary $f \in C_c(X)$. When $X = G/\{e\}$ there exists a non-zero left (right) invariant Radon measure called a *left (right) Haar measure.*

Let $\{\varphi_i\}$ denote a countable orthonormal basis for the separable Hilbert space $\mathcal{H}$. Then the function $\chi_\rho : G \to \mathbf{C}$ defined by

$$\chi_\rho(g) = \mathrm{Tr}\rho(g) = \sum_{i=1}^{\infty}(\varphi_i, \rho(g)\varphi_i) \tag{11}$$

is called the *character of the representation* $(\rho, \mathcal{H})$, where $(\rho, \mathcal{H})$ is finite-dimensional. If $(\rho, \mathcal{H})$ is infinite- dimensional and G is semisimple Lie group, Harish-Chandra [45b,c] has established the following:

If $(\rho, \mathcal{H})$ is a continuous unitary representation of G, then the character of $(\rho, \mathcal{H})$, denoted by $\Theta_\rho(f)$, is defined on the subspace $C_c^\infty(G)$ and is of the form

$$< \chi, f >= \int_G \chi_\rho(g) f(g)\, dg\,, \; f \in C_c^\infty\,, \tag{12}$$

where $\chi_\rho(g)$ is a measurable and totally integrable function on G. Two irreducible unitary representations are equivalent if and only if their characters are the same.

Let $(\rho, \mathcal{H})$ be a unitary representation of G on $\mathcal{H}$. Then the function

$$g \mapsto (x, \rho_g x)\,,\ g \in G \tag{13}$$

is *positive definite* for every $x \in \mathcal{H}$. Conversely, any positive definite function on a topological group G can be expressed as (13) for some unitary representation $(\rho, \mathcal{H})$. Consequently, using Kreĭn - Milman theorem [41,60,69], Gel'fand and Raikov [35] proved that every locally compact group G has *sufficiently many irreducible unitary representations.* The groups having sufficiently many finite-dimensional irreducible unitary representations are called *maximally almost periodic.* Any noncompact connected simple Lie group has no finite-dimensional irreducible unitary representation other than the trivial unit representation.

Let G be a locally compact Lie group satisfying the second countability axiom. Let K be a closed subgroup of G and $(\rho^K, \mathcal{H}_K)$ be a unitary representation of K on a separable Hilbert space $\mathcal{H}_K$. Let δ and Δ be modular fuctions on K and G respectively. (The *modular function* of G is a continuous homomorphism of G into multiplicative $\mathbf{R}_+ - \{0\}$). Then there exists a continuous positive function σ on G such that

$$\sigma(kg) = \frac{\delta(k)}{\Delta(k)}\sigma(g)\,,\ \forall k \in K\,,\, g \in G. \tag{14}$$

Let a group G act on a set X from the left (resp.right). Let $\mathfrak{B}$ be a *completely additive family of subsets of* X (that is, the union of a countable number of elements of $\mathfrak{B}$ is in $\mathfrak{B}$). Then a measure μ on X is said to be a *quasi-invariant measure with respect to* G if the measures $\gamma(g)\mu$ (resp.$\delta(g)\mu$) and μ are *equivalent* (that is, they have the same sets of measure zero) for every $g \in G$, where $\gamma(g)\mu$ (resp.$\delta(g)\mu$) is a measure on $\mathfrak{B}$ defined by

$$\begin{aligned}(\gamma(g)\mu)(gA) &= \mu(A)\ \forall g \in G \text{ and } gA \in \mathfrak{B}.\\ (\delta(g)\mu)(Ag) &= \mu(A)\ \forall g \in G \text{ and } Ag \in \mathfrak{B}.\end{aligned}$$

Let μ be any quasi-invariant measure with respect to G on the right K-coset space $K \setminus G = \{Kg : g \in G\}$. Let $\mathcal{H}_G$ be a vector space of weakly measurable functions $f : G \to \mathcal{H}_K$ such that

1. $f(kg) = \rho^K(k)f(g)\ \forall k \in K\,,\, g \in G.$

2. $||f||^2 = \int_{K\setminus G} ||f(g)||^2\, d\mu(Kg) \ < \infty$, which is well-defined due to the condition 1 above.

Thus $\mathcal{H}_G$ is a Hilbert space equipped with the norm given by the condition 2. In fact $\mathcal{H}_G$ is isomorphic to the Hilbert space $\mathcal{L}^2(K \setminus G,\, \mu,\, \mathcal{H}_K)$ of square integrable vector functions with domain in $K \setminus G$ and values in the Hilbert space $\mathcal{H}_K$. A unitary representation $(\rho^G, \mathcal{H}_G)$ of G on $\mathcal{H}_G$ is defined by, for $g_0, g \in G$,

$$(\rho^G(g_0)f)(g) = \sqrt{\frac{\sigma(gg_0)}{\sigma(g)}} f(gg_0) \in \mathcal{H}_K, \tag{15}$$

where

$$\frac{\sigma(gg_0)}{\sigma(g)} = \frac{d\mu((Kg)g_0)}{d\mu(Kg)} \tag{16}$$

is the *Radon-Nikodym derivative* [69] of the quasi-invariant measure μ on $K \setminus G$. The representation $(\rho^G, \mathcal{H}_G)$ is called *the unitary representation induced by the representation* $(\rho^K, \mathcal{H}_K)$ *of* K *on* $\mathcal{H}_K$ *(in the sense of Mackey* [65b,c]). These induced representations have the following properties:

1. If $(\rho^G, \mathcal{H}_G)$ is irreducible, then $(\rho^K, \mathcal{H}_K)$ is irreducible.

2. Let K and K' be two closed subgroups of G such that $K \subseteq K'$. Let $(\rho^{K'}, \mathcal{H}_{K'})$ be the representation of K' on $\mathcal{H}_{K'}$ induced by the unitary representation $(\rho^K, \mathcal{H}_K)$ of K on $\mathcal{H}_K$. Let $(\rho^G, \mathcal{H}_G)$ and $(\rho^G, \mathcal{H}_G)'$ be two unitary representations of G on $\mathcal{H}_G$ induced by the representations $(\rho^K, \mathcal{H}_K)$ and $(\rho^{K'}, \mathcal{H}_{K'})$, respectively. Then the induced representations $(\rho^G, \mathcal{H}_G)$ and $(\rho^G, \mathcal{H}_G)'$ of G on $\mathcal{H}_G$ are equivalent.

Let $\mathcal{M}$ be the set of all Borel subsets of $K \setminus G$, $M \in \mathcal{M}$, and χ_M the *characteristic function* of the set M. Let $\mathcal{P}(\mathcal{H}_G)$ be the subset of all *projections* in $\mathcal{L}(\mathcal{H}_G)$ - the set of all continuous linear operators on $\mathcal{H}_G$. Let $\mathcal{I} : \mathcal{M} \to \mathcal{P}(\mathcal{H}_G)$ be the mapping defined, for any $f \in \mathcal{H}_G$, by

$$(\mathcal{I}(M)f)(g) = \chi_M(Kg)f(g) \in \mathcal{H}_K\, , \, g \in G. \tag{17}$$

One can observe the following:

1. $\mathcal{I}$ is a weakly measurable function.

2. For $k \in K$

$$(\mathcal{I}(M)f)(kg) = \rho^K(k)(\mathcal{I}(M)f)(g) \in \mathcal{H}_K\, , \, k \in K. \tag{18}$$

3. and

$$\int_{K\setminus G} ||(\mathcal{I}(M)f)(g)||^2 \, d\mu(Kg) = \int_M ||f(g)||^2 \, d\mu(Kg) < \infty. \tag{19}$$

Therefore the function $\mathcal{I}(M)f \in \mathcal{H}_G$ by the definition of $\mathcal{H}_G$ and $\mathcal{I}$ satisfies the following properties:

1. $\mathscr{I}(\emptyset) = 0 \in \mathscr{P}(\mathcal{H}_G)$.

2. $\mathscr{I}(K \setminus G) = I \in \mathscr{P}(\mathcal{H}_G)$.

3. If $M_i \cap M_j = \emptyset$ for $i \neq j$, then

$$\mathscr{I}(\bigcup_{k=1}^{\infty} M_k) = \sum_{k=1}^{\infty} \mathscr{I}(M_k).$$

4. $\mathscr{I}(M_1 \cap M_2) = \mathscr{I}(M_1)\mathscr{I}(M_2) \forall M_1, M_2 \subseteq K \setminus G$.

5. For all $M \subseteq K \setminus G$, $g \in G$,

$$\mathscr{I}(Mg) = \rho^G(g)\mathscr{I}(M)\rho^G(g^{-1}).$$

Properties 1-3 define the function $\mathscr{I}$ as a *projection measure* (or a *spectral measure*) on $K \setminus G$. Consequently one can construct an operator $\mathscr{O}$ on $C_c(K \setminus G)$ as

$$\mathscr{O}(h) = \int_{K \setminus G} h(Kg)\, d\mathscr{I}(\{Kg\}), \ h \in C_c(K \setminus G).$$

Such a projection measure is called a (*the canonical*) *system of imprimitivity* on $K \setminus G$ with respect to $(\rho^G, \mathcal{H}_G)$. Thus as obtained by Mackey [65a] *the criterion of inducibility of the representation* $(\rho, \mathcal{H})$ *of* G *on* $\mathcal{H}$ *from the subgroup* K *of* G *is the existence of a system of imprimitivity on* $K \setminus G$ *with respect to* $(\rho, \mathcal{H})$, or, we say that *the representation* $(\rho, \mathcal{H})$ *admits a system of imprimitivity.*

Let us now consider irreducible representations of a complex semisimple Lie group G. There exist four main series of such representations:

1. Let B be a *Borel subgroup* of G. Let $\Big\{(\rho^B, \mathcal{H}_B)\Big\}$ be the set of all one-dimensional *unitary representations* of B. A *principal series* consists of unitary representations of G induced from the members of $\Big\{(\rho^B, \mathcal{H}_B)\Big\}$ [65d].

2. A *degenerate series* consists of unitary representations of G induced from the set $\{(\rho^{\mathscr{P}}, \mathcal{H}_{\mathscr{P}})\}$ of all one-dimensional *unitary representations* of a *parabolic subgroup* $\mathscr{P} \neq B$ of G.

3. A *complementary principal series* consists of irreducible unitary representations of G induced from the set $\Big\{(\rho_0^B, \mathcal{H}_B^0)\Big\}$ of all one-dimensional *non-unitary representations* of a *Borel subgroup* B of G.

4. A *complementary degenerate series* consists of irreducible unitary representations of G induced from the set $\{(\rho_0^{\mathscr{P}}, \mathcal{H}_{\mathscr{P}}^0)\}$ of all one-dimensional *non-unitary representations* of a *parabolic subgroup* $\mathscr{P} \neq B$ of G.

The four series of representations of complex semisimple Lie groups have been studied extensively by Gel'fand and Naimark [34], and by Stein [83].

In general, the classification of irreducible unitary representations of the real semisimple Lie group is more complicated than in the complex semisimple case. Similar to the case of a complex semisimple Lie group, a connected real semisimple Lie group G has four series of irreducible unitary representations. However, if G has no parabolic subgroup other than a minimal parabolic subgroup (Borel subgroup) B, then clearly G has no degenerate series of representations. Irreducible unitary representations arising from the irreducible decomposition of the regular representation are called representations in the *principal series.* The principal series of G is divided into subseries. A subseries is called the *(principal) continuous series* if G has a *Cartan subgroup* whose vector part has maximal dimension. In this subseries, unitary representations can be obtained as unitary representations induced by finite-dimensional irreducible unitary representations of the Borel subgroup. A connected semisimple Lie group G has a *square integrable representation* if and only if G has a compact Cartan subgroup, or equivalently, if and only if $\mathrm{rank}G = \mathrm{rank}K$, where K is the maximal compact subgroup of G. We say that the representation $(\rho, \mathcal{H})$ is square integrable if any one of the following two mutually equivalent conditions holds:

1. There exist non-zero elements $v_1, v_2 \in \mathcal{H}$ such that

$$\int_G |(v_1, \rho(g)v_2)|^2 \, dg \;<\; \infty.$$

2. There exists a closed subspace V of $\mathcal{L}^2(G)$ which is stable under the right-regular representation $(\rho^r, \mathcal{L}^2(G))$ of G on $\mathcal{L}^2(G)$ such that the representation $(\rho, \mathcal{H})$ is equivalent to $(\rho^{\check{r}}, V)$ where $\rho^{\check{r}}(g) = \rho^r(g)|V$.

The set of all square integrable representations of G is called the *discrete series* of irreducible unitary representations and is a subseries of the principal series. It was Bargmann [5] who first determined certain *continuous* (denoted by C_q^0, $q > 0$) and *discrete* (denoted by $\mathcal{D}_k^-$, $\mathcal{D}_k^+$, $k \geq 2$) series of irreducible representations of $SL(2, \mathbf{R})$ (the three dimensional Lorentz group). The representations $\mathcal{D}_1^-$ and $\mathcal{D}_1^+$ are *not* square-integrable and are known to be *limits of discrete series.* The representations in the discrete series were classified by Harish-Chandra [45d] (see also [29b]). Furthermore, there exist representations in the principal series that are neither continuous nor discrete. Such representations may be constructed by combining a discrete series of representation with an induced representation of a subgroup. For an arbitrary semisimple Lie group, Schmid [81a,b] constructed most of the discrete series of representations by using Langlands' generalization [62a,b] of the Borel-Weil-Bott Theorem. In 1975, he [81c] showed how to characterize the discrete series in terms of their restrictions to a maximal compact subgroup K.

Let $(\rho, \mathcal{H})$ be a continuous representation of a real semisimple Lie group G on $\mathcal{H}$. Let K be the maximal compact subgroup of G. A vector $v \in \mathcal{H}$ is said to be *K-finite* if

$$\dim \operatorname{span}(\rho(K)v) < \infty.$$

Let Ω be the set of all equivalence classes of finite-dimensional irreducible representations of K. Let $(\kappa, V_K) \in \Omega$ be a fixed equivalence class of irreducible representations of K (or a *K-module*). A K-module is *locally finite* if every element in V_K is K-finite. Let $\omega \in \mathcal{R}((\kappa, V_K), (\rho|K, \mathcal{H}))$ be an intertwining operator. Then we say that the representation $(\rho, \mathcal{H})$ is *admissible* (or Harish-Chandra's *permissible*) if

$$\dim \bigcup_{\omega} \omega(V_K) < \infty$$

for all equivalence classes of irreducible representations of K. Harish-Chandra proved that any irreducible unitary representation of G is admissible. Let Θ_ρ be the character of $(\rho, \mathcal{H})$ (see (12)). Then we say that the representation $(\rho, \mathcal{H})$ is a *tempered representation* if the character Θ_ρ is *tempered* (as a distribution in the sense of Schwartz) [45b].

Let $(\rho, \mathcal{H})$ be a representation of G on $\mathcal{H}$. Let $\mathcal{U}$ be a linear subspace of $\mathcal{H}$. We say that $\mathcal{U}$ is *differentiable* if there exists a representation $(\pi^{\mathcal{U}}, \mathcal{U})$ of the algebra $\mathfrak{g}_0$ of G on $\mathcal{U}$ such that

$$\pi^{\mathcal{U}}(\mathfrak{X})(\mathfrak{u}) = \lim_{t \to 0}\left(\frac{1}{t}\right)(\rho(e^{t\mathfrak{X}})\mathfrak{u} - \mathfrak{u}), \quad \mathfrak{X} \in \mathfrak{g}_0, \ \mathfrak{u} \in \mathcal{U}. \tag{20}$$

Since by this definition, the sum of two differentiable spaces is again differentiable, there exists a largest differentiable space in $\mathcal{H}$, namely the union of all differentiable subspaces in $\mathcal{H}$. We say that an element $v \in \mathcal{H}$ is differentiable under ρ if it lies in this union.

Let $\tilde{V}$ be the Garding subspace of $\mathcal{H}$ with respect to ρ. Let $(\pi^{\tilde{V}}, \tilde{V})$ be the (Garding) representation of the *universal enveloping algebra* $\mathfrak{U}(\mathfrak{g})$ of the complexification $\mathfrak{g}$ of the algebra $\mathfrak{g}_0$ of G. The representation $(\rho, \mathcal{H})$ is called *quasi-simple* if

1. $(\rho, \mathcal{H})$ is admissible, and

2. $\exists \chi \in \operatorname{Hom}(\mathcal{Z}(\mathfrak{U}(\mathfrak{g})), \mathbf{C})$ such that $\pi^{\tilde{V}}(\mathfrak{z})v = \chi(\mathfrak{z})v \ \forall \mathfrak{z} \in \mathcal{Z}(\mathfrak{U}(\mathfrak{g}))$, $v \in V$,

where $\mathcal{Z}(\mathfrak{U}(\mathfrak{g}))$ denotes the center of $\mathfrak{U}(\mathfrak{g})$ and χ is called the *infinitesimal character* of $(\rho, \mathcal{H})$.

Let $(\rho, \mathcal{H})$ be a representation of G on $\mathcal{H}$. Let $v \in \mathcal{H}$. We say that v is *well-behaved (an analytic vector)* under ρ if the mapping $G \to \rho(G)(v)$ is analytic (that is, the set of elements in $\mathcal{H}$ which can be written as finite linear combinations of elements of the form $\int_G f(x)\rho(x)v\,dx$ where $v \in \mathcal{H}$, $f \in C_c^\infty(G)$, and dx is the left invariant Haar measure on G). Let $\mathcal{W}$ denote the set of all well-behaved elements in $\mathcal{H}$. By definition $\mathcal{W}$ is differentiable. For any $(\kappa, V_K) \in \Omega$ let $\mathcal{H}_{(\kappa,V_K)}$ be the set of all elements $v \in \mathcal{H}$ with the property that there exists a finite-dimensional linear subspace $\mathcal{V} \subseteq \mathcal{H}$ invareiant under $\rho(K)$ and containing v such that the representation $(\kappa, \mathcal{V})$ of K induced on every simple subspace $\mathcal{V}$ lies in (κ, V_K). That is, $\mathcal{H}_{(\kappa,V_K)}$ is the set (subspace) of all elements in $\mathcal{H}$ which transform under $\kappa(K)$ according to (κ, V_K). One can show [45a] that

$$\dim\mathcal{H}_{(\kappa,V_K)} < \infty \ \forall(\kappa, V_K) \in \Omega.$$

Since $\mathcal{W}$ is stable under $\rho(G)$, one obtains

$$\mathcal{W} \cap \sum_{(\kappa,V_K)\in\Omega} \mathcal{H}_{(\kappa,V_K)} = \sum_{(\kappa,V_K)\in\Omega} \mathcal{W} \cap \mathcal{H}_{(\kappa,V_K)}.$$

This leads to the following theorem due to Harish-Chandra:

Let $(\rho, \mathcal{H})$ be an admissible representation of G on $\mathcal{H}$. Let $\mathcal{W}$ be the set of all well-behaved elements in $\mathcal{H}$. Then the space $\sum_{(\kappa,V_K)\in\Omega} \mathcal{W} \cap \mathcal{H}_{(\kappa,V_K)}$ is dense in $\mathcal{H}$. Let V be a linear subspace of $\mathcal{H}$ invariant under $\kappa(K)$. Then if $\sum_{(\kappa,V_K)\in\Omega} V \cap \mathcal{H}_{(\kappa,V_K)}$ is dense in $\mathcal{H}$, $\mathcal{H}_{(\kappa,V_K)} = \mathrm{Cl}(V \cap \mathcal{H}_{(\kappa,V_K)})$, where Cl *denotes closure in $\mathcal{H}$.*

Let, as for groups, the pair $(\pi, \mathcal{W})$ denote the representation of $\mathfrak{U}(\mathfrak{g})$ on $\mathcal{W}$. Now we consider two admissible representations $(\rho_i, \mathcal{H}_i)$ $i = 1, 2$, of G on $\mathcal{H}_i$. Let $\mathcal{W}_i$ be the respective set of well-behaved elements in $\mathcal{H}_i$. Let $(\pi_i, \mathcal{W}_i)$ be the representations of $\mathfrak{U}(\mathfrak{g})$ on $\mathcal{W}_i$. As seen by Harish-Chandra

$$\mathcal{H}_i^0 = \sum_{(\kappa,V_K)} \mathcal{W}_i \cap (\mathcal{H}_i)_{(\kappa,V_K)}$$

is dense in $\mathcal{H}_i$ and is stable under $\pi_i(\mathfrak{U}(\mathfrak{g}))$. Let $(\pi_i^0, \mathcal{H}_i^0)$ be the representations of $\mathfrak{U}(\mathfrak{g})$ induced on $\mathcal{H}_i^0$. We say that the representations $(\rho_1, \mathcal{H}_1)$ and $(\rho_2, \mathcal{H}_2)$ are *infinitesimally equivalent* if there exists a linear isomorphism $\alpha : \mathcal{H}_1^0 \to \mathcal{H}_2^0$ such that

$$\pi_2^0(\mathfrak{X})\alpha(v) = \alpha(\pi_1^0(\mathfrak{X})v) \ \forall \mathfrak{X} \in \mathfrak{U}(\mathfrak{g}) \ , \ v \in \mathcal{H}_1^0.$$

Indeed if $(\pi_1, \mathcal{H}_1)$ and $(\pi_2, \mathcal{H}_2)$ are equivalent then they are also infinitesimally equivalent although the converse is not generally true. However, for irreducible unitary representations on Hilbert spaces these two concepts of equivalence actually coincide.

2.3 Harish-Chandra and Verma modules

Let us now consider V to be a complex locally convex Hausdorff topological space. A $(\mathfrak{g}, K)$-*module* is a complex vector space, denoted by a triple (π, ρ, V) which is associated with

1. a $\mathfrak{g}$-module - a representation (π, V) of $\mathfrak{g}$ on V, and
2. a locally finite K-module - a representation (ρ, V) of K on V,

such that the operations of $\pi(\mathfrak{g})$ and $\rho(K)$ satisfy the following conditions:

1. $\rho(k)\pi(\mathfrak{X})v = \pi(\mathrm{Ad}(k)(\mathfrak{X}))\rho(k)v$, $k \in K$, $\mathfrak{X} \in \mathfrak{U}(\mathfrak{g})$, $v \in V$.
2. If V_f is a K-stable finite-dimensional subspace of V, then the representation (ρ, V_f) of K on V_f is differentiable (see (20)) and has $\pi|\mathfrak{k}$ as its differential, where $\mathfrak{k}$ is the algebra corresponding to the group K.

Let $\mathcal{H}_{\tilde{K}} = \{v \in \mathcal{H} : v \text{ is } K\text{-finite}\}$. Then the $(\mathfrak{g}, K)$-module $(\pi, \rho, \mathcal{H}_{\tilde{K}})$ associated with an admissible representation $(\rho, \mathcal{H})$ of G on $\mathcal{H}$ is called the *Harish-Chandra module* of $(\rho, \mathcal{H})$. One can prove that every irreducible $(\mathfrak{g}, K)$-module is the Harish-Chandra module of an irreducible admissible representation of G.

Let (π, V) be a $\mathfrak{g}$-module and (π', V') be an irreducible $\mathfrak{g}$-module. The sum of all sub $\mathfrak{g}$-modules of (π, V) isomorphic to (π', V') is a sub $\mathfrak{g}$-module of (π, V) called the *isotypic component of type* (π', V') *of* (π, V). Every irreducible sub $\mathfrak{g}$-module of this sum is isomorphic to (π', V'). A $(\mathfrak{g}, K)$-module is *admissible* if all the isotypic components of the representation (ρ, V) of K on V are finite-dimensional. Two $(\mathfrak{g}, K)$-modules (π', ρ', V') and (π'', ρ'', V'') are said to be *equivalent* if there exists an invertible intertwining operator between them.

Let $\mathfrak{g} = \mathfrak{n}_- \oplus \mathfrak{h} \oplus \mathfrak{n}_+$ be the *triangular decompsition* of $\mathfrak{g}$, [14,48] where $\mathfrak{h} \oplus \mathfrak{n}_+ = \mathfrak{b}$ (*Levi decomposition of* $\mathfrak{b}$) is a solvable subalgebra - the *(canonical) Borel subalgebra* of $\mathfrak{g}$ and $\mathfrak{h}$ is a *Cartan subalgebra* of $\mathfrak{g}$. By the Birkhoff-Witt Theorem we have $\mathfrak{U}(\mathfrak{g}) \cong \mathfrak{U}(\mathfrak{n}_-) \otimes_{\mathbf{C}} \mathfrak{U}(\mathfrak{b})$. Let $\mathfrak{h}^*$ be the *linear dual* of $\mathfrak{h}$ endowed with a lexicographic linear ordering. Let W be the *Weyl group* of $\mathfrak{g}$ relative to $\mathfrak{h}$, that is, W is represented faithfully as a permutation group over the finite set

$$\Delta = \{\lambda \in \mathfrak{h}^* : \lambda \neq 0, \mathfrak{g}_\lambda \neq \{0\}\} \tag{21}$$

called the *root system of* $\mathfrak{g}$ *relative to* $\mathfrak{h}$, where for every $\lambda \in \mathfrak{h}^*$,

$$\mathfrak{g}_\lambda = \{\mathfrak{X}_\mathfrak{g} \in \mathfrak{g} : \mathrm{ad}(\mathfrak{X}_\mathfrak{h})\mathfrak{X}_\mathfrak{g} = \lambda(\mathfrak{X}_\mathfrak{h})\mathfrak{X}_\mathfrak{g} \ \forall \mathfrak{X}_\mathfrak{h} \in \mathfrak{h}\} \tag{22}$$

is the *root subspace* corresponding to λ. Thus,

$$\mathfrak{g} = \mathfrak{h} \oplus \sum_{\lambda \in \Delta} \mathfrak{g}_\lambda \ ; \ \mathfrak{g}_0 = \mathfrak{h}. \tag{23}$$

Let $(\pi^\mathfrak{g}, V)$ be an irreducible representation of $\mathfrak{g}$ (a $\mathfrak{g}$-module). Let $(\pi^{\mathfrak{U}(\mathfrak{g})}, V)$ be a $\mathfrak{U}(\mathfrak{g})$-module. For any $\lambda \in \mathfrak{h}^*$ define as in [23c] the linear subspace

$$V_\lambda = \{v \in V : \pi^\mathfrak{g}(\mathfrak{X}_\mathfrak{h})v = \lambda(\mathfrak{X}_\mathfrak{h})v \ \forall \mathfrak{X}_\mathfrak{h} \in \mathfrak{h}\}$$

which is stable under $\pi|\mathfrak{h}$. If $V_\lambda \neq \{\mathbf{0}\}$, then λ is called a *weight* of the representation $\pi^{\mathfrak{g}}$ relative to $\mathfrak{h}$ and $\lambda\,[\mathfrak{h},\mathfrak{h}] = 0$. The $\dim V_\lambda$ is called the *multiplicity of the weight* λ. The set of all weights of $\pi^{\mathfrak{g}}$ is a finite W-stable subset of $\mathfrak{h}^*$. Denote this set by $\{\lambda_1, \lambda_2, \cdots, \lambda_\ell\}$, ℓ being the rank of $\mathfrak{g}$. Then $V = \oplus_{i=1}^{\ell} V_{\lambda_i}$. The maximal element in the set $\{\lambda_1, \cdots, \lambda_\ell\}$ with respect to the given ordering of $\mathfrak{h}^*$ is denoted by Λ and is called the *highest weight* of $(\pi^{\mathfrak{g}}, V)$. The multiplicity of Λ is one, that is, $\dim V_\Lambda = 1$. Every weight λ appears with finite multiplicity and is of the form

$$\lambda = \Lambda - \sum_{\alpha\in\Delta+} n_\alpha \alpha,$$

where Λ is a highest weight, $\Delta_+ \subset \Delta$ is the set of all positive roots, and n_α is a non-negative integer. We shall call a weight λ_e of $(\pi^{\mathfrak{g}}, V)$ *extreme* if there exists no $\lambda \in \Delta$ and $w \in W$ such that $\lambda_e \pm w(\lambda)$ are both weights of $(\pi^{\mathfrak{g}}, V)$. Clearly a highest weight Λ is extreme.

Let $(\,,\,)$ denote the inverse Killing form on $\mathfrak{h}^*$. Then

$$\mathfrak{h}^*_{\mathbf{Z}} = \left\{\lambda \in \mathfrak{h}^* : \frac{2(\lambda, \alpha)}{(\alpha, \alpha)} \in \mathbf{Z}\ \forall \alpha \in \Delta\right\}$$

is the *lattice of integral elements* in $\mathfrak{h}^*$, and λ is said to be an *integral form* if $\lambda \in \mathfrak{h}^*_{\mathbf{Z}}$. Also, let

$$\mathfrak{h}^*_+ = \{\lambda \in \mathfrak{h}^* : (\lambda, \alpha) \geq 0\ \forall \lambda \in \Delta^+\}.$$

We say that λ is *dominant* in $\mathfrak{h}^*$ if and only if $\lambda \in \mathfrak{h}^*_+$, in other words, if and only if $w(\lambda) \leq \lambda\ \forall w \in W$. As seen by Cartan, Weyl, and Harish-Chandra, there exists an irreducible representation $(\pi(\Lambda), V_\Lambda)$ of $\mathfrak{g}$ with Λ as its highest weight if and only if Λ is a dominant integral form. Thus there exists a bijection between $\mathfrak{h}^*_+$ and isomorphism classes of irreducible $\mathfrak{g}$-modules which are associated with a highest weight. The isomorphism classes are established by the Cartan Theorem: *Two irreducible representations* $(\pi^{\mathfrak{g}}(\Lambda_1), V_{\Lambda_1})$ and $(\pi^{\mathfrak{g}}(\Lambda_2), V_{\Lambda_2})$ *of* $\mathfrak{g}$ *associated with highest weights* Λ_1 *and* Λ_2 *respectively are equivalent if and only if* $\Lambda_1 = \Lambda_2$. Furthermore, Harish-Chandra proved the following:

1. *Let* $(\pi^{\mathfrak{U}(\mathfrak{g})}, V)$ *be an irreducible representation of* $\mathfrak{U}(\mathfrak{g})$ *on a finite-dimensional space* $V \neq \{\mathbf{0}\}$. *Then* $\pi^{\mathfrak{U}(\mathfrak{g})}$ *has a highest weight* Λ *which is a dominant integral form, and* $V = \sum_\lambda V_\lambda$, *where the summation is over all the weights* λ *of* $(\pi^{\mathfrak{U}(\mathfrak{g}}, V)$.

2. *Every finite-dimensional irreducible representation of* $\mathfrak{U}(\mathfrak{g})$ *is equivalent to some representation* $(\pi(\Lambda), V_\Lambda)$ *where* Λ *is a dominant integral form on* $\mathfrak{h}^*$.

Now, let $\lambda \in \mathfrak{h}^*$. We define an action of $\mathfrak{b}$ on $\mathbf{C}$ by

$$\begin{aligned} \mathfrak{X}_{\mathfrak{h}} z &= ((\lambda - \delta)(\mathfrak{X}_{\mathfrak{h}}))z\,,\mathfrak{X}_{\mathfrak{h}} \in \mathfrak{h}\,,\, z \in \mathbf{C}\,,\, \delta = \frac{1}{2}\sum_{\alpha\in\Delta_+} \alpha \in \mathfrak{h}^*_{\mathbf{Z}} \\ \mathfrak{N} z &= 0\,,\ \mathfrak{N} \in \mathfrak{n}_+. \end{aligned} \tag{24}$$

This defines a one-dimensional representation of $\mathfrak{b}$ with highest weight $(\lambda-\delta)$. We denote the respective $\mathfrak{U}(\mathfrak{b})$-module by $\mathbf{C}_{\lambda-\delta}$. Also, under multiplication $\mathfrak{U}(\mathfrak{g})$ is a left $\mathfrak{U}(\mathfrak{g})$-module as well as a right $\mathfrak{U}(\mathfrak{b})$-module. The left $\mathfrak{U}(\mathfrak{g})$-module

$$\begin{aligned} M(\lambda) &= \mathfrak{U}(\mathfrak{g}) \otimes_{\mathfrak{U}_{\mathfrak{b}}} \mathbf{c}_{\lambda-\delta}, \\ &\cong (\mathfrak{U}(\mathfrak{n}_-) \otimes_{\mathbf{C}} \mathfrak{U}(\mathfrak{b})) \otimes_{\mathfrak{U}_{\mathfrak{b}}} \mathbf{C}_{\lambda-\delta}, \\ &\cong \mathfrak{U}(\mathfrak{n}_-) \otimes_{\mathbf{C}} \mathbf{C}_{\lambda-\delta}, \\ &\cong \mathfrak{U}(\mathfrak{n}_-) \end{aligned} \tag{25}$$

is called the *Verma module* associated with $\mathfrak{g}, \mathfrak{h}, (\lambda-\delta)$. Thus if we denote $1_{\mathfrak{g}} \in \mathfrak{U}(\mathfrak{g})$ and $1 \in \mathbf{C}_{\lambda-\delta}$ as the respective identity elements, then $M(\lambda)$ is generated by one *canonical generator* $(1_{\mathfrak{g}} \otimes 1)$ with weight $(\lambda-\delta)$. That is,

$$M(\lambda) \cong \mathfrak{U}(\mathfrak{g})(1_{\mathfrak{g}} \otimes 1).$$

By the Birkhoff-Witt Theorem one obtains

$$M(\lambda) \cong \mathfrak{U}(\mathfrak{n}_-) \otimes 1.$$

In (25) the $\mathfrak{U}(\mathfrak{n}_-)$-module isomorphism is defined by $u \mapsto u \otimes 1 \ \forall u \in \mathfrak{U}(\mathfrak{n}_-)$. In fact, if we set the one-dimensional representation of $\mathfrak{b}$ with highest weight $(\lambda-\delta)$ to be $\rho(\mathfrak{X}_{\mathfrak{h}} + \mathfrak{N}) = (\lambda-\delta)\mathfrak{X}_{\mathfrak{h}}$ for $\mathfrak{X}_{\mathfrak{h}} \in \mathfrak{h}$ and $\mathfrak{N} \in \mathfrak{n}_+$, then the representation induced on $\mathfrak{g}$ by ρ gives the Verma module. Verma [88] proved that *for* $\lambda, \mu \in \mathfrak{h}^*$,

$$\mathrm{Hom}_{\mathfrak{U}(\mathfrak{g})}(M(\lambda), M(\mu)) = 0 \text{ or } \mathbf{C},$$

where in the latter case every nontrivial homomorphism is an embedding. Furthermore, if $\mathfrak{p}$ is a *parabolic subalgebra* of $\mathfrak{g}$ containing a Borel subalgebra $\mathfrak{b}$, then one defines the *generalized Verma module* by

$$M_{\mathfrak{p}}(\lambda) = \mathfrak{U}_{\mathfrak{g}} \otimes_{\mathfrak{U}(\mathfrak{p})} F_{\mathfrak{p}}(\lambda),$$

where $F_{\mathfrak{p}}(\lambda)$ is a finite-dimensional irreducible $\mathfrak{p}$-module with highest weight λ.

Suppose (π, ρ, V) is a $(\mathfrak{g}, K)$-module, and (π', ρ', V') is an irreducible $(\mathfrak{g}, K)$ -module. We say that (π', ρ', V') is a *subquotient (or composition factor)* of (π, ρ, V) if there exist two submodules of (π, ρ, V), denoted by M_1 and M_2 such that $M_1/M_2 \cong (\pi', \rho', V')$. When (π, ρ, V) is an admissible $(\mathfrak{g}, K)$-module, the *multiplicity* of (π', ρ', V') in (π, ρ, V) is the largest integer n such that there exists a chain of submodules:

$$M_0 \subseteq M_1 \subseteq M_2 \subseteq \cdots \subseteq M_s = (\pi, \rho, V) \text{ with } M_i/M_{i-1} \cong (\pi', \rho', V')$$

for n distinct values of i. Two admissible $(\mathfrak{g}, K)$-modules (π_i, ρ_i, V_i), $i = 1, 2$ are said to have *equivalent composition series* if for every irreducible $(\mathfrak{g}, K)$-module (π', ρ', V'), the multiplicity of (π', ρ', V') in (π_1, ρ_1, V_1) is equal to the multiplicity of (π', ρ', V') in (π_2, ρ_2, V_2).

We now go back to the discussion of different series of representations of semisimple Lie groups. Harish-Chandra [45a] proved the subquotient theorem which states that any irreducible admissible representation of a semisimple Lie group must be infinitesimally equivalent to a composition factor of a principal series representation induced by a representation of a Borel subgroup. Želobenko [97] used this theorem in 1974 for the classification of admissible representations of complex semisimple Lie groups. Following the work of Želobenko and of Hirai [47a,b] on $SU(p,1)$, Langlands [62a,b] considered, for real semisimple Lie groups, not only principal series representations induced from a Borel subgroup but also the series induced from tempered representations on parabolic subgroups in general. He proved that any irreducible admissible representation can be realized as a composition factor of such a representation. The work of Knapp and Zuckerman [59a,b] on a one-to-one parametrization of representations combined with Langlands work gives an explicit parametrization of the irreducible admissible representations of a real semisimple Lie group.

Hirai [47b] has computed the characters of the principal series, and has given a method for computing the explicit form of the characters of discrete series. Wolf [95a-d] (see also [79]) has obtained geometric realization of the principal series on flag manifolds.

2.4 Harmonic Analysis

Let M be a C^∞-manifold [20] and $C^\infty(M)$ be the set of all real-valued differentiable functions on M. Let $\mathcal{D}^1(M)$ denote the set of all vector fields on M. An *affine connection* on M is a mapping $X \mapsto \nabla_X$ which assigns to every vector field $X \in \mathcal{D}^1(M)$ a linear mapping $\nabla_X : \mathcal{D}^1(M) \to \mathcal{D}^1(M)$ satisfying

$$\begin{aligned} \nabla_{fX+gY} &= f\nabla_X + g\nabla_Y, \\ \nabla_X(fY) &= f\nabla_X(Y) + (Xf)Y, \end{aligned}$$

where $f, g \in C^\infty(M)$, $X, Y \in \mathcal{D}^1(M)$ and fX and $X + Y$ denote the vector fields

$$\begin{aligned} fX &: g \mapsto f(X(g)), \\ X + Y &: g \mapsto X(g) + Y(g). \end{aligned}$$

Hence $\mathcal{D}^1(M)$ is a $C^\infty(M)$-module (module over $C^\infty(M)$). The operator ∇_X is called *covariant differentiation* with respect to X. Now, let M be a C^∞-manifold with an affine connection ∇. Let $\mathcal{D}_1(M)$ denote the dual of $\mathcal{D}^1(M)$ consisting of all *differential 1-forms* on M. Each differential 1-form is a mapping $\mathcal{D}^1(M) \to C^\infty(M)$. The mapping $\tau : \mathcal{D}_1(M) \times \mathcal{D}^1(M) \times \mathcal{D}^1(M) \to C^\infty(M)$, called the *torsion tensor field* is defined by

$$\tau : (\omega, X, Y) \mapsto \omega(T(X, Y)),$$

where $\omega \in \mathcal{D}_1(M)$ and the *torsion tensor* $T : \mathcal{D}^1(M) \times \mathcal{D}^1(M) \to \mathcal{D}^1(M)$ is given by

$$\begin{aligned} T(X,Y) &= \nabla_X(Y) - \nabla_Y(X) - [X,Y] \ \forall X,Y \in \mathcal{D}^1(M) \\ &= -T(Y,X) \\ T(fX,gY) &= fgT(X,Y) \ ; \ f,g \in C^\infty(M). \end{aligned}$$

In fact τ is a tensor field of type $(1,2)$. Similarly the mapping $\rho : \mathcal{D}_1(M) \times \mathcal{D}^1(M) \times \mathcal{D}^1(M) \times \mathcal{D}^1(M) \to C^\infty(M)$, called the *curvature tensor field*, is defined by

$$\rho : (\omega, X, Y, Z) \mapsto \omega(R(X,Y) \cdot Z) \ \forall X,Y,Z \in \mathcal{D}^1(M),$$

where the *curvature tensor* R is given by ("$\cdot$" indicates the action of the operator $R(X,Y)$ on $Z \in \mathcal{D}^1(M)$)

$$\begin{aligned} R(X,Y) &= \nabla_X \nabla_Y - \nabla_Y \nabla_X - \nabla_{[X,Y]} \\ &= -R(Y,X) \\ R(fX,gY) \cdot hZ &= fghR(X,Y) \cdot Z \ ; \ f,g,h \in C^\infty(M). \end{aligned}$$

ρ is a tensor field of type $(1,3)$. If $T = 0$, then one obtains the *Bianchi identity*

$$R(X,Y) \cdot Z + R(Y,Z) \cdot X + R(Z,X) \cdot Y = 0.$$

Let, as before, M be a C^∞-manifold with an affine connection ∇ which has torsion tensor T and curvature tensor R. Let $m \in M$ and N_m denote an open neighbourhood of m. For every point $p \in N_m$, let $t \mapsto \gamma_p(t)$ be the geodesic in N_m passing through $m = \gamma_p(0)$ and $p = \gamma_p(1)$. Let $p' = \gamma_p(-1)$. The reflection mapping $S_m : N_m \to N_m$ defined by $S_m : p \mapsto p'$ is a local diffeomorphism of N_m onto itself and is called the *geodesic symmetry* with respect to m. M is called *affine locally symmetric* if for each $m \in M$ the local diffeomorphism S_m is an affine transformation. We say that S_m is the *local symmetry around* m. One can show [46c] that M is affine locally symmetric if and only if $T = 0$ and $\nabla_Z R = 0 \ \forall Z \in \mathcal{D}^1(M)$. The manifold M is called an *affine symmetric space* or briefly a *symmetric space* if S_m can be extended to an affine diffeomorphism of M. In this case we say that S_m is the *symmetry around* m.

Let M be as defined above. Let $p \in M$ and $X \in \mathcal{D}^1(M)$. Let $X_p : C^\infty(p) \to \mathbf{R}$ be a linear mapping defined by $X_p : f \mapsto (Xf)(p)$. (Note that one puts $(Xf)(p) = (X\tilde{f})(p)$, where $\tilde{f} \in C^\infty(M)$ such that f and $\tilde{f}$ coincide in a neighbourhood of p. See [46b]). The set $\{X_p : X \in \mathcal{D}^1(M)\} = M_p$ (or $\mathcal{D}^1(p)$) is the tangent space to M at p. A *Riemannian structure* on M is a tensor field $\sigma : \mathcal{D}^1(M) \times \mathcal{D}^1(M) \to C^\infty(M)$ of type $(0,2)$ satisfying

1. $\sigma(X,Y) = \sigma(Y,X) \ \forall X,Y \in \mathcal{D}^1(M)$.

2. For every point $p \in M$, $\sigma_p : M_p \times M_p \to C^\infty(M)$ is a nondegenerate bilinear form.

3. For every $p \in M$, σ_p is positive definite.

If condition (3) is not satisfied then one obtains a *pseudo-Riemannian structure.* A connected C^∞-manifold with a (pseudo)-Riemannian structure is called a *(pseudo)-Riemannian manifold.* If M is analytic and the tensor field σ is analytic, then M is called an *analytic (pseudo)-Riemannian manifold* [29c].

Let $\mathcal{E}^1$ denote the set of all complex vector fields on M, and let the complexification $C^\infty(M) + \sqrt{-1}C^\infty(M)$ be the set of all complex-valued differentiable functions on M. Let $\mathcal{E}_1 : \mathcal{E}^1 \to C^\infty(M) + \sqrt{-1}C^\infty(M)$ be the dual of $\mathcal{E}^1$ consisting of all *complex 1-forms* on M. As in the real case, a *complex tensor field* of type (r, s) is a map

$$\underbrace{\mathcal{E}_1 \times \mathcal{E}_1 \times \cdots \times \mathcal{E}_1}_{r\ times} \times \underbrace{\mathcal{E}^1 \times \mathcal{E}^1 \times \cdots \times \mathcal{E}^1}_{s\ times} \to C^\infty(M) + \sqrt{-1}C^\infty(M).$$

An *almost complex structure* on M (see [49]) is a tensor field $\nu : \mathcal{D}_1(M) \times \mathcal{D}^1(M) \to C^\infty(M)$ of type $(1,1)$ defined by

$$\nu(\omega, X) = \omega(J(x)) \ , \ \omega \in \mathcal{D}_1(M) \ , \ X \in \mathcal{D}^1(M),$$

where $J : \mathcal{D}^1(M) \to \mathcal{D}^1(M)$ is defined by

$$J(X) = \pm\sqrt{-1}X \ \forall X \in \mathcal{D}^1(M). \tag{26}$$

An *almost complex manifold* is denoted by the pair (M, J). A *torsion tensor field of the almost complex structure* J is a tensor field $\psi : \mathcal{D}_1(M) \times \mathcal{D}^1(M) \times \mathcal{D}^1(M) \to C^\infty(M)$ of type $(1,2)$ defined by

$$\psi(\omega, X, Y) = \omega(S(X, Y)),$$

where $S : \mathcal{D}^1(M) \times \mathcal{D}^1(M) \to \mathcal{D}^1(M)$ is defined by

$$\begin{aligned} S(X,Y) &= [X,Y] + J\,[JX,Y] + J\,[X,JY] - [JX,JY] \\ &= -S(Y,X). \end{aligned}$$

S is called the *torsion tensor of the almost complex structure* J and is related to the *Nijenhuis Tensor.* The almost complex structure J is said to be *integrable* if $S = 0$. By the Newlander-Nirenberg Theorem [71] it is known that if $S(X,Y) = 0 \ \forall X, Y \in \mathcal{D}^1(M)$, then M has a unique complex structure associated with J.

A Riemannian structure σ on M is said to be a *Hermitian structure* if

$$\sigma(JX, JY) = \sigma(X, Y) \ \forall X, Y \in \mathcal{D}^1(M),$$

and a *Kählerian structure* if in addition

$$\nabla_X J = J\nabla_X \ ; \ X \in \mathcal{D}^1(M).$$

Let $X, Y, Z \in \mathcal{D}^1(M)$, and $X_p, Y_p, Z_p \in M_p$. The mapping

$$Z_p \mapsto R_p(X_p, Z_p) \cdot Y_p = (R(X,Z) \cdot Y)_p \in M_p$$

is an endomorphism of M_p. The tensor field $\mathfrak{r}$ given by

$$\begin{aligned} (\mathfrak{r}(X,Y))(p) &= \mathfrak{r}_p(X_p, Y_p) \\ &= \mathrm{Trace}(Z_p \mapsto R_p(X_p, Z_p) \cdot Y_p) \end{aligned}$$

is called the *Ricci curvature or Ricci-form* of the affine connection.

Let G be a connected Lie group. Let θ be an involutive automorphism of G, G^θ be the set of all fixed points of θ, and $(G^\theta)_0$ be the identity component of G^θ. Let H be a closed subgroup of G such that $(G^\theta)_0 \subset H \subset G^\theta$. Then the pair (G, H) is called a *symmetric pair.* There exists a unique G-invariant affine connection on the factor space $\mathcal{X} = G/H$ which makes $\mathcal{X}$ a symmetric space. The symmetry around the point $gH \in \mathcal{X}$ is given by

$$S_{gH}(g'H) = g\theta(g^{-1}g')H.$$

Let $\mathfrak{g}$ denote the Lie algebra of G. Let

$$\begin{aligned} \mathfrak{h} &= \{\mathfrak{X} \in \mathfrak{g} : \theta(\mathfrak{X}) = \mathfrak{X}\} \text{ and} \\ \mathfrak{q} &= \{\mathfrak{X} \in \mathfrak{g} : \theta(\mathfrak{X}) = -\mathfrak{X}\}. \end{aligned} \tag{27}$$

Then $\mathfrak{g} = \mathfrak{h} + \mathfrak{q}$ and

$$[\mathfrak{h}, \mathfrak{h}] \subseteq \mathfrak{h}\ ,\ [\mathfrak{h}, \mathfrak{q}] \subseteq \mathfrak{q}\ ,\ [\mathfrak{q}, \mathfrak{q}] \subseteq \mathfrak{h}.$$

$\mathfrak{h}$ is the Lie algebra of H, and $\mathfrak{q}$ may be identified with the tangent space $T_{eH}(\mathcal{X})$. Similarly, under *Cartan involution* θ_c with $[\theta_c, \theta] = 0$ one has $\mathfrak{g} = \mathfrak{k} \oplus \mathfrak{p}$, where

$$\begin{aligned} \mathfrak{k} &= \{\mathfrak{X} \in \mathfrak{g} : \theta_c(\mathfrak{X}) = \mathfrak{X}\} \text{ and} \\ \mathfrak{p} &= \{\mathfrak{X} \in \mathfrak{g} : \theta_c(\mathfrak{X}) = -\mathfrak{X}\} \end{aligned} \tag{28}$$

satisfying

$$[\mathfrak{k}, \mathfrak{k}] \subseteq \mathfrak{k},\ [\mathfrak{k}, \mathfrak{p}] \subseteq \mathfrak{p},\ [\mathfrak{p}, \mathfrak{p}] \subseteq \mathfrak{k}.$$

One also sees that

$$\mathfrak{g} = \mathfrak{h} \cap \mathfrak{k} \oplus \mathfrak{h} \cap \mathfrak{p} \oplus \mathfrak{q} \cap \mathfrak{k} \oplus \mathfrak{q} \cap \mathfrak{p}.$$

The curvature tensor R_{eH} in the neighbourhood of eH satisfies

$$R_{eH}(\mathfrak{X},\mathfrak{Y})\cdot\mathfrak{Z} = -[[\mathfrak{X},\mathfrak{Y}],\mathfrak{Z}]\ ;\ \mathfrak{X},\mathfrak{Y},\mathfrak{Z}\in\mathfrak{q}.$$

It is known [29b,c] that *any connected Lie group is a symmetric space, and every symmetric space is given by* $\mathcal{X} = G/H$.

Let $\mathfrak{q}_1$ be an H-invariant subspace of $\mathfrak{q}$. Then $\mathfrak{g}_1 = [\mathfrak{q}_1,\mathfrak{q}_1]+\mathfrak{q}_1$ is a subalgebra of $\mathfrak{g}$. Let G_1 be the analytic subgroup of G corresponding to $\mathfrak{g}_1$, and let $H_1 = G_1 \cup H$. Then $\mathcal{X}_1 := G_1/H_1$ is a *subsymmetric space* of $\mathcal{X}$. We say that $\mathcal{X}_1$ is an invariant *subsymmetric space* if $\mathfrak{g}_1$ is an ideal in $\mathfrak{g}$. If $\mathcal{X}_1$ is an invariant subsymmetric space, then the *symmetric quotient space* is defined by

$$\begin{aligned}\mathcal{X}/\mathcal{X}_1 &= \{g\mathcal{X}_1 : g\in G\}\\ &= (G/G_1)/(H/H_1).\end{aligned}$$

If $\mathfrak{q}_1$ and $\mathfrak{q}_2$ are two H-invariant subspaces such that $\mathfrak{q} = \mathfrak{q}_1 \oplus \mathfrak{q}_2$ and if also $\mathcal{X}_2 = G_2/H_2$ is an invariant subsymmetric space as defined above, then we say that $\mathcal{X}_2$ is a *complementary subsymmetric space* to $\mathcal{X}_1$.

As in [29c] the symmetric space $\mathcal{X} = G/H$ is said to be

1. *flat* if the curvature tensor vanishes identically,
2. *semisimple* if the Ricci-form $\mathfrak{r}$ is nondegenerate and symmetric,
3. *reductive* if every invariant subsymmetric space has an invariant complementary subsymmetric space, and
4. *simple (or irreducible)* if it has no nontrivial invariant subsymmetric spaces.

Furthermore, an irreducible semisimple symmetric space $\mathcal{X} = G/H$ is said to be of the

1. *compact type* if $G = K$, K being the maximal compact subgroup of G,
2. *noncompact type* if $G \neq K$ and $K = H$, and
3. *non-Riemannian type* if $G \neq K$ and $K \neq H$.

It is known [29c] that an irreducible semisimple symmetric space $\mathcal{X} = G/H$ falls into one of the following nine catagories in terms of the Lie algebra $\mathfrak{g}$ of G:

1. $\mathfrak{g}$ is a compact and simple Lie algebra.
2. $\mathfrak{g} = \mathfrak{g}_1 \oplus \mathfrak{g}_1$, where $\mathfrak{g}_1$ is compact and simple, and $\theta(\mathfrak{X},\mathfrak{Y}) = (\mathfrak{Y},\mathfrak{X})\ \forall \mathfrak{X},\mathfrak{Y}\in\mathfrak{g}_1$, and θ is an involutive automorphism of $\mathfrak{g}$ (as previously for G).
3. $\mathfrak{g}$ is noncompact, simple and with no complex structure, and $\theta = \theta_c$.

4. $\mathfrak{g}$ is simple with a complex structure, and the involution θ is a *conjugation* with respect to a compact real form of $\mathfrak{g}$.

5. $\mathfrak{g}$ is noncompact, simple and with no complex structure, and $\theta \neq \theta_c$.

6. $\mathfrak{g} = \mathfrak{g}_1 \oplus \mathfrak{g}_1$, where $\mathfrak{g}_1$ is noncompact, simple and with no complex structure, and $\theta(\mathfrak{X}, \mathfrak{Y}) = (\mathfrak{Y}, \mathfrak{X})\ \forall \mathfrak{X}, \mathfrak{Y} \in \mathfrak{g}_1$.

7. $\mathfrak{g}$ is simple with a complex structure, and $\mathfrak{h}$ is a noncompact real form of $\mathfrak{g}$.

8. $\mathfrak{g}$ is simple with a complex structure, and the involution θ is complex linear.

9. $\mathfrak{g} = \mathfrak{g}_1 \oplus \mathfrak{g}_1$, where $\mathfrak{g}_1$ is simple with a complex structure, and $\theta(\mathfrak{X}, \mathfrak{Y}) = (\mathfrak{Y}, \mathfrak{X})\ \forall \mathfrak{X}, \mathfrak{Y} \in \mathfrak{g}_1$.

In the above, the catagories (1) and (2) are of the compact type, (3) and (4) are of the noncompact type, and (5)-(9) are of the non-Riemannian type.

Let G be a Lie group and $H \subseteq G$ be a closed subgroup. Let $\mathbf{D}(G/H)$ be the algebra of all differential operators on G/H which are invariant under the *left translations* $\tau : G/H \to G/H$ defined by $\tau(g) : xH \mapsto gxH$, $g \in G$, $xH \in G/H$. As noted by Schwartz and proved by Harish-Chandra [45c] one has the canonical isomorphism

$$\mathbf{D}(G/\{e_G\}) \cong \mathfrak{U}(\mathfrak{g}),$$

where $\mathfrak{U}(\mathfrak{g})$ is the universal enveloping algebra of $\mathfrak{g}$. More generally for $H = K$ where K is the maximal compact subgroup of G, if the coset space G/K is reductive and K is connected then as proved by Helgason [46a]

$$\mathbf{D}(G/K) \cong Z_{\mathfrak{U}(\mathfrak{g})}(\mathfrak{k})/(Z_{\mathfrak{U}(\mathfrak{g})}(\mathfrak{k}) \cap \mathfrak{U}(\mathfrak{g})\mathfrak{k}),$$

where $Z_{\mathfrak{U}(\mathfrak{g})}(\mathfrak{k})$ is the centralizer of $\mathfrak{k}$ in $\mathfrak{U}(\mathfrak{g})$; $\mathfrak{k}$ as in (28) being the Lie algebra of K. Furthermore, in terms of respective complexifications one also has [29c] the isomorphism

$$\mathbf{D}(G/H) \cong Z_{\mathfrak{U}(\mathfrak{g}_{\mathbf{C}})}(\mathfrak{h}_{\mathbf{C}})/(Z_{\mathfrak{U}(\mathfrak{g}_{\mathbf{C}})}(\mathfrak{h}_{\mathbf{C}} \cap \mathfrak{U}(\mathfrak{g}_{\mathbf{C}})\mathfrak{h}_{\mathbf{C}}),$$

where $\mathfrak{h}_{\mathbf{C}}$ is the complexification of $\mathfrak{h}$ in (27). Helgason [46c] showed that *for a symmetric pair (G, H) with G semisimple and H connected, the algebra $\mathbf{D}(G/H)$ is commutative.*

A function on G/H which is an eigenfunction of each $D \in \mathbf{D}(G/H)$ is called a *joint eigenfunction* of $\mathbf{D}(G/H)$. Let $\mu \in \mathrm{Hom}(G/H, \mathbf{C})$ and let E_μ, the corresponding *joint eigenspace*, be given by

$$E_\mu(G/H) = \{f \in C^\infty(G/H) : Df = \mu(D)f\ \forall D \in \mathbf{D}(G/H)\}.$$

Let (ρ_μ, E_μ) denote the natural representation of G on this eigenspace E_μ such that

$$(\rho_\mu(g)f)(xH) = f(g^{-1}xH), \quad g, x \in G, \ f \in E_\mu.$$

These representations are called *eigenspace representations*, and the topic *harmonic analysis on G/H* attempts to

1. decompose an arbitrary function on G/H into joint eigenfunctions of $\mathbf{D}(G/H)$,
2. describe the joint eigenspaces $E_\mu(G/H)$ of $\mathbf{D}(G/H)$,
3. determine for which μ the eigenspace representation (ρ_μ, E_μ) is irreducible.

Let us now consider the symmetric space G/K of noncompact type [29a,46c] Let $\pi \in \mathrm{Epi}(G, G/H)$. Let ϕ be a $C^\infty(G/K)$-function such that $(\phi \circ \pi)(e_G) = 1$. ϕ is called a *spherical function* if

1. $\phi \circ \tau^{-1}(k) = \phi \ \forall k \in K$ (ϕ is K-invariant).
2. $D\phi = \lambda_D \phi \ \forall D \in \mathbf{D}(G/K)$; $\lambda_D \in \mathbf{C}$ (ϕ is a joint eigenfunction on G/K).

Helgason [46c] shows that *ϕ is a spherical function if and only if*

$$\int_K \phi(g_1 k g_2)\, dk = \phi(g_1)\phi(g_2) \ \ \forall g_1, g_2 \in G.$$

Furthermore, $\phi \circ \pi$ *is a spherical function on G if and only if ϕ is a spherical function on G/K.*

A $C^\infty(G/K)$-function f is called *harmonic* if $Df = 0$ for all $D \in \mathbf{D}(G/K)$ which are without a constant term. Godement [39a,b] showed that the harmonic functions on G/K are characterized by the mean-value property

$$f(gK) = \int_K f(gkg'K)\, dk, \ \ g, g' \in G.$$

The investigation of the behaviour of the spherical functions at ∞ led Harish-Chandra [45e,f] to develop the function which came to be known as *Harish-Chandra's* **c**-*function.* We see this function as follows:

Let $\mathfrak{g} = \mathfrak{k} \oplus \mathfrak{p}$ be the Cartan decomposition of $\mathfrak{g}$ as in (28). Let $\mathfrak{a} \subset \mathfrak{p}$ be any maximal abelian subspace, $\mathfrak{a}^*$ be the dual of $\mathfrak{a}$, and $\mathfrak{a}^*_{\mathbf{c}}$ the set of complex-valued linear functions on $\mathfrak{a}$. Let $Z_{\mathfrak{k}}(\mathfrak{a}) := \mathfrak{m}$ be the centralizer of $\mathfrak{a}$ in $\mathfrak{k}$. Let

$$\mathfrak{g}_\alpha = \{\mathfrak{X} \in \mathfrak{g} : [\mathfrak{H}, \mathfrak{X}] = \alpha(\mathfrak{H})\mathfrak{X} \ \forall \mathfrak{H} \in \mathfrak{a}\}, \tag{29}$$

where $\alpha \in \mathfrak{a}_{\mathfrak{c}}^*$ is said to be a *root of the pair* $(\mathfrak{g}, \mathfrak{a})$ (or *restricted root*) if $\alpha \neq 0$ and $\mathfrak{g}_\alpha \neq \{0\}$. Let $m_\alpha = \dim \mathfrak{g}_\alpha$ (also equal to the multiplicity of α). Let Σ denote the set of all restricted roots. Then one obtains a decomposition

$$\mathfrak{g} = \mathfrak{g}_0 \oplus \sum_{\alpha \in \Sigma} \mathfrak{g}_\alpha \,, \; \mathfrak{g}_0 = \mathfrak{a} + \mathfrak{m}. \tag{30}$$

An element $\mathfrak{H} \in \mathfrak{a}$ is called *regular in* $\mathfrak{a}$ if $\alpha(\mathfrak{H}) \neq 0 \; \forall \alpha \in \Sigma$, otherwise it is said to be *singular.* A *Weyl chamber* is a connected component of the set of all regular elements in $\mathfrak{a}$. Each Weyl chamber defines a set Σ^+ of *positive roots of the pair* $(\mathfrak{g}, \mathfrak{a})$. A *simple root* is a positive root which is not a sum of two positive roots. Then one defines

$$\mathfrak{a}_+ = \{\mathfrak{H} \in \mathfrak{a} : \alpha_i(\mathfrak{H}) > 0 \, ; \, 1 \leq i \leq r\},$$

where $\{\alpha_i \, ; \, 1 \leq i \leq r\}$ is the set of simple roots (with respect to the given Weyl chamber). Let $\mathfrak{a}_+^*$ be the dual of $\mathfrak{a}_+$. Let

$$\rho = \frac{1}{2} \sum_{\alpha \in \Sigma^+} (m_\alpha)\alpha.$$

A root $\alpha \in \Sigma$ is called *indivisible* if $c\alpha \in \Sigma$ implies $c = \pm 1, \pm 2$. The set of all indivisible roots is denoted by Σ_0 and $\Sigma_0 \cap \Sigma^+ := \Sigma_0^+$. Let $\overline{N} = \theta_c(N)$, where N is the nilpotent group corresponding to the algebra $\mathfrak{n}$ given by

$$\mathfrak{n} = \sum_{\alpha \in \Sigma^+} \mathfrak{g}_\alpha$$

so that $\mathfrak{g} = \mathfrak{k} \oplus \mathfrak{a} \oplus \mathfrak{n}$ or $G = KAN$ (*Iwasawa decomposition*).

A *horocycle* in $X = G/K$ is an orbit in X of a group of the form gNg^{-1}, $g \in G$. The group G permutes the horocycles transitively. Let $M = Z_K(A)$, the centralizer of A in K, and $B = K/M$. Every horocycle can be written as $ka \cdot (N \cdot e_G K)$, where $k \in K$, $a \in A$, and $N \cdot (e_G K)$ is the horocycle corresponding to the origin $e_G K$ of X. The coset kM is called the *normal* to the horocycle and is unique in B. Also, $a \in A$ is unique in A and is called the *complex distance* from $e_G K$ to the horocycle $ka \cdot (N \cdot e_G K)$. Given $gK \in X$, $kM \in B$ there is a unique horocycle through gK with normal kM. The complex distance from this horocycle to the origin $e_G K$of X is denoted by $a(gK, kM) \in A$. Let $\mathfrak{A}(gK, kM) \in \mathfrak{a}$ such that $\exp(\mathfrak{A}(gK, kM)) = a(gK, kM)$. Given a function f on X its *Fourier transform* is defined by

$$\tilde{f}(\lambda, kM) = \int_K f(gK) e^{(-i\lambda + \rho)A(gK, kM)} \, d(gK) \tag{31}$$

for all $\lambda \in \mathfrak{a}_{\mathfrak{c}}^*$ and $kM \in B$ for which the integral converges.

Harish-Chandra has proved that *for* $\lambda \in \mathfrak{a}_{\mathbf{c}}^*$ *the functions*

$$\phi_\lambda(g) = \int_K e^{(i\lambda+\rho)\mathfrak{A}(kg)}\, dk\,,\ g \in G$$

exhaust the class of spherical functions on a connected semisimple Lie group G *with finite center, where* K *is the maximal compact subgroup of* G *and* $\mathfrak{A}(g) \in \mathfrak{a}$ *denotes the unique element such that*

$$g = n \exp(\mathfrak{A}(g))k\,,\ \ n \in N,\ k \in K.$$

Moreover, two such functions ϕ_ν *and* ϕ_λ *are identical if and only if* $\nu = \omega\lambda$ *for some* $\omega \in W$, *the Weyl group.* If $\mathfrak{H} \in \mathfrak{a}_+ \subset \mathfrak{a}$, then $\mathfrak{H}(g)$ is the unique element in $\mathfrak{a}$ such that

$$g = k' \exp(\mathfrak{H}(g))n'\,,\ \ k' \in K,\ n' \in N.$$

The function $\phi_\lambda(g)$ satisfies

$$\phi_\lambda(g^{-1}) = \phi_{-\lambda}(g).$$

Harish-Chandra's **c**-function is given by

$$\mathbf{c}(\lambda) = \lim_{t\to+\infty} e^{(-i\lambda+\rho)(t\mathfrak{H})}\phi_\lambda(\exp(t\mathfrak{H}))\,,\ \Re e(i\lambda) \in \mathfrak{a}_+^*,\ \mathfrak{H} \in \mathfrak{a}_+. \tag{32}$$

$$= \int_{\overline{N}} e^{(-i\lambda+\rho)\mathfrak{H}(\bar{n})}\, d\bar{n}\,,\ \Re e(i\lambda) \in \mathfrak{a}_+^*. \tag{33}$$

$$= c_0 \prod_{\alpha\in\Sigma_0^+} \frac{2^{-<i\lambda,\alpha_0>}\Gamma(< i\lambda, \alpha_0 >)}{\Gamma(\frac{1}{2}(\frac{1}{2}m_\alpha + 1 + < i\lambda, \alpha_0 >))\Gamma(\frac{1}{2}(\frac{1}{2}m_\alpha + m_{2\alpha} + < i\lambda, \alpha_0 >))}, \tag{34}$$

where

$$\mathfrak{H}(\bar{n}) \in \mathrm{Cl}\,\{\mathfrak{H} \in \mathfrak{a}\ :\ B(\mathfrak{H}, \mathfrak{H}') > 0\ \forall \mathfrak{H}' \in \mathfrak{a}_+\}\,,$$

$d\bar{n}$ is the Haar measure normalized by

$$\int_{\overline{N}} e^{-2\rho(\mathfrak{H}(\bar{n}))}\, d\bar{n} = 1,$$

$\alpha_0 = \frac{\alpha}{<\alpha,\alpha>}$ and c_0 is a constant which is determined by the condition $\mathbf{c}(-i\rho) = 1$. The general validity of a form of formula (34) was proved by Gindikin and Karpelevič [38]. We give the following special cases:

1. If G is complex, then

$$\mathbf{c}(\lambda) = \frac{\prod_{\alpha\in\Sigma^+} < \alpha, \rho >}{\prod_{\alpha\in\Sigma^+} < \alpha, i\lambda >}.$$

2. If G is such that rank$G$ = rank$K$ + rankG/K, then

$$\mathbf{c}(\lambda) = \frac{\prod_{\alpha\in\Sigma^+} \frac{\Gamma(\frac{1}{2}m_\alpha + <\rho,\alpha_0>)}{\Gamma(<\rho,\alpha_0>)}}{\prod_{\alpha\in\Sigma^+} \frac{\Gamma(\frac{1}{2}m_\alpha + <i\lambda,\alpha_0>)}{\Gamma(<i\lambda,\alpha_0>)}}.$$

3. For $G = SU(2,2)$ and $K = S(U(2)\otimes U(2))$, Wehrhahn and Barut [93] obtain

$$\mathbf{c}(\lambda) = \bar{c}_0 \frac{\Gamma(\frac{i}{4}(2\lambda_\alpha + \lambda_\beta))\Gamma(\frac{i}{4}(2\lambda_\beta + \lambda_\alpha))}{\Gamma(\frac{1}{2} + \frac{i}{4}(2\lambda_\alpha + \lambda_\beta))\Gamma(\frac{1}{2} + \frac{i}{4}(2\lambda_\beta + \lambda_\alpha))(\lambda_\alpha^2 - \lambda_\beta^2)}.$$

where $\bar{c}_0$ is some constant, $\lambda_\alpha, \lambda_\beta \in \mathbf{C}$ such that $\lambda = \lambda_\alpha \alpha + \lambda_\beta \beta$, and $< \alpha, \alpha > = < \beta, \beta > = 1$. They relate this $\mathbf{c}(\lambda)$ to the *Jost function* of a scattering problem, and the *scattering operator* $S(\lambda)$ is given by

$$S(\lambda) = \frac{\mathbf{c}(\lambda)}{\mathbf{c}(-\lambda)}.$$

4. If G is of real rank one, then

$$\mathbf{c}(\lambda) = \frac{2^{\frac{1}{2}m_\alpha + m_{2\alpha}}\Gamma(\frac{1}{2}(m_\alpha + m_{2\alpha} + 1))2^{-<i\lambda,\alpha_0>}\Gamma(< i\lambda, \alpha_0 >)}{\Gamma(\frac{1}{2}(\frac{1}{2}m_\alpha + 1 + < i\lambda, \alpha_0 >))\Gamma(\frac{1}{2}(\frac{1}{2}m_\alpha + m_{2\alpha} + < i\lambda, \alpha_0 >))}.$$

The formulas of $\mathbf{c}(\lambda)$ for $G = SL(n, \mathbf{R})$ and $G = Sp(n, \mathbf{R})$ were determined by Bhanu-Murthy [13a,b]. Helgason [46c] proves that *the function* $\mathbf{c}(\lambda)$ (given by (33)) *is holomorphic on the set*

$$\widehat{\mathfrak{a}_\mathbf{c}^*} = \left\{\lambda \in \mathfrak{a}_\mathbf{c}^* : \lambda \notin \omega\sigma_\mu,\ \mu \in \Lambda - \{0\},\ \lambda \notin \tau_\nu(\omega,\omega'),\ \nu \in \tilde{\Lambda},\ \omega,\omega' \in W,\ \omega \neq \omega'\right\},$$

where Λ *is the set of all linear combinations* $\sum_{j=1}^{j=r} n_j\alpha_j$, $n_j \in \mathbf{Z}^+$, $\tilde{\Lambda} = \sum_{j=1}^{j=r} \mathbf{Z}\alpha_j$ *and the hyperplanes* σ_μ *and* $\tau_\nu(\omega,\omega')$ *in* $\mathfrak{a}_\mathbf{c}^*$ *are given by*

$$\begin{aligned} \sigma_\mu &= \{\lambda \in \mathfrak{a}_\mathbf{c}^* : < \mu, \mu > = 2i < \mu, \lambda >\}, \text{ and} \\ \tau_\nu(\omega,\omega') &= \{\lambda \in \mathfrak{a}_\mathbf{c}^* : i(\omega\lambda - \omega'\lambda) = \nu\}. \end{aligned}$$

By definition, the W-invariant set $\widehat{\mathfrak{a}_\mathbf{c}^*}$ is a connected open dense subset of $\mathfrak{a}_\mathbf{c}^*$. The expression (34) gives an extension of $\mathbf{c}(\lambda)$ from a holomorphic function on the set $\widehat{\mathfrak{a}_\mathbf{c}^*}$ to a meromorphic function on $\mathfrak{a}_\mathbf{c}^*$.

The properties and the underlying theory of Harish-Chandra's $\mathbf{c}$-function are discussed in detail by Helgason [*]. The article by Van den Ban, Flensted-Jensen and Schlichtkrull [*] gives an up-to-date survey of the harmonic analysis on the pseudo-Riemannian symmetric space G/H. Various physical applications of the special case $SU(2,2)/S(U(2)\otimes U(2))$ are analysed by Barut [*]. The theory of conformal compactifications of space-time and momentum space along with its physical applications, in particular to large scale correlations of galaxies, are discussed by Budinich [*].

Now we go back to the discussion of spherical functions. The following special cases [46c] (see also [50]) are noteworthy:

1. For $G = \mathbf{R}^2$, $K = \{e_G\}$, $X = G/K = \mathbf{R}^2$, the spherical functions on $\mathbf{R}^2$ are the exponential functions $\mathbf{R}^2 \to \mathbf{C}$ with
$$(x,y) \mapsto e^{\alpha x+\beta y}, \quad \alpha, \beta \in \mathbf{C}.$$

2. For $G = M_e(2)$ (the identity component of M(2) where
$$M(2) = \left\{ \begin{pmatrix} e^{i\alpha} & z \\ 0 & 1 \end{pmatrix} : \alpha \in \mathbf{R},\, z \in \mathbf{C} \right\}$$
is the *motion group* of the Euclidean plane), $K = SO(2)$, $X = G/K = \mathbf{R}^2$, the spherical functions on $\mathbf{R}^2$ are the functions $\mathbf{R}^2 \to \mathbf{C}$ with
$$(x,y) \mapsto \frac{1}{2\pi}\int_0^{2\pi} e^{i\alpha(x\cos\theta + y\sin\theta)}\, d\theta, \quad \alpha \in \mathbf{C}.$$

3. For $G = SO(3)$, $K = SO(2)$, $X = G/K = \mathcal{S}^2$, the spherical functions on $\mathcal{S}^2$ are the functions $\mathcal{S}^2 \to \mathbf{C}$ with
$$p \mapsto \frac{1}{2\pi}\int_0^{2\pi} (\cos r + i \sin r \cos u)^n\, du = P_n(\cos r),$$
where the distance $r = d((0,0,1),p)$, $n \in \mathbf{N}\cup\{0\}$, and $P_n(\cos r)$ is the *Legendre polynomial.*

4. For $G = SL(2,\mathbf{R})$, $K = SO(2)$, $X = G/K = \{(x,y) \in \mathbf{R}^2 : y > 0\} := \mathcal{U}^+$, the spherical functions on $\mathcal{U}^+$ are the functions $\mathcal{U}^+ \to \mathbf{C}$ with
$$p \mapsto \frac{1}{2\pi}\int_0^{2\pi} (\cosh r + \sinh r \cos u)^\rho\, du = P_\rho(\cosh r),$$
where the distance $r = d((0,1),p)$, $\rho \in \mathbf{C}$ and $P_\rho(\cosh r)$ is the *Legendre function.*

In the case for which
$$G = \left\{ \begin{pmatrix} \alpha & 0 & 0 \\ \beta & 1 & \gamma \\ 0 & 0 & 1 \end{pmatrix} \in GL(3,\mathbf{R}) : \alpha > 0 \right\},$$

$$H = \left\{ \begin{pmatrix} 1 & 0 & 0 \\ \beta & 1 & -\beta \\ 0 & 0 & 1 \end{pmatrix} \in G : \beta \in \mathbf{R} \right\} \cong \mathbf{R},$$
$$X = G/H = \left\{ (x,y) \in \mathbf{R}^2 : x > 0 \right\},$$

where H is the stabilizer of the point $(1,0) \in X$, the harmonic analysis on X is studied by Hilgert [*].

As before, let G be a semisimple noncompact Lie group, and K be a maximal compact subgroup of G. Let Γ be a torsion-free discrete subgroup of G which is cocompact in G (that is, G/Γ is compact). [Note that for a Lie group G with only finitely many connected components and with a maximal compact subgroup K, the *cohomological dimension*, denoted by cd, of any torsion-free discrete subgroup Γ of G is cd Γ = dim$G$ - dim$K$ if and only if Γ is cocompact]. The double-coset space $K\backslash G/\Gamma$ is a compact locally symmetric space. The symmetric space $K\backslash G$ is simply connected and thus is the universal covering space of $K\backslash G/\Gamma$ with covering group Γ. Let us assume that G has no factor which is locally isomorphic either to $SL(3,\mathbf{R})$ or to $SO(p,q)$ with pq odd. Let ρ be an orthogonal representation of the *fundamental group (first homotopy group or Poincaré group)* $\pi_1(K\backslash G/\Gamma)$ of (path-connected) $K\backslash G/\Gamma$. Then, as defined by Ray and Singer (see Speh [*]), the *analytic torsion* $\tau_\rho(K\backslash G/\Gamma)$ is in $(0,\infty)$. Let T_π denote an admissible representation (space) of G. Then the *torsion of the representation* T_π, denoted by $tor(T_\pi)$, is defined by

$$tor(T_\pi) = \sum_i (-1)^i i \dim \mathrm{Hom}_K(\Lambda^i \mathfrak{p}, T_\pi),$$

where $\mathfrak{p}$ is as given by the Cartan decomposition (28). If ρ_e denotes the trivial representations of $\Gamma = \pi_1(K\backslash G/\Gamma)$ then by a theorem due to Moscovici and Stanton,

$$\tau_{\rho_e}(K\backslash G/\Gamma) = 1.$$

Speh [*] proves this theorem by relating the analytic torsion of $K\backslash G/\Gamma$ to the torsion of unitary representations in the discrete spectrun of G/Γ.

Let again G be a locally compact topological group and K a maximal compact subgroup. A function f on G is called *K-bi-invariant* if

$$f(kgk') = f(g) \ \forall g \in G, \ k,k' \in K.$$

Similarly, a Borel measure μ on G is called *K-bi-invariant* if

$$\mu(kAk') = \mu(A), \ \ k,k' \in K$$

for all measurable subsets A of G. We say that a complex-valued function on G *vanishes at infinity* if, given $\epsilon > 0$ there exists a compact subset $C_\epsilon \subset G$ such that $|f(g)| < \epsilon$ for $g \notin C_\epsilon$. Let $M(G)$ be the space of all finite Borel measures on G, $C_c(G)$ be the space of all continuous complex-valued functions on G with compact support, $C_\infty(G)$ be the space of all continuous complex-valued functions on G which vanish at infinity, and $\mathcal{L}^p(G)$ be the space of all Lebesgue-measurable complex-valued functions on G with $1 \leq p \leq \infty$. Let $M(K\backslash G/K)$, $C_c(K\backslash G/K)$ ($= C_c^\natural(G)$ of Helgason [46c]), $C_\infty(K\backslash G/K)$ and $\mathcal{L}^p(K\backslash G/K)$, $1 \leq p \leq \infty$ be the respective subspaces of K-bi-invariant functions (or measures). With respect to convolution, defined by

$$(f * f')(x) = \int_G f(g) f'(g^{-1}x)\, dg,$$

$C_c(G) \subset \mathcal{L}^1(G) \subset M(G)$ are associative algebras and $C_c(K\backslash G/K) \subset \mathcal{L}^1(K\backslash G/K) \subset M(k\backslash G/K)$ are the respective subalgebras. One can see that the following conditions are equivalent:

1. The algebra $C_c(K\backslash G/K)$ is commutative,
2. The algebra $\mathcal{L}^1(K\backslash G/K)$ is commutative,
3. The algebra $M(K\backslash G/K)$ is commutative,
4. The double-coset group $K\backslash G/K)$ is commutative.

When these conditions hold, one says that the pair (G, K) is *commutative* (or is a *Gelfand pair*).

For a given Gelfand pair (G, K), a nonzero Radon measure μ on G is called *spherical* if it is K-bi-invariant and $\mu \in \mathrm{Hom}(C_c(K\backslash G/K), \mathbf{C})$ with

$$\mu : f \mapsto \int_G f(g)\, d\mu(g).$$

A complex-valued continuous function ϕ on G is called a *zonal spherical function* (or *elementary spherical function*, or *spherical function of class one*) if $d\mu(g) = \phi(g^{-1})\, d\eta_G(g)$, where η_G is a Haar measure on G, defines a spherical measure. The function ϕ is said to be *positive definite* if

$$\sum_{i,j} \overline{c_i} c_j \phi(g_i^{-1} g_j) \geq 0\,, \; c_i \in \mathbf{C}\,, \, g_i \in G\,, \, 1 \leq i, j \leq n,$$

for all finite sets $\{c_1, \cdots, c_n\}$ and $\{g_1, \cdots, g_n\}$. Let $S(G, K)$ denote the set of all zonal spherical functions and $S^+(G, K)$ be the set of all positive definite zonal spherical functions. The *spherical transform* is the mapping $f \mapsto \widehat{f}$, from K-bi-invariant functions f on G to the functions $\widehat{f}$ on $S(G, K)$ such that

$$\widehat{f}(\phi) = \int_G f(g) \phi(g^{-1})\, d\eta_G(g) = \mu_\phi(f) \;\forall \phi \in S(G, K).$$

The integral above is absolutely convergent if $f \in \mathcal{L}^1(k\backslash G/K)$. The *inverse* mapping $\widehat{f} \mapsto f$ is defined such that

$$f(g) = \int_{S(G,K)} \widehat{f}(\phi)\phi(g)\, d\nu(\phi) := \widehat{f}^{\vee}.$$

According to the Riemann-Lebesgue lemma, *if* $f \in \mathcal{L}^1(K\backslash G/K)$ *then* $\widehat{f} \in C_\infty(S^+(G,K))$. Furthermore, the *Plancherel theorem* and the *Pontrjagin duality theorem* for locally compact abelian groups [59] were extended by Mautner and Godement [39,67] to a Gelfand pair as follows: Let (G,K) be a Gelfand pair. Then there exists a unique positive Radon measure ν on $S^+(G,K)$ (called the *Plancherel measure* for $S^+(G,K)$), such that

if $f \in C_c(K\backslash G/K)$ then $\widehat{f} \in \mathcal{L}^2(S^+(G,K),\nu)$ and $||\widehat{f}||_{\mathcal{L}^2(S^+(G,K),\nu)} = ||f||_{\mathcal{L}^2(G)}$.

Let us now consider the measure space $(S^+(G.K),\nu)$. For every $\phi \in S^+(G,K)$ let $\mathcal{H}_\phi$ be a Hilbert space. We define $\mathcal{H}_p$, $1 \le p \le \infty$ to be the linear span of the mappings

$$s : S^+(G,K) \to \cup_{\phi \in S^+(G,K)} \mathcal{H}_\phi \text{ such that } s(\phi) \in \mathcal{H}_\phi.$$

The functions $\phi \mapsto < s(\phi), s(\phi) >^{\frac{1}{2}}_{\mathcal{H}_\phi}$ are elements of $\mathcal{L}^p(S^+(G,K),\nu)$. Thus $\mathcal{H}_p$ is a Banach space with norm

$$||s||_p = \left(\int_{S^+(G,K)} < s(\phi), s(\phi) >^{\frac{p}{2}}_{\mathcal{H}_\phi} d\nu(\phi)\right)^{\frac{1}{p}} \ , \ 1 \le p < \infty.$$

In terms of zonal spherical functions, we define the *Fourier transform on* G/K (see (31)) to be the mapping

$$\mathcal{F} : \mathcal{L}^1(G/K) \to \int_{S^+(G,K)} \mathcal{H}_\phi\, d\nu(\phi) \text{ with } (\mathcal{F}(f))(\phi) = s_f(\phi).$$

In quantum mechanics, the *Heisenberg uncertainity principle* says that $\Delta x \Delta p \ge \frac{\hbar}{2}$, where $\Delta x = \sqrt{< x^2 > - < x >^2}$ and $\Delta p = \sqrt{< p^2 > - < p >^2}$ are respectively the statistical fluctuations of the result of measurements of *position* x and *momentum* p around the average values $< x >$ and $< p >$, and $h = 2\pi\hbar$ is *Planck's constant.* That is, by this principle, the position and momentum of a moving object cannot be determined simultaneously. In terms of functional analysis this principle says that a function and its Fourier transform cannot both be *mostly concentrated* on short intervals. That is, if $f(t)$ has *most of its support* in an interval of length ℓ, and its Fourier transform $\widetilde{f}(\tau)$ has *most of its support* in an interval of length $\widetilde{\ell}$ then $\ell\widetilde{\ell} \ge 1 - \eta$, where η is a criterion for *most of its support.* A classical extension of the uncertainity principle was given by Donoho and Stark, and by Smith (see Wolf [95e,f]) as follows: Let T be a measurable set and χ_T be its characteristic function. We say that $f \in \mathcal{L}^p$ is *ϵ-concentrated on T* if $||f - \chi_T f||_p \leq \epsilon ||f||_p$. For $p = 2$, let $f, \widehat{f} \in \mathcal{L}^2(\mathbf{R})$, let $\epsilon, \delta \geq 0$, and let $T, W \subset \mathbf{R}$ be measurable subsets, and $\mu(T)$ and $\mu(W)$ the associated Lebesgue measures. Assume that f is ϵ-concentrated on T and $\widehat{f}$ is δ- concentrated on W. Then

$$\mu(T)^{\frac{1}{2}}\mu(W)^{\frac{1}{2}} \geq ||\mathcal{O}_T\mathcal{O}_W||_2 \geq (1 - \epsilon - \delta),$$

where the operators $\mathcal{O}_T$ and $\mathcal{O}_W$ are defined by

$$\mathcal{O}_T f = \chi_T f \text{ and } \mathcal{O}_W f = (\chi_W \widehat{f})^\vee \text{ with } ||\mathcal{O}_T\mathcal{O}_W||_2 \leq \mu(T)^{\frac{1}{2}}\mu(W)^{\frac{1}{2}}.$$

Smith also extended the above to a locally compact abelian group by extending the operator norm inequality given above. Wolf [95e,f] extended the above arguments to Gelfand pairs and thus to Riemann symmetric spaces, compact topological groups and locally compact abelian groups. Wolf's extension goes as follows:

Let (G, K) be a Gelfand pair. We consider two cases:

1. Let $X^\natural := K\backslash G/K$. Let $T = KTK \subset G$ and $U \subset S^+((G,K),\nu)$ be fixed subsets of finite measure. We define the operators $\mathcal{O}_T$ and $\mathcal{O}_U$ such that

$$\mathcal{O}_T f = \chi_T f \text{ and } \mathcal{O}_U f = (\chi_U \widehat{f})^\vee.$$

Given $\epsilon \geq 0$ we say that $f \in \mathcal{L}^p(X^\natural)$ is *ϵ-concentrated on T* if

$$||f - \chi_T f||_p \leq \epsilon ||f||_p.$$

And, given $\delta \geq 0$ we say that $h \in \mathcal{L}_{p'}(S^+(G,K),\nu)$ is *δ-concentrated on U* if

$$||h - \chi_U h||_{p'} \leq \delta ||h||_{p'},\ p' = \frac{p}{p-1}.$$

We say that $f \in \mathcal{L}^p(X^\natural)$ is *δ-bandlimited to U* if there exists $f_U \in \mathcal{L}^p(X^\natural)$ such that $\widehat{f_U}$ has a support in U and

$$||f - f_U||_p \leq \delta ||f||_p.$$

The **scalar uncertainity principle for** $X^\natural$ obtained by Wolf is as follows: *Assume that $0 \neq f \in \mathcal{L}^p(X^\natural)$ with $1 \leq p \leq 2$ and $\epsilon, \delta \geq 0$ such that f is ϵ-concentrated on T* and *δ-bandlimited to U. Then*

$$\mu_G(T)^{\frac{1}{p}}\nu(U)^{\frac{1}{p}} \geq ||\mathcal{O}_T\mathcal{O}_U||_p \geq \frac{1 - \epsilon - \delta}{1 + \delta},$$

where $\mu_G(T)$ is a Haar measure on G and $\nu(U)$ is the Plancherel measure on $S^+((G,K),\nu)$. And for $p = 2$, $||\mathcal{O}_T\mathcal{O}_U||_2 \geq (1 - \epsilon - \delta)$.

2. Let $X^\natural := G/K$. Let $T = TK \subset G$ and $U \subset S^+((G,K),\nu)$ be fixed subsets of finite measure. We define the operators $\mathcal{O}_T$ and $\mathcal{O}_U$ such that for $f \in \mathcal{L}^p(X^\natural)$,

$$\mathcal{O}_T f = \chi_T f \text{ and } \mathcal{O}_U f = \mathcal{F}^{-1}(\chi_U \mathcal{F}(f)).$$

Given $\epsilon \geq 0$ we say that $f \in \mathcal{L}^p(X^\natural)$ is *ϵ-concentrated on T* if

$$||f - \chi_T f||_p \leq \epsilon ||f||_p.$$

And, given $\delta \geq 0$, $h \in \mathcal{H}_{p'}$ is *δ-concentrated on U* if

$$||h - \chi_U h||_{p'} \leq \delta ||h||_{p'}, \ p' = \frac{p}{p-1}.$$

Also, we say that $f \in \mathcal{L}^p(X^\natural)$ is *δ-bandlimited to U* if there exists $f_U \in \mathcal{L}^p(X^\natural)$ such that $\mathcal{F}(f_U)$ has a support in U and

$$||f - f_U||_p \leq \delta ||f||_p.$$

The **vector uncertainity principle** for $X^\natural$ obtained by Wolf is as follows: *Assume that $0 \neq f \in \mathcal{L}^p(X^\natural)$ with $1 \leq p \leq 2$ and $\epsilon, \delta \geq 0$ such that f is ϵ-concentrated on T and δ-bandlimited to U. Then*

$$\mu_G(T)^{\frac{1}{p}} \nu(U)^{\frac{1}{p}} \geq ||\mathcal{O}_T \mathcal{O}_U||_p \geq \frac{1-\epsilon-\delta}{1+\delta}.$$

And for $p = 2$, $||\mathcal{O}_T \mathcal{O}_U||_2 \geq (1-\epsilon-\delta)$.

Wolf announced these remarkable extensions in the Radon Conference in Vienna during December 1992 and in the NATO Workshop in San Antonio during January 1993.

We now turn to an application to a stochastic process. Let Ω be an abstract space, $\mathfrak{B}$ be a σ-algebra of subsets of Ω, and π be a probability measure (probability distribution) on $\Omega(\mathfrak{B})$ (that is, $\pi(\Omega) = 1$). The triple $(\Omega, \mathfrak{B}, \pi)$ is called a *probability space.* A *Markov process* on a topological space with a continuous time parameter is called a *diffusion process* if the *sample function* is continuous in time with probability one. A *diffusion kernal* is a positive linear mapping $\mathcal{P} : M_0 \to M$ which is continuous with respect to weak topologies on M_0 and M, where M_0 is the class of Radon measures on Ω with compact support, and M is a class of measures on Ω which do not necessarily have compact support. Figà-Talamanca [*] defines a natural continuous diffusion process on an infinte, compact, *metrically homogeneous, ultrametric* space X (identified with G/K, G being a compact group of isometries of X and $K = \{g \in G : gx_0 = x_0\}$, where x_0 is a fixed point in X) using spherical functions on X and a K-bi-invariant probability measure on G.

Let G be a connected Lie group. Let $(\rho, \mathcal{H})$ be a nontrivial unitary representation of G on a complex Hilbert space $\mathcal{H}$ ($\dim\mathcal{H} > 1$). By a *coherent state* *orbit* for $(\rho, \mathcal{H})$, we mean a complex orbit of G in the projective space $\mathbf{P}(\mathcal{H})$. The representation $(\rho, \mathcal{H})$ is called a ***coherent state representation*** if it is irreducible, admits a coherent state orbit, and has a discrete kernel. The group G is called a ***coherent state group*** if it has a coherent state representation. In physics, any orbit of G in $\mathbf{P}(\mathcal{H})$ is called a ***system of coherent states.*** The concept of such coherent states was first introduced by Barut and Girardello [9] for $G = SU(1,1)$. It was later generalized by Perelomov [74] for arbitrary Lie groups. Antoine [*] gives a generalized method for the construction of coherent states based on a generalization of the square integrability of a group representation on a homogeneous space. He then applies this method to the Poincaré group.

Let M be a differentiable manifold of class C^∞ and Ω be a system of independent 1-*forms* ω_i, $1 \leq i \leq m$ on M. Then the pair (M, Ω) is called a *symplectic manifold* and Ω is called a *completely integrable system* if for every point $x \in M$

$$d\omega_i = \sum_{j=1}^{m} \theta_{ij} \wedge \omega_j\,, \quad 1 \leq i \leq m,$$

where θ_{ij} are 1-*forms* on a neighbourhood of M. A *Hamiltonian system* is a triple (M, Ω, H), where (M, Ω) is a symplectic manifold called the *phase space* of the system, and H a real differentiable function on M called the *Hamiltonian* of the system. Winternitz, del Olmo and Rodríguez [*] discuss the integrability of the Hamiltonian systems obtained by projecting *free motion* on the homogeneous space $SU(p,q)/U(p-1,q)$ onto the space $O(p,q)/O(p-1,q)$.

Feynman's *path integral* method [28,51] for solving physical problems has been a topic of intense research for several years, although the method does not lead to closed form solutions for a majority of problems - including the case of the hydrogen atom. The use of the Kustaanheimo-Stiefel mapping $\mathbf{R}^3 \to \mathbf{R}^4$ [61,89] in obtaining closed form solutions for certain classes of physical problems was announced in the Munich workshop during the summer of 1978 by Barut and Inomata. A detailed survey of the formulation of the Feynman path integral on the homogeneous space G/H is given by Inomata and Junker [*].

Before we leave our discussion of harmonic analysis on G we point out that several applications of harmonic analysis to invariant differential equations, wave equations, and conformal invariance have been discussed in detail by Helgason [46a,d]. Several interesting applications of harmonic analysis have also been given by Mackey [65e,g] and Wells [91].

Let us now consider a connected, paracompact $C^{(r)}(1 \leq r \leq \infty)$- manifold M. Let *Diff*(M) be the group of all the $C^{(r)}$-diffeomorphisms on M. Let, for $g \in G$,

$$\mathrm{supp}(g) = \mathrm{Cl}\,\{p \in M \,:\, gp \neq p\}$$

and let

$$G = \{g \in \mathit{Diff}(M) \ : \ \mathrm{supp}(g) \text{ compact } \}$$

be a topological group (not locally compact) with a topology of compact uniform convergence of the differentials of the functions $M \to M$ defined by $p \mapsto gp$. Let S_∞ be the infinite symmetry group of all the finite permutations on $\mathbf{N}$ and $\widetilde{S}_\infty$ be the group of all permutations on $\mathbf{N}$. Let

$$X = \prod_{i \in \mathbf{N}} M_i, \ \ M_i = M \ \forall i \in \mathbf{N}.$$

Let $x = (x_i)_{i\in\mathbf{N}} \in X$, $g \in G$ and $\tau \in \widetilde{S}_\infty$. Then the left action of G and the right action of $\widetilde{S}_\infty$ on X are given by

$$\begin{aligned} gx &= (gx_i)_{i\in\mathbf{N}}, \\ x\tau &= (x_{\tau(i)})_{i\in\mathbf{N}}. \end{aligned}$$

A point $x \in X$ is said to be an *ordered configuration of points in* M if the set $\{x_i \ : \ i \in \mathbf{N}\,,\ x_i \neq x_j \text{ if } i \neq j\}$ has no accumulation points in M. Let $\widetilde{X}$ be the set of all ordered configurations. One can show that $\widetilde{X}$ is stable under $G \times \widetilde{S}_\infty$. Hirai [*] constructs a family of unitary irreducible representations of G associated with any irreducible unitary representations of S_∞, by constructing and then using quasi-invariant measures on the spaces of ordered configurations.

Let G be a connected Lie group with a cocompact radical. Let $\mathfrak{l}$ be an admissible linear form on $\mathfrak{g}$. We assume that $\mathfrak{l}$ has a polarization. Let $\mathfrak{g}^*$ denote the dual of $\mathfrak{g}$. The group G acts on $\mathfrak{g}^*$ through the co-adjoint representation. We denote by $\mathfrak{g}(\mathfrak{l})$ the Lie algebra of the group $G(\mathfrak{l})$, the stabilizer of $\mathfrak{l}$. Anoussis [*] has obtained a formula for the character (see (11) and (12)) of the equivalence class of unitary irreducible representations of G associated with $\mathfrak{l}$.

More than 30 years ago Kirillov [57] established the character formula for nilpotent groups. Auslander and Kostant [4] gave the character formula for a large class of solvable groups. Rossmann [78a] shows that for semisimple Lie groups, Kirillov's formula is still valid for characters of irreducible tempered representations, and proposes [78b] an integral character formula for irreducible, unitary or non-unitary representations of semisimple Lie groups. This integral formula expresses the characters as Fourier transforms of *certain cycles* in co-adjoint orbits of the complexified group. Rossmann identified these cycles for a complex group. Schmid and Vilonen [*] identify these cycles for any semisimple Lie group.

2.5 Lie Algebras

We now turn to the discussion of representations of Lie algebras. Angelopoulos [*] describes a method which gives necessary and sufficient conditions, in the form of algebraic inequalities, of unitarizability of representations of a real reductive Lie algebra. Let $\mathcal{M}$ be a simple Harish-Chandra $\mathfrak{g}$-module which is *unitarizable.* ($\mathcal{M}$ is *unitarizable* if and only if its sesquilinear form is positive definite). Let V and W be $\mathfrak{k}$-modules. Let $\mathcal{L}(V)$ be the ring of linear self-mappings of V. Let $\mathfrak{X} \in \mathrm{Hom}_{\mathfrak{g}}(\mathcal{L}(V), \mathfrak{g})$ and $\mathscr{C}$ and $\mathscr{T}$ be its characteristic and Taylor polynomials for $t \in \mathbf{C}$ respectively, satisfying

$$(t\sum_u \mathbf{p}_u - \mathfrak{X}) * \mathscr{T}(t) = \mathscr{T}(t) * (t\sum_u \mathbf{p}_u - \mathfrak{X}) = \sum_u \mathbf{p}_u \cdot \mathscr{C}(t),$$

where $\mathbf{p}_u$ is an idempotent element (see the original paper). Let ξ be a root of $\mathscr{C}$. Let W be a $\mathfrak{k}$-isotypic component of $\mathcal{M}$, and W_u be an isotypic component of type $\overline{W}_u$, that is, $V \otimes W = \oplus_u W_u$. Let m_u^v be real-valued affine expressions of the generators of the center $\overline{Z}(\mathfrak{k})$. Then Angelopoulos proves that *for every component W_u of $V \otimes W$ one must have either*

$$\inf_v(m_u^v) \leq \Re e(\xi) \leq \sup_v(m_u^v)$$

or else all elements $\mathscr{T}(\xi)_c^d(\mathbf{p}_u)_d^a$ vanish on W. Furthermore, if ξ fails to satisfy the above inequality for every u, then elements $\mathscr{T}(\xi)_c^d$ vanish on the entire $\mathfrak{k}$-module $\mathcal{M}$.

We now define *Fock space.* It is a Hilbert space $\mathcal{H}$ consisting of all vectors of the form

$$\varphi = \sum_\nu c_\nu \varphi_\nu,$$

where $\nu = \{\nu_1, \nu_2, \cdots, \nu_f\}$ is a set of nonnegative integers, $c_\nu \in \mathbf{C}$ such that $\sum_\nu |c_\nu|^2 < \infty$, φ_ν are orthonormal vectors given by

$$\varphi_\nu = \frac{(a_1^*)^{\nu_1}(a_2^*)^{\nu_2}\cdots(a_f^*)^{\nu_f}}{\sqrt{\nu_1!\nu_2!\cdots\nu_f!}}\varphi_0$$

with φ_0 the only normalized vector (up to a phase) which satisfies the conditions

$$a_i\varphi_0 = 0\,, \quad i = 1, 2, \cdots, f\,,$$

and with *creation* and *annihilation* operators a_i^* and a_i satisfying the commutation relations

$$\left[a_i^*, a_j^*\right] = 0\,, \ [a_i, a_j] = 0\,, \ \left[a_i, a_j^*\right] = \delta_{ij}\,, \ i, j = 1, 2, \cdots, f.$$

The actions of the operators a_i^* and a_i on φ_ν are given by

$$\begin{aligned} a_i^*\,\varphi_{\nu_1\cdots\nu_i\cdots\nu_f} &= \sqrt{\nu_i+1}\,\varphi_{\nu_1\cdots\nu_i+1\cdots\nu_f} \\ a_i\,\varphi_{\nu_1\cdots\nu_i\cdots\nu_f} &= \begin{cases} \sqrt{\nu_i}\,\varphi_{\nu_1\cdots\nu_i-1\cdots\nu_f} & \text{if } \nu_i > 0 \\ 0 & \text{if } \nu_i = 0 \end{cases}. \end{aligned}$$

The requirement $(\varphi_0, \varphi_0) = 1$ and the commutation relations for a_i^* and a_i ensure that the vectors φ_ν form an orthonormal set:

$$(\varphi_\mu, \varphi_\nu) = \delta_{\mu_1\nu_1} \cdots \delta_{\mu_f\nu_f}.$$

Mackey [*] gives an explicit connection between this Fock space construction and Lie algebra roots and weights.

A brief overview of the way the representations of symplectic group $Sp(3, \mathbf{R})$ and of the way its Lie algebra $sp(3, \mathbf{R})$ are used in the microscopic theory of nuclear collective motion is given by Rowe [*]. He introduces as a consequence the concepts of an *embedded representation of* $sp(3, \mathbf{R})$ and *quasi-spinor representations of* $SU(3)$, and illustrates them in connection with heavy nuclei and *strong coupling.*

Klink [*] introduces the (quartic) nilpotent group defined by

$$G = \left\{ \begin{pmatrix} 1 & b & \frac{b^2}{2} & b_3 \\ 0 & 1 & b & b_2 \\ 0 & 0 & 1 & b_1 \\ 0 & 0 & 0 & 1 \end{pmatrix} : \ b, b_1, b_2, b_3 \in \mathbf{R} \right\},$$

(the familiar *Heisenberg group* is a subgroup of G) and associated with the quartic anharmonic oscillator. He gives the relationship between the quartic anharmonic oscillator Hamiltonian and the irreducible representations of the Lie algebra of G. He then generalizes the results for an arbitrary nilpotent Lie group.

The representations of the Lie algebras of the Lorentz group $SO(3,1) \cong SL(2, \mathbf{C})$ [84,85,85], the deSitter group $SO(4,1)$ (or $SO(3,2)$) [23a] and the conformal group $SU(2,2)$
$\cong SO(4,2)$ have been extensively used in physical applications [11c]. A four-dimensional (non-unitary) *Dirac representation* of the Lorentz group is associated with the *Dirac equation* for an *elementary particle* (for example, an *electron*), given by $(\gamma_\mu p^\mu - m)\psi = 0$, $\mu = 0, 1, 2, 3$. Here the 4×4 matrices γ_μ are related to the generators of the Lie algebra, p^μ is *four-momentum*, m is the *mass* of the particle, and ψ is the *spinor* transforming under the four-dimensional representation of the Lorentz group. A simple generalization of the Dirac equation was the *Majorana equation* $(\Gamma_\mu p^\mu - m)\psi = 0$, where now Γ_μ are related to the generators of the Lie algebra of $SO(3,2)$ and ψ transforms under the infinite-dimensional unitary irreducible representation of $SO(3,2)$. The Majorana equation was considered to be a basic equation for a composite system. However, its inadequacy in explaining a simple composite system like the hydrogen atom led Barut to formulate the theory of *dynamical symmetry groups* [6a-c,7,8,11a,31a-c]. He first announced this concept in the Miami conference [6a] in 1964. This dynamical symmetry group theory has provided the tools to obtain physically acceptable solutions to a certain class of physical problems.

Indecomposable representations of Lie groups have recently attracted the attention of both mathematicians and mathematical physicists [16]. In particular, the indecomposable representations of the Poincaré group are associated with *mass-zero* particles and with physical models related to decaying systems [11b]. Conley [*], using Guichardet's earlier work [43a-c], constructs a unique indecomposable representation of the Poincaré group, composed of n copies of the zero-mass and *zero-helicity* representation in terms of smooth compactly supported functions on the *forward light cone.* His result gives an extension of Wigner's *little group* description of the irreducible representations of the Poincaré group.

Heidenreich [*] computes in a conformal space the *invariant causal propagators* for scalar massless particles and for scalar fields with *anomalous dimension.*

The applications of non-abelian groups to *gauge theory* have been discussed in detail in the monographs of Atiyah [3], O'Raifeartaigh [72] and Polyakov [76]. Streater [*] reviews the arguments leading to the existence of qauge groups, emphasizing the dependence of the group on the representation of the observables. He then shows that the anamolies are *two-cycles* in a non-abelian cohomology.

Gürsey [44] in 1974 suggested that the space of internal degrees of freedom of *leptons* and *quarks*, may correspond to the quantum mechanical space of Jordan, von Neumann and Wigner [54]. Further he suggested that a spontaneously broken gauge field theory based on the exceptional Lie group E_7 is a reasonable candidate for the unification of the *weak, electromagnetic* and *strong interactions.* Consequently, the representations of exceptional Lie groups and of their Lie algebras attracted much attention. Recently, Moody and Patera [*] have studied a family of *quasicrystals* of dimensions $1,2,3,4$ governed by the *root lattice* of E_8. The use of *Coxeter diagrams* clarifies the relationship of E_8 and quasicrystal symmetries and leads to the fundamental chain $E_8 \supset D_6 \supset A_4 \supset A_1 \times A_1$ that underlies the 5-fold symmetry of quasicrystals.

Let G be again a locally compact group with left Haar measure $d\mu$ and right Haar measure $d\mu'$. Let $(\rho, \mathcal{H})$ be a continuous unitary irreducible representation of G in a Hilbert space $\mathcal{H}$. Let $(\rho, \mathcal{H})$ be square integrable. If the function $(\rho(g)\varphi, \varphi)$, $g \in G$ is also square integrable with respect to $d\mu'(g)$ then any such vector φ will be called an *admissible analysing wavelet.* Let $\mathscr{A}(\rho)$ denote the set of admissible analysing wavelets. One can show that $\mathscr{A}(\rho)$ is dense in $\mathcal{H}$. Let $\varphi \in \mathscr{A}(\rho)$ and $\psi \in \mathcal{H}$. Then the function $(\rho(g)\varphi, \psi)$ is square integrable with respect to $d\mu$. That is, $(\rho(g)\varphi, \psi) \in \mathcal{L}^2(G, d\mu(g))$. Wells and Zhou [*] formulate and prove a second order interpolation result for square integrable functions by means of locally finite series of *Daubechies'wavelets* (a class of compactly supported wavelet functions). Using their result they derive some error estimates for the wavelet solutions of certain elliptic partial differential equations.

3 Lie superalgebras and Lie supergroups

Let $\mathfrak{g}$ be a finite-dimensional semisimple Lie algebra of rank ℓ. Let $\{\alpha_j\}$ be the set of simple roots, $1 \leq j \leq \ell$. Let $e_{\alpha_j} := e_j$ be any nonnegetive element in the root space $\mathfrak{g}_{\alpha_j}$. Let $e_{-\alpha_j} := f_j$ and the *coroot* $h_j := h_{\alpha_j} \in \mathfrak{h}$, where $\mathfrak{h}$ is a maximal toral subalgebra (or Cartan subalgebra), such that $[e_j, f_j]$ is a multiple of h_j. Then $\mathfrak{g}_{\mathbf{C}}$ is generated by $\{e_i, f_i, h_i, 1 \leq i \leq \ell\}$, and when e_j and f_j are normalized so that $[e_j, f_j] = h_j$ then the following relations define the Lie algebra $\mathfrak{g}_{\mathbf{C}}$:

$$\left.\begin{array}{rcl} [h_i, h_j] & = & 0 \\ {[e_i, f_j]} & = & \delta_{ij} h_i \\ {[h_i, e_j]} & = & a_{ji} e_j \\ {[h_i, f_j]} & = & -a_{ji} f_j \end{array}\right\} \forall\, 1 \leq i, j \leq \ell, \tag{35}$$

where the integers a_{ij} form a unique $\ell \times \ell$ matrix $A = [a_{ij}]$ called the *Cartan matrix* of $\mathfrak{g}$. The Cartan matrix satisfies

1. $a_{ii} = 2,\ 1 \leq i \leq \ell;\ \ a_{ij} \leq 0$ if $i \neq j\,;\ \ a_{ij} = 0 \Rightarrow a_{ji} = 0.$

2. $\det A \neq 0$, and the group generated by the linear transformations s_i, $1 \leq i \leq \ell$ given by $s_i x_j = x_j + a_{ji} x_i$, x_i, $1 \leq i \leq \ell$ being indeterminates, is finite.

Conversely, given any $\ell \times \ell$ matrix A satisfying conditions (1) and (2), the commutation relations (35) define an abstract *finite-dimensional* semisimple Lie algebra $\widetilde{\mathfrak{g}}(A)$ which is isomorphic to $\mathfrak{g}_{\mathbf{C}}$. However, if the matrix A satisfies only the condition (1) (*generalized Cartan matrix*) then the commutation relations (35) define an *infinite-dimensional* Lie algebra called *Kac-Moody Lie algebra* (*graded Lie algebra*, or *pseudo Lie algebra*).

In fact, the Kac-Moody Lie algebra is one of four classes of infinite-dimensional Lie algebras that have undergone extensive study. These four classes are as follows:

1. Simple infinite-dimensional Lie algebras of vector fields on a finite-dimensional manifold, classified by Cartan.

2. Lie algebras of smooth mappings of a given manifold into a finite-dimensional Lie algebra. Certain central extensions of Lie algebras in this class are referred to as *current algebras.*

3. Lie algebras of operators in a Hilbert space or Banach space. A representation of an infinite-dimensional symplectic Lie algebra in this class, called Segal-Shale-Weil (or *metaplectic*) representation, plays an important role in *quantum field theory.*

4. Kac-Moody algebras.

The algebraic structure and several applications of the Kac-Moody algebras to other topics in mathematics and physics have been discussed in detail in [12,21,55,63,64,68a-c,80]. The Kac-Moody Lie algebras with generalized Cartan matrices A satisfying the condition (1) and the condition: $\det A = 0$ *and all the proper principal minors of A are greater than zero*, are usually called *affine Lie algebras.* All affine Lie algebras admit an invariant symmetric bilinear form, and their root systems are well understood [46b]. Spinor representations of affine Lie algebras are studied by Frenkel [30].

A Lie superalgebra, a Z_2-graded Lie algebra, is a real or complex Z_2-graded vector space $\mathcal{L} = \mathcal{L}_0 \oplus \mathcal{L}_1$ with a fixed *parity* in which a bilinear operation $[,]$ is given such that

1. $[x, y] \in \mathcal{L}_0$ if $x, y \in \mathcal{L}_0$ or $x, y \in \mathcal{L}_1$,
2. $[x, y] \in \mathcal{L}_1$ if $x \in \mathcal{L}_0$, $y \in \mathcal{L}_1$, or *vice versa,*
3. $[x, y] = (-1)^{\alpha(x)\alpha(y)+1}\,[y, x]$, and
4. $[x, [y, z]]\,(-1)^{\alpha(x)\alpha(z)} + [z, [x, y]]\,(-1)^{\alpha(z)\alpha(y)} + [y, [z, x]]\,(-1)^{\alpha(y)\alpha(x)} = 0,$

where $\alpha(t)$ is the degree of t defined by $\alpha(t) = 0$ ($\alpha(t) = 1$) if $t \in \mathcal{L}_0$ ($t \in \mathcal{L}_1$) satisfying $\alpha([x, y]) = \alpha(x) + \alpha(y)$. $\mathcal{L}_0$ is a Lie algebra, called the *even* part of $\mathcal{L}$, and $\mathcal{L}_1$ is called the *odd* part of $\mathcal{L}$. In a matrix representation let $x = \begin{pmatrix} A & B \\ C & D \end{pmatrix} \in \mathcal{L}$, where A and D are square and B and C are rectangular matrices such that $\begin{pmatrix} A & 0 \\ 0 & D \end{pmatrix} \in \mathcal{L}_0$ and $\begin{pmatrix} 0 & B \\ C & 0 \end{pmatrix} \in \mathcal{L}_1$. Then one defines respectively *supertrace* and *superdeterminant* by

$$\begin{aligned} \mathrm{str}x &= \mathrm{tr}A - \mathrm{tr}D. \\ \mathrm{sdet}x &= \det(A - BD^{-1}C)\det D^{-1}. \end{aligned}$$

The generalization of the *Schur lemma* for Lie superalgebras is given as follows: *Let $(\rho, \mathcal{V})$ be an irreducible representation of $\mathcal{L}$ on the Z_2-graded vector space $\mathcal{V} = \mathcal{V}_0 \oplus \mathcal{V}_1$. Let $k \in \mathcal{L}$ be such that $[k, x] = 0\ \forall x \in \mathcal{L}$. Then either k is a multiple of the unit matrix or if* $\dim\mathcal{V}_0 = \dim\mathcal{V}_1$ *then k is a nonsingular matrix such that $k\mathcal{V}_{0(1)} \subseteq \mathcal{V}_{1(0)}$.*

Simple Lie superalgebras are classified generally into two classes depending upon whether the $\mathcal{L}_0$-module $\mathcal{L}_1$ is completely reducible (*classical Lie superalgebras*) or *not* completely reducible (*Cartan Lie superalgebras*). Specific details can be found in [80].

The concept of a Lie supergroup is a generalization of the concept of a Lie group. The idea of this generalization can be found in [12]. Hussin and Nieto [*]

determine, using a matrix realization, a generic element of the classical supergroup $OSP(m/2n)$ by exponentiation of the corresponding superalgebra element.

Lie superalgebras and Lie supergroups have extensive applications in many areas, some of which are number theory (modular forms, Dedekind's η-funtions [55]), topology (loop spaces, loop groups [77]), completely integrable systems (Toda field theories [53]), and quantum field theory (current algebras, supersymmetry, supergravity [33]). Mickelsson [*] discusses a generalized representation theory of current algebras in 3 + 1 dimensions and describes rules for a systematic computation of *vacuum expectation values* of products of currents. Niederle [*] gives a non-linear realization of a supergauge group and its implications for supergravity. O'Raifeartaigh [*] views the non-linear *Toda system* as a linearly constrained system, which enables him to obtain the general solutions and symmetry algebras of the Toda theories.

4 Quantum groups

For this introduction to quantum groups we follow closely the Trieste lectures of Guichardet [43d]. Let X ba a vector space over a field k. Let $\widetilde{X}$ be the space of all formal series

$$\widetilde{x} = \sum_{n\geq 0} x_n h^n \, , \; x_n \in X \; \forall n \geq 0,$$

where h is a variable. By this definition $\widetilde{k}$ is a commutative ring and $\widetilde{X}$ is a $\widetilde{k}$-module with the mapping $\widetilde{k} \times \widetilde{X} \to \widetilde{X}$ defined by

$$(\sum_i \lambda_i h^i, \sum_j x_j h^j) \mapsto \sum_n (\sum_{i+j=n} \lambda_i x_j) \cdot h^n,$$

where $\lambda_i \in k$ and $x_j \in X$. The tensor product space $X \otimes_k \widetilde{k}$ is dense in $\widetilde{X}$, and $X \otimes_k \widetilde{k} = \widetilde{X}$ if and only if X is finite-dimensional. One can show that *given two vector spaces X and Y over k, there exists*

$$f \in \mathrm{Iso}(\widetilde{(\mathrm{Hom}_k(X,Y))}, \mathrm{Hom}_{\widetilde{k}}(\widetilde{X}, \widetilde{Y}))$$

defined by

$$(f(\widetilde{u}))(\widetilde{x}) = \sum_n (\sum_{i+j=n} u_i(x_j)) \cdot h^n \in \widetilde{Y},$$

where

$$\widetilde{u} = \sum_{n\geq 0} u_n h^n \, , \; u_n \in \mathrm{Hom}_k(X,Y) \, , \; \widetilde{x} \in \widetilde{X}.$$

Let us now consider three spaces $\widetilde{X}, \widetilde{Y}$ and $\widetilde{Z}$. Let $S_{\widetilde{k}}$ denote the space of all $\widetilde{k}$-bilinear mappings $\widetilde{X} \times \widetilde{Y} \to \widetilde{Z}$. Since

$$\begin{aligned} S_{\widetilde{k}} &\cong \mathrm{Hom}_{\widetilde{k}}((X \otimes Y)\tilde{}, \widetilde{Z}), \\ &\cong \mathrm{Hom}_k((X \otimes Y), Z), \end{aligned}$$

it is natural that the tensor product of $\widetilde{X}$ and $\widetilde{Y}$ will be $(X \otimes Y)\tilde{}$. Let

$$\widetilde{u}^{(i)} : \widetilde{X}^{(i)} \to \widetilde{Y}^{(i)}, \quad i = 1, 2$$

be $\widetilde{k}$-linear mappings. Then the tensor product of two such mappings

$$\widetilde{u}^{(1)} \otimes \widetilde{u}^{(2)} : (X^{(1)} \otimes X^{(2)})\tilde{} \to (Y^{(1)} \otimes Y^{(2)})\tilde{}$$

is defined by

$$\widetilde{u}^{(1)} \otimes \widetilde{u}^{(2)} : \sum_i \xi_i h^i \mapsto \sum_n \Big(\sum_{i+j+m=n} (u_i^{(1)} \otimes u_j^{(2)})(\xi_m) \Big) \cdot h^n .$$

Let $\mathfrak{A}$ be an algebra over k with unit I. Let $\mu : \mathfrak{A} \otimes \mathfrak{A} \to \mathfrak{A}$ be the *mutiplication* map. The $\widetilde{k}$-module $\widetilde{\mathfrak{A}}$ acquires a canonical algebraic structure over $\widetilde{k}$ by the composition

$$\sum_i a_i h^i \cdot \sum_j b_j h^j = \sum_n \Big(\sum_{i+j=n} (a_i b_j) \Big) \cdot h^n .$$

A *formal deformation* (in the sense of Gerstenhaber [36a,b,37]) of $\mathfrak{A}$ is a pair $(\widetilde{\mathfrak{A}}, \widetilde{\mu})$ such that $\widetilde{\mu} : \widetilde{\mathfrak{A}} \otimes \widetilde{\mathfrak{A}} \to \widetilde{\mathfrak{A}}$ is defined by

$$\widetilde{\mu} : \Big(\sum_i a_i h^i, \sum_j b_j h^j \Big) \mapsto \sum_n \Big(\sum_{i+j+m=n} \mu_m(a_i . b_j) \Big) \cdot h^n ,$$

where $\mu_m : \mathfrak{A} \times \mathfrak{A} \to \mathfrak{A}$ is a sequence of k-bilinear mappings that satisfy

1. $\mu_0 = \mu$ (as $\widetilde{\mathfrak{A}}/h\widetilde{\mathfrak{A}} \cong \mathfrak{A}$).

2. The associative condition:

$$\sum_{i+j=n} \mu_i(\mu_j(a, b), c) = \sum_{i+j=n} \mu_i(a, \mu_j(b, c)) \ \forall n \geq 1, \ a, b, c \in \mathfrak{A}.$$

3. $\mu_n(a, I) = \mu_n(I, a) = 0 \ \forall n \geq 0, \ a \in \mathfrak{A}$.

4. $a \cdot \mu_1(b, c) - \mu_1(a, b) \cdot c = \mu_1(ab, c) - \mu_1(a, bc)$.

5. For $n \geq 2$,

$$\sum_{i=1}^{n-1} \mu_i(\mu_{n-i}(a,b),c)) - \sum_{i=1}^{n-1} \mu_i(a,\mu_{n-i}(b,c)) =$$
$$a \cdot \mu_n(b,c) - \mu_n(a,b) \cdot c + \mu_n(a,bc) - \mu_n(ab,c).$$

A formal deformation is said to be *constant* if and only if $\mu_n = 0 \; \forall n > 0$. Two formal deformations $(\widetilde{\mathfrak{A}}, \widetilde{\mu})$ and $(\widetilde{\mathfrak{A}}, \widetilde{\mu'})$ are called *equivalent* if there exists a sequence of linear mappings $u_n \in \mathrm{End}_k(\mathfrak{A})$ satisfying

1. $u_0 = I_{\mathfrak{A}}$

2. and

$$\sum_{i+j+m=n} \mu_i(u_j(a), u_m(b)) = \sum_{i+j=n} u_i(\mu'_j(a,b)).$$

A formal deformation is called *trivial* if it is equivalent to the constant deformation. That is, if the exact sequence

$$\{0\} \to h\widetilde{\mathfrak{A}} \to \widetilde{\mathfrak{A}} \to \mathfrak{A} \to \{0\}$$

is split.

Let $(\widetilde{\mathfrak{A}}, \widetilde{\mu})$ be a formal deformation of $\mathfrak{A}$. A *representation* of $\widetilde{\mathfrak{A}}$ is a pair $(\widetilde{\rho}, \widetilde{V})$, where V is a vector space over k, and

$$\widetilde{\rho} : \widetilde{\mathfrak{A}} \to \mathrm{End}_{\widetilde{k}}(\widetilde{V}) \cong (\mathrm{End}_k(V))\widetilde{\;}$$

is a $\widetilde{k}$-linear and multiplicative mapping. Let

$$\rho_n \in \mathrm{Hom}_k(\mathfrak{A}, \mathrm{End}_k(V))$$

be a sequence of mappings such that

$$\sum_{i+j=n} \rho_i(\mu_j(a,b)) = \sum_{i+j=n} \rho_i(a) \cdot \rho_j(b)$$

and (ρ_0, V) is a representation of $\mathfrak{A}$. We say that $\widetilde{\rho}$ is a *formal deformation* of ρ_0. The representation $(\widetilde{\rho}, \widetilde{V})$ has *finite rank n* if V has finite dimension n. For a constant deformation $(\widetilde{\mathfrak{A}}, \widetilde{\mu})$ one obtains

$$\rho_n(a,b) = \sum_{i+j=n} \rho_i(a) \cdot \rho_j(b).$$

The *constant deformation* is defined by $\rho_n = 0 \; \forall n > 0$. We say that the representations $(\widetilde{\rho}, \widetilde{V})$ and $(\widetilde{\sigma}, \widetilde{W})$ are *equivalent* if there exists a $\widetilde{k}$-isomorphism $\widetilde{u} : \widetilde{V} \to \widetilde{W}$

transforming $\widetilde{\rho}$ to $\widetilde{\sigma}$. That is to say, there exists $u_n \in \mathrm{Hom}(V, W)$ such that u_0 is bijective, with

$$\sum_{i+j=n} u_i \circ \rho_j(a) = \sum_{i+j=n} \sigma_j(a) \circ u_i.$$

A formal deformation $\widetilde{\rho}$ is said to be *trivial* if it is equivalent to the constant one. Let $(\widetilde{\rho^{(i)}}, \widetilde{V^{(i)}})$ be a family of representations of $\widetilde{\mathfrak{A}}$. Then there exists a representation $(\widetilde{\rho}, \widetilde{V})$ of $\widetilde{\mathfrak{A}}$ in $\widetilde{V} = \oplus_i \widetilde{V^{(i)}}$ such that

$$\widetilde{\rho}(\widetilde{a}) = \oplus_i \widetilde{\rho^{(i)}}(\widetilde{a}).$$

Let $[\mathfrak{A}]$ (resp. $\left[\widetilde{\mathfrak{A}}\right]$) denote the set of equivalence classes of finite-dimensional representations of $\mathfrak{A}$ (resp. finite rank representations of $\widetilde{\mathfrak{A}}$). The following theorem can be proved [24a,b,43d] using some of the concepts in Lie group cohomology: *Let $\mathfrak{g}$ be a semisimple Lie algebra over a field k of characteristic zero. Let $\mathfrak{U}(\mathfrak{g}) := \mathfrak{U}$ be the universal enveloping algebra of $\mathfrak{g}$. Then*

1. *Every formal deformation $(\widetilde{\mathfrak{U}}, \widetilde{\mu})$ is trivial.*
2. *There is a bijection $\left[\widetilde{\mathfrak{U}}\right] \to [\mathfrak{U}]$ which preserves direct sums and indecomposablity by associating with every finite rank representation $(\widetilde{\rho}, \widetilde{V})$ of $\widetilde{\mathfrak{U}}$ its component ρ_0.*
3. *Every finite rank representation of $\widetilde{\mathfrak{U}}$ is a direct sum of indecomposable representations.*

We now define *Hopf algebra.* It is an associative algebra $\mathfrak{A}$ over k with unit I and multiplication $\mu : \mathfrak{A} \otimes \mathfrak{A} \to \mathfrak{A}$, equipped with a linear mapping called the *comultiplication* $\Delta : \mathfrak{A} \to \mathfrak{A} \otimes \mathfrak{A}$ defined by

$$\Delta : a \mapsto \sum_i a'_i \otimes a''_i \quad \forall a \in \mathfrak{A}$$

(the pair $(\mathfrak{A}, \Delta)$ is called a *coalgebra*) and a linear mapping ϵ called the *counit* where $\epsilon : \mathfrak{A} \to k$, such that the following conditions are satisfied:

1. Δ is a morphism such that

$$\Delta \circ \mu = (\mu \otimes \mu) \circ \pi_{23} \circ (\Delta \otimes \Delta),$$

where π_{23} is the (permutation) automorphism of $\mathfrak{A} \otimes \mathfrak{A} \otimes \mathfrak{A} \otimes \mathfrak{A}$ defined by $\pi_{23} : (a_1 \otimes a_2 \otimes b_1 \otimes b_2) \mapsto a_1 \otimes b_1 \otimes a_2 \otimes b_2$.

2. Δ is *coassociative*, that is,

$$(\Delta \otimes I_{\mathfrak{A}}) \circ \Delta = (I_{\mathfrak{A}} \otimes \Delta) \circ \Delta : \mathfrak{A} \to \mathfrak{A} \otimes \mathfrak{A} \otimes \mathfrak{A}.$$

3. ϵ is a morphism such that

$$\sum_i \epsilon(a_i') \cdot a_i'' = \sum_i \epsilon(a_i'') \cdot a_i' = a \in \mathfrak{A}.$$

4. There exists a linear mapping called the *antipode* $S : \mathfrak{A} \to \mathfrak{A}$ satisfying

$$\mu \circ (S \otimes I_{\mathfrak{A}}) \circ \Delta = \mu \circ (I_{\mathfrak{A}} \otimes S) \circ \Delta = I \circ \epsilon.$$

That is,

$$\sum_i S(a_i') \cdot a_i'' = \sum_i a_i' \cdot S(a_i'') = \epsilon(a) I \ \ \forall a \in \mathfrak{A}.$$

One can show that S is unique and is an anti-endomorphism, that is $S(ab) = S(b) \cdot S(a)$. A Hopf algebra is therefore denoted by a sextuplet $(\mathfrak{A}, \mu, I, \Delta, \epsilon, S)$ whereas a *bialgebra* is denoted by the quintuplet $(\mathfrak{A}, \mu, I, \Delta, \epsilon)$.

The formal deformation of a Hopf algebra denoted (in short) by $\mathfrak{A}$ is defined as a formal deformation $(\widetilde{\mathfrak{A}}, \widetilde{\mu})$ of the algebra $(\mathfrak{A}, \mu)$ equipped with a sequence of linear mappings $\Delta_n : \mathfrak{A} \to \mathfrak{A} \otimes \mathfrak{A}$ satisfying

$$\begin{aligned}
\Delta_0 &= \Delta, \\
0 &= \sum_{i+j=n} (\Delta_i \otimes I_{\mathfrak{A}} - I_{\mathfrak{A}} \otimes \Delta_i) \circ \Delta_j, \\
\sum_{i+j=n} \Delta_i \circ \mu_j &= \sum_{i+j+m+r=n} (\mu_i \otimes \mu_j) \circ \pi_{23} \circ (\Delta_m \otimes \Delta_r), \\
(\epsilon \otimes I) \circ \Delta_n &= (I \otimes \epsilon) \circ \Delta_n = 0 \ \ \forall n > 0.
\end{aligned}$$

Let (ρ, V) and (σ, W) be two representations of a Hopf algebra $\mathfrak{A}$. Let $(\widetilde{\rho}, \widetilde{V})$ and $(\widetilde{\sigma}, \widetilde{W})$ be representations of a formal deformation $\widetilde{\mathfrak{A}}$ of the Hopf algebra $\mathfrak{A}$. Then the *tensor product representation* $(\rho \otimes \sigma, V \otimes W)$ is defined by

$$(\rho \otimes \sigma)(a) = (\rho \times \sigma)(\Delta(a)) = \sum_i \rho(a_i') \otimes \sigma(a_i'')$$

and the components of the tensor product representation $(\widetilde{\rho} \otimes \widetilde{\sigma}, \widetilde{(V \otimes W)})$ are given by

$$\begin{aligned}
(\widetilde{\rho} \otimes \widetilde{\sigma})_n &= \sum_{i+j+m=n} (\rho_i \times \sigma_j) \circ \Delta_m, \\
(\widetilde{\rho} \otimes \widetilde{\sigma})_0 &= (\rho_0 \times \sigma_0) \circ \Delta_0 = \rho_0 \otimes \sigma_0.
\end{aligned}$$

It is proved [43d] that *the bijection* $\left[(\widetilde{\mathfrak{U}},\widetilde{\mu})\right] \to [\mathfrak{U}]$ *is compatible with tensor products.*

We now define a *Poisson bracket* $\{,\} : \mathfrak{A} \times \mathfrak{A} \to \mathfrak{A}$ such that the mapping $\mathfrak{A} \to \mathfrak{A}$ defined by $a \mapsto \{a,c\}$ $\forall c \in \mathfrak{A}$ is a *derivation* on the commutative algebra $(\mathfrak{A}, \mu, I)$. Then the quartet $(\mathfrak{A}, \mu, I, \{,\})$ is called a *Poisson algebra.* The algebra morphism Δ is called a *Poisson algebra morphism* if it is compatible with the Poisson bracket $\{,\}$. Then the object $(\mathfrak{A}, \mu, I, \Delta, \epsilon, S, \{,\})$ is called a *Poisson- Hopf algebra.*

Poisson-Hopf algebras with the formal deformation (quantum) parameter h (Planck's constant) are often referred to as *Quantum groups.* Manin [66] and Woronowicz [96] introduced an alternative notion of a *q-deformation* of the commutative **C**-Hopf algebra into a noncommutative **C**-Hopf algebra. Gugenheim [42] studied the extensions of Hopf algebras. Guichardet [*] discusses in detail the definitions and construction of the quantum group $\mathfrak{U}_h(sl(2,k))$. Representations of quantum groups have also been studied, for example, in [1,40,75,87]. Pusz and Woronowich [*] give their recent results concerning representation theory of the quantum Lorentz group. Irreducible representations of the quantum group Gl_n are discussed in [22,25,73]. Fronsdal [*] discusses in detail the *Universal T-matrix* for multi-parameter (*twisted*) quantum algebra $gl(N)$. As immediate applications, Giachetti [*] gives a contraction of $SU_q(2)$ and applies it to harmonic excitations of a crystal and spin chains. Biedenharn and Tarlini [*] discuss a quantum Poincaré group and apply it to the Dirac-Coulomb problem.

In conclusion we would just like to mention that this introduction to the study of Lie groups and Lie algebras is by no means complete. There are several other intriguing areas related to this field which are certainly worth study. Although the inclusion of such a comprehensive introduction is not appropriate to this volume of lectures, we would like to encourage the reader to consult other recent literature in this field.

BIBLIOGRAPHY

1. H. H. Anderson, P.Polo and W. Kexin, *Representations of quantum algebras*, Invent. Math. **104** (1991), 1-59.

2. E. Artin, *Geometric Algebra*, John Wiley & Sons, New York, 1957.

3. M. F. Atiyah, *Geometry of Yang-Mills Field*, Scuola Normale Superiore, Pisa, 1979.

4. L. Auslander and B. Kostant, *Polarization and unitary representations of solvable Lie groups*, Invent. Math. **14** (1971), 255-354.

5. V. Bargmann, *Irreducible unitary representations of Lorentz group*, Ann. of Math. **48** (1947), 568-640.

6. A. O. Barut, (a) *Dynamical symmetry groups and mass spectrum for elementary particles*, Proceedings of Conference on Symmetry at High Energy, University of Miami, Florida, 1964.
(b) *Some unusual applications of Lie algebra representations in quantum theory*, SIAM J. Appl. Math. **25** (1973), 247-259.
(c) *Dynamical group quantization*, Rept. Math. Phys. **11** (1977), 401-413.
(d) *Geometry and Physics*, Bibliopolis, Napoli, 1989.

7. A. O. Barut, P. Budini and C. Fronsdal, *Two examples of covariant theories with internal symmetries involving spin*, Proc. Roy. Soc. London **291** (1966), 106-112.

8. A. O. Barut and C. Fronsdal, *On noncompact groups II. The* $2+1$ *Lorentz group*, Proc. Roy. Soc. London **287** (1965), 532-548.

9. A. O. Barut and L. Girardello, *New coherent states associated with noncompact groups*, Commun. Math. Phys. **21** (1971), 41-55.

10. A. O. Barut and E. Phillips, *Indecomposable representations of Lie algebras and Lie groups*, Symposia Mathematica **31** (1989), 197-213.

11. A. O. Barut and R. Raczka, (a) *On noncompact groups I. Classification of noncompact real simple Lie groups and groups containing the Lorentz group*, Proc. Roy. Soc. London **287** (1965), 519-531.
(b) *Properties of nonunitary zero-mass induced representations of the Poincaré group on the space of tensor-valued functions*, Ann. Inst. H. Poincaré **A17** (1972), 111-118.
(c) *Theory of Group Representations and Applications*, World Scientific, Singapore, 1987.

12. F. A. Berezin, *Introduction to Superanalysis*, D. Reidel Pub. Company, Dordrecht, 1987.

13. T. S. Bhanu-Murthy, (a) *Plancherel's measure for the factor space* $SL(n,\mathbf{R})/SO(n)$, Dokl. Akad. Nauk SSSR **133** (1960), 503-506.
(b) *The asymtotic behavior of zonal spherical functions on the Siegel upper-half plane*, Dokl. Akad. Nauk SSSR **135** (1960), 1027-1030.

14. A. Borel, *Linear Algebraic Groups*, Springer-Verlag, New York, 1991.

15. F. Bruhat, *Sur les reprèsentations induites der groupe de Lie*, Bull. Soc. Math. France **84** (1956), 97-205.

16. V. Cantoni, (editor) *Indecomposable Representations of Lie Groups and their Physical Applications*, Academic Press, London, 1989.

17. E. Cartan, (a) *Über die einfachen transformationsgruppen, Sur la structure des groupes de transformations finis et continus, Les groupes réels simples, finis et continus*, Oeuvres Complètes, Gauthier-Villars, Paris, 1952.
(b) *On Manifolds with an Affine Connection and the Theory of General Relativity*, Bibliopolis, Napoli, 1986.

18. E. Cartan and A. Einstein, *Letters on Absolute Parallelism 1929-1932*, Princeton University Press, Princeton, 1979.

19. C. Chevalley, *Theory of Lie Groups*, Princeton University Press, Princeton, 1946.

20. G. de Rham, *Differentiable Manifolds*, Springer-Verlag, Berlin, 1984.

21. B. DeWitt, *Supermanifolds*, Cambridge University Press, Cambridge, 1984.

22. R. Dipper and S. Donkin, *Quantum* GL_n, Proc. Roy. Soc. London **63** (1991), 165-211.

23. J. Dixmier, (a) *Représentations intégrables du groupe deSitter*, Bull. Soc. Math. France **89** (1961), 9-41.
(b) *Les Algèbres d'Opérateurs dans L'espace Hilbertien (Algebres de von Neumann)*, Gauthier-Villars, Paris, 1969.
(c) *Enveloping Algebras*, North-Holland, Amsterdam, 1977.
(d) C^**-Algebras*, North-Holland, Amsterdam, 1982.

24. V. G. Drinfel'd, (a) *Quantum groups*, J. of Soviet Math. **41** (1988), 898-915.
(b) *On almost cocommtative Hopf algebras*, Lenningrad Math. J. **1** (1990), 321-342.

25. J. Du, *Canonical bases for irreducible representations of quantum* Gl_n, Bull. London Math. Soc. **24** (1992), 325-334.

26. L. P. Eisenhart, *Continuous Groups of Transformations.* Dover Publications, New York, 1961.

27. J. A. Ernest, *A decomposition theory for unitary representations of locally compact groups*, Trans. Amer. Math. Soc. **104** (1962), 252-277.

28. R. P. Feynman and A. R. Hibbs, *Quantum Mechanics and Path Integrals*, McGraw-Hill, New York, 1965.

29. M. Flensted-Jensen, (a) *Spherical functions on a real semisimple Lie group. A method of reduction to the complex case*, J. Funct. Anal. **30** (1978), 106-146.

(b) *Discrete series for semisimple symmetric spaces*, Ann. of Math. **111** (1980), 253-311.
(c) *Analysis on Non-Riemannian Symmetric Spaces*, CBMS Regional Conference series no.**61**, Amer. Math. Soc., Providence, 1980.

30. I. B. Frenkel, *Spinor representations of affine Lie algebras*, Proc. Natl. Acad. Sci. USA **77** (1980), 6303-6306.

31. C. Fronsdal, (a) *Infinite multiplets and local fields*, Phys. Rev. **156** (1967), 1653-1664.
(b) *Infinite multiplets and the Hydrogen atom*, Phys. Rev. **156** (1967), 1665-1677.
(c) *Relativistic Lagrangian field theory for composite systems*, Phys. Rev. **171** (1968), 1811-1824.

32. F. Gantmacher, *On the classification of real simple Lie groups*, Math. Sbornik **5** (1939), 217-249.

33. K. Gawedzki, *Supersymmetries - Mathematics of supergeometry*, Ann. Inst. H. Poincaré **27** (1977), 335-366.

34. I. M. Gel'fand and M. A. Naimark, *Unitäre Darstellungen der Klassischen Gruppen*, Akademische Verlag, Berlin, 1957.

35. I. M. Gel'fand and D. A. Raikov, *Irreducible unitary representations of locally compact groups*, Math. Sbornik **55** (1943), 301-316.

36. M. Gerstenhaber, (a) *On the deformation of rings and algebras*, Ann. of Math. **79** (1964), 59-103.
(b) *On the deformation of rings and algebras III*, Ann. of Math. **88** (1968), 1-34.

37. M. Gerstenhaber and S. D. Schack, *Algebras, bialgebras, quantum groups, and algebraic deformations*, Contemporary Math. **134** (1992), 51-92.

38. S. G. Gindikin and F. I. Karpelevič, *Plancherel measure of Riemann symmetric spaces of nonpositive curvature*, Soviet Math. **3** (1962), 962-965.

39. R. Godement, (a) *Une généralisation du théoréme de la moyenne pour les fonctions harmoniques*, C. R. Acad. Sci. Paris **234** (1952), 2137-2139.
(b) *Introduction aux Travaux de A. Selberg*, Séminaire Bourbaki no.**144**, Paris, 1957.

40. E. C. Gootman and A. J. Lazar, *Quantum groups and duality*, Rev. in Math. Phys. **5** (1993), 417-451.

41. A. Grothendieck, *Topological Vector Spaces*, Gordon and Breach, London, 1973.

42. V. K. A. M. Gugenheim, *On extensions of algebras, coalgebras and Hopf algebras I*, Amer. J. Math. **84** (1964), 349-382.

43. A. Guichardet, (a) *Sur un problème posé par G. Mackey*, C. R. Acad. Sci. Paris **250** (1960), 962-963.
(b) *Extensions de représentations induites des produits semidirects*, J. für reine angew. Math. **310** (1979), 7-32.
(c) *Représentations de longeur finie des groups de Lie inhomogènes*, Astérisque **124-125** (1985), 212-252.
(d) *Introduction aux groupes quantiques (Point de vue formel)*, Trieste lectures, 1993.

44. F. Gürsey, in *Johns Hopkins Workshop on Current Problems in High Energy Particle Theory*, page 15, John Hopkins University Press, Baltimore, 1974.

45. Harish-Chandra, (a) *Representations of a semisimple Lie group on a Banach space I-VI*, Trans. Amer. Math. Soc. **75** (1953), 185-243; **76** (1954), 26-65; **76** (1954), 234-253; Amer. J. Math. **77** (1955), 743-777; **78** (1956), 1-41; **78** (1956), 564-628.
(b) *On characters of semisimple Lie groups*, Bull. Amer. Math. Soc. **61** (1955), 389-396.
(c) *The characters of semisimple Lie groups*, Trans. Amer. Math. Soc. **83** (1956), 98-163.
(d) *Discrete series for semisimple Lie groups I,II*, Acta Math. **113** (1965), 241-318; **116** (1966), 1-111.
(e) *Invariant eigendistributions on a semisimple Lie group*, Trans. Amer. Math. Soc. **119** (1965), 457-508.
(f) *Two theorems on semisimple Lie groups*, Ann. of Math. **83** (1966), 74-128.

46. S. Helgason, (a) *Differential operators on homogeneous spaces*, Acta Math. **102** (1959), 239-299.
(b) *Differential Geometry, Lie Groups and Symmetric Spaces*, Academic Press, New York, 1978.
(c) *Groups and Geometric Analysis*, Academic Press, Orlando, 1984.
(d) *Wave equations on homogeneous spaces*, Lecture Notes in Math. (Springer-Verlag), **1077** (1984), 254-287.

47. T. Hirai, (a) *Classification and the characters of irreducible representations of* $SU(p,q)$, Proc. Japan Acad. **42** (1966), 907-912.
(b) *The characters of some induced representations of semisimple Lie groups*, J. Math. Kyoto University **8** (1968), 313-363.

48. G. P. Hochschild, *Basic Theory of Algebraic Groups and Lie Algebras*, Springer-Verlag, New York, 1981.

49. L. Hörmander, *An Introduction to Complex Analysis in Several Variables*, North-Holland, Amsterdam, 1979.

50. L. K. Hua, *Harmonic Analysis of Functions of Several Complex Variables in the Classical Domains*, Amer. Math. Soc., Providence, 1963.

51. A. Inomata, H. Kuratsuji and C. C. Gerry, *Path Integrals and Coherent States of $SU(2)$ and $SU(1,1)$*, World Scientific, Singapore, 1992.

52. N. Jacobson, *Lie Algebras*, Dover Publications, New York, 1979.

53. M. Jimbo and T. Miwa, *Solitons and infinite dimensional Lie algebras*, Publ. RIMS Kyoto University **19** (1983), 943-1001.

54. P. Jordan, J. von Neumann and E. P. Wigner, *On the algebraic generalization of the quantum mechanical formalism*, Ann. of Math. **35** (1935), 29-64.

55. V. G. Kac, *Infinite Dimensional Lie Algebras*, Birkhäuser, Boston, 1983.

56. W. Killing, *Die Zussammensetzung der stetigen endlichen transformations-gruppen I,II,II*, Math. Ann. **31** (1888), 252-290; **33** (1888), 1-48; **34** (1889), 57-122.

57. A. A. Kirillov, *Elements of the Theory of Representations*, Springer-Verlag, Berlin, 1976.

58. F. Klein, *Vergleichende Betrachtungen über neuere geometrische Forschungen*, Math. Ann. **43** (1893), 63-100.

59. A. W. Knapp, (a) *Representation Theory of Semisimple Groups*, Princeton University Press, Princeton, 1986.
(b) *Lie Groups, Lie Algebras, and Cohomology*, Princeton University Press, Princeton, 1988.

60. G. Köthe, *Toplogical Vector Spaces I*, Springer-Verlag, New York, 1969.

61. P. Kustaanheimo and E. Stiefel, *Perturbation theory of Kepler motion based on spinor regularization*, J. Reine Angew. Math. **218** (1965), 204-219.

62. R. P. Langlands, (a) *Dimension of spaces of automorphic forms* in *Algebraic Groups and Discontinuous Subgroups*, Proc. Symp. in Pure Math. **9** (1966), 253-257.
(b) *On the classification of irreducible representations of real algebraic groups* in *Representation Theory and Harmonic Analysis on Semisimple Lie Groups*, Mathematical Surveys and Monographs no.**31**, Amer. Math. Soc., Providence, 1989.

63. D. A. Leites, *Introduction to the theory of supermanifolds*, Russian Math. Surveys **35** (1980), 1-64.

64. I. G. Nacdonald, *Kac-Moody algebra* in *Lie Algebras and Related Topics*, CMS Conference Proceedings no.**5** (1986), 69-135.

65. G. W. Mackey, (a) *Imprimitivity for representations of locally compact groups, I*, Proc. Nat. Acad. Sci. USA **35** (1949), 537-545.
(b) *On induced representation of groups*, Amer. J. Math. **73** (1951), 576-592.
(c) *Induced representations of locally compact groups, I,II*, Ann. of Math. **55** (1952), 101-139; **58** (1953), 193-221.
(d) *Borel structure in groups and their duals*, Trans. Amer. Math. Soc. **85** (1957), 134-165.
(e) *Unitary representations of group extensions, I*, Acta Math. **99** (1958), 265-311.
(f) *Infinite-dimensional group representations*, Bull. Amer. Math. Soc. **69** (1963), 628-686.
(g) *The Scope and History of Commutative and Noncommutative Harmonic Analysis*, Amer. Math. Soc., Providence, 1992.

66. Yu. I. Manin, *Quantum Groups and Noncommutative Geometry*, Les Publications du Centre de Recherches Mathématiques, Montréal, 1988.

67. F. I. Mautner, *Unitary representations of locally compact groups, I,II*, Ann. of Math. **51** (1950), 1-25; **52** (1950), 528-556.

68. R. V. Moody, (a) *Lie algebras associated with generalized Cartan matrices*, Bull. Amer. Math. Soc. **73** (1967), 217-221.
(b) *A new class of Lie algebras*, J. of Algebra **10** (1968), 211-230.
(c) *Euclidean Lie algebras*, Canad. J. Math. **21** (1969), 1432-1454.

69. M. A. Naimark, *Normed Algebras*, Wolters-Noordhoff, Groningen, 1972.

70. M. A. Naimark and A. I. Štern, *Theory of Group Representations*, Springer-Verlag, New York, 1982.

71. A. Newlander and L. Nirenberg, *Complex analytic coordinates in almost complex manifolds*, Ann. of Math. **65** (1957), 391-404.

72. L. O'Raifeartaigh, *Group Structure of Gauge Theories*, Cambridge University Press, Cambridge, 1986.

73. B. Parshall and J-P. Wang, *Quantum linear group*, Mem. Amer. Math. Soc. no.**439**, 1991.

74. A. Perelomov, *Generalized Coherent States and their Applications*, Springer-Verlag, Berlin, 1986.

75. S. Piunikhin, *State sum models for trivalent knotted graph invariants using quantum group $SL_q(2)$*, J. of Knot Theory and its Ramifications **1** (1992), 253-278.

76. A. M. Polyakov, *Gauge Fields and Strings*, Harwood Academic Publishers, Chur, 1987.

77. A. Pressley and G. Segal, *Loop Groups*, Oxford University Press, Oxford, 1986.

78. W. Rossmann, (a) *Kirillov's character formula for reductive Lie groups*, Invent. Math. **48** (1978), 207-220.
(b) *Characters as contour integrals*, Lecture Notes in Math. (Spinger-Verlag) **1077** (1984), 375-388.

79. I. Satake, *Algebraic Structures of Symmetric Domains*, Princeton University Press, Princeton, 1980.

80. M. Scheunert, *The Theory of Lie Superalgebras*, Lecture Notes in Math. (Springer-Verlag) **716**, 1979.

81. W. Schmid, (a) *On the realization of the discrete series of a semisimple Lie group*, Rice University Studies **56** (1970), 99-108.
(b) *On a conjecture of Langlands*, Ann. of Math. **93** (1971), 1-42.
(c) *Some properties of square integrable representations of semisimple Lie groups*, Ann. of Math. **102** (1975), 535-564.

82. J-P. Serre, (a) *Lie Algebras and Lie Groups*, Benjamin Publishing Company, Reading, 1965.
(b) *Complex Semisimple Lie Algebras*, Springer-Verlag, New York, 1987.

83. E. M. Stein, *Analysis in matrix spaces and some new representations of $SL(N, \mathbf{C})$*, Ann. of Math. **86** (1967), 461-490.

84. M. Toller, *Three dimensional Lorentz group and harmonic analysis of the scattering amplitude*, Nuovo Cimento **37** (1965), 631-657.

85. M. Toller and A. Bassetto, *Harmonic analysis on the one-sheet hyperboloid and multiperipheral inclusive distributions*, Ann. Inst. H. Poincaré **18** (1973), 1-38.

86. M. Toller and A. Sciarrino, *Decomposition of the unitary representations of the group $SL(2, \mathbf{C})$ restricted to the subgroup $SU(1,1)$*, J. Math. Phys. **8** (1967), 1252-1265.

87. L. L. Vaksman and Ya. S. Soibel'man, *Algebra of functions on the quantum group $SU(2)$*, Funct. Analy. ans its Appl. **22** (1989), 170-181.

88. D. N. Verma, *Structure of certain induced representations of complex semisimple Lie algebras*, Bull. Amer. Math. Soc. **74** (1968), 160-166,628.

89. M. D. Vivarelli, *The KS-transformation in hupercomplex form*, Celestial Mechanics **29** (1983), 45-50.

90. G. Warner, *Harmonic Analysis on Semisimple Lie Groups, I,II*, Springer-Verlag, Berlin, 1972.

91. R. O. Wells, *Complex manifolds and mathematical physics*, Bull. Amer. Math. Soc. **1** (1979), 296-336.

92. H. Weyl, *Classical Groups*, Princeton University Press, Princeton, 1973.

93. R. F. Wehrhahn and A. O. Barut, *Symmetry scattering for* $SU(2,2)$ *and its applications*, Deutsches Elektronen-Synchrotron preprint, DESY 93-037 (March 1993).

94. E. P. Wigner, *On unitary representations of the inhomogeneous Lorentz group*, Ann. of Math. **40** (1939), 149-204.

95. J. A. Wolf, (a) *The action of a real semisimple group on a complex flag manifold.I: Orbit structure and holomorphic arc components*, Bull. Amer. Math. Soc. **75** (1969), 1121-1237.
(b) *The action of a real semisimple Lie group on a complex flag manifols.II: Unitary representations on partially holomorphic cohomology spaces*, Mem. of Amer. Math. Soc. no,**138**, 1974.
(c) *Unitary representations of maximal parabolic subgroups of the classical groups*, Mem. Amer. of Math. Soc. no.**180**, 1976.
(d) *Classification and Fourier inversion for parabolic subgroups with square integrable nilradical*, Mem. of Amer. Math. Soc. **225**, 1979.
(e) *The uncertainity principle for Gelfand pairs*, Nova J. of Algebra and Geometry **1** (1992), 383-396.
(f) *Uncertainity principles for Gelfand pairs and Cayley complexes*, Proceedings of the Radon Conference, Vienna, 1992.

96. S. L. Woronowicz, *Compact matrix pseudogroups*, Commun. Math. Phys. **111** (1987), 613-665.

97. D. P. Želobenko, *Harmonic Analysis on Complex Semisimple Lie Groups*, Nauka, Moscow, 1974.

Division of Mathematics, University of Texas
San Antonio, Texas 78249

Harish-Chandra's c-Function. A Mathematical Jewel

Sigurdur Helgason

Abstract

In his work on spherical functions on semisimple Lie groups G Harish- Chandra showed that these functions' asymptotic behaviour is governed by a certain meromorphic function $\mathbf{c}$. While he showed that this function determines the Plancherel measure for the spherical transform on G it has later turned out that this c-function plays many other roles in the representation theory of G and in analysis on various homogeneous spaces of G; see particularly Theorems 6.1, 8.1, and 9.1.

1 Introduction

In his paper [4a], published in 1958, Harish-Chandra associated a certain function $\mathbf{c}$ to an arbitrary noncompact connected semisimple Lie group G. He showed later that this function determines the Plancherel measure for the spherical transform on G (see [4b,c]).

In this expository paper we shall describe the $\mathbf{c}$-function for G and then show how it enters in the solution of various problem in analysis on the symmetric space X associated with G. For the most part the results will be stated in full generality rather than for examples because only then will the relation with the $\mathbf{c}$-function emerge clearly.

The real- and complex numbers are denoted by $\mathbf{R}$ and $\mathbf{C}$, respectively, $\operatorname{Re} z$ is the real part of $z \in \mathbf{C}$. The dual of a vector space E is denoted by E^*. The adjoint representation of a Lie group G (respectively Lie algebra $\mathfrak{g}$) is denoted by Ad (respectively ad). I am indebted to Henrik Schlichtkrull for helpful remarks.

2 Semisimple Lie Groups

In this section we summarize some basic facts from the structure theory of semisimple Lie groups. Let G be a noncompact connected semisimple Lie group with finite center and K a maximal compact subgroup (all such are conjugate). Then as proved by Cartan there exists an involutive automorphism θ (actually unique) with

E. A. Tanner and R. Wilson (eds.), Noncompact Lie Groups and Some of Their Applications, 55–67.

fixed point set K. Let θ also denote the corresponding automorphism of the Lie algebra $\mathfrak{g}$ of G and

$$\mathfrak{g} = \mathfrak{k} + \mathfrak{p} \tag{1}$$

the decomposition into +1 and -1 eigenspaces of θ. Then if $X \in \mathfrak{p}$ the endomorphism ad$X$ of $\mathfrak{g}$ is diagonalizable (over $\mathbf{R}$). To capitalize on this one chooses a maximal abelian subspace $\mathfrak{a} \subset \mathfrak{p}$ and diagonalizes the endomorphisms ad$H$ ($H \in \mathfrak{a}$) simultaneously. Thus

$$\mathfrak{g} = \sum_{\alpha} \mathfrak{g}_\alpha + \mathfrak{g}_o\,, \tag{2}$$

where for each $\alpha \in \mathfrak{a}^*$,

$$\mathfrak{g}_\alpha = \{X \in \mathfrak{g} : \mathrm{ad}H(X) = \alpha(H)X \; for \; H \in \mathfrak{a}\}.$$

If $\alpha \neq 0$ and $\mathfrak{g}_\alpha \neq 0$ then α is called a *root* (of $(\mathfrak{g}, \mathfrak{a})$). Let Σ denote the set of all roots. The Killing form of $\mathfrak{g}$ induces a positive definite inner product $<,>$ on $\mathfrak{a}$ and its dual $\mathfrak{a}^*$.

Example: Let G=$\mathbf{SL}(n, \mathbf{R})$. Then $\mathfrak{g}$ consists of the real $n \times n$ matrices of trace 0, $\mathfrak{k} = \mathfrak{so}(n)$ (the algebra of skew symmetric matrices) and $\mathfrak{p}$ the space of symmetric matrices. For $\mathfrak{a}$ we can take the space of diagonal matrices. Then $\mathfrak{g}_o$=$\mathfrak{a}$ and (2) becomes

$$\mathfrak{sl}(n, \mathbf{R}) = \sum_{i \neq j} \mathbf{R} E_{ij} + \mathfrak{a}.$$

In general let $\mathfrak{a}'$ denote the subset of $\mathfrak{a}$ where all roots are $\neq 0$. Then $\mathfrak{a}'$ is the complement of a finite union of hyperplanes. Fix a component $\mathfrak{a}^+$ of $\mathfrak{a}'$. By connectedness each α has a fixed sign on $\mathfrak{a}^+$. We write $\alpha > 0$ if α is positive on $\mathfrak{a}^+$. Let $\Sigma^+ = \{\alpha \in \Sigma : \alpha > 0\}$. The space

$$\mathfrak{n} = \sum_{\alpha \in \Sigma^+} \mathfrak{g}_\alpha \tag{3}$$

is a subalgebra of $\mathfrak{g}$. Let A and N be the analytic subgroups of G corresponding to $\mathfrak{a}$ and $\mathfrak{n}$, respectively. Then one has the following result (the Iwasawa decomposition):

Theorem 2.1: *We have the direct vector space sum*

$$\mathfrak{g} = \mathfrak{k} + \mathfrak{a} + \mathfrak{n} \tag{4}$$

and unique decompositions

$$G = KAN\ ,\ G = NAK. \tag{5}$$

By (5) we write

$$g = k_1 exp\, H(g) n_1 = n_2 exp\, A(g) k_2 \ ; \ H(g), A(g) \in \mathfrak{a} \tag{6}$$

where all factors are unique.

Since $\mathfrak{a} \subset \mathfrak{p}$ was chosen freely and since $\mathfrak{a}^+ \subset \mathfrak{a}'$ was chosen freely it is important to recall that all such choices are conjugate under the adjoint action of K on $\mathfrak{p}$. The linear transformations of $\mathfrak{a}$ induced by those members of K which leave $\mathfrak{a}$ invariant constitute the *Weyl group* W.

Since $\theta(\mathfrak{g}_\alpha) = \mathfrak{g}_{-\alpha}$, the set Σ is invariant under $\alpha \to -\alpha$. Also, if $\alpha, c\alpha \in \Sigma^+, (c \in \mathbf{R})$ then $c = \frac{1}{2}$ or 2. Let

$$\Sigma_o^+ = \left\{ \alpha \in \Sigma^+ : \frac{1}{2}\alpha \notin \Sigma^+ \right\},$$

the set of *indivisible* positive roots. The integer m_α=dim$\mathfrak{g}_\alpha$ is called the *multiplicity* of α and the function $m : \alpha \to m_\alpha$ the multiplicity function. The function $\rho = \frac{1}{2}\sum_{\alpha>0} m_\alpha \alpha$ will appear frequently in the sequel.

Theorem 2.2: *The triple* $(\mathfrak{a}, \Sigma, m)$ *determines* $\mathfrak{g}$ *up to isomorphism.*

Using Cartan's classification of all $\mathfrak{g}$ this theorem follows by verification ([5e], p.535). However, mathematicians would prefer an *a priori* proof.

3 Spherical Functions and Spherical Transforms

In this section we recall some basic facts about spherical functions and introduce the **c**-function.

A continuous function $\varphi \not\equiv 0$ on G is said to be a *spherical function* if

$$\int_K \varphi(xky)\, dk \equiv \varphi(x)\varphi(y), \tag{7}$$

dk denoting normalized Haar measure on K. Then $\varphi(e) = 1$ and φ is *bi-invariant* under K. Let Φ be the set of spherical functions and let $C_c(K \setminus G/K)$ denote the space of continuous functions of compact support on G bi-invariant under K. If $f \in C_c(K \setminus G/K)$ its *spherical transform* is the function $\tilde{f}$ on Φ defined by

$$\tilde{f}(\varphi) = \int_G f(g)\varphi(g^{-1})\, dg, \quad \varphi \in \Phi, \tag{8}$$

dg denoting Haar measure. By a somewhat weakened form of Plancherel theorem of Godement [3] there exists a measure ν on Φ such that

$$\int_G | f(g) |^2 dg = \int_\Phi | \tilde{f}(\varphi) |^2 d\nu(\varphi), \quad f \in C_c(K \setminus G/K). \tag{9}$$

In [4a] Harish-Chandra set himself the problem of relating the measure $d\nu$ to the rich structure of G. His first result is the following parametrization of Φ as $\mathfrak{a}_{\mathbf{c}}^*/W$.

Theorem 3.1: *As λ runs through the space $\mathfrak{a}_{\mathbf{c}}^*$ of complex-valued linear functions on $\mathfrak{a}$, the functions*

$$\varphi_\lambda(g) = \int_K e^{(i\lambda-\rho)(H(gk))}\, dk\,, \quad g \in G\,, \tag{10}$$

run through the set Φ. Also $\varphi_\mu \equiv \varphi_\lambda$ if and only if $\mu \in W \cdot \lambda$.

The spherical functions can also be characterized as eigenfunctions of certain invariant differential operators on G. This led Harish-Chandra to precise results about their asymptotic behaviour and this is where the **c**-function enters. Let $\mathfrak{a}_+^* \subset \mathfrak{a}^*$ denote the subset corresponding to $\mathfrak{a}^+$ under the natural identification of $\mathfrak{a}$ and $\mathfrak{a}^*$.

Theorem 3.2: *Let* $\mathrm{Re}(i\lambda) \in \mathfrak{a}_+^*$ *and $H \in \mathfrak{a}^+$. Then the limit*

$$\mathbf{c}(\lambda) = \lim_{t\to+\infty} e^{(-i\lambda+\rho)(tH)}\varphi_\lambda(exp(tH)) \tag{11}$$

exists (and is independent of the choice of H).

Suppose now G has real rank one, i.e. dim$A$=1. If Ω denotes the Casimir operator on G, φ_λ satisfies the differential equation

$$\Omega\varphi_\lambda = -(<\lambda,\lambda> + <\rho,\rho>)\varphi_\lambda. \tag{12}$$

When expressed on A this becomes a singular second order ordinary differential equation which, by substitution $z = -\sinh^2(\alpha(\log a))$ goes over in the hypergeometric equation. Thereby

$$\varphi_\lambda(h) = F(a,b,c;-\sinh^2(\alpha(\log h)))\,, \quad h \in A\,, \tag{13}$$

where F is the hypergeometric function and the parameters a, b, c depend on λ and the multiplicities $m_\alpha, m_{2\alpha}$. From the knowledge of the hypergeometric function at ∞, (11) and (13) give the following formula for $\mathbf{c}(\lambda)$ (cf.[4a]):

Theorem 3.3: (G of real rank one). *Then $\mathbf{c}(\lambda)$ extends to a meromorphic function on $\mathfrak{a}_{\mathbf{c}}^*$ given by*

$$\mathbf{c}(\lambda) = c_o \frac{2^{-<i\lambda,\alpha_o>}\Gamma(<i\lambda,\alpha_o>)}{\Gamma(\frac{1}{2}(\frac{1}{2}m_\alpha + 1 + <i\lambda,\alpha_o>))\Gamma(\frac{1}{2}(\frac{1}{2}m_\alpha + m_{2\alpha} + <i\lambda,\alpha_o>))}\,, \tag{14}$$

where $\alpha_o = \frac{\alpha}{<\alpha,\alpha>}$ and c_o is given by $\mathbf{c}(-i\rho) = 1$.

On the other hand, for G complex, $\mathbf{c}(\lambda)$ is given by

$$\mathbf{c}(\lambda) = \prod_{\alpha>0} \frac{<\rho,\alpha>}{<i\lambda,\alpha>}. \tag{15}$$

In his papers [1a,b] Bhanu-Murthy determined the $\mathbf{c}$-function for the cases $G = \mathbf{SL}(n,\mathbf{R})$, $G = \mathbf{Sp}(n,\mathbf{R})$. The formulas exhibited a product structure reminiscent of (15). This led Gindikin and Karpelevič [2] to the following general product formula for the $\mathbf{c}$-function. Let $\beta \in \Sigma_o^*$ and $\mathfrak{g}_{(\beta)}$ be the subalgebras of $\mathfrak{g}$ generated by the subspaces $\mathfrak{g}_\beta$ and $\mathfrak{g}_{-\beta}$. Then $\mathfrak{g}_{(\beta)}$ is semisimple,

$$\mathfrak{g}_{(\beta)} = \mathfrak{k}_\beta + \mathfrak{p}_\beta, \tag{16}$$

where $\mathfrak{k}_\beta = \mathfrak{k} \cap \mathfrak{g}_{(\beta)}, \mathfrak{p}_\beta = \mathfrak{p} \cap \mathfrak{g}_{(\beta)}$, and the subspace $\mathfrak{a}_\beta = \mathfrak{a} \cap \mathfrak{p}_\beta$ is one-dimensional and maximal abelian in $\mathfrak{p}_\beta$. The product formula quoted is

$$\mathbf{c}(\lambda) = c \prod_{\beta\in\Sigma_o^+} \mathbf{c}_\beta(\lambda_\beta), \tag{17}$$

where c is a constant, $\mathbf{c}_\beta$ is the $\mathbf{c}$-function for $\mathfrak{g}_{(\beta)}$ and λ_β is the restriction $\lambda \mid \mathfrak{a}_\beta$. Combining (14) and (17) we get finally

$$\mathbf{c}(\lambda) = c_o \prod_{\alpha\in\Sigma_o^+} \frac{2^{-<i\lambda,\alpha_o>}\Gamma(<i\lambda,\alpha_o>)}{\Gamma(\frac{1}{2}(\frac{1}{2}m_\alpha + 1 + <i\lambda,\alpha_o>))\Gamma(\frac{1}{2}(\frac{1}{2}m_\alpha + m_{2\alpha} + <i\lambda,\alpha_o>))}, \tag{18}$$

where the constant c_o is determined by $\mathbf{c}(-i\rho) = 1$.

The association $G \to \mathbf{c}$ being natural and canonical, formula (18) gives evidence for the Gamma functions' appearance in nature. Note that if G is complex then $m_{2\alpha} = 0$ and $m_\alpha = 2$. Using the Legendre duplication formula for the Gamma function, (18) reduces to (15) in this case.

Guided by the spectral theory of ordinary differential equations, Harish-Chandra [4b,c] related the measure $d\nu$ in (9) to the $\mathbf{c}$-function. More precisely, the following result holds:

Theorem 3.4: *For a suitable normalization of dg we have (writing $\tilde{f}(\lambda) = \tilde{f}(\varphi_\lambda)$),*

$$\int_G |f(g)|^2\, dg = \int_{\mathfrak{a}^*/W} |\tilde{f}(\lambda)|^2 |\mathbf{c}(\lambda)|^{-2}\, d\lambda, \quad f \in C_c(K\backslash G/K)$$

and the map $f \to \tilde{f}$ extends to an isometry of $L^2(K\backslash G/K)$ onto $L^2(\mathfrak{a}^*/W, |\mathbf{c}(\lambda)|^{-2} d\lambda)$.

All the details in formula (18) are significant; for example the majoration of $\mathbf{c}^{-1}(\lambda)$ by a polynomial and the localization of its singularities is important for the

Paley-Wiener theorem, that is the characterization of the range $C_c^\infty(K\backslash G/K)$ (see [5f], Ch.IV and the references there). We shall also see later that the numerator and the denominator in (18) have their individual significance.

Remark: Theorem 2.2 suggests that, at least in principle, the Plancherel measure $d\nu$ should be expressible in terms of the triple $(\mathfrak{a}, \Sigma, m)$. Theorem 3.4 in connection with (18) is a striking manifestation of this.

4 The Fourier Transform on G/K and the c-Function

The maximal compact subgroups K of G all being conjugate the symmetric space $X = G/K$ is canonically determined by G. The functions in $C_c(K \backslash G/K)$ give only K-invariant functions on X, but for a genuine Fourier transform on X we have to consider "arbitrary" functions on X.

Let M denote the centralizer of A in K and consider the homogeneous space

$$B = K/M = G/MAN\,,$$

and let db be the normalized K-invariant measure on B. Consider the vector-valued inner product $A : X \times B \to \mathfrak{a}$ given by

$$A(gK, kM) = A(k^{-1}g) \tag{19}$$

in terms of the second formula in (6). If f is a function on X its *Fourier transform is defined by* (cf.[5c])

$$\tilde{f}(\lambda, b) = \int_X f(x)e^{(-i\lambda+\rho)(A(x,b))}\, dx\,,\ \lambda \in \mathfrak{a}_c^*\,,\ b \in B\,, \tag{20}$$

dx being "the" G-invariant measure on X.

Theorem 4.1: *For a suitable normalization of dx the following Plancherel formula holds:*

$$\int_X |\, f(x)\,|^2\, dx = \int_{\mathfrak{a}^*}\int_B |\, \tilde{f}(\lambda, b)\,|^2\, |\, \mathbf{c}(\lambda)\,|^{-2}\, d\lambda\, db. \tag{21}$$

Also $f \to \tilde{f}$ is an isometry of $L^2(X)$ onto $L^2((\mathfrak{a}^*/W) \times B)$.

If f is K-invariant then $\tilde{f}(\lambda, b)$ is independent of b and (21) reduces to Theorem 3.4. The spherical transform (8) has the same relation to the Fourier transform in (20) above as the Hankel transform (with Bessel functions) has to the Euclidean Fourier transform. The proof of the identity (21) is based on (9) and the following symmetry property of the spherical function ([5d,I],Ch.III,§5):

Theorem 4.2: *For all $g, h \in G$ we have*

$$\varphi_\lambda(g^{-1}h) = \int_K e^{(-i\lambda+\rho)(A(kg))}e^{(i\lambda+\rho)(A(kh))}\, dk. \tag{22}$$

The range property in Theorem 4.1 requires additional arguments.

We shall now explain an analog of (20) and (21) for compact symmetric spaces (where the analog of the group N is missing). Denoting complexification by superscript $\mathbf{C}$ we consider the (compact) Lie algebra

$$\mathfrak{u} = \mathfrak{k} + i\mathfrak{p} \subset \mathfrak{g}^{\mathbf{C}}.$$

If U and G are the analytic subgroups of the simply connected Lie group $G^{\mathbf{C}}$ then U is compact and U/K is a symmetric space (dual to G/K). Complexifying the decomposition (4) we get

$$\mathfrak{g}^{\mathbf{C}} = \mathfrak{n}^{\mathbf{C}} + \mathfrak{a}^{\mathbf{C}} + \mathfrak{k}^{\mathbf{C}}.$$

It follows that there exists a neighborhood U_o of e in U (which can be taken invariant under conjugation by members of K) and a mapping $u \to A(u)$ of U_o into $i\mathfrak{a}$ such that

$$u \in N^{\mathbf{C}} exp(A(u)) K^{\mathbf{C}}.$$

Consider the subset of $\mathfrak{a}^*$ given by

$$\Lambda = \left\{ \mu \in \mathfrak{a}^* : \frac{< \mu, \alpha >}{< \alpha, \alpha >} \in \mathbf{Z}^+ \; for\, \alpha \in \Sigma^+ \right\}.$$

It is known ([5d,I],Ch.III) that this is the set of highest restricted weights of the irreducible spherical representations of U (that is representations which have a fixed vector under K). Let

$$S = U/K \; , \; S_o = \{uK \, : \, u \in U_o\}.$$

Let F be a continuous function on S with support inside S_o. Its *Fourier transform* $\tilde{F}$ is defined by

$$\tilde{F}(\mu, kM) = \int_{U/K} F(uK) e^{(\mu+2\rho)(A(k^{-1}u))} \, du_K \; , \; \mu \in \Lambda \; , \; kM \in K/M. \tag{23}$$

Thus the Fourier transform is a function on $\Lambda \times B$. The following result was proved by Sherman [7a,b]:

Theorem 4.3: *For a suitable normalization of the invariant measures ds and db,*

$$\int_S | F(s) |^2 \, ds = \sum_{\mu \in \Lambda} d_\mu \int_B | \tilde{F}(\mu, b) |^2 \, db \tag{24}$$

for $F \in C(S)$, supp$F \subset S_o$, d_μ *denoting the degree of the spherical representation with highest restricted weight* μ. *If* dim$\mathfrak{a} = 1$ *the support condition can be removed.*

The proof (under the support condition) is based on an analytic continuation of formula (22).

Since the definitions (20) and (23) are related by analytic continuation the analogy between (21) and (24) would be complete if we had a relation between d_μ and $|\,\mathbf{c}(\lambda)\,|^{-2}$. Such a relation had been proved by Vretare [9]:

$$d_\mu = \left\{ \frac{\mathbf{c}(\lambda + i\mu)\mathbf{c}(-\lambda - i\mu)}{\mathbf{c}(\lambda)\mathbf{c}(-\lambda)} \right\}_{\lambda=-i(\mu+\rho)}. \tag{25}$$

In the case when G is complex, U/K is a group and Harish-Chandra's product formula (15) gives by (25) Weyl's product formula for the degrees of the irreducible representations of U.

5 The Radon Transform and the c-Function

Again consider the symmetric space $X = G/K$ and the decomposition (5). A *horocycle* in X is by definition an orbit ξ of a group gNg^{-1} conjugate to N. The space Ξ of all horocycles is again a homogeneous space of G,

$$\Xi = G/MN = (K/M) \times A. \tag{26}$$

If f is a function on X its *Radon transform* is the function $\widehat{f}$ on Ξ defined by

$$\widehat{f}(\xi) = \int_\xi f(x)\, dm(x), \tag{27}$$

dm being the volume element on ξ induced by the Riemannian structure of X. There is a *dual transform* associated to a function φ on Ξ a point function $\check{\varphi}$ given by

$$\check{\varphi}(x) = \int_{\xi \ni x} \varphi(\xi)\, d\mu(\xi), \tag{28}$$

$d\mu$ being the measure invariant under the isotropy group at x. Viewing φ by (26) as a function on $(K/M) \times A$ we consider its Fourier transform φ^* in the A-variable,

$$\varphi^*(kM, \lambda) = \int_A \varphi(kM, a) e^{-i\lambda(\log(a))}\, da. \tag{29}$$

Let Λ^o be the operator on $C_c^\infty(\Xi)$ defined by

$$(\Lambda^o \varphi)^*(kM, \lambda) = |\,\mathbf{c}(\lambda)\,|^{-2} \varphi^*(kM, \lambda) \tag{30}$$

and put $\Lambda = e^{-\rho} \circ \Lambda^o \circ e^\rho$. The Radon transform is then inverted by the following result ([5b],[5d,I]):

Theorem 5.1: *For a suitable normalization of* μ

$$f = (\Lambda \widehat{f})^{\vee}, \quad f \in C_c^\infty(X).$$

6 Huygens' Principle and the c-Function

Let Y be a Lorentzian manifold with pseudo-Riemannian structure of signature $(1, n-1)$. Given $y \in Y$ the isotropic geodesics through y make up the light cone C_y. Let L be the Laplace-Beltrami operator on Y and c a constant. Let $S \subset Y$ be a spacelike hypersurface and consider the Cauchy problem for the equation $(L+c)u = 0$ with u having given initial data on S. Huygens' principle is said to hold if the value $u(y)$ only depends on the initial data on an arbitrary neighborhood of the edge $s = C_y \cap S$. Huygens' principle holds if the fundamental solution has support in C_y.

Theorem 6.1: *The following conditions* (1)-(4) *on* G *are equivalent:*

1. *G has all its Cartan subgroups conjugate.*
2. *All multiplicities m_α are even and each $m_{2\alpha} = 0$.*
3. *The function $\mathbf{c}^{-1}(\lambda)$ is a polynomial.*
4. *The operator Λ in Theorem 5.1 is a differential operator.*
5. *If these conditions are satisfied and if* $\dim X$ *is odd then the differential equation*

$$\frac{\partial^2 u}{\partial t^2} = (L + |\rho|^2)u\,, \quad u(x,0) = f_o(x)\,, \quad u_t(x.0) = f_1(x)$$

satisfies Huygens' principle.

The equivalence of (1)-(4) is proved in [5e],Ch.IX,§6 and [5f],Ch.IV,§6. For part (5), see [5g], [8], [6] and [5h].

7 Orbital Integrals and Huygens' Principle

A Lorentzian manifold of constant curvature $\kappa = 0, -1, +1$ is locally isometric to the respective spaces

$$\mathbf{O}(1, n-1)\mathbf{R}^n/\mathbf{O}(1, n-1)\,, \quad \mathbf{O}(1, n)/\mathbf{O}(1, n-1)\,, \quad \mathbf{O}(2, n-1)/\mathbf{O}(1, n-1). \quad (31)$$

Let $X = G/H$ be one of the spaces (31), where we replace the groups by their identity component. Then H acts transitively on the retrograde cone $C_o - \{0\}$, with the vertex removed. Fix $y \in X$ and a point x inside the cone at distance r from y. We consider the *orbital integral*

$$(M^r u)(y) = \int_H u(gh \cdot x)\, dh\,, \qquad (32)$$

dh being a fixed invariant measure on H. The following results are proved in [5a] (see also [5f],Ch.I):

Theorem 7.1:

1. *If u has compact support then*

$$\lim_{r\to 0} r^{n-2}(M^r u)(y) \text{ exists and is } \not\equiv 0. \tag{33}$$

2. $M^r(Lu) = L_r(M^r u)$, *where* L_r *is the radial part of the Laplacian.*

3. *Suppose* n *is even and* > 2. *Put*

$$Q(L) = (L - \kappa(n-3)2)(L - \kappa(n-5)4)\cdots(L - \kappa(1)(n-2)).$$

Then for a constant c,

$$u(y) = c \lim_{r\to 0} r^{n-2} Q(L_r)((M^r u)(y)). \tag{34}$$

The limit (33) being a gHg^{-1}-invariant measure with support in C_y it is given by a constant multiple of $\int_{C_y} u(z)\, d\mu(z)$ where $d\mu$ is the invariant measure on C_y. (It is not hard to see that the delta function at y does not appear in the limit).

Remark: (Schlichtkrull) Because of part(2) of Theorem 7.1, expression (34) thus equals

$$c \lim_{r\to 0} r^{n-2} M^r(Q(L)u) = c \int_{C_y} (Q(L)u)(z)\, d\mu(z)$$

so $\delta_y = (L - \kappa(n-3)2)\cdots(L - \kappa(1)(n-2))(c\mu)$. Thus each operator $L_m = L - \kappa(n-m)(m-1)$, for $m = 3, 5, \cdots, n-1$ satisfies Huygens' principle. Because of the conformal invariance this was known for the operator $L - \frac{\kappa}{4}n(n-2)$ (cf.[10]) which equals L_m for $m = 1 + 2[\frac{n}{4}]$. For $n = 4$ it was known that no additional operator could satisfy Huygens' principle but for $n = 6$ there are two such operators.

8 Conical Distributions and the c-Function

The horocycle space $\Xi = G/MN$ considered in §5 has been found to have many features analogous to $X = G/K$ (cf.[5d,I]). This leads to the consideration of the analogs for Ξ of the spherical function on X. They would be the MN-invariant eigendistributions of each $D \in \mathbf{D}(G/MN)$, the algebra of G-invariant differential operators on G/MN. These are by definition the *conical distributions* on Ξ (cf.[5d,I]).

For each $s \in W$ let $m_s \in K$ be an element such that $\mathrm{Ad}(m_s)/\mathfrak{a} = s$. If Ξ_s denotes the orbit $MNAm_s \cdot \xi_o$ (ξ_o=origin in Ξ) then the Bruhat decomposition for G implies

$$\Xi = \bigcup_{s\in W} \Xi_s \quad (\textit{disjoint union}).$$

If $\xi \in \Xi$ then $\xi = mna(\xi)m_s \cdot \xi_o$ where $a(\xi) \in A$ is unique. For a certain natural measure $d\nu$ on Ξ_s which is MN- invariant we consider for $\lambda \in \mathfrak{a}_c^*$ the functional on $C_c^\infty(\Xi)$ given by

$$\Psi'_{\lambda,s}(\varphi) = \int_{\Xi_s} \varphi(\xi) e^{(is\lambda+s\rho)(\log a(\xi))}\, d\nu(\xi).$$

This integral converges absolutely if λ lies in a certain open tube in $\mathfrak{a}_c^*$ and for those λ, $\Psi'_{\lambda,s}$ is a conical distribution.

Now let $\mathbf{c}_s(\lambda)$ denote the product (18) where α only ranges over $\Sigma_o^+ \cap s^{-1}\Sigma_o^-$ and let $\mathbf{d}_s(\lambda)$ denote the *numerator* in the formula for $\mathbf{c}_s(\lambda)$. Then we have the following result ([5d,I],Ch.III,§4):

Theorem 8.1: *Let $s \in W$. Then the mapping*

$$\lambda \to \mathbf{d}_s^{-1}(\lambda)\Psi'_{\lambda,s}$$

extends to a holomorphic function, denoted $\lambda \to \Psi_{\lambda,s}$, on $\mathfrak{a}_c^*$, with values in the space of distributions on Ξ. For each $\lambda \in \mathfrak{a}_c^*$, $s \in W$, $\Psi_{\lambda,s}$ is a conical distribution.

Roughly speaking (cf.[5d,I],Ch.III,Theorem 4.9) these $\Psi_{\lambda,s}$ constitute all the conical distributions on Ξ. This parametrization of the set of conical distributions by the product $\mathfrak{a}_c^* \times W$ had been expected on the basis of the analogy with spherical functions where

$$K \backslash G/K \approx A/W\,, \quad \Phi \approx \mathfrak{a}_c^*/W$$

combined with the identification

$$MN \backslash G/MN \approx A \times W.$$

While the conical distributions were defined purely by a formal analogy they turn out to give the intertwining operators for the spherical principal series for G ([5d,I],Ch.III, Theorem 6.1).

9 Eigenspace Representations and the c-Function

Let L/H be a coset space of a Lie group L and a closed subgroup H and let $\mathbf{D}(L/H)$ be the algebra of invariant differential operators on L/H. Let $\chi : \mathbf{D}(L/H) \to \mathbf{C}$ be any homomorphism, let E_χ be the corresponding *joint eigenspace*

$$E_\chi = \{f \in C^\infty(L/H) \,:\, Df = \chi(D)f\,,\ for\, D \in \mathbf{D}(L/H)\}$$

and let S_χ be the representation of L on E_χ given by

$$(S_\chi(\ell)f)(xH) = f(\ell^{-1}xH)\,, \ \ \ell \in L\,,\ f \in E_\chi.$$

These representations S_χ were introduced in [5d,I] under the name *eigenspace representations* and the problem was posed to find for which χ the representation S_χ was irreducible. We shall now see how this question is answered for the symmetric space $X = G/K$ in terms of the **c**-function. Each joint eigenspace of $\mathbf{D}(G/K)$ contains a K-invariant joint eigenfunction which by Theorem 3.1 is a constant multiple of φ_λ for some $\lambda \in \mathfrak{a}_{\mathbf{c}}^*$. Let $c_\lambda(D) \in \mathbf{C}$ be determined by

$$D\varphi_\lambda = c_\lambda(D)\varphi_\lambda\,, \quad D \in \mathbf{D}(G/K)$$

and let $\mathcal{E}_\lambda(X)$ denote the joint eigenspace

$$\mathcal{E}_\lambda(X) = \{f \in C^\infty(X) \,:\, Df = c_\lambda(D)f\,,\ for\, D \in \mathbf{D}(G/K)\}\,.$$

Of course $\mathcal{E}_{s\lambda}(X) = \mathcal{E}_\lambda(X)$ for $s \in W$ and each joint eigenspace has the form $\mathcal{E}_\lambda(X)$ for some $\lambda \in \mathfrak{a}_{\mathbf{c}}^*$. Let T_λ denote the corresponding eigenspace representation. With $\mathbf{c}(\lambda)$ given by (18) let $\Gamma_X(\lambda)$ denote the denominator of $\mathbf{c}(\lambda)\mathbf{c}(-\lambda)$. This could appropriately be called the Gamma function of X. We have then the following irreducibility criterion (cf.[5d,II],Theorem 9.1).

Theorem 9.1: *Let $\lambda \in \mathfrak{a}_{\mathbf{c}}^*$, T_λ the corresponding eigenspace representation of G on $\mathcal{E}_\lambda(X)$. Then*

$$T_\lambda \text{ is irreducible} \Leftrightarrow \frac{1}{\Gamma_X(\lambda)} \neq 0.$$

BIBLIOGRAPHY

1. T. S. Bhanu-Murthy, (a) *Plancherel's measure for the factor space* $\mathbf{SL}(n,\mathbf{R})/\mathbf{SO}(n,\mathbf{R})$, Dokl. Akad. Nauk SSSR **133** (1960), 503-506.
(b) *The asymptotic behaviour of zonal spherical functions on the Siegel upper half-plane*, ibid. **135** (1960), 1027-1030.

2. S. G. Gindikin and F. I. Karpelevič, *Plancherel measure of Riemannian symmetric spaces of nonpositive curvature*, Sov. Math. **3** (1962), 962-965.

3. R. Godement, *Introduction aux Travaux de A.Selberg*, Séminaire Bourbaki, *n°* **144** (1957), Paris.

4. Harish-Chandra, (a) *Spherical functions on a semisimple Lie group I*, Amer. J. Math. **80** (1958), 241-310.
(b) *Spherical functions on a semisimple Lie group II*, Amer. J. Math. **80** (1958), 553-613.
(c) *Discrete series for semisimple Lie groups II*, Acta Math. **116** (1966), 1-111.

5. S. Helgason, (a) *Differential operators on homogeneous spaces*, Acta Math. **102** (1959), 239-299.
(b) *A duality in integral geometry; some generalizations of the Radon transform*, Bull. Amer. Math. Soc. **70** (1964), 435-446.
(c) *Radon-Fourier transforms on symmetric spaces and related group representations*, Bull. Amer. Math. Soc. **71** (1965), 757-763.
(d) *A duality for symmetric spaces with applications to group representations I, II*, Advan. Math. **5** (1970), 1-154; Advan. Math. **22** (1976), 187-219.
(e) *Differential Geometry, Lie Groups and Symmetric Spaces*, Academic Press, New York, 1978.
(f) *Groups and Geometric Analysis*, Academic Press, Orlando, 1984.
(g) *Wave equation on homogeneous spaces*, in *Lie Group Representations III*, Lecture Notes in Math. **1077**, pp.254-287, Springer-Verlag, New York, 1984.
(h) *Huygens' principle for wave equations on symmetric spaces*, J. Funct. Anal. **107** (1992), 279-288.

6. G. 'Olafsson and H. Schlichtkrull, *Wave propagation on Riemannian symmetric spaces*, J. Funct. Anal. **107** (1992), 270-278.

7. T. O. Sherman, (a) *Fourier analysis on compact symmetric space*, Bull. Amer. Math. Soc. **83** (1977), 378-380.
(b) *The Helgason-Fourier transform for compact Riemannian symmetric spaces of rank one*, Acta Math. **164** (1990), 73-144.

8. L. E. Solomatina, *Translation representation and Huygens' principle for an invariant wave equation in a Riemannian symmetric space*, Soviet Math. Iz. **30** (1986), 108-111.

9. L. Vretare, *Elementary spherical functions on symmetric spaces*, Math. Scand. **39** (1976), 343-358.

10. B. Oersted, *The conformal invariance of Huygens' principle*, J. Diff. Geometry **16** (1981), 1-9.

Department of Mathematics, Massachusetts Institute of Technology,
Cambridge, Massachusetts 02139

BASIC HARMONIC ANALYSIS ON PSEUDO–RIEMANNIAN SYMMETRIC SPACES

E. VAN DEN BAN, M. FLENSTED–JENSEN and H. SCHLICHTKRULL

Abstract. We give a survey of the present knowledge regarding basic questions in harmonic analysis on pseudo–Riemannian symmetric spaces G/H, where G is a semisimple Lie group: The definition of the Fourier transform, the Plancherel formula, the inversion formula and the Paley–Wiener theorem.

1. Introduction

The rich and beautiful theory of harmonic analysis on $\mathbb{R}^n$ and $\mathbb{T}^n = (\mathbb{R}/\mathbb{Z})^n$ has become a powerful tool, widely used in other branches of mathematics, in physics, engineering etc. From our point of view all the basic questions are completely and explicitly solved: The Fourier transform is defined, there exists a Plancherel formula and an inversion formula for it, and (for $\mathbb{R}^n$) there is a Paley–Wiener theorem, describing the image of the space of compactly supported functions.

There exist many generalizations of this theory. Let us mention a few directions, based on various ways of viewing $\mathbb{R}^n$ and $\mathbb{T}^n$.

$\mathbb{R}^n$ and $\mathbb{T}^n$ are locally compact groups:

- Fourier analysis on locally compact Abelian groups.
- The Peter–Weyl theory for Fourier analysis on compact groups.
- Representation theory for locally compact groups and rings of operators on Hilbert spaces (C^*–algebras, von Neumann algebras, etc.).

$\mathbb{R}^n$ and $\mathbb{T}^n$ are Lie groups:

- The representation theory for compact Lie groups (the Cartan–Weyl classification, Weyl's character formula etc.).
- Representation theory for general Lie groups (semisimple, reductive, nilpotent, solvable etc.).

$\mathbb{R}^n$ and $\mathbb{T}^n$ are smooth homogeneous manifolds:

- Harmonic analysis related to homogeneous spaces and their transformation groups.

Here we take the last mentioned viewpoint. We claim that inside the class of

E. A. Tanner and R. Wilson (eds.), Noncompact Lie Groups and Some of Their Applications, 69–101.

smooth manifolds the class of (not necessarily Riemannian) symmetric spaces constitutes an appropriate framework for generalization of harmonic analysis: On the one hand this class of manifolds is wide enough to contain very many important spaces of relevance in other branches of mathematics and in physics. On the other hand it is restrictive enough to make feasible a theory of harmonic analysis, with explicit parametrizations and descriptions of representations, explicit Plancherel formulae, etc.

2. Symmetric spaces

2.1. Definition and structure

We define a (affine) symmetric space as follows:

Definition. *A connected smooth manifold M with an affine connection is called a* symmetric space *if for every x in M the local reflection in x along geodesics extends to a global affine diffeomorphism, S_x, of M.*

Without going into technicalities we shall need a few facts about symmetric spaces (for details, see [21], [37], [38] and the references cited there):

The group $G = G(M)$ generated by the transformations $S_x \circ S_y$, $(x, y \in M)$, is a connected Lie group acting transitively on M. Therefore, choosing a base point $x_o \in M$, we may identify M with the homogeneous space G/H, where H is the stabilizer of x_o. If we define $\sigma(g) = S_{x_o} \circ g \circ S_{x_o}$ for $g \in G$, then σ is an involution of G, i.e. an automorphism whose square is the identity. It easily follows that H is an open subgroup of the group G^σ of σ–fixed points in G. On the other hand, if G is a connected Lie group with an involution σ, and H is an open subgroup of G^σ, then the homogeneous space G/H is a symmetric space on which G acts by affine transformations (but G differs in general from $G(G/H)$).

Let r be the Ricci curvature tensor on M. This is a covariant tensor of degree 2, which is canonically associated with the affine connection on M, and therefore G-invariant. A symmetric space has the special feature that its affine connection is torsion free, and that the Ricci curvature tensor r is covariantly constant. In particular, if r is symmetric and non-degenerate, then it defines a pseudo-Riemannian structure on M whose associated connection is the original affine connection.

Theorem 1. *Let $M = G/H$ be a symmetric space with $G = G(M)$.*

(i) *M is flat if and only if G is Abelian.*

(ii) *(M, r) is pseudo–Riemannian (that is, r is symmetric and non–degenerate) if and only if G is semisimple.*

(iii) *(M, r) is Riemannian (that is, r is symmetric and definite) if and only if G is semisimple and H is compact.*

(iv) *If M is irreducible then either $\dim(M) = 1$, or G is simple, or M is a simple Lie group G_1.*

In the last mentioned case, when $M = G_1$, we have that G is the product $G_1 \times G_1$ with the left times right action on G_1. In this case, the reflection $S_x\colon G_1 \to G_1$ in an element $x \in G_1$ is given by $S_x(g) = xg^{-1}x$. Choosing the identity element of G_1 as our base point we get that H is the diagonal $d(G_1)$ and that the involution of G is given by $\sigma(x,y) = (y,x)$. We call this the *group case.*

Our goal in this paper is to describe the state of the art for harmonic analysis on *semisimple symmetric spaces*, i.e. the spaces of case (ii) above. From now on we assume that $M = G/H$ is such a space, with $G = G(M)$ semisimple. Notice that this is a stronger assumption than just requiring M to be equipped with some pseudo-Riemannian structure which is compatible with the given affine connection ($\mathbb{R}^n$ with any pseudo–norm is an example – here $r = 0$). However by (iv), if M is irreducible and of dimension at least 2 then G is semisimple.

For simplicity of exposition we assume (which we may up to coverings of M) that G is a closed subgroup of $\mathrm{GL}(n,\mathbb{R})$ for some n, and that G is stable under transposition. Let $K = G \cap \mathrm{SO}(n)$, or equivalently $K = G^\theta$, where $\theta(x) = {}^t x^{-1}$, then K is a maximal compact subgroup of G. We may choose the base point such that $\theta(H) = H$, or equivalently, such that $\sigma \circ \theta = \theta \circ \sigma$.

We shall distinguish between the following 3 types of irreducible semisimple symmetric spaces:

- M is of the *compact type* if $G = K$, or equivalently if all geodesic curves have compact closures.
- M is of the *non–compact type* if $H = K$, or equivalently if all geodesic curves have non–compact closures.
- M is of the *non–Riemannian type* if $G \neq K$ and $K \neq H$, or equivalently if there exist geodesic curves of both types.

If M is of one of the first two types we say that it is of the Riemannian type (cf. Thm. 1(iii)). Notice that a simple group G_1, considered as a symmetric space, is either of the compact type or of the non–Riemannian type.

2.2. EXAMPLES

The irreducible symmetric spaces have been classified by M. Berger [10]. Compared with the list of Riemannian symmetric spaces (see [27, Ch.X]), Berger's list is considerably longer.

There is (up to coverings) one two–dimensional space of each of the three types:

- The compact type: The *2–sphere* $\mathrm{S}^2 = \mathrm{SO}(3)/\mathrm{SO}(2)$.
- The non–compact type: The *hyperbolic 2–space* $M = \mathrm{H}^2$. This has several isomorphic realizations: As $\mathrm{SL}(2,\mathbb{R})/\mathrm{SO}(2)$, as $\mathrm{SU}(1,1)/\mathrm{S}(\mathrm{U}(1)\times\mathrm{U}(1))$, or as $\mathrm{SO}_e(2,1)/\mathrm{SO}(2)$, corresponding to, respectively, the upper half plane in $\mathbb{C}$, the unit disk in $\mathbb{C}$, or a sheet of the two–sheeted hyperboloid in $\mathbb{R}^3$.
- The non–Riemannian type: The *one–sheeted hyperboloid* in $\mathbb{R}^3$, $H^{1,1} = \mathrm{SO}_e(2,1)/\mathrm{SO}_e(1,1)$, which can also be realized as $\mathrm{SL}(2,\mathbb{R})/\mathrm{SO}(1,1)$. It has the two–fold cover $\mathrm{SL}(2,\mathbb{R})/\mathrm{SO}_e(1,1)$.

In higher dimensions there exist several 'families' of symmetric spaces, many of which have one of the spaces above as their lowest dimensional member. For example we could mention:

The *n-spheres*: $S^n = SO(n+1)/SO(n)$.

The spaces of ***positive definite quadratic forms*** in $\mathbb{R}^n$: $SL(n,\mathbb{R})/SO(n)$.

The spaces of *quadratic forms of signature* (p,q) in $\mathbb{R}^n$, (where $n = p+q$): $SL(n,\mathbb{R})/SO(p,q)$.

The *hyperboloids* in $\mathbb{R}^{n+1}$: $H^{p,q} = \{x \in \mathbb{R}^{n+1} \mid x_1^2+\cdots+x_p^2-x_{p+1}^2-\cdots-x_{p+q+1}^2 = -1\}$ where $p+q=n$ (if $q=0$ one must take a connected component). Here $M = SO_e(p,q+1)/SO_e(p,q)$.

Similarly one can take the corresponding spaces over the complex numbers or over the quaternions.

2.3. Some basic notation

Let G, H, K, σ and θ be as above. Let $\mathfrak{g}$ be the (real) Lie algebra of G, and let $\mathfrak{h}$ and $\mathfrak{k}$ be the subalgebras corresponding to H and K, and $\mathfrak{q}$ and $\mathfrak{p}$ their respective orthocomplements with respect to the Killing form. Then

$$\mathfrak{g} = \mathfrak{h} \oplus \mathfrak{q} = \mathfrak{k} \oplus \mathfrak{p}$$

is the decomposition of $\mathfrak{g}$ into the ± 1 eigenspaces for σ and θ respectively. Since θ and σ commute we also have the joint decomposition

$$\mathfrak{g} = \mathfrak{h}\cap\mathfrak{k} \oplus \mathfrak{h}\cap\mathfrak{p} \oplus \mathfrak{q}\cap\mathfrak{k} \oplus \mathfrak{q}\cap\mathfrak{p}. \tag{1}$$

Notice that there is a natural identification of $\mathfrak{q}$ with the tangent space $T_{x_o}(M)$. We denote by $\mathfrak{g}_{\mathbb{C}}, \mathfrak{h}_{\mathbb{C}}$ etc. the complexifications of $\mathfrak{g}$, $\mathfrak{h}$ etc.

A *Cartan subspace* $\mathfrak{b}$ for G/H is a maximal Abelian subspace of $\mathfrak{q}$, consisting of semisimple elements. (If we assume, as we may in the following, that $\mathfrak{b}$ is θ–invariant, then all its elements are automatically semisimple, once $\mathfrak{b}$ is maximal Abelian). All Cartan subspaces have the same dimension, which we call *the rank of* M. The number of H–conjugacy classes of Cartan subspaces is finite. Geometrically, a Cartan subspace is the tangent space of a maximally flat regular subsymmetric space.

We say that a Cartan subspace $\mathfrak{b}$ is *fundamental* if the intersection $\mathfrak{b}\cap\mathfrak{k}$ is maximal Abelian in $\mathfrak{q}\cap\mathfrak{k}$, and that it is *split* if the intersection $\mathfrak{b}\cap\mathfrak{p}$ is maximal Abelian in $\mathfrak{q}\cap\mathfrak{p}$. There is, up to conjugation by $K\cap H$, a unique fundamental and a unique split Cartan subspace. If the fundamental Cartan subspace is contained in $\mathfrak{k}$ it is called a *compact* Cartan subspace. The dimension of the $\mathfrak{p}$–part of a split Cartan subspace is called the *split rank* of M.

Let $\mathbb{D}(G/H)$ denote the algebra of G–invariant differential operators on G/H. There is a natural isomorphism (the *Harish-Chandra isomorphism*) χ of this algebra with the algebra $S(\mathfrak{b})^W$ of W–invariant elements in the symmetric algebra of any

Cartan subspace $\mathfrak{b}_{\mathbb{C}}$. Here W is the reflection group of the root system of $\mathfrak{b}_{\mathbb{C}}$ in $\mathfrak{g}_{\mathbb{C}}$. In particular, $\mathbb{D}(G/H)$ is commutative, and its characters are parametrized up to W-conjugation by $D \mapsto \chi_\lambda(D) = \chi(D)(\lambda) \in \mathbb{C}$. It is known (see [2]) that the symmetric elements of $\mathbb{D}(G/H)$ have self-adjoint closures as operators on $L^2(G/H)$.

3. Basic harmonic analysis

3.1. Harmonic analysis on $\mathbb{R}^n$

We want to generalize the basic notions and results from harmonic analysis on $\mathbb{R}^n$. These are:

The *Fourier transform:* $f \mapsto f^\wedge(\lambda) = (2\pi)^{-n/2} \int_{\mathbb{R}^n} f(t) e^{-i\lambda \cdot t}\, dt$, $f \in C_c^\infty(\mathbb{R}^n)$.

The *Plancherel theorem:* $f \mapsto f^\wedge$ extends to an isometry of $L^2(\mathbb{R}^n)$ onto $L^2(\mathbb{R}^n)$.

The *inversion formula:* If $f \in C_c^\infty(\mathbb{R}^n)$ then

$$f(x) = (2\pi)^{-n/2} \int_{\mathbb{R}^n} f^\wedge(\lambda) e^{i\lambda \cdot x}\, d\lambda.$$

The *Paley–Wiener theorem:* $f \mapsto f^\wedge$ is a bijection of $C_c^\infty(\mathbb{R}^n)$ onto $\mathrm{PW}(\mathbb{R}^n)$, where $\mathrm{PW}(\mathbb{R}^n)$ is the space of rapidly decreasing entire functions of exponential type. More precisely, a complex function ψ on $\mathbb{R}^n$ belongs to $\mathrm{PW}(\mathbb{R}^n)$ if and only if it extends to an entire function on $\mathbb{C}^n$ for which there exists $R > 0$ such that the following holds for all $N \in \mathbb{N}$:

$$\sup_{\lambda \in \mathbb{C}} (1 + |\lambda|)^N e^{-R|\operatorname{Im} \lambda|} |\psi(\lambda)| < +\infty. \tag{2}$$

The aim of the basic harmonic analysis on G/H is to obtain analogues of these notions and results.

3.2. The 'abstract' harmonic analysis on a semisimple symmetric space

Let G and H be as above, then $M = G/H$ has an invariant measure, and the action of G by translations gives a unitary representation $\mathcal{L}$ in the associated Hilbert space $L^2(G/H)$. From general representation theory it is known (since G is 'type 1') that this representation can be decomposed as a direct integral of irreducible unitary representations:

$$\mathcal{L} \simeq \int_{G^\wedge}^{\oplus} m_\pi\, \pi\, d\mu(\pi), \tag{3}$$

where the measure $d\mu$ (whose class is uniquely determined) is called the *Plancherel measure*, and m_π (which is unique almost everywhere) the *multiplicity* of π. Moreover, only the so-called *H-spherical representations* can occur in this decomposition. By definition, an irreducible unitary representation $(\pi, \mathcal{H}_\pi)$ of G is H-spherical if the space $(\mathcal{H}_\pi^{-\infty})^H$ of its H–fixed distribution vectors is non–trivial. Here we denote by $\mathcal{H}_\pi^\infty$ and $\mathcal{H}_\pi^{-\infty}$, respectively the C^∞ and the distribution vectors for $\mathcal{H}_\pi$, such that $\mathcal{H}_\pi^\infty \subset \mathcal{H}_\pi \subset \mathcal{H}_\pi^{-\infty}$. We write

$$\mathcal{V}_\pi = (\mathcal{H}_\pi^{-\infty})^H.$$

It is known (see [2]) that $m_\pi \leq \dim \mathcal{V}_\pi < +\infty$, in particular, all multiplicities are finite. Denote by $G_H^\wedge$ the set of (equivalence classes) of H–spherical representations, then it follows that the Plancherel measure $d\mu$ is carried by $G_H^\wedge$.

The *'abstract' Fourier transform* $f \mapsto f^\wedge(\pi)$ for G/H is now defined by

$$f^\wedge(\pi)(\eta) = \pi(f)\eta = \int_{G/H} f(x)\pi(x)\eta \, dx \in \mathcal{H}_\pi^\infty$$

for $\pi \in G_H^\wedge, \eta \in \mathcal{V}_\pi$ and $f \in C_c^\infty(G/H)$. Thus

$$f^\wedge(\pi) \in \mathrm{Hom}_{\mathbb{C}}(\mathcal{V}_\pi, \mathcal{H}_\pi^\infty) \simeq \mathcal{H}_\pi^\infty \otimes \mathcal{V}_\pi^*$$

(notice that the integral over G/H only makes sense because η is H–invariant). One can prove (using [35] and [40]) that there exists for almost all $\pi \in G_H^\wedge$ a subspace $\mathcal{V}_\pi^o$ (of dimension m_π) of $\mathcal{V}_\pi$, equipped with the structure of a Hilbert space, such that if $f^\wedge(\pi)$ is restricted to $\mathcal{V}_\pi^o$ for almost all π, then $f \mapsto f^\wedge$ extends to an isometry of $L^2(G/H)$ onto $\int_{G_H^\wedge}^\oplus \mathrm{Hom}_{\mathbb{C}}(\mathcal{V}_\pi^o, \mathcal{H}_\pi) d\mu(\pi)$. Here the norm on $\mathrm{Hom}_{\mathbb{C}}(\mathcal{V}_\pi^o, \mathcal{H}_\pi)$ is given by

$$\|\varphi\|_\pi^2 = \mathrm{Tr}(\varphi^* \circ \varphi) = \sum_i \|\varphi(v_i)\|^2, \qquad \varphi \in \mathrm{Hom}_{\mathbb{C}}(\mathcal{V}_\pi^o, \mathcal{H}_\pi),$$

where φ^* is the adjoint of φ and $\{v_i\}_{i=1,\ldots,m_\pi}$ is an orthonormal basis in $\mathcal{V}_\pi^o$.

We thus have the *Plancherel formula*

$$\|f\|_2^2 = \int_{G_H^\wedge} \|f^\wedge(\pi)\|_\pi^2 \, d\mu(\pi), \qquad f \in L^2(G/H).$$

Similarly, there is the *inversion formula* (for suitably nice functions f)

$$f(e) = \int_{G_H^\wedge} \sum_{i=1}^{m_\pi} \langle f^\wedge(\pi)v_i | v_i \rangle \, d\mu(\pi). \tag{4}$$

(Here $\langle \cdot | \cdot \rangle$ denotes the inner product on $\mathcal{H}_\pi$, as well as the naturally associated pairing $\mathcal{H}_\pi^\infty \times \mathcal{H}_\pi^{-\infty} \to \mathbb{C}$.) Consequently we also have, for suitable f

$$f(x) = \int_{G_H^\wedge} \sum_{i=1}^{m_\pi} \langle f^\wedge(\pi)v_i | \pi(x)v_i \rangle \, d\mu(\pi).$$

The basic problems are now

(a) Describe (parametrize) $G_H^\wedge$, or at least μ–almost all of it.
(b) For μ–almost all $\pi \in G_H^\wedge$ describe (parametrize) $\mathcal{V}_\pi^o$ and its Hilbert space structure.
(c) Determine $d\mu$ explicitly.

A *Paley–Wiener theorem* would amount to an intrinsic description of the Fourier image of $C_c^\infty(G/H)$ in terms of $G_H^\wedge$. We add this as a fourth basic problem:

(d) Describe $C_c^\infty(G/H)^\wedge$ in terms of the parametrizations and possible holomorphic extensions.

For each $\pi \in G_H^\wedge$ we have that $\mathcal{V}_\pi$ is a $\mathbb{D}(G/H)$–module in a natural way. Using that the symmetric elements of $\mathbb{D}(G/H)$ are essentially selfadjoint operators on $L^2(G/H)$ one can show (with the arguments in [40]) that $\mathcal{V}_\pi^o$ can be chosen to be invariant and diagonalizable for this action. Thus $\mathcal{V}_\pi^o$ is spanned by its joint eigenvectors for $\mathbb{D}(G/H)$. Let $\mathfrak{b} \subset \mathfrak{q}$ be a Cartan subspace. Then such an eigenvector satisfies

$$\pi(D)v = \chi_\lambda(D)v, \qquad D \in \mathbb{D}(G/H),$$

for some $\lambda \in \mathfrak{b}_{\mathbb{C}}^*$. We say that v is a *spherical vector of type* λ, and that the orthonormal basis $\{v_i\}_{i=1,\ldots,m_\pi}$ in $\mathcal{V}_\pi^o$ is spherical if its members are spherical.

The maps $\xi_{\pi,i}\colon f \mapsto \langle f^\wedge(\pi)v_i|v_i\rangle$ in (4) are H–invariant distributions on G/H. As distributions on G they are positive definite and extreme (see [40]). With a spherical basis $\{v_i\}$ each $\xi_{\pi,i}$ is also a *spherical distribution*, that is an H–invariant eigendistribution for $\mathbb{D}(G/H)$. The solution of Problem (b) is then closely related to the study of the spherical distributions.

3.3. Results valid for specific classes of symmetric spaces

Here we give some brief remarks concerning the status of the above problems for some specific classes of semisimple symmetric spaces.

3.3.1. *The compact type.* For a homogeneous space G/H with a compact group G the abstract formulation above follows easily from the Peter–Weyl theorem and the Schur orthogonality relations. In particular, $\mathcal{V}_\pi^o = \mathcal{V}_\pi = \mathcal{H}_\pi^H$, and if we give $\mathcal{V}_\pi^o$ the subspace norm from $\mathcal{H}_\pi$, we have $d\mu(\pi) = \dim(\pi)$. For the symmetric spaces of compact type we then have the following explicit solutions to the above problems (see [28, § V.4]):

(a) $G_H^\wedge$ is parametrized by a subset of the set of dominant weights.
(b) $\dim \mathcal{V}_\pi^o = 1$ for $\pi \in G_H^\wedge$.
(c) $d\mu$ is given by Weyl's dimension formula.
(d) The smooth functions are determined by a certain growth condition on the Fourier transforms (see [39]).

3.3.2. *The non-compact type.* We write M as G/K. The four questions are settled beautifully by the work of Harish–Chandra, Helgason and others. See [28, § IV.7]. Let $\mathfrak{a}$ be a maximal Abelian subspace of $\mathfrak{p}$.

(a) A sufficient subset of $G_K^\wedge$ is parametrized (up to conjugacy by the Weyl group W of $\mathfrak{a}$ in $\mathfrak{g}$) by means of the spherical functions $\varphi_\lambda, \lambda \in i\mathfrak{a}^*$ (see (23)) and the corresponding spherical principal series representations $(\pi_\lambda, \mathcal{H}_\lambda)$.

(b) For $\pi = \pi_\lambda \in G_K^\wedge$ we have $\mathcal{V}_\pi^o = \mathcal{H}_\lambda^K$ and $\dim(\mathcal{V}_\pi^o) = 1$. We can then use the subspace norm from $\mathcal{H}_\lambda$.

(c) The Plancherel measure is given by $d\mu(\pi_\lambda) = |\mathbf{c}(\lambda)|^{-2} d\lambda$ on $i\mathfrak{a}^*/W$. Here $\mathbf{c}(\lambda)$ is Harish–Chandra's c–function, which is explicitly given in terms of the structure of G/K by the formula of Gindikin–Karpelevic.

(d) We have $C_c^\infty(K\backslash G/K)^\wedge = \mathrm{PW}(\mathfrak{a})^W$. Here $\mathrm{PW}(\mathfrak{a})^W$ is the space of W–invariant functions in the image space $\mathrm{PW}(\mathfrak{a})$ for the Fourier transform

$$f \mapsto f^\wedge(\lambda) = \int_{\mathfrak{a}} f(X) e^{-\lambda(X)} dX, \quad \lambda \in \mathfrak{a}_{\mathbb{C}}^*, f \in C_c^\infty(\mathfrak{a}), \tag{5}$$

that is, the space of rapidly decreasing entire functions of exponential type on $\mathfrak{a}_{\mathbb{C}}^*$ (see Sect. 3.1, but note that since the imaginary unit i is not present in the exponent in (5) one has to replace $\operatorname{Im}\lambda$ by $\operatorname{Re}\lambda$ in (2)). Helgason has extended the Paley–Wiener theorem to the space $C_c^\infty(K; G/K)$ of K–finite functions in $C_c^\infty(G/K)$, and also to the full space $C_c^\infty(G/K)$.

3.3.3. *The group case,* $M = G_1$. This case is almost completely settled by the work of Harish–Chandra ([23]) and others.

(a) The map $\pi_1 \mapsto \pi_1 \otimes \pi_1^*$ is a bijective correspondence from the unitary dual $G_1^\wedge$ onto $G_H^\wedge$. A sufficient subset of $G_1^\wedge$ is described by the discrete series and different families of (cuspidal) principal series.

(b) For $\pi_1 \in G_1^\wedge$ and $\pi = \pi_1 \otimes \pi_1^*$ we have $\mathcal{V}_\pi = (\mathcal{H}_\pi^{-\infty})^H = \mathbb{C} 1_{\pi_1}$, where 1_{π_1} is the identity operator on $\mathcal{H}_{\pi_1}$. Notice however that in this case $\mathcal{V}_\pi \not\subset \mathcal{H}_\pi$, since the latter space can be identified with the space of Hilbert–Schmidt operators on $\mathcal{H}_{\pi_1}$. We take $\mathcal{V}_\pi^o = \mathcal{V}_\pi$, and use on it the Hilbert space structure obtained from the identification with $\mathbb{C}$ in which $1_{\pi_1} = 1$.

(c) With the above choice one can give $d\mu$ explicitly in terms of the formal degrees of discrete series and certain c–functions.

(d) A Paley–Wiener theorem for the K–finite functions on G_1 has been established in [14] (in split rank one) and [1] (in general). In particular, the Paley–Wiener space is determined by the minimal principal series only. For the full space $C_c^\infty(G_1)$ a Paley–Wiener theorem has not been established.

3.3.4. *The non-Riemannian type, rank one.* There is a vast literature dealing with the questions (a)–(c) on specific classes of rank one symmetric spaces of the non–Riemannian type. See for example [19], [40], [31]. Common for all these spaces is that the decomposition of $L^2(G/H)$ contains a discrete series as well as a continuous part. Problem (d) is solved in [9] (see below) for the K–finite functions, under the more general assumption that the *split* rank is one.

3.3.5. $G/H = \mathrm{SL}(n,\mathbb{C})/\mathrm{SU}(p,q)$. See [12].

4. A survey of results valid for general semisimple symmetric spaces

Even though the basic problems have been solved for many specific classes of semisimple symmetric spaces, there are still few final answers known which hold in complete generality. On the other hand, very much is known about the representations connected with these problems, and there is hope for the general answers in a not too distant future.

By analogy with the group case one expects in general that the left regular representation $\mathcal{L}$ on $L^2(G/H)$ can be decomposed in several 'series' of representations, one series for each H–conjugacy class of Cartan subspaces for $\mathfrak{q}$. The most extreme of these would then be the 'most continuous' part, corresponding to the conjugacy class of Cartan subspaces with maximal $\mathfrak{p}$–part (the split Cartan subspaces) and the 'most discrete' part (sometimes called the fundamental series), corresponding to the conjugacy class of Cartan subspaces with maximal $\mathfrak{k}$–part (the fundamental Cartan subspaces). If the fundamental Cartan subspaces are compact, then this 'most discrete' part is in fact the *discrete series*, that is, the irreducible subrepresentations of $\mathcal{L}$. For both of these 'extreme' parts of $L^2(G/H)$ very much is known about the Problems (a), (b) and (c); below (in Subsections 4.1, 4.3 and 4.4) we shall review some details and give precise references.

With respect to Problem (d) we want to mention two results of a general nature: One ([17], see Subsect. 4.2) which exhibits a large class (though too small to be 'the Paley–Wiener space' in general) of functions which are Fourier transforms of K–finite functions in $C_c^\infty(G/H)$, and another ([9], see Subsect. 4.5) which shows that the Fourier transform of a function in $C_c^\infty(G/H)$ is determined by its restriction to the meromorphic extension of the unitary principal series (the 'most continuous' part mentioned above). Along with the latter result goes a conjectural description of the K–finite Paley–Wiener space. The conjecture can be confirmed in the above mentioned cases 3.3.1, 3.3.2 and 3.3.3, and it also holds when G/H has split rank one.

4.1. The discrete series

The basic existence theorem is the following, where we preserve the notions from above. Let $L^2_d(G/H) \subset L^2(G/H)$ be the closed linear span of the irreducible subrepresentations of $\mathcal{L}$.

Theorem 2, [20], [33]. *Let G/H be a semisimple symmetric space. Then the discrete series space $L^2_d(G/H)$ is non–zero if and only if*

$$\operatorname{rank}(G/H) = \operatorname{rank}(K/K \cap H). \tag{6}$$

The condition (6) means that G/H has a compact Cartan subspace. An equivalent more geometric formulation is that it has a compact maximally flat subsymmetric space.

We shall now discuss Problems (a), (b) and (c) for the discrete series. Assume (as we may by the above theorem) that (6) holds, and let $\mathfrak{t}$ be a compact Cartan subspace of $\mathfrak{q}$. Let Σ be the root system of $\mathfrak{t}_\mathbb{C}$ in $\mathfrak{g}_\mathbb{C}$ and Σ_c the subsystem of $\mathfrak{t}_\mathbb{C}$ in $\mathfrak{k}_\mathbb{C}$. Let W and W_c be the corresponding reflection groups.

A rough classification of the discrete series is obtained by means of the commutative algebra $\mathbb{D}(G/H)$. Recall that the characters of $\mathbb{D}(G/H)$ are parametrized by $\mathfrak{t}^*_\mathbb{C}/W$ via the Harish–Chandra isomorphism $\chi\colon \mathbb{D}(G/H) \to S(\mathfrak{t})^W$. Let $\mathcal{E}_\lambda(G/H)$ denote the joint eigenspace for $\mathbb{D}(G/H)$ in $C^\infty(G/H)$ corresponding to the character χ_λ, where $\lambda \in \mathfrak{t}^*_\mathbb{C}$. Then $\mathcal{E}_{w\lambda}(G/H) = \mathcal{E}_\lambda(G/H)$ for all $w \in W$. Since $\mathbb{D}(G/H)$ is commutative and its symmetric elements act as essentially selfadjoint operators on $L^2(G/H)$, there is a joint spectral resolution of $L^2(G/H)$ for this algebra. The resulting decomposition is G–invariant because of the invariance of the elements in $\mathbb{D}(G/H)$. It follows (see [2]) that $L^2_d(G/H)$ admits an orthogonal G–invariant decomposition

$$L^2_d(G/H) = \widehat{\bigoplus_\lambda} L^2_\lambda(G/H),$$

where $L^2_\lambda(G/H)$ is the closure in $L^2(G/H)$ of $L^2(G/H) \cap \mathcal{E}_\lambda(G/H)$, and where the sum extends over the W–orbits in the set of those $\lambda \in \mathfrak{t}^*_\mathbb{C}$ for which $L^2_\lambda(G/H)$ is non–trivial. In order to parametrize the discrete series we must then determine this set of λ's, and for each λ therein the irreducible subrepresentations of $L^2_\lambda(G/H)$.

Let $\Lambda \subset i\mathfrak{t}^*$ denote the set of elements $\lambda \in i\mathfrak{t}^*$ satisfying the following conditions (i)–(iii).

(i) $\langle \lambda, \alpha \rangle \neq 0$ for all $\alpha \in \Sigma$.

Given that (i) holds, let

$$\Sigma^+ = \{\alpha \in \Sigma \mid \langle \lambda, \alpha \rangle > 0\}, \tag{7}$$

then this is a positive system for Σ. Put $\Sigma^+_c = \Sigma^+ \cap \Sigma_c$, and let ρ, resp. ρ_c, be defined as half the sum of the Σ^+–roots, resp. Σ^+_c–roots, counted with multiplicities.

(ii) $\lambda + \rho$ is a weight for T_H, i.e. $e^{\lambda+\rho}$ is well defined on T_H. Here T_H denotes the torus in G/H corresponding to $\mathfrak{t}$ (that is, $T_H = T/(T \cap H)$ where $T = \exp \mathfrak{t}$).

(iii) $\langle \lambda - \rho, \beta \rangle \geq 0$ for each compact simple root β in Σ^+.

(that β is compact means that the root space $\mathfrak{g}_{\mathbb{C}}^{\beta}$ is contained in $\mathfrak{k}_{\mathbb{C}}$). Notice that (ii) implies that Λ is a discrete subset of it^*.

Under the assumption that $\lambda \in \Lambda$ there is a rather simple construction (which we shall outline below) of a $\mathfrak{g}$–invariant subspace $\mathcal{U}_{\lambda,K}$ of $C^\infty(K;G/H)$ (the space of K–finite functions in $C^\infty(G/H)$), which can be shown to be contained in $L^2_\lambda(G/H)$. Let $\mathcal{U}_\lambda$ denote the closure of $\mathcal{U}_{\lambda,K}$ in $L^2(G/H)$, then $\mathcal{U}_\lambda$ is a subrepresentation of $L^2_\lambda(G/H)$. Let π_λ denote this subrepresentation.

For 'large' $\lambda \in \Lambda$, or more precisely if $\langle \lambda + \rho - 2\rho_c, \alpha \rangle \geq 0$ for all $\alpha \in \Sigma_c^+$, it can be shown by elementary methods that $\mathcal{U}_\lambda \neq \{0\}$. For the remaining λ's one has to add a more technical assumption in order to ensure that $\mathcal{U}_\lambda \neq \{0\}$. We shall not state this condition here (the condition is stated in [30] together with a proof of its necessity for the non–vanishing of $\mathcal{U}_\lambda$).

Theorem 3, [33], [41]. *The discrete series space $L^2_d(G/H)$ is spanned by the $\mathcal{U}_\lambda$'s with $\lambda \in \Lambda$. Moreover for each $\lambda \in \Lambda$ either the representation π_λ is irreducible or $\mathcal{U}_\lambda$ is zero, and if $\lambda, \lambda' \in \Lambda$ we have $\mathcal{U}_{\lambda'} = \mathcal{U}_\lambda$ if and only if $\lambda' = w\lambda$ for some $w \in W_c$.*

It follows that if $\lambda \in \mathfrak{t}^*_{\mathbb{C}}$ then $L^2_\lambda(G/H)$ is the sum of those $\mathcal{U}_{w\lambda}$ for which $w \in W$ and $w\lambda \in \Lambda$. In particular it has at most as many components as the order of the quotient W/W_c.

With this result, Problem (a) is solved as regards to the discrete series. It is conjectured that $\pi_{\lambda'}$ is unitarily equivalent to π_λ if and only if $\mathcal{U}_{\lambda'} = \mathcal{U}_\lambda$, or equivalently in view of the above, that the discrete series have multiplicity one in the Plancherel formula. The conjecture is proved for all classical groups G, and is only open for a few exceptional cases for very special values of λ (see [11]).

Evaluation at the base point in G/H gives rise to an H–fixed distribution vector η_λ for $\mathcal{U}_\lambda$, for which it is easily seen that we have

$$f^\wedge(\pi_\lambda)(\eta_\lambda) = \mathrm{P}_\lambda f, \qquad f \in C_c^\infty(G/H),$$

where P_λ is the orthogonal projection of $L^2(G/H)$ onto $\mathcal{U}_\lambda$. It follows that if we take $\mathcal{V}^o_{\pi_\lambda} = \mathbb{C}\eta_\lambda$ and use on it the Hilbert space structure obtained from the identification with $\mathbb{C}$ in which $\eta_\lambda = 1$, then $d\mu(\pi_\lambda) = 1$. In other words, the Plancherel measure restricts to the counting measure on the discrete series. This provides the solution to Problems (b) and (c) for the discrete series.

At this point it is however interesting to note the following. Though the discrete series has been parametrized as above, it seems to be an open problem to determine an explicit expression for the spherical distribution $\xi_\lambda : f \mapsto \langle f^\wedge(\pi_\lambda)\eta_\lambda | \eta_\lambda \rangle$ on G/H associated to η_λ (or equivalently, for the projection operator P_λ, which is given by convolution with ξ_λ). In the group case one knows that ξ_λ is given by $d_\lambda \Theta_\lambda$, where d_λ is the formal degree and Θ_λ the character of π_λ (see [22, §5]), but there is no obvious generalization of this formula.

We shall not try to describe the proof of the above theorems. However as the construction of $\mathcal{U}_{\lambda,K}$ can be described by quite elementary methods we would like to indicate it.

Let the notation be as above, and recall the decomposition (1) of $\mathfrak{g}$. Let $\mathfrak{g}^d$ be the real form of $\mathfrak{g}_{\mathbb{C}}$ given by

$$\mathfrak{g}^d = \mathfrak{h} \cap \mathfrak{k} \oplus i(\mathfrak{h} \cap \mathfrak{p}) \oplus i(\mathfrak{q} \cap \mathfrak{k}) \oplus \mathfrak{q} \cap \mathfrak{p},$$

where i is the imaginary unit. Assume (again for simplicity of exposition) that G is a real form of a linear complex Lie group $G_{\mathbb{C}}$, and let G^d be the real form of $G_{\mathbb{C}}$ whose Lie algebra is $\mathfrak{g}^d$. Then the subgroup $K^d = G^d \cap H_{\mathbb{C}}$ is a maximal compact subgroup. The symmetric space G^d/K^d is called the *non-compact Riemannian form* of G/H. The subgroup $H^d = G^d \cap K_{\mathbb{C}}$ of G^d is a (in general non-compact) real form of $K_{\mathbb{C}}$. Let $(G \cap G^d)_e$ denote the identity component of $G \cap G^d$. Then both G and G^d are contained in the set $K_{\mathbb{C}}(G \cap G^d)_e H_{\mathbb{C}}$. The K–finite functions on G/H extend naturally to left $K_{\mathbb{C}}$–finite and right $H_{\mathbb{C}}$–invariant functions on this set (and so do the H^d–finite functions on G^d/K^d, provided the H^d–action admits a holomorphic extension to $K_{\mathbb{C}}$). We call this *partial holomorphic extension.* Let $C^\infty(K; G/H)$ and $C^\infty(H^d; G^d/K^d)$ be the spaces of K–finite, resp. H^d–finite smooth functions on G/H, resp. G^d/K^d. There is a natural action of $\mathfrak{g}_{\mathbb{C}}$ on both of these spaces.

Proposition 4, [20]. *Partial holomorphic extension defines a $\mathfrak{g}_{\mathbb{C}}$–equivariant linear injection $f \to f^r$ of $C^\infty(K; G/H)$ into $C^\infty(H^d; G^d/K^d)$, the image of which is the set of functions in $C^\infty(H^d; G^d/K^d)$ for which the H^d–action extends holomorphically to $K_{\mathbb{C}}$. Moreover, f is a joint eigenfunction for $\mathbb{D}(G/H)$ if and only if f^r is a joint eigenfunction for $\mathbb{D}(G^d/K^d)$.*

The construction of G^d/K^d and this proposition hold independently of (6). However, this assumption is crucial for the following construction.

Since G^d/K^d is a Riemannian symmetric space the joint eigenfunctions for the algebra $\mathbb{D}(G^d/K^d)$ can be described by means of the so-called generalized Poisson transform. This is defined as follows. It follows from the fact that $\mathfrak{t}$ is a maximal Abelian subspace of $\mathfrak{q}$, that $\mathfrak{t}^r = i\mathfrak{t}$ is a maximal Abelian split subspace for $\mathfrak{g}^d$. Hence there is an Iwasawa decomposition

$$G^d = K^d T^r N^d \tag{8}$$

of G^d with $T^r = \exp \mathfrak{t}^r$, which corresponds to a given Σ^+. Let $P^d = M^d T^r N^d$ be the corresponding minimal parabolic subgroup in G^d, and for $\lambda \in \mathfrak{t}^*_{\mathbb{C}}$ let $\mathcal{D}'_\lambda = \mathcal{D}'_\lambda(G^d/P^d)$ be the space of $(\lambda - \rho)$–homogeneous distributions on G^d/P^d, that is the space of generalized functions f on G^d satisfying

$$f(gman) = a^{\lambda-\rho} f(g), \qquad g \in G^d, m \in M^d, a \in T^r, n \in N^d.$$

The group G^d acts from the left on this space. The Poisson transform $\mathcal{P}_\lambda\colon \mathcal{D}'_\lambda \to C^\infty(G/H)$ is defined by

$$\mathcal{P}_\lambda f(x) = \int_{K^d} f(xk)\,dk = \int_{K^d} p_\lambda(x,k) f(k)\,dk, \quad x \in G^d.$$

Here the 'Poisson kernel' $p_\lambda \in C^\infty(G^d \times K^d)$ is defined by $p_\lambda(x,k) = a^{-\lambda-\rho}$, where $a \in T^r$ is the T^r-part of $x^{-1}k$ in the decomposition (8). It is known that $\mathcal{P}_\lambda$ is a G^d-equivariant transformation into a joint eigenspace for $\mathbb{D}(G^d/K^d)$ in $C^\infty(G^d/K^d)$, and that it is injective if Σ^+ is given by (7) (see for example [7, Thm. 12.2]).

Let $\mathcal{D}'_{\lambda,H^d}$ be the set of H^d-finite elements in $\mathcal{D}'_\lambda$, and let $\mathcal{D}'_{\lambda,H^d}(H^dP^d)$ denote the subset of elements supported on the H^d-orbit H^dP^d in G^d/P^d (which is closed, cf. [37, Prop. 7.1.8]). Let now $\lambda \in \Lambda$. Then condition (ii) implies that the H^d-finite action on $\mathcal{D}'_{\lambda,H^d}(H^dP^d)$ extends to a holomorphic $K_{\mathbb{C}}$-action. The space $\mathcal{U}_{\lambda,K}$ is now defined by

$$\mathcal{U}_{\lambda,K} = \{f \in C^\infty(K; G/H) \mid f^r \in \mathcal{P}_\lambda(\mathcal{D}'_{\lambda,H^d}(H^dP^d))\}.$$

The proof that $\mathcal{U}_{\lambda,K} \subset L^2_\lambda(G/H)$ can be found in [33] (see also [7, Thm. 19.1]).

4.2. A partial Paley–Wiener theorem

We now return to the general case, where condition (6) is not necessarily fulfilled. We shall see that a variation of the ideas going into Prop. 5 provides us with a construction of the inverse Fourier transform for a large family of 'nice' functions on $G^\wedge_H$.

Recall that for $f \in C_c^\infty(G/H)$ we have defined the Fourier transform $f^\wedge$ on $G^\wedge_H$ such that

$$f^\wedge(\pi) \in \mathrm{Hom}(\mathcal{V}^o_\pi, \mathcal{H}_\pi) = \mathcal{H}_\pi \otimes (\mathcal{V}^o_\pi)^*.$$

Let $\mathfrak{b} \subset \mathfrak{q}$ be a θ-invariant Cartan subspace. Let $\{v_i\}_{i=1,\dots,m_\pi}$ be a spherical basis for $\mathcal{V}^o_\pi$ (for a given π), and let $\lambda_i \in \mathfrak{b}^*_{\mathbb{C}}$ be the type of v_i (determined up to conjugation by W).

As in the previous section let G^d/K^d be the non-compact Riemannian form of G/H. In analogy with the definition of $\mathfrak{t}^r$ we define

$$\mathfrak{b}^r = \mathfrak{b} \cap \mathfrak{p} + i(\mathfrak{b} \cap \mathfrak{k}) = \mathfrak{b}_{\mathbb{C}} \cap \mathfrak{g}^d, \tag{9}$$

then $\mathfrak{b}^r$ is a maximal Abelian split subspace for $\mathfrak{g}^d$. Hence the roots of $\mathfrak{b}_{\mathbb{C}}$ in $\mathfrak{g}_{\mathbb{C}}$ are real valued on $\mathfrak{b}^r$, and $\mathfrak{b}^r$ is W-invariant.

Let $PW(\mathfrak{b}^r)^W$ be the space of W invariant entire rapidly decreasing functions of exponential type on $\mathfrak{b}^*_{\mathbb{C}}$. By the classical Paley–Wiener theorem this is the image of the space $C_c^\infty(\mathfrak{b}^r)^W$ under the Fourier transform on the Euclidean space $\mathfrak{b}^r$ (defined as in (5)), and by Helgason's Paley–Wiener theorem it is also the image of $C_c^\infty(K^d\backslash G^d/K^d)$ under the spherical Fourier transform (see 3.3.2 (d)).

Let $K^\wedge_{K\cap H}$ be the set of (equivalence classes) of irreducible representations of K with non-trivial $K\cap H$-fixed vectors. For any $\psi \in PW(\mathfrak{b}^r)^W$, $\mu \in K^\wedge_{K\cap H}$ and $\pi \in G^\wedge_H$ we define $F_{\psi,\mu}(\pi) \in \mathrm{Hom}(\mathcal{V}^o_\pi, \mathcal{H}_\pi)$ by

$$F_{\psi,\mu}(\pi)v_i = \psi(\lambda_i)P_\mu v_i, \qquad (i = 1, \ldots, m_\pi),$$

where $P_\mu\colon \mathcal{H}_{-\infty} \to \mathcal{H}_\infty$ is the K-equivariant extension to $\mathcal{H}_{-\infty}$ of the orthogonal projection of $\mathcal{H}_\pi$ onto its μ-component (given by the convolution with the normalized character of $\mu^\vee$). Notice that $F_{\psi,\mu}$ is independent of the choice of the spherical basis $\{v_i\}$ for $\mathcal{V}^o_\pi$.

Theorem 5, [17]. *Let $\psi \in PW(\mathfrak{b}^r)^W$ and $\mu \in K^\wedge_{K\cap H}$ be given, and let $F_{\psi,\mu}$ be as above. There exists a unique function $f = f_{\psi,\mu}$ in $C_c^\infty(G/H)$ such that $f^\wedge = F_{\psi,\mu}$, or equivalently, for any $\pi \in G^\wedge_H$ and any spherical vector $v \in \mathcal{V}^o_\pi$ of type $\lambda \in \mathfrak{b}^*_{\mathbb{C}}$ we have*

$$f^\wedge(\pi)v = \psi(\lambda)P_\mu v. \tag{10}$$

Moreover, the function f is $K\cap H$-invariant and K-finite of type μ, and the equation (10) holds more generally with v a spherical vector of type λ in $\mathcal{V}_\pi$.

Notice that it follows from (10) that the spherical distributions given by $\xi_{\pi,i}\colon \varphi \mapsto \langle \varphi^\wedge(\pi)v_i | v_i\rangle$, $i = 1, \ldots, m_\pi$, satisfy

$$\xi_{\pi,i}(f) = \psi(\lambda_i)\langle P_\mu v_i | v_i\rangle$$

for all $\pi \in G^\wedge_H$.

In order to indicate the proof of Thm. 5 we shall need the following proposition, which is closely related to Prop. 4. Let the spaces $K\backslash G/H$ and $H^d\backslash G^d/K^d$ be given the measures inherited from the invariant measures on G/H and $H^d\backslash G^d$, respectively.

Proposition 6, [20]. *Partial holomorphic extension defines a norm-preserving isomorphism $f \to f^r$ of $L^1(K\backslash G/H)$ onto $L^1(H^d\backslash G^d/K^d)$.*

Indication of the proof of Thm. 5. The uniqueness of f follows easily from the abstract Plancherel theory discussed earlier. The existence is established as follows.

Let $\psi \in PW(\mathfrak{b}^r)^W$ and $\mu \in K^\wedge_{K\cap H}$ be given, and let V_μ be the representation space for μ, equipped with an inner product for which μ is unitary. By the Paley-Wiener theorem for the spherical Fourier transform on G^d/K^d there exists a K^d-invariant function $F \in C_c^\infty(G^d/K^d)$ such that

$$\int_{G^d/K^d} F(x)\varphi_\lambda(x)\,dx = \psi(\lambda)$$

for all $\lambda \in \mathfrak{b}^*_{\mathbb{C}}$, where φ_λ is the elementary spherical function on G^d/K^d. Let $e_o \in V_\mu$ be a $K\cap H$-fixed unit vector and define

$$F_\mu(x) = \dim(\mu)\int_{H^d} F(hx)\langle \mu(h)e_o | e_o\rangle dh, \qquad x \in G^d/K^d.$$

Here μ is defined on H^d by the holomorphic extension from K to $K_{\mathbb{C}}$. It follows that F_μ is H^d-finite of this type. Let $f \in C^\infty(K; G/H)$ be the element such that $f^r = F_\mu$ by Prop. 4.

It follows easily from this construction that f has compact support. To finish we must calculate $f^\wedge(\pi)v$ for all spherical $v \in \mathcal{V}_\pi$. Let $\lambda \in \mathfrak{b}_{\mathbb{C}}^*$ be the type of v. Since f is K-finite of type μ and $K \cap H$-fixed it suffices to calculate $\langle f^\wedge(\pi)v|v'\rangle$, where v' is a $K \cap H$-fixed vector in V_μ. Now $\langle f^\wedge(\pi)v|v'\rangle$ can be written as an integral over K followed by an integral over $K\backslash G/H$, and by Prop. 6 the latter can be transferred to an integral over $H^d\backslash G^d/K^d$ involving F_μ. After some rewriting one ends up by finding

$$\langle f^\wedge(\pi)v|v'\rangle = \int_{G^d/K^d} F(x)\varphi_\lambda(x)\,dx\,\langle P_\mu v|v'\rangle = \psi(\lambda)\langle P_\mu v|v'\rangle,$$

from which the result follows.

4.3. A Plancherel formula for the most continuous part of $L^2(G/H)$

In this subsection we discuss Problems (a), (b) and (c) for the 'most continuous part' of $L^2(G/H)$ (to be defined below). The main reference is [9].

Let notation be as in Sect. 2 (in [9] the assumptions on G/H are somewhat more general, but we shall not discuss that here). The representations $\pi_{\xi,\lambda}$ that occur in the most continuous part of $L^2(G/H)$ are constructed as follows. Let $P = MAN$ be a parabolic subgroup of G, with the indicated Langlands decomposition, satisfying $\sigma\theta P = P$ and being minimal with respect to this condition. Then M and A are σ-stable. Let $\mathfrak{a}_\mathfrak{q} = \mathfrak{a} \cap \mathfrak{q}$, where $\mathfrak{a}$ is the Lie algebra of A, then it follows that $\mathfrak{a}_\mathfrak{q}$ is a maximal Abelian subspace of $\mathfrak{p} \cap \mathfrak{q}$, and that the Levi part MA of P is the centralizer of $\mathfrak{a}_\mathfrak{q}$ in G. Let $(\xi, \mathcal{H}_\xi) \in M^\wedge_{\mathrm{fu}}$, the set of (equivalence classes of) finite dimensional irreducible unitary representations of M, and let $\lambda \in i\mathfrak{a}^*$. We require that $\lambda \in i\mathfrak{a}_\mathfrak{q}^*$, that is that λ vanishes on $\mathfrak{a} \cap \mathfrak{h}$. Then by definition $\pi_{\xi,\lambda}$ is the induced representation $\pi_{P,\xi,\lambda} = \mathrm{Ind}_{P=MAN}^G\, \xi \otimes e^\lambda \otimes 1$ (the 'principal series' for G/H), that is, the representation space $\mathcal{H}_{\xi,\lambda}$ consists of (classes of) $\mathcal{H}_\xi$ valued measurable functions f on G, square integrable on K and satisfying

$$f(gman) = a^{-\lambda-\rho}\xi(m)^{-1}f(g), \qquad (g \in G, m \in M, a \in A, n \in N), \tag{11}$$

and G acts from the left. Here $\rho = \frac{1}{2}\,\mathrm{Tr}\,\mathrm{Ad}_\mathfrak{n} \in \mathfrak{a}_\mathfrak{q}^*$. (The convention in (11) differs from the above cited references: The induction takes place on the opposite side.)

The Plancherel decomposition for the most continuous part of $L^2(G/H)$ is obtained by realizing the abstract Fourier transform explicitly for the principal series. This realization is then a partial isometry of $L^2(G/H)$ onto the direct integral

$$\int_{\xi,\lambda}^{\oplus} m_\xi\, \pi_{\xi,\lambda}\, d\mu(\xi,\lambda). \tag{12}$$

The multiplicities m_ξ (which happen to be independent of λ) and the measure $d\mu(\xi,\lambda)$ are explicitly described below. The most continuous part of $L^2(G/H)$, denoted $L^2_{\mathrm{mc}}(G/H)$, is then by definition the orthocomplement of the kernel of this partial isometry. Its Plancherel decomposition is exactly given by (12).

In order to realize the Fourier transform we must first discuss the space $\mathcal{V}_{\xi,\lambda} = (\mathcal{H}^{-\infty}_{\xi,\lambda})^H$. Let $\mathcal{W} \subset N_K(\mathfrak{a}_\mathrm{q})$ be a fixed set of elements such that $w \mapsto HwP$ parametrizes the open $H \times P$ orbits on G (it is known (see [36] or [29]) that any set of representatives for the double quotient $N_{K\cap H}(\mathfrak{a}_\mathrm{q})\backslash N_K(\mathfrak{a}_\mathrm{q})/Z_K(\mathfrak{a}_\mathrm{q})$ can be used as $\mathcal{W}$ – in particular, $\mathcal{W}$ is finite). Viewing an element $f \in \mathcal{H}^{-\infty}_{\xi,\lambda}$ as an $\mathcal{H}_\xi$-valued distribution on G, satisfying appropriate conditions of homogeneity for the right action of P, it is easily seen that if f is H-invariant then f must restrict to a smooth function on each open $H \times P$ orbit. Hence it makes sense to evaluate f in the elements of $\mathcal{W}$, and in fact its restriction to the open orbit HwP will be uniquely determined from the value at w. We denote this value by $\mathrm{ev}_w(f)$. It is easily seen that ev_w maps $\mathcal{V}_{\xi,\lambda}$ into the space $\mathcal{H}_\xi^{w^{-1}(M\cap H)w}$ of $w^{-1}(M\cap H)w$-fixed elements in $\mathcal{H}_\xi$ (note that $w^{-1}Mw = M$, but $w^{-1}Hw$ may differ from H). Let $V(\xi)$ denote the formal direct sum

$$V(\xi) = \bigoplus_{w\in\mathcal{W}} \mathcal{H}_\xi^{w^{-1}(M\cap H)w}. \tag{13}$$

provided with the direct sum inner product (thus, by definition the summands are mutually orthogonal, even though this may not be the case in $\mathcal{H}_\xi$). Furthermore, let

$$\mathrm{ev}\colon \mathcal{V}_{\xi,\lambda} \to V(\xi)$$

denote the direct sum of the maps ev_w. The construction of the induced representations $\pi_{\xi,\lambda}$ and of the map ev makes sense for $\lambda \in \mathfrak{a}^*_{\mathrm{q}\mathbb{C}}$, the complex linear dual of $\mathfrak{a}_\mathrm{q}$ (though the representations need not be unitary for λ outside $i\mathfrak{a}^*_\mathrm{q}$). We now have

Theorem 7, [3]. *The map ev is bijective for generic $\lambda \in \mathfrak{a}^*_{\mathrm{q}\mathbb{C}}$.*

For generic λ, let

$$j(\xi,\lambda)\colon V(\xi) \to \mathcal{V}_{\xi,\lambda}$$

be the inverse of ev, then by definition we have for $\eta \in V(\xi)$ that the restriction of the distribution $j(\xi,\lambda)(\eta)$ to the open $H \times P$ orbit HwP, $w \in \mathcal{W}$, is the smooth $\mathcal{H}_\xi$-valued function given by

$$j(\xi,\lambda)(\eta)(hwman) = a^{-\lambda-\rho}\xi(m^{-1})\eta_w. \tag{14}$$

(Here η_w denotes the w-component of η, viewed as an element of $\mathcal{H}_\xi$.) Notice that if G/H is a Riemannian symmetric space, so that $H = K$, then we have $G = HP$ by the Iwasawa decomposition. Hence we can take $\mathcal{W} = \{e\}$, and since $M \subset K = H$ we have $V(\xi) = \{0\}$ unless ξ is the trivial representation $\mathbf{1}$, in which case $V(\mathbf{1}) = \mathbb{C}$. Then $j(\mathbf{1}, \lambda)$ is completely determined by (14); in fact we have

$$j(\mathbf{1}, \lambda)(x) = e^{-(\lambda+\rho)H(x)},$$

where $H: G \to \mathfrak{a}$ is the Iwasawa projection (since $V(\mathbf{1}) = \mathbb{C}$ we can omit η). Thus the kernel $P_\lambda(x, k) = j(\mathbf{1}, \lambda)(x^{-1}k)$ on $G/K \times K$ is the generalized Poisson kernel. For general G/H we can supplement (14) as follows: If $\mathrm{Re}\langle\lambda + \rho, \alpha\rangle < 0$ for all α in the set Σ^+ of positive roots (the $\mathfrak{a}$-roots of $\mathfrak{n} = \mathrm{Lie}(N)$), then $j(\xi, \lambda)(\eta)$ is the continuous function on G given by (14) on HwP for all $w \in \mathcal{W}$ and vanishing on the complement of these sets (the condition on λ ensures the continuity). For elements λ outside the above region the distribution $j(\xi, \lambda)$ can be obtained from the above by meromorphic continuation. (See [34], [32], [3]. These results have been generalized to other principal series representations in [13], [15].)

Having constructed the H-invariant distribution vectors $j(\xi, \lambda)\eta$ as above we can now attempt to write down a Fourier transform for the principal series. For $f \in C_c^\infty(G/H)$ we consider the map

$$(\xi, \lambda) \mapsto f^\wedge(\pi_{\xi,\lambda})j(\xi, \lambda) = \pi_{\xi,\lambda}(f)j(\xi, \lambda) \in \mathcal{H}^\infty_{\xi,\lambda} \otimes V(\xi)^*. \tag{15}$$

In the Riemannian case this is exactly the Fourier transform, as defined by Helgason (see [24]). However when G/H is not Riemannian a new phenomenon may occur: by the above definitions (15) is a meromorphic function in λ, which may have singularities on the set $i\mathfrak{a}_\mathfrak{q}^*$ of interest for the Plancherel decomposition, and thus it may not make sense for some singular $\lambda \in i\mathfrak{a}_\mathfrak{q}^*$. This unpleasantness is overcome by a suitable normalization of $j(\xi, \lambda)$, which removes the singularities. The normalization is carried out by means of the standard intertwining operators $A(\bar{P}, P, \xi, \lambda)$ from $\pi_{P,\xi,\lambda}$ to $\pi_{\bar{P},\xi,\lambda}$, where $\bar{P}$ is the parabolic subgroup opposite to P. Let

$$j^\circ(\xi, \lambda) = A(\bar{P}, P, \xi, \lambda)^{-1} j(\bar{P}, \xi, \lambda),$$

where $j(\bar{P}, \xi, \lambda)$ is constructed as $j(\xi, \lambda)$ above, but with P replaced by $\bar{P}$. Since the intertwining operator $A(\bar{P}, P, \xi, \lambda)$ is bijective for generic λ, it follows that

$$j^\circ(\xi, \lambda): V(\xi) \to \mathcal{V}_{\xi,\lambda}$$

is again a bijection, for generic λ. Moreover, we now have

Theorem 8, [9]. *The meromorphic function $\lambda \mapsto j^\circ(\xi, \lambda)$ is regular on $i\mathfrak{a}_\mathfrak{q}^*$.*

We can now define the Fourier transform $f \mapsto f^\wedge$ for the principal series properly by (15), but with j replaced by j°:

$$f^\wedge(\xi, \lambda) = \pi_{\xi,\lambda}(f)j^\circ(\xi, \lambda) \in \mathcal{H}^\infty_{\xi,\lambda} \otimes V(\xi)^*.$$

Notice that when G/H is Riemannian the normalization makes our Fourier transform different from that of Helgason – in this case the normalization amounts to a division by Harish–Chandra's c–function $\mathbf{c}(\lambda)$. See [8] for the determination of j° in the group case.

We can now give the solution to Problem (b) for this part of $L^2(G/H)$: We take $\mathcal{V}^o_{\xi,\lambda} = \mathcal{V}_{\xi,\lambda}$, and give it the Hilbert space structure that makes $j^\circ(\xi,\lambda)$ an isometry. The solution to Problem (c) is as follows. Let $\mathcal{H}$ be the Hilbert space given by

$$\mathcal{H} = \int_{\xi,\lambda}^{\oplus} \mathcal{H}_{\xi,\lambda} \otimes V(\xi)^* \, d\mu(\xi,\lambda), \tag{16}$$

with the measure $d\mu(\xi,\lambda) = \dim(\xi)\, d\lambda$, where $d\lambda$ is Lebesgue measure on $i\mathfrak{a}_{\mathrm{q}}^*$ (suitably normalized). Here ξ runs over $M_{\mathrm{fu}}^\wedge$ (notice however that some of them may disappear because $V(\xi)$ is trivial), and λ runs over an open chamber $i\mathfrak{a}_{\mathrm{q}}^{*+}$ in $i\mathfrak{a}_{\mathrm{q}}^*$ for the Weyl group $W_{\mathrm{q}} = N_K(\mathfrak{a}_{\mathrm{q}})/Z_K(\mathfrak{a}_{\mathrm{q}})$.

Theorem 9, [9]. *Let $f \in C_c^\infty(G/H)$. Then $f^\wedge \in \mathcal{H}$ and $\|f^\wedge\| \le \|f\|_2$. Moreover, the map $f \mapsto f^\wedge$ extends to an equivariant partial isometry $\mathfrak{F}$ of $L^2(G/H)$ onto $\mathcal{H}$. In particular, we thus have the multiplicities $m_\xi = \dim V(\xi)$.*

We define the most continuous part $L^2_{\mathrm{mc}}(G/H)$ of $L^2(G/H)$ as the orthocomplement of the kernel of $\mathfrak{F}$. Then $\mathfrak{F}$ restricts to an isometry of this space onto $\mathcal{H}$. In [9] it is shown that $L^2_{\mathrm{mc}}(G/H)$ is 'large' in $L^2(G/H)$ in a certain sense – in particular its orthocomplement (the kernel of $\mathfrak{F}$) has trivial intersection with $C_c^\infty(G/H)$ (thus $f \mapsto f^\wedge$ is injective, even though the extension $\mathfrak{F}$ need not be). Moreover, if G/H has split rank one, that is if $\dim \mathfrak{a}_{\mathrm{q}} = 1$, then there are at most two conjugacy classes of Cartan subspaces, and hence one expects from the analogy with the group case as mentioned earlier that only the corresponding two 'series' of representations will be present. Indeed this is the case; it is shown in [9] that the kernel of $\mathfrak{F}$ decomposes discretely when the split rank is one. Thus, in this case the Plancherel decomposition of $L^2(G/H)$ can be determined from Thm. 9 together with the description of the discrete series (see Sect. 4.1 above), except for the explicit determination of the Hilbert space structure on $\mathcal{V}^o_\pi$ for the discrete series representations π.

On the other hand, when G/H is Riemannian then $\mathfrak{F}$ is injective and Thm. 9 gives the complete Plancherel decomposition of $L^2(G/H)$ (in the formulation of Harish–Chandra and Helgason the Plancherel measure is $|\mathbf{c}(\lambda)|^{-2}\, d\lambda$, but here the factor $|\mathbf{c}(\lambda)|^{-2}$ disappears because of the normalization of j°).

A further discussion of the multiplicities m_π can be found in [8].

4.4. THE K–FINITE CASE

The isomorphism of (16) onto $L^2_{\mathrm{mc}}(G/H)$ (the 'inverse Fourier transform') can be given more explicitly when one restricts to K-finite functions. In this subsection we shall discuss this restriction, which happens to be crucial in the proofs of Thms. 8 and 9.

4.4.1. *Eisenstein integrals.* Let (μ, V_μ) be a fixed, irreducible unitary representation of K. Taking μ-components in (16) we have

$$\mathcal{H}^\mu = \int_{\xi,\lambda}^{\oplus} \mathcal{H}^\mu_{\xi,\lambda} \otimes V(\xi)^* \, d\mu(\xi,\lambda). \tag{17}$$

Moreover, by Frobenius reciprocity we have

$$\mathcal{H}^\mu_{\xi,\lambda} \simeq \mathrm{Hom}_{M\cap K}(V_\mu, \mathcal{H}_\xi) \otimes V_\mu \tag{18}$$

as K-modules (where K acts on the second component in the tensor product), for all $\xi \in M^\wedge_{\mathrm{fu}}$, $\lambda \in \mathfrak{a}^*_{\mathrm{q}\mathbb{C}}$. Note that since each representation $\xi \in M^\wedge_{\mathrm{fu}}$ is trivial on the non-compact part of M, we have that $\xi|_{M\cap K}$ is irreducible, and that $\mathrm{Hom}_{M\cap K}(V_\mu, \mathcal{H}_\xi)$ is non-trivial if and only if this restriction occurs as a subrepresentation of $\mu|_{M\cap K}$. We use the notation $\xi \uparrow \mu$ to indicate this occurrence; it happens only for finitely many ξ. Thus by taking K-types the integral over ξ in (17) becomes a finite sum, hence more manageable. In analogy with the earlier definition of the space $V(\xi)$ we now define the space $\mathcal{V}(\mu)$ to be the formal direct sum

$$\mathcal{V}(\mu) = \bigoplus_{w\in\mathcal{W}} V_\mu^{w^{-1}(K\cap M\cap H)w}.$$

It is easily seen from the above that

$$\mathcal{V}(\mu) \simeq \bigoplus_{\xi\uparrow\mu} \mathrm{Hom}_{M\cap K}(\mathcal{H}_\xi, V_\mu) \otimes V(\xi). \tag{19}$$

Hence in view of (18) we have

$$\mathcal{V}(\mu)^* \otimes V_\mu \simeq \bigoplus_{\xi\uparrow\mu} \mathcal{H}^\mu_{\xi,\lambda} \otimes V(\xi)^* \tag{20}$$

for all $\lambda \in \mathfrak{a}^*_{\mathrm{q}\mathbb{C}}$. From (17) and (20) we finally obtain

$$\mathcal{H}^\mu \simeq \int_\lambda^{\oplus} \mathcal{V}(\mu)^* \otimes V_\mu \, d\lambda \simeq L^2(i\mathfrak{a}^{*+}_{\mathrm{q}}) \otimes \mathcal{V}(\mu)^* \otimes V_\mu. \tag{21}$$

This isomorphism indicates that the Fourier transform, when restricted to K-finite functions of type μ, can be considered as a map into the $\mathcal{V}(\mu)^* \otimes V_\mu$-valued functions on $i\mathfrak{a}^*_{\mathrm{q}}$.

Instead of working with K-finite scalar-valued functions on G/H, it is convenient to consider 'μ-spherical' functions f on G/H, that is, V_μ-valued functions satisfying

$$f(kx) = \mu(k)f(x), \qquad k \in K,\ x \in G/H.$$

Let $L^2(G/H;\mu)$ denote the space of square integrable such functions, then by contraction we have a K–equivariant isomorphism

$$\gamma_\mu\colon L^2(G/H;\mu^\vee)\otimes V_\mu \xrightarrow{\sim} L^2(G/H)^\mu. \tag{22}$$

(Again K acts on the second component in the tensor product. The map $\dim(\mu)\gamma_\mu$ is an isometry.) Notice that when passing from K–finite functions to spherical functions one must also pass from μ to its contragradient $\mu^\vee$. Since $\mathcal{V}(\mu)^* = \mathcal{V}(\mu^\vee)$ we are led to the search, for each μ, of a Fourier transform, which is a partial isometry of $L^2(G/H;\mu)$ onto $L^2(i\mathfrak{a}_\mathrm{q}^{*+})\otimes\mathcal{V}(\mu)$. Going through the above isomorphisms in detail, we are led to the following construction culminating in (26), which essentially is the 'projection' of the construction of $f\mapsto f^\wedge$ to functions of type μ.

For $\psi\in\mathcal{V}(\mu)$ and $\lambda\in\mathfrak{a}_{\mathrm{q}\mathbb{C}}^*$ with $\mathrm{Re}\langle\lambda+\rho,\alpha\rangle<0$ for all $\alpha\in\Sigma^+$, let $\tilde{\psi}_\lambda$ be the V_μ–valued function on G defined by

$$\tilde{\psi}_\lambda(x)=\begin{cases} a^{-\lambda-\rho}\mu(m^{-1})\psi_w & \text{if } x=hwman\in Hw(M\cap K)AN, w\in\mathcal{W},\\ 0 & \text{if } x\notin\cup_{w\in\mathcal{W}}HwP.\end{cases}$$

(It is to be noted that $M=w^{-1}(M\cap H)w(M\cap K)$, and hence $Hw(M\cap K)AN=HwMAN$.) It can be shown that $\tilde{\psi}_\lambda$ is continuous as a function of x, and has a distribution–valued meromorphic continuation in $\lambda\in\mathfrak{a}_{\mathrm{q}\mathbb{C}}^*$. Let $E_\mu(\psi,\lambda)$ be the smooth μ–spherical function on G/H defined by

$$E_\mu(\psi,\lambda)(x)=\int_K\mu(k)\tilde{\psi}_\lambda(x^{-1}k)\,dk.$$

(Even when $\tilde{\psi}_\lambda$ is only a distribution, the convolution with μ makes $E_\mu(\psi,\lambda)$ smooth.) We call these functions *Eisenstein integrals* for G/H. When G/H is Riemannian and μ is the trivial K–type **1**, the construction produces the spherical functions

$$\varphi_\lambda(x)=\int_K e^{-(\lambda+\rho)H(x^{-1}k)}\,dk, \tag{23}$$

and for other K–types we get the generalized spherical functions of [26]. In the group case the Eisenstein integrals defined in this manner coincide, up to normalization, with Harish–Chandra's Eisenstein integrals associated to the minimal parabolic subgroup. It can be seen that the vector components of the Eisenstein integral $E_\mu(\psi,\lambda)$ are linear combinations of generalized matrix coefficients formed by the $j(\xi,\lambda)\eta$, ($\eta\in V(\xi)$, $\xi\uparrow\mu^\vee$), with K–finite vectors of type μ.

The spherical functions are eigenfunctions for the invariant differential operators on G/K – in analogy we have

$$DE_\mu(\psi,\lambda)=E_\mu(\chi_\mu(D,\lambda)\psi,\lambda) \tag{24}$$

for all $D \in \mathbb{D}(G/H)$. Here $\chi_\mu(D)$ is an End($\mathcal{V}(\mu)$)-valued polynomial in λ. Just as it is the case for the spherical functions, one can derive an asymptotic expansion from this 'eigenequation'. Here we have to recall the 'KAH'-decomposition of G,

$$G = \mathrm{cl} \bigcup_{w \in \mathcal{W}} KA_{\mathrm{q}}^+ w^{-1} H,$$

where A_{q}^+ is the exponential of the positive chamber in $\mathfrak{a}_{\mathrm{q}}$ corresponding to Σ^+, and where the union inside the closure operator cl is disjoint. Since the Eisenstein integrals are K-spherical, we have to consider their behavior on $A_{\mathrm{q}}^+ w^{-1}$, for all $w \in \mathcal{W}$. Notice that when G/H is Riemannian there is only one 'direction' to control, since the KAH-decomposition then specializes to the Cartan decomposition $G = \mathrm{cl}\, KA^+K$. The expansion is essentially as follows (see [4] and the remark below):

$$E_\mu(\psi, \lambda)(aw^{-1}) = \sum_{s \in W_{\mathrm{q}}} a^{s\lambda - \rho}[C(s, \lambda)\psi]_w + \text{ lower order terms in } a, \tag{25}$$

for $a \in A_{\mathrm{q}}^+$, $w \in \mathcal{W}$, where W_{q} is as defined above Thm. 9, and the 'c-function' $\lambda \mapsto C(s, \lambda)$ is a meromorphic function on $\mathfrak{a}_{\mathrm{q}\mathbb{C}}^*$ with values in End($\mathcal{V}(\mu)$) (it follows easily from the μ-sphericality that we have $E_\mu(\psi, \lambda)(aw^{-1}) \in V_\mu^{w^{-1}(K \cap M \cap H)w}$ for $a \in A_{\mathrm{q}}$). The expansion converges for $a \in A_{\mathrm{q}}^+$; the 'lower order terms' involve powers of the form $a^{s\lambda - \rho - \nu}$ where ν is a sum of positive roots.

Remark. The definition of the Eisenstein integral given here does require μ to be finite dimensional, but not necessarily irreducible, and therefore remains valid for an arbitrary finite dimensional unitary representation of K. In fact such an Eisenstein integral E_τ is introduced in [4] for a representation τ defined as follows. Let $\mu \in K^\wedge$, and let $C(K)_{\mu^\vee}$ denote the space of continuous functions $K \to \mathbb{C}$ which are finite and isotypical of type $\mu^\vee$ for the right regular representation R of K. Put $V_\tau = C(K)_{\mu^\vee}$ and let τ be the restriction of the right regular representation R to V_τ. Then by the Peter-Weyl theorem we have natural isomorphisms $V_\tau \simeq V_{\mu^\vee} \otimes V_\mu$, and $\tau \simeq \mu^\vee \otimes I_{V_\mu}$. It now follows easily that the Eisenstein integrals E_τ and $E_{\mu^\vee}$ are related as follows. One has $\mathcal{V}(\tau) \simeq \mathcal{V}(\mu^\vee) \otimes V_\mu$, and accordingly, for $\psi \in \mathcal{V}(\mu^\vee)$, $v \in V_\mu$:

$$E_\tau(\psi \otimes v, \lambda)(x) = E_{\mu^\vee}(\psi, \lambda)(x) \otimes v.$$

It follows from this that the corresponding c-functions are related by $C_\tau(s, \lambda) = C_{\mu^\vee}(s, \lambda) \otimes I$. From these remarks it should be clear how the results of [4] carry over to the present situation.

4.4.2. *The Fourier transform.* It would now be natural to define the Fourier transform $\mathcal{F}_\mu f$ of a function $f \in C_c^\infty(G/H; \mu)$, the space of compactly supported and smooth μ-spherical functions on G/H, as the $\mathcal{V}(\mu)$-valued function φ on $\mathfrak{a}_{\mathrm{q}\mathbb{C}}^*$ given by

$$\langle \varphi(\lambda) | \psi \rangle = \int_{G/H} \langle f(x) | E_\mu(\psi, -\bar{\lambda})(x) \rangle \, dx, \qquad \psi \in \mathcal{V}(\mu).$$

where the inner products $\langle\cdot|\cdot\rangle$ are the sesquilinear Hilbert space inner products on $\mathcal{V}(\mu)$ and V_μ, respectively. Via the isomorphisms in (21) and (22) this would essentially correspond to the Fourier transform in (15). However, as with $j(\xi,\lambda)$ we have the problem that $E_\mu(\psi,\lambda)$, which is meromorphic in λ, may have singularities on $i\mathfrak{a}_q^*$. Again we have to carry out a normalization: the *normalized Eisenstein integral* is defined by

$$E_\mu^\circ(\psi,\lambda)=E_\mu(C(1,\lambda)^{-1}\psi,\lambda).$$

In other words, the Eisenstein integral is normalized by its asymptotics, so that we have $E_\mu^\circ(\psi,\lambda)(aw^{-1})\sim a^{\lambda-\rho}\psi_w$ for $a\in A^+$, $w\in\mathcal{W}$ and $\mathrm{Re}\,\lambda$ strictly dominant. It can be shown that this normalization corresponds to the one on $j(\xi,\lambda)$, in the sense that the vector components of $E_\mu^\circ(\psi,\lambda)$ are linear combinations of matrix coefficients formed by the $j^\circ(\xi,\lambda)\eta$, $(\eta\in V(\xi),\ \xi\uparrow\mu^\vee)$, with K-finite vectors of type μ. Moreover, it can be shown that the statement of Thm. 8 is equivalent with the following 'K-finite version':

Theorem 10, [9]. *The meromorphic function $\lambda\mapsto E_\mu^\circ(\psi,\lambda)$ is regular on $i\mathfrak{a}_q^*$, for every $\mu\in K^\wedge$ and $\psi\in\mathcal{V}(\mu)$.*

With this in mind we define the *μ-spherical Fourier transform* $\mathcal{F}_\mu f$ as above, but with E_μ replaced by E_μ°, that is, by

$$\langle\mathcal{F}_\mu f(\lambda)|\psi\rangle=\int_{G/H}\langle f(x)|E_\mu^\circ(\psi,-\bar\lambda)(x)\rangle\,dx,\qquad \psi\in\mathcal{V}(\mu).\tag{26}$$

Then $\mathcal{F}_\mu f$ corresponds to $f^\wedge$ via the isomorphisms in (20) and (22). For completeness the precise correspondence is given as follows. Let $\gamma_\mu\colon C_c^\infty(G/H;\mu^\vee)\otimes V_\mu\to C_c^\infty(G/H)^\mu$ be the contraction (as in (22)) and let $\mathrm{pr}_{\xi,\lambda}\colon\mathcal{V}(\mu)^*\otimes V_\mu\to\mathcal{H}_{\xi,\lambda}\otimes V(\xi)^*$ be the projection corresponding to (20). Then for all $f\in C_c^\infty(G/H)^\mu$ we have

$$\dim(\xi)\,f^\wedge(\xi,\lambda)=\mathrm{pr}_{\xi,\lambda}\left[\left((\mathcal{F}_{\mu^\vee}\otimes I_{V_\mu})\,(\gamma_\mu^{-1}f)\right)(-\lambda)\right],\tag{27}$$

for $\lambda\in\mathfrak{a}_{q\mathbb{C}}^*$, $\xi\uparrow\mu$, and $f^\wedge(\xi,\lambda)=0$ for all other ξ.

When G/H is Riemannian and $\mu=\mathbf{1}$, the normalization again amounts to division by $\mathbf{c}(\lambda)$, and thus $\mathcal{F}_\mu f$ is in this case related to the spherical Fourier transform of f as follows:

$$\mathcal{F}_\mu f(\lambda)=\mathbf{c}(-\lambda)^{-1}\int_{G/K}f(x)\varphi_{-\lambda}(x)\,dx,$$

where φ_λ is the elementary spherical function in (23). If G/H is Riemannian and μ is non-trivial there is a similar relation, also involving $\mathbf{c}(\lambda)^{-1}$, to the Fourier transform in [26].

Let $C^\circ(s,\lambda)=C(s,\lambda)C(1,\lambda)^{-1}$, then we have from (25)

$$E_\mu^\circ(\psi,\lambda)(aw)=\sum_{s\in W_q}a^{s\lambda-\rho}[C^\circ(s,\lambda)\psi]_w+\text{ lower order terms in }a.\tag{28}$$

The following theorem generalizes results of Helgason and Harish–Chandra for the Riemannian case and the group case, respectively (see [25, Thm. 6.6], [23, Lemma 17.6]).

Theorem 11, [4], [5]. *For every $s \in W_{\mathrm{q}}$ we have the following identity of meromorphic functions:*

$$C^\circ(s,\lambda)C^\circ(s,-\bar{\lambda})^* = I_{\mathcal{V}(\mu)} \qquad (\lambda \in \mathfrak{a}^*_{\mathrm{q}\mathbb{C}}).$$

In particular, for $\lambda \in i\mathfrak{a}^*_{\mathrm{q}}$, the endomorphism $C^\circ(s,\lambda)$ of $\mathcal{V}(\mu)$ is unitary.

Notice that by Riemann's boundedness theorem it follows from the above result that the meromorphic function $\lambda \mapsto C^\circ(s,\lambda)$ has no singularities on $i\mathfrak{a}^*_{\mathrm{q}}$. Therefore the possible singularities of $E^\circ_\mu(\psi,\lambda)$ must occur in the lower order terms of (28). This observation plays a crucial role in the proof of Thm. 10.

On G/K the spherical functions satisfy the functional equation $\varphi_{s\lambda} = \varphi_\lambda$, for all $s \in W_{\mathrm{q}}$. The analog for the normalized Eisenstein integral on G/H is

$$E^\circ_\mu(C^\circ(s,\lambda)\psi, s\lambda) = E^\circ_\mu(\psi,\lambda) \tag{29}$$

(see [4, Prop. 16.4]. For the group case, see also [23, Lemma 17.2]).

Though $E^\circ_\mu(\psi,\lambda)$ by Thm. 10 is regular on $i\mathfrak{a}^*_{\mathrm{q}}$, it will in general have singularities elsewhere on $\mathfrak{a}^*_{\mathrm{q}\mathbb{C}}$. It is remarkable, though, that in a certain direction only finitely many singularities occur. To be more precise, one has the following. Let

$$(\mathfrak{a}^*_{\mathrm{q}\mathbb{C}})_+ = \{\lambda \in \mathfrak{a}^*_{\mathrm{q}\mathbb{C}} \mid \mathrm{Re}\langle\lambda,\alpha\rangle \geq 0,\ \alpha \in \Sigma^+\},$$

and put $(\mathfrak{a}^*_{\mathrm{q}\mathbb{C}})_- = -(\mathfrak{a}^*_{\mathrm{q}\mathbb{C}})_+$.

Theorem 12, [4]. *There exists a polynomial π' on $\mathfrak{a}^*_{\mathrm{q}\mathbb{C}}$, which is a product of linear factors of the form $\lambda \mapsto \langle\lambda,\alpha\rangle$+constant, with α a root, such that $\pi'(\lambda)E^\circ_\mu(\psi,\lambda)$ is holomorphic on a neighborhood of $(\mathfrak{a}^*_{\mathrm{q}\mathbb{C}})_+$.*

Notice that π' depends on the K–type μ. Notice also that when G/H is Riemannian we actually have that $E^\circ_\mu(\psi,\lambda)$ itself is holomorphic on $(\mathfrak{a}^*_{\mathrm{q}\mathbb{C}})_+$. Indeed, the spherical functions are everywhere holomorphic, and the normalizing divisor $\mathbf{c}(\lambda)$ has no zeros on this set. Thus, for this case one can take $\pi' = 1$.

It follows from Thm. 12 and (26) that if we put

$$\pi(\lambda) = \overline{\pi'(-\bar{\lambda})} \tag{30}$$

then $\lambda \mapsto \pi(\lambda)\mathcal{F}_\mu f(\lambda)$ is holomorphic on a neighborhood of $(\mathfrak{a}^*_{\mathrm{q}\mathbb{C}})_-$.

4.4.3. ***Wave packets.*** For the μ–spherical Fourier transform a 'partial inversion formula' is given in [9] as follows. For a $\mathcal{V}(\mu)$–valued function φ on $i\mathfrak{a}^*_{\mathrm{q}}$ of suitable decay one can form a 'wave packet', which is the superposition of normalized Eisenstein integrals, with amplitudes given by φ, that is

$$\mathcal{J}_\mu\varphi(x) = \int_{i\mathfrak{a}_\mathrm{q}^*} E_\mu^\circ(\varphi(\lambda),\lambda)(x)\,d\lambda.$$

It is easily seen that the transform $\mathcal{J}_\mu$ is the transposed of $\mathcal{F}_\mu$. For Euclidean Fourier transform (and more generally for the spherical Fourier transform on a Riemannian symmetric space) this transform is also the inverse of $\mathcal{F}_\mu$; the inversion formula states that $\mathcal{J}_\mu\mathcal{F}_\mu$ is the identity operator (when measures are suitably normalized). In the non–Riemannian generality of G/H this cannot be expected, because of the possible presence of discrete series. However we do have

Theorem 13, [9]. *There exists an invariant differential operator D (depending on μ) on G/H satisfying the following:*

(i) *As an operator on $C_c^\infty(G/H)$, D is injective and symmetric.*
(ii) $\mathcal{J}_\mu\mathcal{F}_\mu f = f$ *for all* $f \in D(C_c^\infty(G/H;\mu))$.

From (24) one can derive that $\mathcal{J}_\mu\mathcal{F}_\mu D = \mathcal{J}_\mu\chi_\mu(D)\mathcal{F}_\mu = D\mathcal{J}_\mu\mathcal{F}_\mu$. Hence it follows from (ii) that $D(\mathcal{J}_\mu\mathcal{F}_\mu f - f) = 0$ for all $f \in C_c^\infty(G/H;\mu)$. Nevertheless, one cannot then conclude from (i) that in fact $\mathcal{J}_\mu\mathcal{F}_\mu f = f$ because $\mathcal{J}_\mu\mathcal{F}_\mu f$ is not compactly supported in general. The presence of D is important, for example it annihilates all the discrete series in $L^2(G/H;\mu)$.

The proof of Thm. 13 is very much inspired by Rosenberg's proof (see [28, Ch. IV, §7]) of the inversion formula for the spherical Fourier transform on G/K (in which case one can take $D = 1$). A key step in both proofs is the use of a 'shift argument', originally used by Helgason for the proof of the Paley–Wiener theorem, where the integration in $\mathcal{J}_\mu$ (after use of (28)) is moved away from $i\mathfrak{a}_\mathrm{q}^*$ in the direction of $(\mathfrak{a}_{\mathrm{q}\mathbb{C}}^*)_-$, using Cauchy's theorem. It can be seen that one only meets a finite number of singular hyperplanes in this shift. The purpose of the operator D is to remove these singularities (among other things this means that π should be a divisor in $\chi_\mu(D)$), so that no residues are present. The shift allows one to conclude that $\mathcal{J}_\mu\mathcal{F}_\mu Df$ is compactly supported whenever f is, which is an important step in the proof of the theorem.

Thm. 13 is crucial in the proof of Thm. 9. Via the isomorphism (22) one obtains with $\mathcal{J}_{\mu^\vee}$ an explicit formula for the restriction to $\mathcal{H}^\mu$ of the isomorphism of $\mathcal{H}$ onto $L^2_{\mathrm{mc}}(G/H)$.

4.5. A Paley–Wiener theorem for G/H

Let π' be the minimal polynomial satisfying the conclusion of Thm. 12, and as before let π be given by (30). We define the *pre-Paley-Wiener space*, $\mathcal{M}_\mu$ as the space of $\mathcal{V}(\mu)$-valued meromorphic functions φ on $\mathfrak{a}_{\mathrm{q}\mathbb{C}}^*$, satisfying the following conditions:

(i) $\varphi(s\lambda) = C^\circ(s,\lambda)\varphi(\lambda)$, *for all* $s \in W_\mathrm{q}$, $\lambda \in \mathfrak{a}_{\mathrm{q}\mathbb{C}}^*$.

(ii) *$\pi(\lambda)\varphi(\lambda)$ is holomorphic on a neighborhood of* $(\mathfrak{a}^*_{\mathrm{q}\mathbb{C}})_-$.
(iii) *There exists a constant $R > 0$ and for every $n \in \mathbb{N}$ a constant $C > 0$ such that*

$$\|\pi(\lambda)\varphi(\lambda)\| \leq C(1 + |\lambda|)^{-n} e^{R|\operatorname{Re}\lambda|}$$

for all $\lambda \in (\mathfrak{a}^*_{\mathrm{q}\mathbb{C}})_-$.

It can be seen that $\mathcal{F}_\mu$ maps $C_c^\infty(G/H;\mu)$ into $\mathcal{M}_\mu$ (properties (i) and (ii) are straightforward consequences of (29) and Thms. 11 and 12, whereas (iii) requires a more difficult estimate for $E^\circ_\mu(\psi,\lambda)$). It follows from the Paley–Wiener theorem of Helgason and Gangolli (see [28, Ch. IV, §7]), that when G/H is Riemannian and μ the trivial K–type then $\mathcal{F}_\mu$ is a surjection onto the pre–Paley–Wiener space, as defined above for this special case. However in general one has to require further conditions on a function $\varphi \in \mathcal{M}_\mu$ before it belongs to $\mathcal{F}_\mu(C_c^\infty(G/H;\mu))$. Briefly put, the extra condition is that any existing relation between the normalized Eisenstein integrals and their derivatives (with respect to λ) should be reflected by a similar condition on φ. More precisely, we require that:

For all finite collections of $\partial_1,\ldots,\partial_k \in S(\mathfrak{a}^*_{\mathrm{q}})$ (that is, constant coefficient differential operators on $\mathfrak{a}^*_{\mathrm{q}}$), $\psi_1,\ldots,\psi_k \in \mathcal{V}(\mu)$ and $\lambda_1,\ldots,\lambda_k \in (\mathfrak{a}^*_{\mathrm{q}\mathbb{C}})_-$, for which the relation

$$\sum_{i=1}^{k} \partial_i \big[\, \pi(\lambda)\, \langle \psi | E^\circ_\mu(\psi_i, -\bar{\lambda})(x)\rangle \big]_{\lambda=\lambda_i} = 0 \tag{31}$$

holds for every $\psi \in \mathcal{V}(\mu)$, $x \in G/H$, we also have the relation

$$\sum_{i=1}^{k} \partial_i \big[\, \pi(\lambda)\, \langle \varphi(\lambda) | \psi_i \rangle \big]_{\lambda=\lambda_i} = 0. \tag{32}$$

The space of functions $\varphi \in \mathcal{M}_\mu$ satisfying this requirement is denoted PW_μ. It is clear from the definition (26) of $\mathcal{F}_\mu f$, that $\mathcal{F}_\mu f$ belongs to this space for $f \in C_c^\infty(G/H;\mu)$.

Theorem 14, [9]. *The μ–spherical Fourier transform $\mathcal{F}_\mu$ maps $C_c^\infty(G/H;\mu)$ into the Paley–Wiener space PW_μ. Moreover*

(a) *$\mathcal{F}_\mu$ is injective.*
(b) *If $\dim \mathfrak{a}_{\mathrm{q}} = 1$ then $\mathcal{F}_\mu$ is surjective.*

The injectivity of $\mathcal{F}_\mu$ is an immediate corollary of Thm. 13: If $\mathcal{F}_\mu f = 0$ then $\mathcal{F}_\mu Df = \chi_\mu(D)\mathcal{F}_\mu f = 0$, hence $Df = 0$ by (ii), and hence $f = 0$ by (i). The injectivity of $f \mapsto f^\wedge$ asserted earlier (below Thm. 9) is a consequence, by density of the K–finite functions in $C_c^\infty(G/H)$. The surjectivity statement in (b) is a by-product of the proof of Thm. 13.

For the Riemannian symmetric spaces the surjectivity of $\mathcal{F}_\mu$ (with an arbitrary K-type μ) is a consequence of the Paley–Wiener theorem in [26], and for the group G itself considered as a symmetric space it follows from the results in [15] and [1], as mentioned earlier. In [9] it is conjectured that $\mathcal{F}_\mu$ is surjective for general G/H as well.

We are now going to extend this theory to distributions, or more precisely, to generalized functions. Let $C_c^{-\infty}(G/H)$ denote the space of compactly supported generalized functions on G/H. Multiplication with the invariant measure dx induces an isomorphism of this locally convex space with the topological linear dual of $C^\infty(G/H)$, i.e. with the space of compactly supported distributions on G/H. If $u \in C_c^{-\infty}(G/H)$, and $f \in C^\infty(G/H)$, then we shall write accordingly:

$$\langle u , f \rangle = \int_{G/H} u(x) f(x)\, dx = u\, dx(f).$$

We have a natural embedding $C_c^\infty(G/H) \to C_c^{-\infty}(G/H)$; accordingly there is a natural extension of the Fourier transform $f \mapsto f^\wedge$ to the space of compactly supported generalized functions.

Let $C_c^{-\infty}(G/H;\mu)$ denote the space of compactly supported μ-spherical generalized functions $G/H \to V_\mu$. The μ-spherical Fourier transform $\mathcal{F}_\mu$ allows a natural extension to the space $C_c^{-\infty}(G/H;\mu)$ with values in the space of meromorphic functions $\mathfrak{a}^*_{q\mathbb{C}} \to \mathcal{V}(\mu)$.

A classical extension of the Paley–Wiener theorem for $\mathbb{R}^n$ states that the Fourier transform maps the space $C_c^{-\infty}(\mathbb{R}^n)$ of compactly supported generalized functions bijectively onto the space $\mathrm{PW}^*(\mathbb{R}^n)$ of entire functions on $\mathbb{C}^n$ for which there exists $R > 0$ such that (2) holds for some $N \in \mathbb{Z}$ (such functions are said to have *slow growth of exponential type*). We shall now state a conjectural analog of this result for $C_c^{-\infty}(G/H;\mu)$.

Let $\mathcal{M}^*_\mu$ be the pre-Paley–Wiener space of meromorphic functions $\varphi\colon \mathfrak{a}^*_{q\mathbb{C}} \to \mathcal{V}(\mu)$ satisfying conditions (i) and (ii) of the definition of $\mathcal{M}_\mu$ and moreover the following condition:

(iii)' *There exist constants $R, C > 0$ and $n \in \mathbb{N}$ such that*

$$\|\pi(\lambda)\varphi(\lambda)\| \le C\,(1+|\lambda|)^n e^{R|\operatorname{Re}\lambda|}$$

for all λ in $(\mathfrak{a}^*_{q\mathbb{C}})_-$.

Furthermore, let PW^*_μ be the space of functions $\psi \in \mathcal{M}^*_\mu$ satisfying the Paley–Wiener relations given in (32).

It can be seen that $\mathcal{F}_\mu$ maps $C_c^{-\infty}(G/H;\mu)$ into PW^*_μ (properties (i) and (ii) are obtained by the same arguments that were used to establish these facts for smooth u, and (iii)' follows from the estimates for the derivatives of $E^\circ_\mu(\psi,\lambda)$ obtained in [4]). In analogy with Thm. 14 we now have:

Theorem 15, [6]. *The μ-spherical Fourier transform $\mathcal{F}_\mu$ maps $C_c^{-\infty}(G/H;\mu)$ into the Paley–Wiener space PW^*_μ. Moreover*

(a) *$\mathcal{F}_\mu$ is injective.*
(b) *If $\dim \mathfrak{a}_\mathfrak{q} = 1$ then $\mathcal{F}_\mu$ is surjective.*

Moreover, we conjecture that the surjectivity of $\mathcal{F}_\mu$ holds in general. When G/H is Riemannian the surjectivity is established in [18].

4.6. A MULTIPLIER THEOREM

A linear operator

$$M\colon C_c^\infty(K;G/H) \to C_c^\infty(K;G/H)$$

is called a *multiplier* if it is equivariant for the actions of $\mathfrak{g}$, K and $\mathbb{D}(G/H)$ on this space, and has a continuous restriction to $C_c^\infty(G/H)^\mu$ for each $\mu \in K^\wedge$. If M is a multiplier, then it can be seen from the Fourier theory discussed in Sect. 4.4 that for almost every principal series representation $\pi = \pi_{\xi,\lambda}$ there exists an endomorphism Ψ_π of the space $\mathcal{V}_\pi^o$ such that

$$(Mf)^\wedge(\pi) = f^\wedge(\pi) \circ \Psi_\pi. \tag{33}$$

Moreover, Ψ_π will respect the eigenspace decomposition of $\mathcal{V}_\pi^o$ for $\mathbb{D}(G/H)$.

Simple examples of multipliers are the elements of $\mathbb{D}(G/H)$. If M is given by such an operator $D \in \mathbb{D}(G/H)$, then (33) can be written as follows:

$$(Mf)^\wedge(\pi)v = \chi(D)(\lambda) f^\wedge(\pi)v,$$

for any spherical vector $v \in \mathcal{V}_\pi^o$ of type $\lambda \in \mathfrak{b}_\mathbb{C}^*$. Of course an operator M thus defined extends to $C_c^\infty(G/H)$, but this will not be the case in general.

In [6] we give a simple construction of a large algebra of multipliers, containing the algebra $\mathbb{D}(G/H)$. The result is stated below. The existence of these multipliers is a generalization of Arthur's result [1, Thm. III.4.2] for the group case. Arthur's proof rests on his Paley–Wiener theorem (the generalization of which was conjectured in Sect. 4.5); a simpler construction using the correspondence $\varphi \mapsto \varphi'$ in Prop. 4 was later given in [16]. Our construction for the general case is similar in that it also uses this correspondence.

Let $\mathfrak{b}$ be a θ-invariant maximally split Cartan subspace of $\mathfrak{q}$, and recall from Sect. 4.2 that $\mathrm{PW}(\mathfrak{b}^r)^W$ is the space of W-invariant entire rapidly decreasing functions of exponential type on $\mathfrak{b}_\mathbb{C}^*$. Let $\mathrm{PW}^*(\mathfrak{b}^r)^W$ be the space of W-invariant entire functions with slow growth of exponential type on $\mathfrak{b}_\mathbb{C}^*$ (see Sect. 4.5).

Theorem 16, [6]. *For every $\psi \in \mathrm{PW}^*(\mathfrak{b}^r)^W$ there exists a unique linear operator*

$$M_\psi\colon C_c^\infty(K;G/H) \to C_c^\infty(K;G/H)$$

such that for any $\pi \in G_H^\wedge$ and any spherical vector $v \in \mathcal{V}_\pi^o$ of type $\lambda \in \mathfrak{b}_\mathbb{C}^*$ we have

$$(M_\psi f)^\wedge(\pi)v = \psi(\lambda)f^\wedge(\pi)v, \qquad f \in C_c^\infty(K; G/H). \tag{34}$$

If $D \in \mathbb{D}(G/H)$, then $M_{\chi(D)} = D|C_c^\infty(K; G/H)$. Moreover, the map $\psi \mapsto M_\psi$ is an algebra homomorphism from $\mathrm{PW}^*(\mathfrak{b}^r)^W$ into the algebra of multipliers. Finally, for every $\psi \in \mathrm{PW}^*(\mathfrak{b}^r)^W$ the equation (34) holds more generally with v a spherical vector of type λ in $\mathcal{V}_\pi$.

Indication of the proof. The uniqueness of M_ψ follows easily from the abstract Plancherel theory discussed in Sect. 3.2. The existence is established as follows.

Let $\psi \in \mathrm{PW}^*(\mathfrak{b}^r)^W$ be given, and let $F \in C_c^{-\infty}(G^d/K^d)$ be the K^d–invariant generalized function such that

$$\int_{G^d/K^d} F(x)\varphi_\lambda(x)\,dx = \psi(\lambda)$$

for all $\lambda \in \mathfrak{b}_\mathbb{C}^*$. Then for $f \in C_c^\infty(K; G/H)$ it is easily seen that the convolution product

$$f^r * F(x) = \int_{G^d/K^d} f^r(y)F(y^{-1}x)\,dy, \qquad x \in G^d/K^d, \tag{35}$$

is smooth and H^d–finite, and that Prop. 4 allows us to define $M_\psi f \in C_c^\infty(K; G/H)$ by

$$(M_\psi f)^r = f^r * F.$$

By [21, Cor. II.4] there exists a natural isomorphism of algebras $D \mapsto D^r$ from $\mathbb{D}(G/H)$ onto $\mathbb{D}(G^d/K^d)$, such that $(Df)^r = D^r f^r$ for all $f \in C_c^\infty(K; G/H)$. Moreover, we have $\chi^r(D^r) = \chi(D)$, where $\chi^r\colon \mathbb{D}(G^d/K^d) \to S(\mathfrak{b}^r)^W$ is the Harish-Chandra isomorphism. Let now $D \in \mathbb{D}(G/H)$, and let F be associated to $\psi = \chi(D)$ as above. Then for all $g \in C^\infty(G^d/K^d)$ we have $g * F = D^r g$. It follows from this that $(M_{\chi(D)}f)^r = f^r * F = D^r f^r = (Df)^r$, for $f \in C_c^\infty(K; G/H)$. Hence $M_{\chi(D)} = D$.

It is easily seen that the map $\psi \mapsto M_\psi$ is additive and multiplicative. Hence, if $\psi \in \mathrm{PW}^*(\mathfrak{b}^r)^W$ and $D \in \mathbb{D}(G/H)$, then $M_\psi \circ D = M_\psi \circ M_{\chi(D)} = M_{\psi\chi(D)} = M_{\chi(D)} \circ M_\psi = D \circ M_\psi$, and one readily checks that M_ψ is a multiplier.

Finally (34) is seen by an argument similar to the proof of Thm. 5. □

Remark. It follows from the injectivity statements in Theorem 14 that M_ψ is already uniquely determined by the requirement that equation (34) should hold for all principal series representations $\pi = \pi_{\xi,\lambda}$, where $\xi \in M_{\mathrm{fu}}^\wedge$ and $\lambda \in i\mathfrak{a}_\mathrm{q}^*$.

The multipliers of Thm. 16 do actually extend to the space $C^{-\infty}(K; G/H)$ of K–finite generalized functions on G/H. Let this space be equipped with the direct sum of the usual strong dual topologies on the subspaces $C^{-\infty}(G/H)^\mu$, $\mu \in K^\wedge$.

Theorem 17, [6]. *Let $\psi \in \mathrm{PW}^*(\mathfrak{b}^r)^W$. Then the operator M_ψ of Theorem 16 extends to a continuous linear operator*

$$M_\psi \colon C^{-\infty}(K;G/H) \to C^{-\infty}(K;G/H).$$

This extension is equivariant for the actions of $\mathfrak{g}$, K and $\mathbb{D}(G/H)$ and preserves the subspace $C_c^{-\infty}(K;G/H)$ of compactly supported generalized functions. Moreover, for any $\pi \in G_H^\wedge$ and any spherical vector $v \in \mathcal{V}_\pi$ of type $\lambda \in \mathfrak{b}_{\mathbb{C}}^*$ we have

$$(M_\psi f)^\wedge(\pi)v = \psi(\lambda) f^\wedge(\pi)v, \qquad f \in C_c^{-\infty}(K;G/H). \tag{36}$$

Finally, if $\psi \in \mathrm{PW}(\mathfrak{b}^r)^W$ then M_ψ maps $C^{-\infty}(K;G/H)$ into $C^\infty(K;G/H)$.

Remark. Notice that (36) for $f \in C_c^{-\infty}(K;G/H)$ is an equation of elements in $\mathcal{H}_\pi^{-\infty}$. Notice also that the existence statement in Thm. 5 can be obtained from the final statement of Thm. 17, by applying M_ψ to the K–isotypical component $P_\mu\delta$ of type μ of the Dirac function δ supported at the origin.

Let $\mu \in K^\wedge$ and $\psi \in \mathrm{PW}^*(\mathfrak{b}^r)^W$ be given. We shall now discuss the relation of the multiplier M_ψ to the μ–spherical Fourier transform $\mathcal{F}_\mu$. We first note that the construction of M_ψ is easily extended to μ–spherical functions $f \in C_c^\infty(G/H;\mu)$: The partial holomorphic extension $\varphi \mapsto \varphi^r$ makes sense for vector valued functions, and so does the convolution product in (35). We denote the resulting linear operator $C_c^\infty(G/H;\mu) \to C_c^\infty(G/H;\mu)$ by M_ψ^μ. It is easily seen that we have the following relation between the operators M_ψ and M_ψ^μ:

$$\gamma_\mu \circ (M_\psi^{\mu^\vee} \otimes I_{V_\mu}) = M_\psi \circ \gamma_\mu,$$

where $\gamma_\mu \colon C_c^\infty(G/H;\mu^\vee) \otimes V_\mu \to C_c^\infty(G/H)^\mu$ as earlier is the contraction isomorphism.

Since $\mathfrak{b}$ is maximally split, we may as well assume that $\mathfrak{a}_{\mathrm{q}} = \mathfrak{b} \cap \mathfrak{p}$. Put $\mathfrak{b}_{\mathrm{k}} = \mathfrak{b} \cap \mathfrak{k}$, then (9) becomes

$$\mathfrak{b}^r = \mathfrak{a}_{\mathrm{q}} \oplus i\mathfrak{b}_{\mathrm{k}}.$$

Via this direct sum decomposition we identify $\mathfrak{a}_{\mathrm{q}}^*$ and $\mathfrak{b}_{\mathrm{k}}^{r*} := i\mathfrak{b}_{\mathrm{k}}^*$ with subspaces of $\mathfrak{b}^{r*}$.

Let $\xi \in M_{\mathrm{fu}}^\wedge$, and recall from Sect. 4.3 that $j^\circ(\xi,\lambda)$ for generic $\lambda \in \mathfrak{a}_{\mathrm{q}\mathbb{C}}^*$ is a linear bijection of the space $V(\xi)$ onto the space $\mathcal{V}_{\xi,\lambda} = (\mathcal{H}_{\xi,\lambda}^{-\infty})^H$. Via j° we transfer the $\mathbb{D}(G/H)$–module structure of $\mathcal{V}_{\xi,\lambda}$ to $V(\xi)$. Thus for every $D \in \mathbb{D}(G/H)$ we define $\chi_\xi(D,\lambda) \in \mathrm{End}(V(\xi))$ by

$$Dj^\circ(\xi,\lambda) = j^\circ(\xi,\lambda)\,\chi_\xi(D,\lambda) \tag{37}$$

for generic λ. It is known that $\chi_\xi(D,\lambda)$ is in fact an $\mathrm{End}(V(\xi))$–valued polynomial in λ (cf. [4, Sect. 4]). It allows an eigenspace decomposition which is independent of λ. More precisely, if $\nu \in \mathfrak{b}_{\mathrm{k}}^{r*}$, define

$$V(\xi)_\nu = \{\eta \in V(\xi) \mid \chi_\xi(D,\lambda)\eta = \chi(D,\nu+\lambda)\eta,\ D \in \mathbb{D}(G/H),\ \lambda \in \mathfrak{a}_{\mathrm{q}\mathbb{C}}^*\}$$

(as before χ denotes the Harish–Chandra isomorphism from $\mathbb{D}(G/H)$ onto $S(\mathfrak{b}^r)^W$). Then for $\nu_1, \nu_2 \in \mathfrak{b}_{\mathrm{k}}^{r*}$ with $V(\xi)_{\nu_1} \neq 0$ we have $V(\xi)_{\nu_1} = V(\xi)_{\nu_2}$ if and only if ν_1 and ν_2 are conjugate under the centralizer W_{M} of $\mathfrak{a}_{\mathrm{q}}$ in W. Moreover, let $\mathcal{N}(\xi)$ denote the set of $\nu \in \mathfrak{b}_{\mathrm{k}}^{r*}/W_{\mathrm{M}}$ for which $V(\xi)_\nu \neq 0$. Then we have the direct sum decomposition

$$V(\xi) = \bigoplus_{\nu \in \mathcal{N}(\xi)} V(\xi)_\nu$$

(for details, see [4]). Notice that it follows from the above that $j^\circ(\xi,\lambda)$ maps $V(\xi)_\nu$ onto the space of $\mathbb{D}(G/H)$–spherical vectors of type $\nu+\lambda$ in $\mathcal{V}_{\xi,\lambda}$, for generic λ.

We now define, for a given W_{M}–invariant complex function ψ on $\mathfrak{b}_{\mathbb{C}}^*$ an endomorphism $\mathrm{M}(\psi,\xi,\lambda)$ of $V(\xi)$ by

$$\mathrm{M}(\psi,\xi,\lambda) = \psi(\nu+\lambda)I \quad \text{on} \quad V(\xi)_\nu \tag{38}$$

for $\lambda \in \mathfrak{a}_{\mathrm{q}\mathbb{C}}^*$. It follows that we have

$$(M_\psi f)^\wedge(\xi,\lambda) = f^\wedge(\xi,\lambda) \circ \mathrm{M}(\psi,\xi,\lambda)$$

for all $\xi \in M_{\mathrm{fu}}^\wedge$, $\lambda \in i\mathfrak{a}_{\mathrm{q}}^*$.

Recall the orthogonal decomposition (19) of $\mathcal{V}(\mu)$. According to this decomposition we define for each $\nu \in \mathfrak{b}_{\mathrm{k}}^{r*}$ a subspace $\mathcal{V}(\mu)_\nu$ of $\mathcal{V}(\mu)$ by

$$\mathcal{V}(\mu)_\nu = \bigoplus_{\xi \uparrow \mu} \mathrm{Hom}_{M\cap K}(V_\mu, \mathcal{H}_\xi) \otimes V(\xi)_\nu.$$

Then $\mathcal{V}(\mu)_\nu$ only depends on the W_{M}–conjugacy class of ν and writing $\mathcal{N}(\mu) = \cup_{\xi\uparrow\mu}\mathcal{N}(\xi)$ we have the following finite direct sum of non–trivial vector spaces:

$$\mathcal{V}(\mu) = \bigoplus_{\nu \in \mathcal{N}(\mu)} \mathcal{V}(\mu)_\nu. \tag{39}$$

The maps $\chi_\xi(\lambda)\colon \mathbb{D}(G/H) \to \mathrm{End}(V(\xi))$ and $\chi_\mu(\lambda)\colon \mathbb{D}(G/H) \to \mathrm{End}(\mathcal{V}(\mu))$ are closely related; in fact it follows from [4, Sect. 4] that $\chi_\mu(D,\lambda)$ corresponds to the direct sum of the maps $I \otimes \chi_\xi(D,\lambda)$ in the decomposition (19), or equivalently, to the direct sum of the maps $\chi(D,\nu+\lambda)I_{\mathcal{V}(\mu)_\nu}$ in the decomposition (39), for all $D \in \mathbb{D}(G/H)$, $\lambda \in \mathfrak{a}_{\mathrm{q}\mathbb{C}}^*$.

Let $\mathrm{M}_\mu(\psi,\lambda) \in \mathrm{End}(\mathcal{V}(\mu))$ be defined by the requirement

$$\mathrm{M}_\mu(\psi,\lambda) = I \otimes \mathrm{M}(\psi,\xi,\lambda) \quad \text{on} \quad \mathrm{Hom}_{M\cap K}(\mathcal{H}_\xi, V_\mu) \otimes V(\xi)$$

in the direct sum decomposition (19). We shall view $\mathrm{M}_\mu(\psi)$ as a multiplication operator on $\mathcal{V}(\mu)$-valued functions on $\mathfrak{a}_{q\mathbb{C}}^*$. It follows from the remarks made above that

$$\mathcal{F}_\mu(M_\psi^\mu f) = \mathrm{M}_\mu(\psi^\vee)\mathcal{F}_\mu f \tag{40}$$

for all $f \in C_c^\infty(G/H;\mu)$.

If the surjectivity conjectures for $\mathcal{F}_\mu$ stated in Sect. 4.5 are valid for G/H, then it follows from (40) that multiplication by $\mathrm{M}_\mu(\psi)$ leaves the spaces PW_μ and PW_μ^* invariant. This is indeed true in general:

Proposition 18, [6]. *Let $\psi \in \mathrm{PW}^*(\mathfrak{b}^r)^W$. Then multiplication by $\mathrm{M}_\mu(\psi)$ leaves the spaces PW_μ and PW_μ^* invariant. Moreover, if $\psi \in \mathrm{PW}(\mathfrak{b}^r)^W$ then multiplication by $\mathrm{M}_\mu(\psi)$ maps PW_μ^* to PW_μ.*

References

1. J. Arthur, *A Paley–Wiener theorem for real reductive groups*, Acta Math. **150** (1983), 1-89.
2. E. P. van den Ban, *Invariant differential operators on a semisimple symmetric space and finite multiplicities in a Plancherel formula*, Ark. för Mat. **25** (1987), 175-187.
3. E. P. van den Ban, *The principal series for a reductive symmetric space I. H–fixed distribution vectors*, Ann. sci. Éc. Norm. Sup. **4, 21** (1988), 359–412.
4. E. P. van den Ban, *The principal series for a reductive symmetric space II. Eisenstein integrals*, J. Funct. Anal. **109** (1992), 331-441.
5. E. P. van den Ban, *The action of intertwining operators on H-fixed generalized vectors in the minimal principal series of a reductive symmetric space*, in preparation.
6. E. P. van den Ban, M. Flensted–Jensen and H. Schlichtkrull, *Multipliers on semisimple symmetric spaces*, in preparation.
7. E. P. van den Ban and H. Schlichtkrull, *Asymptotic expansions and boundary values of eigenfunctions on Riemannian symmetric spaces*, J. reine und angew. Math. **380** (1987), 108–165.
8. E. P. van den Ban and H. Schlichtkrull, *Multiplicities in the Plancherel decomposition for a semisimple symmetric space*, Contemporary Math. **145** (1993), 163-180.
9. E. P. van den Ban and H. Schlichtkrull, *The most continuous part of the Plancherel decomposition for a reductive symmetric space*, in preparation.
10. M. Berger, *Les espaces symétriques non compacts*, Ann. Sci. École Norm. Sup. **74** (1957), 85–177.
11. F. Bien, *$\mathcal{D}$–modules and spherical representations*, Princeton U. P., Princeton, N. J., 1990.
12. N. Bopp and P. Harinck, *Formule de Plancherel pour* $\mathrm{GL}(n,\mathbf{R})/\mathrm{U}(p,q)$., J. reine und angew. Math. **428** (1992), 45-95.
13. J.-L. Brylinski and P. Delorme, *Vecteurs distributions H–invariants pour les séries principales géneralisées d'espaces symétriques réductifs et prolongement méromorphe d'integrales d'Eisenstein*, Invent. math. (1992), 619–664.
14. O. Campoli, *The complex Fourier transform for rank–1 semisimple Lie groups*, Thesis, Rutgers University (1976).
15. J. Carmona and P. Delorme, *Base méromorphe de vecteurs distributions H–invariants pour les séries principales géneralisées d'espaces symétriques réductifs. Equation fonctionelle*, preprint (1992).
16. P. Delorme, *Multipliers for the convolution algebra of left and right K–finite compactly supported smooth functions on a semi–simple Lie group*, Invent. Math. **75** (1984), 9-23.

17. P. Delorme and M. Flensted–Jensen, *Towards a Paley–Wiener theorem for semisimple symmetric spaces*, Acta Math. **167** (1991), 127–151.
18. M. Eguchi, M. Hashizume and K. Okamoto, *The Paley–Wiener theorem for distributions on symmetric spaces*, Hiroshima M. Jour. **3** (1973), 109-120.
19. J. Faraut, *Distributions sphériques sur les espaces hyperboliques*, J. Math. Pures Appl. **58** (1979), 369–444.
20. M. Flensted–Jensen, *Discrete series for semisimple symmetric spaces*, Ann. of Math **111** (1980), 253–311.
21. M. Flensted–Jensen, *Analysis on Non–Riemannian Symmetric Spaces*, Regional Conference Series in Math. 61, Amer. Math. Soc., Providence, 1986.
22. Harish–Chandra, *Representations of semisimple Lie groups VI*, Amer. J. Math. **78** (1956), 564–628.
23. Harish–Chandra, *Harmonic analysis on real reductive groups, III. The Maass–Selberg relations*, Ann. of Math. **104** (1976), 117–201.
24. S. Helgason, *A duality for symmetric spaces with applications to group representations*, Adv. in Math. **5** (1970), 1–154.
25. S. Helgason, *The surjectivity of invariant differential operators on symmetric spaces I*, Ann. of Math. **98** (1973), 451–479.
26. S. Helgason, *A duality for symmetric spaces with applications to group representations, II. Differential equations and eigenspace representations*, Adv. in Math. **22** (1976), 187–219.
27. S. Helgason, *Differential Geometry, Lie Groups, and Symmetric Spaces*, Academic Press, New York, San Francisco, London, 1978.
28. S. Helgason, *Groups and Geometric Analysis*, Academic Press, Orlando, 1984.
29. T. Matsuki, *The orbits of affine symmetric spaces under the action of minimal parabolic subgroups*, J. Math. Soc. Japan **31** (1979), 331–357.
30. T. Matsuki, *A description of discrete series for semisimple symmetric spaces II*, Adv. Studies in Pure Math **14** (1988), 531–540.
31. V. F. Molchanov, *Plancherel's formula for pseudo-Riemannian symmetric spaces of rank 1*, Sov. Math., Dokl. **34** (1987), 323–326.
32. G. Ólafsson, *Fourier and Poisson transformation associated to a semisimple symmetric space*, Invent. math. **90** (1987), 605–629.
33. T. Oshima and T. Matsuki, *A description of discrete series for semisimple symmetric spaces*, Adv. Studies in Pure Math. **4** (1984), 331–390.
34. T. Oshima and J. Sekiguchi, *Eigenspaces of invariant differential operators on an affine symmetric space*, Invent. math. **57** (1980), 1–81.
35. R. Penney, *Abstract Plancherel theorems and a Frobenius reciprocity theorem*, J. Funct. Anal. **18** (1975), 177-190.
36. W. Rossmann, *The structure of semisimple symmetric spaces*, Canad. J. Math. **31** (1979), 157–180.
37. H. Schlichtkrull, *Hyperfunctions and harmonic analysis on symmetric spaces*, Birkhäuser, Boston, 1984.
38. H. Schlichtkrull, *Harmonic analysis on semisimple symmetric spaces. Lectures for the European School of Group Theory*, 1992.
39. M. Sugiura, *Fourier series of smooth functions on compact Lie groups*, Osaka Math. J. **8** (1971), 33–47.
40. G. van Dijk and M. Poel, *The Plancherel formula for the pseudo–Riemannian space* $\mathrm{GL}(n,\mathbf{R})/\mathrm{GL}(n-1,\mathbf{R})$, Compos. Math. **58** (1986), 371–397.
41. D. Vogan, *Irreducibility of discrete series representations for semisimple symmetric spaces*, Adv. Studies in Pure Math. **14** (1988), 191–221.

Department of Mathematics
University of Utrecht
P. O. Box 80010
NL 3508 TA Utrecht
The Netherlands

Department of Mathematics and Physics
The Royal Veterinary and Agricultural University
Thorvaldsensvej 40
DK 1871 Frederiksberg C
Denmark

Department of Mathematics and Physics
The Royal Veterinary and Agricultural University
Thorvaldsensvej 40
DK 1871 Frederiksberg C
Denmark

The Extensions of Space-Time. Physics in the 8-dimensional Homogeneous Space $\mathcal{D} = SU(2,2)/K$

Asim O. Barut

Abstract

The Minkowski space-time is only a boundary of a bigger homogeneous space of the conformal group. The conformal group is the symmetry group of our most fundamental massless wave equations. These extended groups and spaces have many remarkable properties and physical implications.

1 Preliminaries

1.1 Physical Considerations

(a) Although physical phenomena happen in ordinary real three- dimensional space, some dynamical information can be described in geometric terms if physical theories are formulated in generalized spaces, in particular, in complex spaces. For example, the scattering amplitude in momentum space when continued analytically to complex energy and momentum contains all the information about the bound states of the system. This is the slow but continuous program of geometrization of physics [2c].

In this work we are interested in complex domains associated with the conformal group of space-time. The homogeneous space $\mathcal{D} = SU(2,2)/K$, $K =$ maximal compact subgroup of the conformal group, is a complex domain with ordinary space-time as its boundary, thus forms an *extension* of space-time. There is no hermitian symmetric domain associated with the Lorentz group $SL(n, \mathbf{C})$ (Cartan).

(b) Although the conformal group of space-time $\mathbf{M}^4$, $C(\mathbf{M}^4)$, has a long history, it goes back to the symmetries of free Maxwell's equations, it has not yet found a basic permanent place in physics as did the Poincaré group of space-time. Bessel-Hagen [10] deduced early the fifteen conservation laws of Maxwell's equations with matter present from the conformal group, but the applications beyond that of the conformal symmetry remained sporadic. The reasons for

E. A. Tanner and R. Wilson (eds.), Noncompact Lie Groups and Some of Their Applications, 103–121.

that is that firstly our basic particle equations (e.g. Dirac equation) contain phenomenological mass terms which formally break the conformal symmetry. Secondly the action of the conformal group on the Minkowski space $\mathbf{M}^4$ is singular. These impediments may change, however. Firstly, the *mass* of quantum particles may be induced by localized oscillating solutions of the massless, hence conformally invariant equations, the internal frequency Ω of the solution replacing the role of the mass [2d,e]. Secondly, instead of the Minkowski space, we may take the *compactified Minkowski space*, or the space $\mathcal{D} = SU(2,2)/K$, or the group space of $SU(2,2)$, as the actual physical or kinematical space, on which the action of the conformal group is non-singular.

(c) The Poincaré group of special relativity related to Minkowski space $\mathbf{M}^4$ is not powerful enough to describe fully the fundamental entities of physics, contrary to a generally held opinion. Wigner's program of defining elementary particles as unitary irreducible representations of the Poincaré group is incomplete even for the absolutely stable entities of physics: electron, neutrino and photon. Thus, electron does not come alone, but always in the form of an electron-positron complex $(e^+ - e^-)$; for electromagnetic field A_μ and for the $(\nu - \bar{\nu})$ complex we have to use peculiar indecomposable representations of the Poincaré group; the higher spins in nature occur in sequences and not singly. Furthermore, the 4-complex $(e^-\bar{\nu}, e^+\nu)$ enters the weak interactions, and repeats itself at least three times: $(e\nu_e), (\mu\nu_\mu), (\tau\nu_\tau)$.

The conformal group does much better for some of these problems. For example, it quite naturally accounts for the fact that electron and neutrino occurs in pairs [6,9].

(d) The conformal group of space-time has been used in physics in at least three different roles:

1. as an (approximate) symmetry in ordinary special relativity when mass effects can be neglected at high energies.
2. as a genuine new and more basic kinematical group replacing the Poincaré group $\mathcal{P}$. It turns out that one can introduce a new conformally invariant mass m_{00}, and transform the usual rest mass m_0 together with coordinates under conformal transformations. This is similar to the situation when the Newtonian mass m is transformed under Lorentz transformations, but m_0 is not. With this the conformal group becomes an exact symmetry at least in a compactified Minkowski space to avoid the singularities. Then the conformal group is indeed a more fundamental group than the Poincaré group, because it accounts for the origin of $(e^+ - e^-)$-complex and further $(e\nu)$-system. It is the group of electrodynamics, of Green's functions and of the *space of light rays* on which really the electromagnetic interactions should be formulated, for two particles interact

only if they are connected by a light ray.

The conformal group also occurs as a *dynamical group* of the ubiquitous $\frac{1}{r}$-interactions (Coulomb, gravitation, hadrons) but this time as the conformal group in momentum space. There seems to be a remarkable symmetry between the conformal transformations in coordinate and momentum spaces.

3. finally, as the conformal group playing an important role in understanding the structure of space and time in the large, in cosmology.

(e) Because an 8-dimensional space is a natural arena for the conformal group, I list the occurrence of various 8-dimensional spaces in physics to which the action of the conformal group may be contemplated.

- The 8-dimensional phase space of spinless particles (x^μ, p^μ).
- The 8-dimensional configuration space of spinning particles (x^μ, ψ_α), where ψ_α is a classical 8-component spinor.
- Basic stable leptons (for each generation) - the space of two spinors (ψ_e, ψ_ν).
- Conformally invariant Dirac operator must involve 8-component Clifford algebra

$$D = \beta^a \partial_a \, , \quad a = 1, \cdots, 6$$

related to the linear representation of the conformal group in a 6-dimensional space.

- Configuration space of 2-particles (x_1^μ, x_2^μ), or center-of-mass and relative coordinates (X^μ, x^μ).
- Complex Minkowski space, $z^\mu = x^\mu + iy^\mu$, in particular the tube domain of the forward light-cone.
- The 4-dimensional defining representation of $SU(2,2)$ acts on the complex space with

$$| z_1 |^2 + | z_2 |^2 - | z_3 |^2 - | z_4 |^2 = \text{invar.}$$

- Coadjoint orbit of the Poincaré group.

1.2 Some Mathematical Preliminaries

Originally introduced on physical grounds by nonlinear transformations of space-time coordinates representing scale and inversion invariance of Maxwell's equations, the conformal algebra is one of the basic Lie algebras, a real form of $A_3 \sim D_3$, i.e. isomorphic to $su(2,2)$, as a real form of $su(4) \sim A_3$, or to $so(4,2)$, as a real form of $so(6) \sim D_3$.

The different Lie algebra bases used in different applications, which we shall refer to are

1. The basis $\{L_{\mu\nu}, P_\mu, K_\mu, D\}$ corresponding to the nonlinear action on the Minkowski space; $\mu, \nu = 0 = 5, 1, 2, 3$.
2. The $so(4,2)$ basis $\{L_{ab}\ , a, b = 1, ..., 6\}$, corresponding to the linear action in a 6-dimensional space,

$$[L_{ab}, L_{ad}] = ig_{aa} L_{bd}\ ; \ g_{aa} = (+ - - - - +), 0 = 5, 1, 2, 3, 4, 6.$$

3. The $su(2,2)$-basis.
4. The Cartan basis $\{H_i, E_\alpha\}$.

The three *Casimir operators* of the conformal group can be written in the $so(4,2)$-basis as

$$\begin{aligned} C_2 &= L_{ab} L^{ab} \\ C_3 &= \epsilon_{abcdef} L^{ab} L_{cd} L^{ef} \\ C_4 &= L_{ab} L^{bc} L_{cd} L^{da}. \end{aligned}$$

The representations are determined by the eigenvalues of the three Casimir operators. In addition we need 6 more labels to specify the states

Algebra	A_ℓ	D_ℓ
Dimension	$\ell(\ell+2)$	$2\ell^2 - \ell$
Dimension of Cartan subalgebra	ℓ	ℓ
Additional quantum numbers	$\frac{1}{2}\ell(\ell-1)$	$\ell(\ell-2)$

The total of nine labels can be seen by the Casimir operators of the subgroup chains, for example: $O(6) \supset O(5) \supset O(4) \supset O(3) \supset O(2)$, giving ($N$ denotes the number of Casimir operators)

Subgroup	N
$O(6)$	3
$O(5)$	2
$O(4)$	2
$O(3)$	1
$O(2)$	1
Total	9

i.e. in general

$$D_\ell \supset B_{\ell-1} \supset D_{\ell-1} \supset B_{\ell-2} \supset \cdots \supset B_1 \supset D_1$$
$$A_\ell \supset A_{\ell-1} \supset A_{\ell-2} \supset \cdots \supset A_2 \supset A_1 .$$

For general theorems about complex domains and unitary irreducible representations that we make use of we refer to Cartan [11a-c], Gel'fand and Naimark [13,33a,b], Godement [14], Harish-Chandra [17a-e], Helgason [18a-c], Hörmander [19], Hua [20], Krein and Smul'jan [25,26], Langlands [27], Vilenkin and Klimyk [38], Wolf [39a,b] and Zelobenko [41]. For physical purposes the explicit construction of representations of $SU(2,2) \sim SO_0(4,2)$ were studied by: Angelopolous [1a,b], Barut and Böhm [3], Esteve and Sona [12], Graev [15a,b], Gross [16], Jakobsen and Vergne [21], Kihlberg, Müller and Halbwachs [22], Knapp and Speh [23a-c,24], Limic, Niederle and Raczka [28a,b], Mack and Todorov [29,30], Mickelsson and Niederle [31], Murai [32a,b], Newton [34], Rühl [35a,b], Thomas [37] and Yao [40a-c].

2 Action of Conformal Group on the Homogeneous space $SU(2,2)/K$

I now discuss the results of a joint work with Wilson [8] on the explicit construction of the most general action of the conformal group on the manifold $SU(2,2)/K = \mathcal{D}$, where K is the maximal compact subgroup, and the explicit construction of the Lie algebra of $SU(2,2)$ as differential operators. Principal discrete series of (unitary) representations of $SU(p,q)$ has been studied in detail by Tanner and Wilson in [36].

2.1 Multicomponent Wave Functions *versus* Internal Coordinates

One of the motivations is to construct internal coordinates for spin and other quantum numbers of particles or composite systems.

We recall that angular momentum operators are differential operators acting on $SU(2)/SO(2) \sim S^2(\theta,\varphi)$, generators of the rotation group, and lead to integer values of ℓ. But on $SU(2)$ the Lie algebra elements as differential operators in terms of Euler angles include also half integer ℓ-values. However, in relativistic theories for the differential operators representing the Lie algebra of the Poincaré group on $\mathbf{M}^4$, there is no room for internal coordinates corresponding to spin. For this reason the spin is introduced without reference to coordinates by using multicomponent functions on $\mathbf{M}^4$.

This is a very general and fundamental situation in physics: Should one use multicomponent fields to describe internal quantum numbers (spin, isospin, etc.) or should we use a larger manifold than the Minkowski space-time and only one-component fields?. It is clear that the second direction is more general leading to wider consequences. Consider a field $f(x^\mu;\xi^\alpha)$, where ξ^α are internal coordinates in addition to the space-time coordinates x^μ. Expanding the function $f(x^\mu;\xi^\alpha)$ around $\xi_\alpha = 0$,

$$f(x^\mu;\xi^\alpha) = f(x^\mu) + \xi^\alpha f_\alpha(x^\mu) + \xi^\alpha\xi^\beta f_{\alpha\beta}(x^\mu) + \cdots$$

we see that the coefficients $f_\alpha, f_{\alpha\beta}\cdots$ are multicomponent spinor, vector, tensor,...fields on space-time. If further f satisfies an equation

$$\mathscr{F}(\frac{\partial}{\partial x},\frac{\partial}{\partial \xi})\, f(x,\xi) = \lambda f(x,\xi)$$

then by comparing coefficients of ξ^α, one obtains equations for the multicomponent fields. For example, if

$$\mathscr{F}(\frac{\partial}{\partial x},\frac{\partial}{\partial \xi}) = \xi^\alpha A^\mu_{\alpha\beta}(\frac{\partial}{\partial x^\mu})\frac{\partial}{\partial \xi^\beta}$$

we have

$$A^\mu_{\alpha\beta}(\frac{\partial}{\partial x^\mu})f_\beta = \lambda f_\alpha.$$

The Dirac equation belongs to this example. But clearly one can so obtain other more grneral interesting systems.

A limiting case of multicomponent equations is the class of infinite component wave equations. For example, let $\Phi(X,\mathbf{r})$ be the wave function of a composite system with center-of-mass coordinates X and relative coordinates $\mathbf{r}$, and

$$(H(X) + H_{rel}(\mathbf{r}))\Phi(X,\mathbf{r}) = E\Phi(x,\mathbf{r})$$

is the Hamiltonian, then in a basis of functions $\varphi_i(\mathbf{r})$ of the relative coordinates we have the equation

$$(H(X) + M)\psi(X) = E\psi(X)$$

where M is an infinite matrix $(\varphi_j, H_{rel}\varphi_k)$. Another set of important infinte component wave equations have the form

$$(\Gamma^\mu\frac{\partial}{\partial X^\mu} + M)\psi(X) = 0\,,$$

where Γ^μ, M are infinite dimensional matrices [2a,b,5,7,10]. With this motivation about higher dimensional spaces we now turn to the homogeneous space of the conformal group.

2.2 The Homogeneous Space $\mathcal{D}$

The group $SU(2,2)$ is parametrized as follows:

$$SU(2,2)=\left\{g=\begin{pmatrix}\alpha & \beta\\ \gamma & \delta\end{pmatrix}\in SL(4,\mathbf{C})\,:\,gJ_0g^*=J_0=\begin{pmatrix}I_2 & 0\\ 0 & -I_2\end{pmatrix}\right\} \tag{1}$$

where $\alpha,\beta,\gamma,\delta\in M_{2\times 2}(\mathbf{C})$, I_2 is 2×2 unit matrix and g^* is the transpose conjugate of g.

Graev in [15a,b] has shown that $SU(2,2)$ has three conjugacy classes of Cartan subgroups (maximal tori) and obtained the corresponding three principal series of irreducible representations denoted by d_0,d_1,d_2. The series d_0 is characterized by three integer parameters, d_1 by two integer and one continuous parameters, and d_2 by one integer and two continuous parameters.

From the condition: $gJ_0g^*=J_0$ we have

$$\alpha\alpha^*\geq I\,,\;\;\delta\delta^*\geq I. \tag{2}$$

Let $Z=\delta^{-1}\gamma$, then it follows that $Z^*=\alpha^{-1}\beta$, and consequently $I-ZZ^*$ and $I-Z^*Z$ are hermitian, positive definite matrices, with

$$det(I-ZZ^*)=det(I-Z^*Z). \tag{3}$$

The maximal compact subgroup is

$$\begin{aligned}K &= \left\{\begin{pmatrix}u & 0\\ 0 & v\end{pmatrix}\,:\,u\in U(2)\,,\,v\in U(2)\,,\,det(u)det(v)=1\right\}\\ &= SU(2,2)\cap U(4)\\ &= S(U(2)\otimes U(2)).\end{aligned} \tag{4}$$

The complex domain

$$\mathcal{D}=SU(2,2)/K=\{Z\in M_{2\times 2}(\mathbf{C})\,:\,I-Z^*Z>0\} \tag{5}$$

has many remarkable properties:

1. $\mathcal{D}$ is an irreducible homogeneous bounded symmetric space of hyperbolic non-compact type.

2. $\mathcal{D}$ is geometrically convex.

3. $\mathcal{D}$ is a domain of holomorphy.

4. $\mathcal{D}$ has compact closure.

5. $\mathcal{D}$ has rank 2.

6. $\mathcal{D}$ contains S^4.

7. $\mathcal{D}$ is an 8-dimensional manifold.

8. $\mathcal{D}$ has a 7-dimensional boundary $\partial\mathcal{D}$.

9. $\mathcal{D}$ has a 4-dimensional Bergman-Šilov boundary $\check{\mathcal{S}}$:

$$\check{\mathcal{S}} \sim S^3 \times S^1 = \{Z \in M_{2\times 2}(\mathbf{C}) : Z^*Z = ZZ^* = I\} . \tag{6}$$

i.e. $Z \in U(2) = S(U(2) \times U(1))$.

10. $\mathcal{D}$ carries a transitive action of $SU(2,2)$.

11. $\mathcal{D}$ is the *generalized upper helf plane* and can be mapped one-to-one into the tube domain.

12. $\mathcal{D}$ has six orbits, three open orbits and the only closed one being $\check{\mathcal{S}}$.

13. $\mathcal{D}$ is such that the modulus of every holomorphic function regular in $\mathcal{D}$ and bounded on the closure $\overline{\mathcal{D}}$ attains its maximum on $\check{\mathcal{S}}$.

2.3 Action of $SU(2,2)$ on $\mathcal{D}$

In order to find the action of $SU(2,2)$ on $\mathcal{D}$ we need a decomposition of $g_0 \in SU(2,2)$ in which $Z \in \mathcal{D}$ occurs. Then the group law

$$gg_0(Z) \equiv g'(g(Z))$$

determines the action of g on $\mathcal{D}$ by finding the same decomposition of $g' \in SU(2,2)$. We use two decompositions:

$$\begin{pmatrix} \alpha & \beta \\ \gamma & \delta \end{pmatrix} = \begin{pmatrix} I & 0 \\ \gamma\alpha^{-1} & I \end{pmatrix} \begin{pmatrix} \alpha & 0 \\ 0 & \delta - \gamma Z^* \end{pmatrix} \begin{pmatrix} I & Z^* \\ 0 & I \end{pmatrix} \tag{7}$$

$$\begin{pmatrix} \alpha & \beta \\ \gamma & \delta \end{pmatrix} = \begin{pmatrix} I & \beta\delta^{-1} \\ 0 & I \end{pmatrix} \begin{pmatrix} \alpha - \beta Z & 0 \\ 0 & \delta \end{pmatrix} \begin{pmatrix} I & 0 \\ Z & I \end{pmatrix}, \tag{8}$$

where $Z = \delta^{-1}\gamma$ and $Z^* = \alpha^{-1}\beta$.

The left action of an element $g = \begin{pmatrix} A & B \\ C & D \end{pmatrix}$ is obtained by taking the product

$$g' = \begin{pmatrix} I & 0 \\ Z & I \end{pmatrix} \begin{pmatrix} A & B \\ C & D \end{pmatrix} = \begin{pmatrix} A & B \\ ZA + C & ZB + D \end{pmatrix} \tag{9}$$

and decomposing it according to (8)

$$g' = \begin{pmatrix} I & B(ZB+D)^{-1} \\ 0 & I \end{pmatrix} \begin{pmatrix} A - Bg(Z) & 0 \\ 0 & ZB+D \end{pmatrix} \begin{pmatrix} I & 0 \\ g(Z) & I \end{pmatrix}$$

where

$$g(Z) = (ZB+D)^{-1}(ZA+C). \tag{10}$$

Similarly the right action is obtained from the decomposition of the product according to again (8)

$$\begin{aligned} g' &= \begin{pmatrix} A & B \\ C & D \end{pmatrix} \begin{pmatrix} I & Z \\ 0 & I \end{pmatrix} = \begin{pmatrix} A & AZ+B \\ C & CZ+D \end{pmatrix} \\ &= \begin{pmatrix} I & g(Z) \\ 0 & I \end{pmatrix} \begin{pmatrix} A - g(Z)C & 0 \\ 0 & CZ+D \end{pmatrix} \begin{pmatrix} I & 0 \\ (CZ+D)^{-1}C & I \end{pmatrix}, \end{aligned}$$

where

$$g(Z) = (AZ+B)(CZ+D)^{-1}. \tag{11}$$

The full group space $SU(2,2)$ can be parametrized, in addition to $Z \in \mathcal{D}$ by two other 2×2 matrices X and Y with the action of $SU(2,2)$ as follows:

$$\begin{aligned} X &\longrightarrow X(C^*Z + A^*)^{-1} \\ Y &\longrightarrow Y(B^tZ^t + D^t)^{-1} \\ Z &\longrightarrow (ZB+D)^{-1}(ZA+C) \end{aligned} \tag{12}$$

where superscript t denotes matrix transposition. Only the first two rows of X and Y enter as effective coordinates: x_{11}, x_{12} and y_{11}, y_{12}.

2.4 Determination of Lie algebra Differential Operators

Here we determine the generators L_{ab} of the conformal group in the eight coordinates: $(x_{11}, x_{12}, y_{11}, y_{12}, z_{11}, z_{12}, z_{21}, z_{22})$. The method is as follows:

(a) Write down the one-parameter subgroups $\{R_{ab}\}$ as 6×6 matrices corresponding to *rotations* in the *ab-planes* by an angle α.

(b) Using the homomorphism between $SO(4,2)$ and $SU(2,2)$ determine the corresponding 4×4 matrices $U_{ab}(\alpha)$ for each series of Graev's representations d_0, d_1, d_2.

(c) Express $U_{ab}(\alpha)$ in terms of $\begin{pmatrix} A(\alpha) & B(\alpha) \\ C(\alpha) & D(\alpha) \end{pmatrix}$ as in (9).

(d) Consider a function $\varphi(X,Y,Z)$ and transformed function, according to (12),

$$\varphi(X(C^*Z+A^*)^{-1}, Y(B^tZ^t+D^t)^{-1}, (ZB+D)^{-1}(ZA+C)). \qquad (13)$$

Because $\frac{d}{d\alpha}U_{ab}(\alpha)\mid_{\alpha=0}=\ell_{ab}$ is the four-dimensional representation of the Lie algebra, we can evaluate explicitly

$$\frac{d}{d\alpha}\varphi\mid_{\alpha=0}=\ell_{ab}\varphi. \qquad (14)$$

This is a lengthy computation - we give below the results:

$$\begin{aligned}
\ell_{12} &= \frac{i}{2}\left[-x_{11}\frac{\partial}{\partial x_{11}}+x_{12}\frac{\partial}{\partial x_{12}}+y_{11}\frac{\partial}{\partial y_{11}}-y_{12}\frac{\partial}{\partial y_{12}}+2(z_{12}\frac{\partial}{\partial z_{12}}-z_{21}\frac{\partial}{\partial z_{21}})\right].\\
\ell_{13} &= \frac{1}{2}\left[-x_{12}\frac{\partial}{\partial x_{11}}+x_{11}\frac{\partial}{\partial x_{12}}-y_{12}\frac{\partial}{\partial y_{11}}+y_{11}\frac{\partial}{\partial y_{12}}-(z_{12}+z_{21})\frac{\partial}{\partial z_{11}}\right.\\
&\left.+(z_{11}-z_{22})(\frac{\partial}{\partial z_{12}}+\frac{\partial}{\partial z_{21}})+(z_{12}-z_{21})\frac{\partial}{\partial z_{22}}\right].\\
\ell_{23} &= \frac{i}{2}\left[-x_{12}\frac{\partial}{\partial x_{11}}-x_{11}\frac{\partial}{\partial x_{12}}+y_{12}\frac{\partial}{\partial y_{11}}+y_{11}\frac{\partial}{\partial y_{12}}-(z_{12}-z_{21})\frac{\partial}{\partial z_{11}}\right.\\
&\left.-(z_{11}-z_{22})(\frac{\partial}{\partial z_{12}}-\frac{\partial}{\partial z_{21}})+(z_{12}-z_{21})\frac{\partial}{\partial z_{22}}\right].\\
\ell_{14} &= \frac{i}{2}\left[x_{12}\frac{\partial}{\partial x_{11}}+x_{11}\frac{\partial}{\partial x_{12}}+y_{12}\frac{\partial}{\partial y_{11}}+y_{11}\frac{\partial}{\partial y_{12}}+(z_{12}+z_{21})(\frac{\partial}{\partial z_{11}}+\frac{\partial}{\partial z_{22}})\right.\\
&\left.+(z_{11}+z_{22})(\frac{\partial}{\partial z_{12}}+\frac{\partial}{\partial z_{21}})\right].\\
\ell_{24} &= \frac{1}{2}\left[-x_{12}\frac{\partial}{\partial x_{11}}+x_{11}\frac{\partial}{\partial x_{12}}+y_{12}\frac{\partial}{\partial y_{11}}-y_{11}\frac{\partial}{\partial y_{12}}-(z_{12}-z_{21})(\frac{\partial}{\partial z_{11}}+\frac{\partial}{\partial z_{22}})\right.\\
&\left.+(z_{11}+z_{22})(\frac{\partial}{\partial z_{12}}-\frac{\partial}{\partial z_{21}})\right].\\
\ell_{34} &= \frac{i}{2}\left[x_{11}\frac{\partial}{\partial x_{11}}-x_{12}\frac{\partial}{\partial x_{12}}+y_{11}\frac{\partial}{\partial y_{11}}-y_{12}\frac{\partial}{\partial y_{12}}+2(z_{11}\frac{\partial}{\partial z_{11}}-z_{22}\frac{\partial}{\partial z_{22}})\right].\\
\ell_{15} &= \frac{1}{2}\left[(x_{12}z_{11}+x_{11}z_{21})\frac{\partial}{\partial x_{11}}+(x_{11}z_{22}+x_{12}z_{12})\frac{\partial}{\partial x_{12}}+(y_{12}z_{11}+y_{11}z_{12})\frac{\partial}{\partial y_{11}}\right.\\
&+(y_{11}z_{22}+y_{12}z_{21})\frac{\partial}{\partial y_{12}}+z_{11}(z_{12}+z_{21})\frac{\partial}{\partial z_{11}}+(z_{11}z_{22}+z_{12}z_{12}-1)\frac{\partial}{\partial z_{12}}\\
&\left.+(z_{11}z_{22}+z_{21}z_{21}-1)\frac{\partial}{\partial z_{21}}+z_{22}(z_{12}+z_{21})\frac{\partial}{\partial z_{22}}+(z_{12}+z_{21})\rho\right].\\
\ell_{25} &= \frac{i}{2}\left[(x_{12}z_{11}-x_{11}z_{21})\frac{\partial}{\partial x_{11}}+(-x_{11}z_{22}+x_{12}z_{12})\frac{\partial}{\partial x_{12}}+(-y_{12}z_{11}+y_{11}z_{12})\frac{\partial}{\partial y_{11}}\right.\\
&+(y_{11}z_{22}-y_{12}z_{21})\frac{\partial}{\partial y_{12}}+z_{11}(z_{12}-z_{21})\frac{\partial}{\partial z_{11}}+(1-z_{11}z_{22}+z_{12}z_{12})\frac{\partial}{\partial z_{12}}\\
&\left.+(-1+z_{11}z_{22}-z_{21}z_{21})\frac{\partial}{\partial z_{21}}+z_{22}(z_{12}-z_{21})\frac{\partial}{\partial z_{22}}+(z_{12}-z_{21})\rho\right].
\end{aligned}$$

$$\begin{aligned}
\ell_{35} &= \frac{1}{2}\Bigg[(x_{11}z_{11}-x_{12}z_{21})\frac{\partial}{\partial x_{11}}+(x_{11}z_{12}-x_{12}z_{22})\frac{\partial}{\partial x_{12}}+(y_{11}z_{11}-y_{12}z_{12})\frac{\partial}{\partial y_{11}}\\
&\quad+(y_{11}z_{21}-y_{12}z_{22})\frac{\partial}{\partial y_{12}}+(-1+z_{11}z_{12}-z_{12}z_{21})\frac{\partial}{\partial z_{11}}+z_{12}(z_{11}-z_{22})\frac{\partial}{\partial z_{12}}\\
&\quad+z_{21}(z_{11}-z_{22})\frac{\partial}{\partial z_{21}}+(1+z_{21}z_{12}-z_{22}z_{22})\frac{\partial}{\partial z_{22}}+(z_{11}-z_{22})\rho\Bigg].\\
\ell_{45} &= \frac{i}{2}\Bigg[(x_{12}z_{21}+x_{11}z_{11})\frac{\partial}{\partial x_{11}}+(x_{11}z_{12}+x_{12}z_{22})\frac{\partial}{\partial x_{12}}+(y_{12}z_{12}+y_{11}z_{11})\frac{\partial}{\partial y_{11}}\\
&\quad+(y_{11}z_{21}+y_{12}z_{22})\frac{\partial}{\partial y_{12}}+(1+z_{12}z_{21}+z_{11}z_{11})\frac{\partial}{\partial z_{11}}+z_{12}(z_{11}+z_{22})\frac{\partial}{\partial z_{12}}\\
&\quad+z_{21}(z_{11}+z_{22})\frac{\partial}{\partial z_{21}}+(1+z_{21}z_{12}+z_{22}z_{22})\frac{\partial}{\partial z_{22}}+(z_{11}+z_{22})\rho\Bigg].\\
\ell_{16} &= \frac{i}{2}\Bigg[(x_{12}z_{11}+x_{11}z_{21})\frac{\partial}{\partial x_{11}}+(x_{11}z_{22}+x_{12}z_{12})\frac{\partial}{\partial x_{12}}+(y_{12}z_{11}+y_{11}z_{12})\frac{\partial}{\partial y_{11}}\\
&\quad+(y_{11}z_{22}+y_{12}z_{21})\frac{\partial}{\partial y_{12}}+z_{11}(z_{12}+z_{21})\frac{\partial}{\partial z_{11}}+(1+z_{11}z_{22}+z_{12}z_{12})\frac{\partial}{\partial z_{12}}\\
&\quad+(1+z_{11}z_{22}+z_{21}z_{21})\frac{\partial}{\partial z_{21}}+z_{22}(z_{12}+z_{21})\frac{\partial}{\partial z_{22}}+(z_{12}+z_{21})\rho\Bigg].\\
\ell_{26} &= \frac{1}{2}\Bigg[(x_{11}z_{21}-x_{21}z_{11})\frac{\partial}{\partial x_{11}}+(x_{11}z_{22}-x_{12}z_{12})\frac{\partial}{\partial x_{12}}+(y_{12}z_{11}-y_{11}z_{12})\frac{\partial}{\partial y_{11}}\\
&\quad+(y_{12}z_{21}-y_{11}z_{22})\frac{\partial}{\partial y_{12}}+z_{11}(z_{21}-z_{12})\frac{\partial}{\partial z_{11}}+(1+z_{11}z_{22}-z_{12}z_{12})\frac{\partial}{\partial z_{12}}\\
&\quad+(-1-z_{11}z_{22}+z_{21}z_{21})\frac{\partial}{\partial z_{21}}+z_{22}(z_{21}-z_{12})\frac{\partial}{\partial z_{22}}-(z_{12}-z_{21})\rho\Bigg].\\
\ell_{36} &= \frac{i}{2}\Bigg[(x_{11}z_{11}-x_{12}z_{21})\frac{\partial}{\partial x_{11}}+(x_{11}z_{12}-x_{12}z_{22})\frac{\partial}{\partial x_{12}}+(y_{11}z_{11}-y_{12}z_{12})\frac{\partial}{\partial y_{11}}\\
&\quad+(y_{11}z_{21}-y_{12}z_{22})\frac{\partial}{\partial y_{12}}+(1-z_{12}z_{21}+z_{11}z_{11})\frac{\partial}{\partial z_{11}}+z_{12}(z_{11}-z_{22})\frac{\partial}{\partial z_{12}}\\
&\quad+z_{21}(z_{11}-z_{22})\frac{\partial}{\partial z_{21}}+(-1+z_{21}z_{12}-z_{22}z_{22})\frac{\partial}{\partial z_{22}}+(z_{11}-z_{22})\rho\Bigg].\\
\ell_{46} &= \frac{1}{2}\Bigg[-(x_{11}z_{11}+x_{12}z_{21})\frac{\partial}{\partial x_{11}}-(x_{11}z_{12}+x_{12}z_{22})\frac{\partial}{\partial x_{12}}-(y_{12}z_{12}+y_{11}z_{11})\frac{\partial}{\partial y_{11}}\\
&\quad-(y_{11}z_{21}+y_{12}z_{22})\frac{\partial}{\partial y_{12}}+(1-z_{12}z_{21}-z_{11}z_{11})\frac{\partial}{\partial z_{11}}-z_{12}(z_{11}+z_{22})\frac{\partial}{\partial z_{12}}\\
&\quad-z_{21}(z_{11}+z_{22})\frac{\partial}{\partial z_{21}}+(1-z_{21}z_{12}-z_{22}z_{22})\frac{\partial}{\partial z_{22}}-(z_{11}+z_{22})\rho\Bigg].\\
\ell_{56} &= -\frac{i}{2}\Bigg[x_{11}\frac{\partial}{\partial x_{11}}+x_{12}\frac{\partial}{\partial x_{12}}+y_{11}\frac{\partial}{\partial y_{11}}+y_{12}\frac{\partial}{\partial y_{12}}\\
&\quad+2(z_{11}\frac{\partial}{\partial z_{11}}+z_{12}\frac{\partial}{\partial z_{12}}+z_{21}\frac{\partial}{\partial z_{21}}+z_{22}\frac{\partial}{\partial z_{22}})-2\rho\Bigg].
\end{aligned} \tag{15}$$

where ρ is a representation parameter. As a special case, for functions $\varphi(X,Y)$ independent of the z-coordinates, the Lie algebra can be written in the physicist's favourite boson operator formalism with

$$a^+ = \begin{pmatrix} x_{11} \\ x_{12} \end{pmatrix}, b^+ = \begin{pmatrix} y_{11} \\ y_{12} \end{pmatrix}, a = \begin{pmatrix} \frac{\partial}{\partial x_{11}} \\ \frac{\partial}{\partial x_{12}} \end{pmatrix}, b = \begin{pmatrix} \frac{\partial}{\partial y_{11}} \\ \frac{\partial}{\partial y_{12}} \end{pmatrix}$$

as

$$\begin{aligned} L_i &= \frac{1}{2}\left[a^+\sigma_i a - b^+\sigma_i^* b\right] \\ L_{i4} &= -\frac{1}{2}\left[a^+\sigma_i a + b^+\sigma_i^* b\right] \\ L_{i5} &= \frac{i}{2}\left[a^+(\sigma_i Z)a + b^+(\sigma_i^* Z^t)b\right] \\ L_{i6} &= -\frac{1}{2}\left[a^+(\sigma_i Z)a - b^+(\sigma_I^* Z^t b\right] \\ L_{45} &= -\frac{1}{2}\left[a^+(\sigma_0 Z)a + b^+(\sigma_0 Z^t)b\right] \\ L_{46} &= -\frac{i}{2}\left[a^+(\sigma_0 Z)a - b^+(\sigma_0 Z^t)b\right] \\ L_{56} &= \frac{1}{2}\left[a^+ a + b^+ b\right] \end{aligned} \tag{16}$$

which is similar but different than the other well known realization of the conformal group used in dynamical algebra formalism [7]. This also shows that X, Y-coordinates may represent additional internal coordinates of the system, whereas the Z-coordinates are related to space-time. In order to see this more precisely we introduce new coordinates related to Z's:

$$\begin{aligned} \mathfrak{y}_1 &= \frac{i}{2}\kappa(z_{12} + z_{21}) \\ \mathfrak{y}_2 &= -\frac{1}{2}\kappa(z_{12} - z_{21}) \\ \mathfrak{y}_3 &= \frac{i}{2}\kappa(z_{11} - z_{22}) \\ \mathfrak{y}_4 &= -\frac{1}{2}\kappa(z_{11} + z_{22}) \end{aligned} \tag{17}$$

where κ is a scale factor. Then for functions $\varphi(Z)$ dependent only on the Z-coordinates, the Lie algebra can be written in the simple form:

$$L_{ab} = \mathfrak{y}_a \partial_b - \mathfrak{y}_b \partial_a; \ a, b = 1, ..., 6 \tag{18}$$

where in addition to (17) we have introduced two new coordinates

$$\begin{aligned} \mathfrak{y}_5 &= \frac{i}{2}\kappa(1 - \det Z) \equiv \mathfrak{y}_0 \\ \mathfrak{y}_6 &= -\frac{1}{2}\kappa(1 + \det Z) \end{aligned} \tag{19}$$

in order to linearize the representation. It follows from (17) and (19) that

$$\mathfrak{y}_1^2 + \mathfrak{y}_2^2 + \mathfrak{y}_3^2 + \mathfrak{y}_4^2 = \kappa^2 \det Z = \mathfrak{y}_5^2 + \mathfrak{y}_6^2 \tag{20}$$

from which we see the $O(4,2)$-invariant metric in the $\mathfrak{y}$-coordinates. In (18) the conjugate variables are

$$\begin{aligned}
\partial_1 &= -\frac{i}{\kappa}\left(\frac{\partial}{\partial z_{12}} + \frac{\partial}{\partial z_{21}}\right) \\
\partial_2 &= -\frac{1}{\kappa}\left(\frac{\partial}{\partial z_{12}} - \frac{\partial}{\partial z_{21}}\right) \\
\partial_3 &= -\frac{i}{\kappa}\left(\frac{\partial}{\partial z_{11}} - \frac{\partial}{\partial z_{22}}\right) \\
\partial_4 &= -\frac{1}{\kappa}\left(\frac{\partial}{\partial z_{11}} + \frac{\partial}{\partial z_{22}}\right) \\
\partial_5 &= -\frac{i}{\kappa}\left(z_{11}\frac{\partial}{\partial z_{11}} + z_{22}\frac{\partial}{\partial z_{22}} + z_{12}\frac{\partial}{\partial z_{12}} + z_{21}\frac{\partial}{\partial z_{21}} + \rho\right) \\
\partial_6 &= -i\partial_5
\end{aligned} \tag{21}$$

with

$$[\mathfrak{y}_5, \partial_5] = -\det Z = [\mathfrak{y}_6, \partial_6].$$

2.5 Mapping from $\mathcal{D}$ to the Tube Domain T

The tube domain is a complex extension of the forward light cone, the space-time therefore is its boundary. We map these two eight - dimensional manifolds by the transformation

$$W = i(I - Z)(I + Z)^{-1} \tag{22}$$

with the inverse

$$Z = (I - iW)^{-1}(I + iW).$$

If we parametrize the coordinates in W as

$$W = w^\mu \sigma_\mu = \begin{pmatrix} w^0 + w^3 & w^1 - iw^2 \\ w^1 + iw^2 & w^0 - w_3 \end{pmatrix} \tag{23}$$

we obtain

$$w^\mu = \frac{2}{\kappa \det(I + Z)} \mathfrak{y}^\mu \tag{24}$$

where $\mathfrak{y}^\mu$ were introduced in Eqns.(17) and (19).

Define further

$$\begin{aligned} w^4 &= \frac{2}{\kappa \det(I+Z)} \mathfrak{y}^4 \\ w^6 &= \frac{2}{\kappa \det(I+Z)} \mathfrak{y}^6 \end{aligned} \tag{25}$$

we have

$$w_1^2 + w_2^2 + w_3^2 + w_4^2 = 4\frac{\det Z}{[\det(I+Z)]^2} = w_0^2 + w_4^2. \tag{26}$$

2.6 Representations of the Group

From the Lie algebra representations we go over to the representations of the group defined by (see Eqns.(12) and (13))

$$\begin{aligned} T_g\varphi(X,Y,Z) &= \varphi\left[X(g_{21}^* Z + g_{11}^*)^{-1}, Y(g_{12}^t Z^t + g_{22}^t)^{-1}, (Zg_{12}+g_{22})^{-1}(Zg_{11}+g_{21})\right] \cdot \\ & \det{}^{-\lambda}(Zg_{12}+g_{22}). \end{aligned} \tag{27}$$

where

$$g = \begin{pmatrix} g_{11} & g_{12} \\ g_{21} & g_{22} \end{pmatrix} \in SU(2,2)$$

such that

- For fixed Y, Z: $\varphi(X)$ is a homogeneous polynomial of degree μ_1 in x_{11}, x_{12}.
- For fixed X, Z: $\varphi(Y)$ is a homogeneous polynomial of degree μ_2 in y_{11}, y_{12}.
- For fixed X, Y: $\varphi(Z)$ is holomorphic in $\mathcal{D}$.

Then we can expand $\varphi(X,Y,Z)$ as follows:

$$\varphi(X,Y,Z) = \sum x_{11}^{\mu_1 - m_1} x_{12}^{m_1} y_{11}^{\mu_2 - m_2} y_{12}^{m_2} \psi_{\mu_2, m_2}^{\mu_1, m_1}(Z). \tag{28}$$

We see again here the appearance of multicomponent (or multipole) fields over *space-time* coordinates labelled by the internal quantum numbers.

3 Discussion

Most physical applications of the conformal group have made use of the simple degenerate representations with one or two of the Casimir operators vanishing. On the other hand, the most general representations of the Lie algebra and the group we have discussed here involve nine quantum numbers. As we have discussed, we believe these should be significant in the description of quantum systems with space-time and internal quantum numbers. The Lie algebra representation in the eight-dimensional homogeneous space that we have exhibited is quite rich in this respect.

For composite system, e.g. a system of two particles, the conformal group can be applied as an external group to the center-of-mass coordinates, also as an internal group to the relative coordinates. It turns out that the action of the conformal group in the relative coordinates is in the relative momentum space. This is the well-known momentum space conformal group representations which accounts, for example, for all states of the Hydrogen atom, showing the remarkable geometry of the Kepler problem. Quite generally, the requirement that the coordinates of each particle as well as the center-of-mass coordinates lie on the five-dimensional cone, Eq.(20), leads almost on kinematical grounds, to nontrivial discrete mass spectra for the composite system [4]. This theory can now be generalized to other systems with the present more general representations.

BIBLIOGRAPHY

1. E. Angelopoulos, (a) *Sur les représentations unitaires irréductibles de* $\overline{SO}_0(p,2)$, C.R. Acad.Sci.,Paris I, **292** (1981), 469-471.
   (b) *The unitary irreducible representations of* $\overline{SO}_0(4,2)$, Commun. Math. Phys., **89** (1983), 41-57.

2. A. O. Barut, (a) There are a number of volumes on the various aspects of physical applications of the conformal group: *De Sitter and Conformal Groups and Their Applications*, Colorado Assoc. University Press, Boulder, 1971 (edited by A. O. Barut).
   (b) *Conformal Group and Related Symmetries*, Lecture Notes in Physics, **261**, Springer-Verlag, New York, 1986 (edited by A. O. Barut and H. Doebner).
   (c) *Geometry and Physics*, Bibliopolis, Napoli, (1989).
   (d) *Quantum theory of single events*, Lecture Notes in Physics, **379** (1991), 241-256.
   (e) *Quantum theory of single events.* Found. of Phys., **20** (1990), 1233-1240.

3. A. O. Barut and A. Böhm, *Reduction of a class of* $O(4,2)$ *representations with respect to* $SO(4,1)$ *and* $SO(3,2)$, J. Math. Phys., **11** (1970), 2938-2945.

4. A. O. Barut and G. Bornzin, *Unification of the external conformal symmetry group and the internal conformal dynamical group*, J. Math. Phys., **15** (1974), 1000-1006.

5. A. O. Barut, P. Budinich, J. Niederle and R. Raczka, *Wave structures in conformally compact space-times*, (to be published).

6. A. O. Barut and R. Haugen, *Solutions of the conformally invariant spinor equations and theory of the electron-muon system*, Lett. Nuovo Cimento, **7** (1973), 625-628.

7. A. O. Barut and R. Raczka, *Theory of Group Representations and Applications*, World Scientific, Singapore, 1987.

8. A. O. Barut and R. Wilson, *Principal discrete series representations of* $SU(2,2)$, (to be published).

9. A. O. Barut and B. wei Xu, *The wave equation for the electron-muon system and their neutrinos*, Phys. Letts., **101B** (1981), 437-438.

10. E. Bessel-Hagen, *Über die Erhaltungssätze der Elektrodynamik*, Math. Ann., **84** (1921), 258-276.

11. E. Cartan, (a) *Sur les domaines bornés homogénes de l'espace de n variables complexes*, Abh. Math. Sem. Univ. Hamburg, **11** (1935), 116-162.
(b) *Les groupes projectifs qui ne laissent invariante aucune multiplicité plane*, Bull. Soc. Math. de France, **41** (1913), 53-96.
(c) *Les groupes projectifs continus réels qui ne laissent invariante aucune multiplicité plane*, J. Math. Pures et Appliquées, **10** (1914), 149-186.

12. A. Esteve and P. G. Sona, *Conformal group in Minkowski space. Unitary irreducible representations*, Il Nuovo Cimento, **32** (1964), 473-485.

13. I. M. Gel'fand and M. A. Naimark, *Unitäre Darstellungen der Klassischen Gruppen*, Akademie-Verlag, Berlin, 1957.

14. R. Godement, *A theory of spherical functions-I*, Trans. Amer. Math. Soc., **73** (1952), 496-556.

15. M. I. Graev, (a) *Unitary representations of real simple Lie groups*, Amer. Math. Soc. Transl. (2), **16** (1960), 393-396.
(b) *Unitary representations of real simple Lie groups*, Amer. Math. Soc. Transl. (2), **66** (1968), 1-62.

16. L. Gross, *Norm invariances of mass-zero equations under the conformal group*, J. Math. Phys., **5** (1964), 687-695.

17. Harish-Chandra, (a) *Representations of semisimple Lie groups on a Banach space. I*, Trans. Amer. Math. Soc., **75** (1953), 185-243.
(b) *Representations of semisimple Lie groups. II*, Trans. Amer. Math. Soc., **76** (1954), 26-65.
(c) *Representations of semisimple Lie groups. III*, Trans. Amer. Math. Sco., **76** (1954), 234-253.

(d) *Discrete series for semisimple Lie groups. I*, Acta Math., **113** (1965), 242-318.
(e) *Discrete series for semisimple Lie groups. II*, Acta Math., **116** (1966), 1-111.

18. S. Helgason, (a) *Differential Geometry, Lie Groups and Symmetric spaces*, Academic Press, New York, 1978.
(b) *Groups and Geometric Analysis*, Academic Press, Orlando, 1984.
(c) *Topics in Harmonic Analysis on Homogeneous Spaces*, Birkhäuser, Basel-Boston, 1981.

19. L. Hörmander, *An Introduction to Complex Analysis in Several Variables*, North-Holland, Amsterdam, 1979.

20. L. K. Hua, *Harmonic Analysis of Functions of Several Complex Variables in the Classical Domains*, Amer. Math. Soc., Providence, 1963.

21. H. P. Jakobsen and M. Vergne, *Wave and Dirac operators and representations of the conformal group*, J. Funct. Anal., **24** (1977), 52-106.

22. A. Kihlberg, V. F. Müller and F. Halbwachs, *Unitary irreducible representations of* $SU(2,2)$, Commun. Math. Phys., **3** (1966), 194-217.

23. A. W. Knapp, (a) *Langlands classification and unitary dual of* $SU(2,2)$, Lectures in Applied Math., **21** (1985), 209-217.
(b) *Representation Theory of Semisimple Lie Groups*, Princeton University Press, Princeton, 1986.
(c) *Lie Groups, Lie Algebras and Cohomology*, Princeton University Press, Princeton, (1988).

24. A. W. Knapp and B. Speh, *Irreducible Unitary representations of* $SU(2,2)$, J. Funct. Anal., **45** (1982), 41-73.

25. M. G. Krein, *Introduction to the geometry of indefinite J-spaces and to the theory of operators in those spaces*, Amer. Math. Soc. Transl. (2), **93** (1970), 103-176.

26. M. G. Krein and Ju. L. Smul'jan, *On linear fractional transformations with operator coefficients*, Amer. Math. Soc. Transl. (2), **103** (1974), 125-152.

27. R. P. Langlands, *On the classification of irreducible representations of real algebraic groups*, in *Representation Theory and Harmonic Analysis on Semisimple Lie Groups*, Amer. Math. Soc., Providence, 1989 (edited by P. J. Sally, Jr. and D. A. Vogan, Jr.).

28. N. Limic, J. Niederle and R. Raczka, (a) *Discrete degenerate representations of noncompact rotation groups. I*, J. Math. Phys., **7** (1967), 1861-1876.
(b) *Discrete degenerate representations of noncompact rotation groups. II*, J. Math. Phys., **7** (1967), 2026-2035.

29. G. Mack, *All unitary ray representations of the conformal group $SU(2,2)$ with positive energy*, Commun. Math. Phys, **55** (1977), 1-28.

30. G. Mack and I. Todorov, *Irreducibility of the ladder representations of $U(2,2)$ when restricted to the Poincaré subgroup*, J. Math. Phys., **10** (1969), 2078-2085.

31. J. Mickelsson and J. Niederle, *On representations of the conformal group which when restricted to its Poincaré or Weyl subgroups remain irreducible*, J. Math. Phys., **13** (1972), 23-27.

32. Y. Murai, (a) *On the group of transformations in six-dimensional space*, Prog. Theor. Phys., **9** (1953), 147-168.
(b) *On the group of transformations in six-dimensional space. II*, Prog. Theor. Phys., **11** (1954), 441-448.

33. M. A. Naimark, (a) *Linear representations of the Lorentz group*, Amer. Math. Soc. Transl. (2), **6** (1957), 379-458.
(b) *Decomposition of a tensor product of irreducible representations of the proper Lorentz group into irreducible representations*, Amer. Math. Soc. Transl. (2), **36** (1964), 101-242.

34. T. D. Newton, *A note on the representations of the de Sitter group*, Ann. of Math., **51** (1950), 730-733.

35. W. Rühl, (a) *Distributions of Minkowski space and their connection with analytic representations of the conformal group*, Commun. Math. Phys., **27** (1972), 53-86.
(b) *Field representations of the conformal group with continuous mass spectrum*, Commun. Math. Phys., **30** (1973), 287-302.

36. E. A. Tanner and R. Wilson, *Unitary representations of $SU(p,q)$ .I : Principal discrete series of representations* (to be published).

37. L. H. Thomas, *On unitary representations of the group of de Sitter space*, Ann. of Math., **42** (1941), 113-126.

38. N. Ja. Vilenkin and A. V. Klimyk, *Representations of Lie Groups and Special Functions*, Vols. I,II,III, Kluwer Academic Publishers, Dordrecht, 1991.

39. J. A. Wolf, (a) *Fine structure of Hermitian symmetric spaces*, in *Symmetric Spaces*, Marcel Deckker, New York, 1972 (edited by W. M. Boothby and G. L. Weiss).
(b) *The action of a real semisimple group on a complex flag manifold. I: Orbit structure and holomorphic arc components*, Bull. Amer. Math. Soc., **75** (1969), 1121-1237.

40. T. Yao, (a) *Unitary irreducible representations of* $SU(2,2).I$, J. Math. Phys., **8** (1967), 1931-1954.
(b) *Unitary irreducible representations of* $SU(2,2).II$, J. Math. Phys., **9** (1968), 1615-1626.
(c) *Unitary irreducible representations of* $SU(2,2).III$: *Reduction with respect to an iso-Poincaré subgroup*, J. Math. Phys., **12** (1971), 315-342.

41. D. P. Želobenko, *The theory of linear representations of complex and real Lie groups*, Trans. Moscow Math. Soc., **12** (1963), 57-110.

International Centre for Theoretical Physics, P.O.Box 586,
Miramare Grignano, 34100 Trieste, Italy.
and
Department of Physics, University of Colorado,
Boulder, Colorado 80309.

ORDINARY- AND MOMENTUM-SPACE CONFORMAL COMPACTIFICATIONS: SOME POSSIBLE OBSERVABLE CONSEQUENCES

P. BUDINICH

ABSTRACT. It is suggested that space-time and momentum space should be both conformally compactified, for the latter the motivations may be found in simple-spinors geometry. The two compact manifolds, both isomorphic to $(S_3 \times S_1)/Z_2$, may be identified with the two 4-dimensional homogeneous spaces of the conformal group, transformed in each other by conformal inversion, and possibly correlated by Radon transforms [20], [21].

It is shown that these conjectures might find confirmations already in existing non relativistic natural phenomena: for ordinary space in recently discovered [4], [5] large-scale correlations of galaxies (still in the course of verification through further observations), for momentum space in the S0(4) symmetry of the Hydrogen atom. The structure of both systems may in fact be derived from the same equation

$$\Delta (S_3)\, Y_{n\ell m} = -n\,(n+2)\, Y_{n\ell m}$$

defining the S_3-sphere spherical harmonics $Y_{n\ell m}$, interpreted in ordinary-space for the first system (universe) and in momentum-space for the second (H-atom).

Simultaneous ordinary- and momentum-space conformal compatifications poses problems which are both difficult and challenging, whose solutions might be relevant for several fields of physics.

1. Introduction

This paper represents a status report of a line of research motivated by the hypothesis of the fundamental role of conformal symmetry in physics and of the consequent conjecture of conformal compactification not only of ordinary space-time, but also of momentum-space(*).

(*) Adopting the conventional language of quantum field theoretists, I mean by "momentum space" the Fourier dual of Minkowski space-time ($\mathbb{R}^{3,1}$ or $\mathbb{R}^{1,3}$) which, outside the domain of quantum mechanics should be named, more appropriately, "wave-number space". This abuse of names is milder if one accepts the convention $\hbar = 1$, often adopted.

E. A. Tanner and R. Wilson (eds.), Noncompact Lie Groups and Some of Their Applications, 123–139.

Some partial results of this line of research, to which several people participate, have been published [1], and some are in the course of preparation [2], [3]. In this paper I present, together with a summary of some of the results already published [1], some of the unsolved problems which are being found on the way and some further results and conjectures. The main emphasis of the paper is layed on the study of the possibility that some confirmation of those hypothesis may be already found in existing natural phenomena and precisely in large scale correlations of galaxies (chapter 2) discovered some years ago [4] and recently confirmed [5] and in the most common element of our universe: the hydrogen atom (chapter 3). The possible geometrical-dynamical correlation of the "very large" with the "very small" was already advanced [6] and seems suggestive. But even more appealing are some of the possibilities together with the challenging problems which this line of research might open (chapter 4, 5).

2. Space-time compactification

Maxwell equations may be considered to represent one of the best and most successful thoeries of physics. In particular they represent one of the best instruments we have at disposal for the study of space-time where natural phenomena occur. The recognition of their Lorentz- and Poincaré-covariance brought to the discovery of the $\mathbb{R}^{3,1}$-structure of Minkowski space-time M, which may be identified as a homogeneous space of the form

$$M = \mathbb{R}^{3,1} = \frac{P}{L} \,,$$

the quotient of the (10 parameter) Poincaré group P by its (6 parameter) Lorentz subgroup L.

It is known since 1909 [7] that Maxwell equations are conformally covariant. From this covariance one can also identify a homogeneous space $\overline{M}$ of the form:

$$\overline{M} = \frac{G}{G_1} \tag{2.1}$$

where G stands for the (15 parameter) conformal group and $G_1 = L \otimes D \rtimes k^{(4)}$ for its (11 parameter) subgroup where $k^{(4)}$ represent special conformal transformations and D dilatations.

It is well known that $\overline{M}$ is compact, diffeomorphic to $(S_3 \times S_1)/Z_2$:

$$\overline{M} = \frac{S_3 \times S_1}{Z_2} \,. \tag{2.2}$$

$\overline{M}$ is generally named conformally compactified space-time, where the conformal group G acts transitively and without singularities, at difference with ordinary space-time $M = \mathbb{R}^{3,1}$ where the conformal group acts non linearly presenting singularities: it may bring any point of $M = R^{3,1}$ to infinity; more precisely to the so-called light cone at infinity J [8]. One has:

$$\overline{M} = M + J ,$$

where J, a submanifold of $\overline{M}$, has dimension 3 (while $\overline{M}$ and M have dimension 4). This means that M is densely contained in $\overline{M}$.

Therefore, while ordinary Minkowski space-time M is enough for the description of phenomena involving the local action of the conformal group G, $\overline{M}$ is necessary for the description of those phenomena where the global action of G may be involved, as electromagnetic phenomena and several others in high energy physics. Consequently, it is conformally compactified space-time $\overline{M}$ which should be considered as the real structure of empty space-time (in so far general relativity may be ignored) where physical phenomena occur. The geometrical properties of $\overline{M}$ have been extensively studied and are well known [9], [10]. Here we are interested in studying how the various space-times adopted by physics for the description of particular classes of phenomena may be densely imbedded in $\overline{M}$.

In fact it is precisely the modality of this imbedding which will determine on one side the way to deal with space and time when the regions involved are very large (cosmology), and on the other the so-called conformal factors which will become important in field theories when integrations on space-time extending to infinity are involved.

2.1. IMBEDDING OF SPACE-TIMES IN $\overline{M}$

The general problem of space-times imbedding in $\overline{M}$ may be group-theoretically classified [11] and is dealt with in a separate paper [2].

Here I wish first to examine the space-times needed for the description of some particular classes of physical phenomena and the way these spaces imbed in $\overline{M}$.

A first class contains non relativistic, local phenomena like slow motions on the earth or in our planetary system. For them the newtonian space-time $\{\mathbb{R}^3, t\}$ where t represents absolute time is adequate; and its imbedding in $\overline{M}$ which is somehow trivial, is not expected to give any new interesting insight for the description of the phenomenology.

A second class is that of relativistic local phenomena like the motion of elementary particles of quantum mechanics, or q.e.d., for which Minkowski space-time $\mathbb{R}^{3,1}$ will be appropriate. Its imbedding in $\overline{M}$ may be of interest for the virtual processes whose study may involve integrations on space-time coordinates x_μ "up to infinity", which is rather meaningless, since: either the integral is convergent, and then we could as well cut the integration at an invariant region of $\mathbb{R}^{3,1}$ space-time including say the nearest galaxy, with no appreciable change, or, the integral is divergent and then we cannot ignore that $\mathbb{R}^{3,1}$ cannot be the space-time structure of our universe and $\mathbb{R}^{3,1}$ has to be imbedded in $\overline{M}$ (possibly via a Robertson-Walker or De Sitter space-time) and then $\mathbb{R}^{3,1}$ will result conformally flat: that is the volume-element d^4x will have to be substituted by $\Omega^2(x)\, d^4x$ where $\Omega(x)$ is the conformal factor which for any imbedding [2] satisfies the equation:

$$\lim_{x \to \infty} \Omega(x) = 0$$

which, in general, will regularize possible infrared divergences. We will come back to this class of phenomena when dealing with conformally compactified momentum space.

A third class of phenomena of particular interest for the present paper is that of non relativistic phenomena involving large regions of space and time like those which are dealt with by cosmology. For this class we may again consider the separation space-absolute time, however we cannot ignore that, for a closed universe, the space part is represented by S^3, and we have the Robertson-Walker space

$$\text{R.W.}: \{S^3, t\} \tag{2.3}$$

where S^3 represents a sphere with time-dependent radius R(t).

It is well known [12] that Robertson-Walker space-time, with line element:

$$ds^2 = R^2(t)\,[d\chi^2 + \sin^2\chi\,(d\theta^2 + \sin^2\theta\ d\phi^2)] - c^2\,dt^2 \tag{2.4}$$

where (χ, θ, ϕ) are angular coordinates on S_3, may be uniquely determined from the cosmological principle stating space-isotropy and -homogeneity of the universe.

For large scale relativistic phenomena instead, it is more appropriate to consider De Sitter space which may be simply obtained from R.W. one assuming

$$R^2(t) = R_0^2 + c^2t^2$$

The local spaces (R^3, t) and $\mathbb{R}^{3,1}$ considered before may be imbedded in $\overline{M}$ via R.W. and De Sitter spaces respectively [2].

2.2. POSSIBLE LARGE-SCALE WAVE STRUCTURE OF THE UNIVERSE

In the frame of the above hypothesis, the evolution of any dynamical system should take place in $\overline{M}$ or in space-times densely contained in it. In $\overline{M}$ should be formulated and solved the conformally covariant equations like Maxwell ones (and perhaps also massless Dirac ones [13], [14]) as well as those representing the early evolution of the universe(*). For the latter a second order scalar field equation is often adopted which indeed admits exact solutions as will be shown and discussed in a forthcoming paper [2].

Since we are interested in the large scale structure of the expanding universe as may be observed at present time when non relativistic motions of massive systems prevail, it is enough to formulate the equation for a scalar field Ψ in Robertson-Walker (or De Sitter) space-time densely contained in $\overline{M}$. The equation, for a closed universe is [1], [2]:

$$\{R^{-3}\,\partial_t\,(R^3\,\partial_t) - R^{-2}\,[\Delta\,(S_3) - 1]\}\,\Psi = 0\,, \tag{2.5}$$

(*) Dealing with the universe, as in the Big Bang cosmology, then $\overline{M} = G/G_1$, should be strictly considered as a homogeneous group-space, space-time instead is represented by the solution of field equations on $\overline{M}$, thus it may spontaneously break the symmetry of $\overline{M}$ (space-isotropy and -homogeneity).

where $\Delta(S_3)$ is the Laplace-Beltrami operator on S_3/Z_2. It may be solved by separation of variables and elementary solutions have the form:

$$\Psi_{n\ell m}(x) = f_n(t)\; Y_{n\ell m}(\chi, \theta, \phi) \tag{2.6}$$

where $Y_{n\ell m}(\chi, \theta, \phi)$ are harmonics(*) on S_3/Z_2, defined by

$$\Delta(S_3)\, Y_{n\ell m} = -n\,(n+2)\, Y_{n\ell m}, \tag{2.7}$$

and $f_n(t)$ satisfies

$$R^{-3}\frac{d}{dt}\left(R^3\frac{df_n}{dt}\right) + n\,(n+2)\,R^{-2}\,f_n(t) = 0. \tag{2.8}$$

One may now identify Ψ as the inflaton field of the inflationary model [17] (but other models of universe evolution could serve as well) and then the energy density H(x), identified as the time component of the energy momentum tensor, may be computed [1b], for each elementary solution, to be of the form

$$H_{n\ell m}(x) = K_n(t)\,|\,Y_{n\ell m}(\chi, \theta, \phi)|^2 \tag{2.9}$$

where $K_n(t)$ is a time-dependent factor depending on $f_n(t)$ (and R(t)).
The harmonics $Y_{n\ell m}(\chi, \theta, \phi)$ have the form

$$Y_{n\ell m}(\chi, \theta, \phi) = K_{n\ell}\; C^{\ell+1}_{n-\ell}(\cos\chi)\,\sin^\ell\chi\; Y_{\ell m}(\theta, \phi) \tag{2.10}$$

where $K_{n\ell}$ are normalization constants, $C_{n-\ell}^{\ell+1}(\cos\chi)$ are Gegenbauer polynomials defined by

$$C^{\ell+1}_{n-\ell}(\cos\chi) = \frac{d^{\ell+1}\,[\cos(n+1)\,\chi]}{d(\cos\chi)^{\ell+1}} \tag{2.11}$$

and $Y_{\ell m}(\theta, \phi)$ are harmonics of S_2.
Under the simplifying hypothesis that only one eigenmode and precisely the most symmetric one with ℓ=0=m dominates the solution of (2.5) the Gegenbauer polynomial $C_n^1(\cos\chi)$ assumes the simple form

$$C^1_n(\cos\chi) = \frac{\sin(n+1)\,\chi}{\sin\chi} \tag{2.12}$$

(*) It is interesting to remind that the possibility that the universe may present a structure determined by the solutions of eq.(2.5) was considered by E. Schrödinger in 1937 [15] following an idea by A.S. Eddington [16].

and the energy density becomes

$$H_n(x) = K_n^2(t) \frac{\sin^2 (n+1)\chi}{\sin^2 \chi} \tag{2.13}$$

It may be compared with large scale distribution of galaxies in the direction of south and north galactic poles published in 1990 [4] and recently confirmed [5] by first considering the relative density $\rho(x, x_0)$

$$\rho (x, x_0) = \frac{H (x)}{H (x_0)} \tag{2.14}$$

where $x = (t, \chi, \theta, \phi)$ and $x_0 = (t, \chi_0, \theta_0, \phi_0)$ which eliminates the unknown factor K(t), and then identifying the distance $d_\chi = R\chi$. Assuming $D = R\,\pi/2 = 3000$ Mpc one obtains the comparison reported in Fig. 1 (Fig. 3 of ref. [1b]), for H(x) given by (2.13) with n = 46.

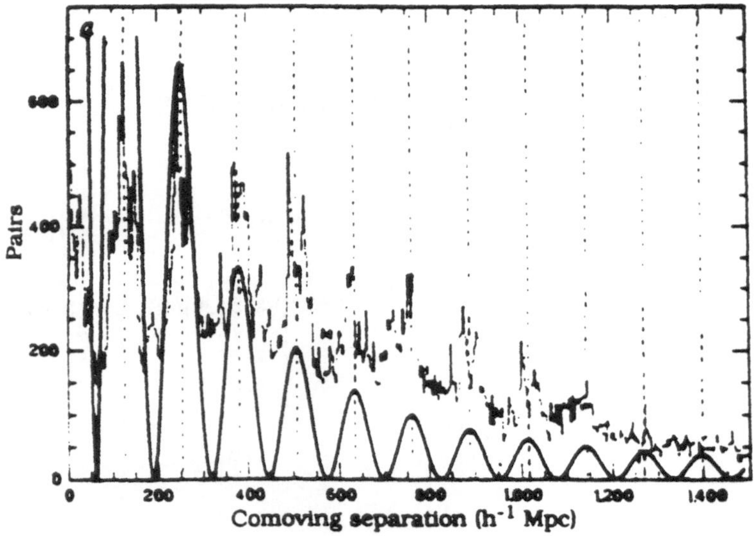

Figure 1. Observational result on large scale distribution of galaxies in the direction of south and north galactic poles reproduced from ref. [4], Fig. 2a) (broken line). The continuous curve represents the theoretical energy density $\rho(x, x_0)$ given by eqs.(2.14) and (2.13) with n = 46. It is normalized to the third peak of the observational data.

It is seen then that one eigenmode and the most symmetric one could explain the 128 Mpc periodicity of galaxies distribution at large distances.

Subsequent astronomical observations [5] have confirmed with more data the periodicity in the south-north galactic directions while have found less striking regularities in other directions of the celestial sphere. In the frame of the present interpretation this could be the case if:

a) keeping the hypothesis of one or few dominant modes one allow $\ell \neq 0 \neq m$.
b) The observer is at some distance $d_0 = R\,\chi_0$ from the center (North pole) of the oscillatory system.

Computations are in progress in order to determine, through a best fit with the most recent observational data, the dominating eigenfunctions contained in $\Psi(x)$, from which to compute an optimized H(x) apt to allow the prediction of further possible observations which, when performed, could then either contradict or give more confidence to the proposed interpretation.

In the latter case our universe would be closed and classical wave field theory could play an important role in the study of dynamical cosmology.

I wish here only to hint to its possible impact on one problem: the one of the horizon. In fact if the large scale structure of the universe may be represented in its broad structure by the solution of a wave field equation, then its regularity and uniformity at very large distances (at the horizon in opposite directions) is determined everywhere by the eigenfunctions, solutions of the dynamical equations, in a similar way as it is the Schrödinger wave function of a decaying system: say, the neutron-decaying in proton-electron-neutrino, which determines the probability of observing the polarization-direction of the decaying particles even when they are measured at space-time separations which cannot be correlated by a light signal. In other words: the horizon paradox would have some aspects in common with the Einstein-Rosen-Podolsky paradox of quantum mechanics.

3. Momentum-space conformal compactification

Fourier transforms and computations in momentum-space are familiar tools for physics. S-matrix theory (in q.e.d., say) is often formulated in momentum space $P = \mathbb{R}^{3,1}$, the Fourier dual of ordinary space-time $M = \mathbb{R}^{3,1}$. If space-time is conformally compactified and represented by $\overline{M}$ as advocated in the preceeding chapter, then M imbedded in $\overline{M}$ results conformally flat. This modification of M will not, by itself, modify the pseudoeuclidean flat nature of $P = \mathbb{R}^{3,1}$. At the most one can expect that, for integrations in momentum space, the compactification of M may originate a cut off at low values of $p \in P = \mathbb{R}^{3,1}$ corresponding to $p_0 \sim 2\pi/R$ where R is the radius of the S_3 and S_1 spheres of M. An asymmetry results then between M and P: the first is conformally flat, while the second is simply flat. This might appear cumbersome in a conformal world where dilatation covariance (and conformal inversion) should be valid. In that world one would rather expect both spaces M and P conformally flat which would then imply conformal compactification of both ordinary and momentum space.

Besides this generic, somehow aestetical motivation for conformal compactification of momentum-space(*) there are also some more concrete ones.

The first comes from spinor geometry. In fact for a pseudo-euclidean space $\mathbb{R}^{n,n}$ (for simplicity we choose the neutral signature) a simple spinor ζ is defined by the equation [18]:

$$p_\alpha \gamma^\alpha \zeta = 0 \,, \qquad \alpha = 1, 2, \dots 2n \tag{3.1}$$

(*) One should keep in mind that, at difference with ordinary space-time, coordinates of momentum space represent relative momenta of physical systems; the absolute ones laying on hyperboloids (mass-shells).

where $p_\alpha \in \mathbb{R}^{n,n}$ and γ^α are the generators of the corresponding Clifford algebra $Cl(n,n)$ obeying $[\gamma_\alpha, \gamma_\beta]_+ = g_{\alpha\beta}$, from which follows: $p_\alpha p^\alpha \zeta = 0$ and, for $\zeta \neq 0$:

$$p_\alpha p^\alpha = 0 \quad , \qquad \alpha = 1, 2, \ldots 2n \tag{3.2}$$

that is p_α are null, they lie on the light-cone of $\mathbb{R}^{n,n}$. But then they define in the projective light-cone a (2n-2) compact manifold (in the considered case isomorphic to: $(S_{n-1} \times S_{n-1})/Z_2$).

The definition may be extended to $\mathbb{R}^{n+1,\,n-1}$. In $\mathbb{R}^{3,1}$, from (3.1) one easily obtains both Weyl and Maxwell equations: however in momentum space [19]. For twistors: Weyl spinors of $\mathbb{R}^{4,2}$, p_α must be complex, however for Weyl spinors of $\mathbb{R}^{5,3}$ one may obtain equations which are significant for physics in real momentum space [19]; in this way compactified momentum space in $\mathbb{R}^{4,2}$ is obtained, and then, naturally, in its lightcone conformally compactified $\overline{P}$:

$$\overline{P} = \frac{S_3 \times S_1}{Z_2} \tag{3.3}$$

is obtained.

Another motivation [1a] may be found in the structure of the conformal group G itself, which, besides G_1, introduced in (2.1) has another eleven-parameter subgroup G_2:

$$G_2 = L \otimes D \rtimes P^{(4)} \tag{3.4}$$

where $P^{(4)}$ indicate the abelian group of Poincaré translations. The subgroups G_1 and G_2 are transformed in each other by conformal inversion J defined in M by (see next paragraph):

$$J\,(x_\mu) = \frac{x_\mu}{x^2} \;, \qquad \mu = 1, 2, 3, 4$$

one has precisely:

$$J\, G_1\, J^{-1} = G_2$$

The homogeneous space

$$\overline{P} = \frac{G}{G_2} = \frac{S_3 \times S_1}{Z_2}$$

is then obtained from $\overline{M} = G/G_1$ by conformal inversion, and the transitive action of G on $\overline{P}$ may be easily defined [1a].

One is tempted to identify the homogeneous space $\overline{P}$ as compactified momentum space. In order to justify this identification one has to show that there is a $P = \mathbb{R}^{3,1}$ densely contained in $\overline{P}$ which is the Fourier dual of $M = R^{3,1}$ densely contained in $\overline{M}$, which is indeed possible as I will try to briefly indicate.

Fourier transforms may only be defined in flat pseudoeuclidean spaces. For the compact manifold $(S_3 \times S_1)/Z_2$ one may define, by solving Laplace-Beltrami equations, complete sets of orthonormal eigenfunctions $Y_{n\ell m\tilde{m}}$ such that any (regular) function on the manifold may be expressed as a discrete series on them. The imbedding of $R^{3,1}$ on $(S_3 \times S_1)/Z_2$ may be realized as a stereographic projection of the latter on a flat $\mathbb{R}^{3,1}$ tangent. In $\mathbb{R}^{3,1}$, in the vicinity of the contact with the compact manifold, one may show that the discrete series might be identified with the Fourier integrals (as Stilties integrals). One may repeat the same operation with P which justifies the identification of $\overline{P}$ as compactified momentum-space, once one identifies the two flat tangents as Fourier dual [3].

What remains to be shown is the connection between the two discrete series on $\overline{M}$ and $\overline{P}$ both isomorphic to $(S_3 \times S_1)/Z_2$ in general. For that some theorems due to Helgason [20] may help which show that: complex-valued continuous functions defined on homogeneous spaces determined by two different closed subgroups of the same group are connected by integral transforms. These in turn are derived from Radon transforms [21].

Then we have that the two orthonormal sets $Y_{n\ell m\tilde{m}}$ on $\overline{M}$ and $\overline{P}$ are connected by integral transforms, but then these transforms may also connect the two discrete Fourier series on $\overline{M}$ and $\overline{P}$ if we show that these two series identify with the Fourier transforms in $M \subset \overline{M}$ and $P \subset \overline{P}$; In this way we may substitute the usual concept of Fourier transforms in flat $\mathbb{R}^{3,1}$ spaces with the one of discrete series in compact manifolds which might imply the existence both of a maximal and a minimal conceivable distance both in $\overline{M}$ and in $\overline{P}$. This will be dealt with in a subsequent paper [3].

Let us now consider the easier problem of imbedding M in $\overline{M}$ and P in $\overline{P}$. They will be both conformally flat. The conformal factors will depend on how the imbedding is performed. To keep the manifest relativistic covariance it will be convenient to imbed them via De Sitter spaces. Then the conformal factors become [1a], [2]

$$\Omega(x) = \frac{1}{1 + x^2/4R^2} \qquad \Omega(p) = \frac{1}{1 + p^2/4\mathcal{M}^2} \tag{3.5}$$

where R and $\mathcal{M}$ are the radiuses of the S_3 spheres in $\overline{M}$ and $\overline{P}$ respectively. Observe that $\Omega(p)$ is identical with the regularization factor introduced by Pauli and Villars [22].

When integrating up to infinity, obviously being now both M and P conformally flat the conformal factors have to be introduced and infrared and ultraviolated divergences will be in general both regularized (and afterwards possibly renormalized).

3.1. A POSSIBLE ROLE OF THE CONFORMAL INVERSION

Traditionally the imbedding of $M = R^{3,1}$ in $\overline{M} = (S_3 \times S_1)/Z_2$ is obtained by the stereographic projection:

$$x_\mu = \frac{\eta_\mu}{\eta_5 + \eta_6}, \quad \eta_5 + \eta_6 \neq 0 \quad ; \; x^2 = \frac{\eta_6 - \eta_5}{\eta_6 + \eta_5} \tag{3.6}$$

where $x_\mu \in M$, $\eta_a \in PQ$, where PQ is the projective null quadric: $\eta_\alpha\eta^\alpha = 0$ in $\mathbb{R}^{4,2}$. The conformal inversion J: $\eta_5 \to -\eta_5$, induces in M:

$$J : x_\mu \to \frac{x_\mu}{x^2}, \quad \text{or} \quad x^2 \to \frac{1}{x^2}; \tag{3.7}$$

For x^2 spacelike J maps in M (or, better, in $\mathbb{R}^3$) each inner point of a sphere S_2 of radius 1 to an outer one, in particular the center to infinity (corresponding to the lightcone at infinity $J \subset \overline{M}$), and viceversa. Loosely speaking one could say that such S_2 sphere is the border between "small": $x^2<1$ and "large": $x^2>1$.

However in a conformal world these notions do not have the common meaning since $x_\mu \in M$ are adimensional like the common radius of the spheres S_3 and S_1 in $\overline{M}$ (conventionally set = 1), and x_μ may be subject to covariant dilatations. The conventional meaning of small and large may be recovered if we introduce by hand a unit of length ℓ in (3.6):

$$x_\mu = \frac{\ell\,\eta_\mu}{\eta_5+\eta_6}, \quad \eta_5+\eta_6 \neq 0; \quad x^2 = \ell^2\,\frac{\eta_6-\eta_5}{\eta_6+\eta_5} \tag{3.6'}$$

by which (3.7) becomes:

$$J: \; x_\mu \to \frac{x_\mu\,\ell^2}{x^2} \quad \text{or} \quad x^2 \to \frac{\ell^4}{x^2} \tag{3.7'}$$

and then the S_2 sphere in M (or in $\mathbb{R}^3$) has the radius ℓ. But then the strict conformal covariance is lost. In our human world then (if, say, ℓ = one meter) small and large mapped to each other by J have the conventional meaning(*):

$$\begin{aligned} &\text{small}: \quad x^2 < \ell^2 \\ &\text{large}: \quad x^2 > \ell^2 \end{aligned} \tag{3.8}$$

We may repeat the same argumentation for P, imbedded in $\overline{P}$

$$k_\mu = \frac{\pi_\mu}{\pi_5+\pi_6}, \quad \pi_5+\pi_6 \neq 0; \quad k^2 = \frac{\pi_6-\pi_5}{\pi_6+\pi_5} \qquad \mu = 1, 2, 3, 4, \tag{3.9}$$

(*) One natural unit for ℓ would be the common radius R of the spheres S_3 and S_1, but then, if R is thought to represent the radius of the universe, large would be meaningless in our world.

where the conformal inversion induces the transformation [1a]:

$$J:\ k_\mu \to \frac{k_\mu}{k^2}\ ,\quad \text{or}\quad k^2 \to \frac{1}{k^2} \tag{3.10}$$

Here again for k_μ spacelike we have in $P = R^{3,1}$ (or in $\mathbb{R}^3$) a sphere of radius 1. If we now introduce a unit of lengths, since as usual we may adopt the convention (for k spacelike):

$$k^2 = \frac{1}{\lambda^2}$$

with λ a wavelength, we will have:

$$k_\mu = \frac{\ell^{-1}\pi_\mu}{\pi_5 + \pi_6}\ ;\quad \pi_6 + \pi_6 \neq 0 \qquad \ell^2 k^2 = \frac{\pi_6 - \pi_5}{\pi_6 + \pi_5} \tag{3.9'}$$

and therefore (3.10) becomes:

$$J:\ k_\mu \to \frac{k_\mu}{k^2\ell^2}\quad \text{or}\quad k^2 \to \frac{1}{k^2\ell^4} \tag{3.10'}$$

Therefore this time the inside of S_2 sphere will be large (in terms of length) and the outside small.

Let us now consider the scalar product $x_\mu k^\mu$; it remains dimensionless even after the introduction of a unit of length, therefore if we subject it to a conformal inversion we have:

$$J:\ x_\mu k^\mu \to \frac{x_\mu k^\mu}{x^2 k^2} \tag{3.11}$$

and again:

$$\begin{aligned} &\text{small:}\quad x^2 k^2 < 1 \\ &\text{large:}\quad x^2 k^2 > 1 \end{aligned} \tag{3.12}$$

and this is consistent with conformal covariance even after introduction of a unit of length.

Adopting now the De Broglie relation, starting point of quantum mechanics:

$$p_\mu = h\, k_\mu$$

where p_μ are the components of ordinary linear momentum and h the Planck constant, then (3.11) becomes:

$$J:\ x_\mu\, p^\mu \rightarrow \frac{x_\mu\, p^\mu}{x^2\, p^2}\, h^2 \tag{3.13}$$

and (3.12) becomes:

$$\begin{aligned} &\text{small:} \quad x^2\, p^2 < h^2 \\ &\text{large:} \quad x^2\, p^2 > h^2 \end{aligned} \tag{3.14}$$

In this way we have then that, the familiar separation of the world (small) of quantum mechanics where actions are smaller (or of the order of) h, from that (large) of classical mechanics, where actions are larger than h, is determined by h as it should. Conformal inversion, transforms $x_\mu p^\mu$ of the one world into $x_\mu p^\mu$ of the other, and viceversa, up to dimensionless factors.

Obviously we have introduced this separation by hand(*), inserting as a postulate the De Broglie equation. Nevertheless it shows that the correlation between "large" and "small" potentially implicit in the conformal inversion may be made significant but independent from the arbitrary choice of a unit of length ℓ and instead dependent from the universal constant h. In this way conformal inversion would have a role in bringing for the macroscopic world of classical mechanics to the microscopic one of quantum mechanics and viceversa. If we now accept, for the moment, the interpretation of the observed large scale structure of galaxies as due to (non relativistic) eigenwaves on $S_3 \subset \overline{M}$, then conformal inversion should presumably bring us from this large classical system to a quantum system. Furthermore, since conformal inversion brings $G/G_1 = \overline{M}$ to $G/G_2 = \overline{P}$, such quantum system should be described in momentum space. I wish to show that such system might perhaps exist in nature and could in fact be identified with the hydrogen atom.

4. The hydrogen atom

We have to deal then with a non-relativistic system in momentum space to be imbedded in $\overline{P} = G/G_2$. It will be the dual of Robertson-Walker space $\{S_3, t\}$ that is:

$$P = \{S_3, E\} \tag{4.1}$$

where E stands for energy, Fourier dual of absolute time t ($\hbar = 1$). Coherently, with our hypothesis the equation to be considered will be (2.5) with t substituted by E, admitting solutions of the form (2.6) that is:

$$\Psi_{n\ell m}(p) = f_n(E)\, Y_{n\ell m}(\chi, \theta, \phi) \tag{4.2}$$

(*) Ideally it would be preferable to get it from pure geometry. For the moment this is just a dream, but it is not excluded that it may be realized (e.g. in the frame of simple-spinor-geometry in $\mathbb{R}^{5,3}$).

where χ, θ, ϕ are angular variables of a unit sphere S_3:

$$S_3 : \pi_1^2 + \pi_2^2 + \pi_3^2 + \pi_5^2 = 1$$

and

$$\begin{aligned} \pi_1 &= \sin\chi \ \sin\theta \ \cos\phi & \pi_3 &= \sin\chi \ \cos\theta \\ \pi_2 &= \sin\chi \ \sin\theta \ \sin\phi & \pi_5 &= \cos\chi \end{aligned} \tag{4.3}$$

Observe in particular that in the maximally symmetric case ($\ell = 0 = m$) we will have

$$Y_{n,0,0} = \frac{\sin(n+1)\chi}{\sin\chi}, \tag{4.4}$$

that is the same eigen wave (2.12) as in the case of wave structure of the universe, which seems to represent the large scale correlation of galaxies (Fig. 1), however now it is to be interpreted in momentum space.

The S_3 spherical harmonics build up an orthonormal set:

$$\frac{1}{2\pi^2} \int Y^*_{n\ell m}(\pi) \ Y_{n'\ell' m'}(\pi) \ d\Omega = \delta_{nn'} \ \delta_{\ell\ell'} \ \delta_{mm'} , \tag{4.5}$$

where π is a unit vector of $\mathbb{R}^4$ with components π_α ($\alpha = 1, 2, 3, 5$).

The equation (2.7) may be expressed in integral form:

$$Y_{n\ell m}(\pi) = \frac{\lambda}{2\pi^2} \int \frac{Y_{n\ell m}(\pi')}{(\pi - \pi')^2} \ d\Omega' \tag{4.6}$$

with $\lambda = n + 1$, as shown(*) by V. Fork {23}.

Let us now project S_3 on a flat $\mathbb{R}^3$ space $p : \{p_1, p_2, p_3\}$, according to:

$$p_j = \frac{p_0 \, \pi_j}{1 + \pi_5} , \quad \pi_5 \neq -1 , \quad j = 1, 2, 3 \tag{4.7}$$

where p_0 is a, for the moment arbitrary, unit of momentum, necessary since π_α are dimensionless as shown by eq.(4.3).

(*) I adopt here the same procedure as that of V. Fock [23] however in reversed order. He discovered (together with W. Pauli) the S0(4) symmetry of the H-atom. Therefore in this line of thought that symmetry is not accidental, as suggested by W. Pauli, but derived from the supposed compactification of both ordinary and momentum space.

To express the integral equation (4.6) in non-relativistic momentum space $p \in \mathbb{R}^3$ one has to first express $d\Omega = \sin^2\chi\, d\chi\, \sin\theta\, d\theta\, d\phi$ in terms of $d^3 p = dp_1, dp_2, dp_3$ as follows

$$d\,\Omega = F^2(p)\, d^3 p \tag{4.8}$$

where $F(p) = 2\,p_0^2\,(p^2 + p_0^2)^{-2}$ is a form-factor easily obtained from (4.7). (It is not a conformal factor which instead could be easily obtained by imbedding in turn the space (4.1) in $P = G/G_2$). Subsequently if one substitutes the wavefunction $\Psi_{n\ell m}(\pi)$ with [23]:

$$\psi_{n\ell m}(p) = \frac{\sqrt{8}}{\pi}\, p_0^{5/2}\,(p_0^2 + p^2)^{-2}\,\Psi_{n\ell m}(\pi)\;, \tag{4.9}$$

such that, because of (4.5), $\psi_{n\ell m}(p)$ are subject to the standard orthonormal conditions in ordinary momentum space, one obtains after setting in equation (4.6), following V. Fock [23],

$$\lambda = \frac{m e^2}{\hbar\, p_0} = n+1 \;\; ; \;\; p_0 = \sqrt{-2mE} \;\; ; \;\; E < 0\;, \tag{4.10}$$

that it becomes:

$$\frac{1}{2m}\,p^2\,\psi_{n\ell m}(p) - \frac{e^2}{2\pi\hbar^2}\int \frac{\psi_{n\ell m}(p)}{|p - p'|^2}\, d^3p' = E_n\,\psi_{n\ell m}$$

where:

$$E_n = -\frac{m e^4}{2\hbar\,(n+1)^2}\;,$$

that is the standard Schrödinger equation for the Hydrogen eigenstates in momentum space.

It is amusing to observe that $p^2_0 = 2mE$ is characteristic of non relativistic relation between energy and momentum ($E = p^2/2m$). In a corresponding relativistic theory, which could be obtained by substituting the space (4.1) with a De Sitter space, one would have rather set as a unit of momentum $p_0 = mc$ by which one would have obtained for λ in eq.(4.6), instead of (4.10): $\lambda = e^2/\hbar c$, the dimensionless, universal fine structure constant(*).

(*) I remember Wolfgang Pauli, during the last years of his life in Zürich repeating to his friends and students: "the number 137 (the inverse of the fine structure constant) has to have a deep meaning, it must hide the clue for finally understanding the deep connection which certainly exists between relativity electrodynamics and quantum mechanics, it has to be obtained from pure space-time geometry". Could the relativistic version of equation (4.6) with consequent identification of λ with the fine structure constant be an indication in that direction?

5. Conclusion and outlooks

It appears then that, if we admit, as conformal covariance of Maxwell equations seems to suggest, that space-time is conformally compactified and represented by the homogeneous space G/G_1 and if we furthermore admit that also the other homogeneous space G/G_2 of the conformal group, obtained from the first through conformal inversion, is realized in nature, and if we interpret it, as simple-spinor-geometry seems to suggest, as compactified (relative) momentum space, then it appears that we could find in nature, already at the level of non relativistic dynamics at least two, at first sight uncorrelated phenomena, which could be consequences of these hypothesis and be therefore explained by them.

The first, deriving from space-time compactification, is represented by the recently discovered large-scale correlation of galaxies in the universe (to be still verified and possibly confirmed by further observations), the second, deriving from momentum-space compactification could be the H-atom. The structure of both systems may be obtained as eigensolutions of the unique Laplace-Beltrami equation:

$$\Delta (S_3) Y_{n\ell m} = - n (n+2) Y_{n\ell m} ,$$

interpreted for the first cosmological system as a classical wave equation in ordinary, compactified space, and, for the second,as quantum wave equation in compactified (relative) momentum space, which after being set in integral form, and stereographically projected to $\mathbb{R}^3$ and, subsequently Fourier transformed, may be identified as the Schrödinger equation for the H-atom.

They are correlated by conformal inversion and have in common the S0(4) isometry group of S_3, common compact submanifold of both $\overline{M}$ and $\overline{P}$ which is appropriate to represent space and momentum respectively for non relativistic phenomena.

It is well known that there are further non-relativistic classical systems in nature which present S0(4) symmetry: the Kepler two body systems. Their dynamics also may be deduced by solving classical wave equations: those of general relativity in ordinary space-time (in the non-relativistic limit). One could then hope to be able to consider them also consequences of the above hypothesis. For the moment their dynamics may be obtained algebraically from the so(4) subalgebra of the so(4,2) conformal algebra expressed in relative momentum space [24]. These classical systems and the H-atom are correlated by the conformal inversion, they could then represent a realization of the connection between classical- and quantum-systems, mediated by the conformal inversion, as suggested in paragraph 3.1.

In this frame then the common r^{-2} dependence (in the static limit) of the two main known long-range natural forces: the gravitational and the Coulomb ones, which uniquely generate the S0(4) symmetry of the simplest composite systems, could not be accidental but just another consequence of conformal compactification.

These hypothesis then, besides presenting some suggestive possibility of unifying natural systems which, at first sight, seem far apart, outline a programme of hard work and pose some challenging and difficult problems which, if solved, could perhaps suggest the answer to some at least of the since long pending questions of physics.

ACKNOWLEDGEMENTS

The author wishes to thank A. Barut, L. Dabrowski, G.F. Dell'Antonio, G. Ellis, S. Helgason, J. Niederle, R. Raczka, D.W. Sciama for helpful discussions.

REFERENCES

1a. P. Budinich and R. Raczka (1992) 'Global Properties of Conformally Flat Momentum Space and their Implications', ISAS Preprint Trieste and Foundations of Physics, in press.

1b. P. Budinich and R. Raczka (1992) 'Eigen Vibrations of the Expanding Universe', ISAS Preprint Trieste and Foundations of Physics, in press.

2. A.O. Barut, P. Budinich, J. Niederle and R. Raczka, in preparation.

3. P. Budinich and L. Dabrowski, in preparation.

4. T.J. Broadhurst, R.S. Ellis, D.C. Koo and A.S. Szalay (1990) Nature 343, 726.

5. D.C. Koo, N. Ellman, R.G. Kron, J.A. Munn, A.S. Szalay, T.J. Broadhurst and R.S. Ellis (1992) 'Deep Redshift Surveys as Probes of 100 MPC Structures', Preprint and Proceedings of VI Marcel Grossman Meeting, World Scientific, in press.

6. D.W. Sciama (1953) 'On the Origin of Inertia', Mon. Not. Roy. Ast. Soc. 113, 34.

7. E. Cunningham (1909) Proc. London Math. Soc. 8, 77.
H. Bateman (1909) Proc. London Math. Soc. 8, 223.

8. R. Penrose (1964) 'Conformal Treatment of Infinity in Relativity, Groups and Topology', ed. by C.M. de Witt, 565 (Gordon and Breach, New York).

9. R. Penrose and W. Rindler (1984) 'Spinor and Space-Time' (Cambridge Univ. Press).

10. I.E. Segal (1976) 'Mathematical Cosmology and Extragalactic Astronomy' (Academic Press, N.Y.).

11. J.F. Plebanski (1987) 'Gravitation and Geometry', 363 ed. by W. Rinder and A. Trautman, Bibliopolis, Naples.

12. S. Weinberg (1972) 'Gravitation and Cosmology', 412 (J. Wiley, New York).

13. L. Dabrowski (1988) 'Group Actions on Spinors' (Bibliopolis, Naples).

14. M. Cahen, S. Gutt and A. Trautman (1992) 'Spin Structures on Real Projective Quadrics', Preprint.

15. E. Schrödinger (1937) Pontificia Academia Scintiarum, Commentationes II, 321.
E. Schrödinger (1939) Physica VI, 272.

16. A.S. Eddington (1936) 'Relativity Theory of Protons and Electrons' (Cambridge Univ. Press).

17. J.V. Narlikar and T. Padmanabhan (1991) Ann. Rev. Astron. Astrophysics, 29, 352.
A.D. Linde (1990) 'Inflation and Quantum Cosmology' (Academic Press, N.Y.).

18. E. Cartan (1938) 'Lecons sur la Théorie des Spineurs' (Hermann, Paris).

19. P. Budinich (1992) 'Conformal Compactifications from Spinor Geometry', ISAS Preprint and Foundations of Physics, in press.
D. Hestenes (1987) 'Space-time Algebra', Gordon and Breach, New York.

20. S. Helgason (1965) 'A Duality in Integral Geometry on Symmetric Spaces', Proc. U.S.-Japan Seminar in Diff. Geometry, Kyoto, Nippon Hironsha, Tokyo.
S. Helgason (1992) 'Radon Transforms for Double Fibrations Example and Viewpoints', Proc. 75 Years of Radon Transforms, Vienna, in press.

21. J. Radon (1917) Ber. Verh. Sächs. Akad. 69, 262.

22. W. Pauli and F. Villars (1949) Rev. Modern Phys., 21, 434.

23. V. Fock (1935) Zeitschrist f. Physik, 98, 145.

24. A.O. Barut and G.L. Borzin (1974) J. Math. Phys., 15, 1000.

Interdisciplinary Laboratory for Natural and Humanistic Sciences
Strada Costiera, 11
34014 Trieste
Italy

Radon transform on halfplanes via group theory

Joachim Hilgert*

1 A suitable double fibration

Consider the halfplane $X = \{(a,b) \in \mathbf{R}^2 | a > 0\}$ as a subset of $\{(a,b,1) \in \mathbf{R}^3\}$ and the group

$$G = \left\{ (\alpha,\beta,\gamma) := \begin{pmatrix} \alpha & 0 & 0 \\ \beta & 1 & \gamma \\ 0 & 0 & 1 \end{pmatrix} \in GL(3,\mathbf{R}) | \alpha > 0 \right\}$$

which acts transitively on X via

$$(\alpha,\beta,\gamma) \odot (a,b) = (\alpha a, a\beta + b + \gamma).$$

The stabilizer of $x_0 := (1,0)$ is

$$H_X = \{(1,\beta,-\beta) \in G | \beta \in \mathbf{R}\} \cong \mathbf{R}.$$

Let Ξ be the set of halflines in X which end in $\partial X \cong \{0\} \times \mathbf{R}$. Such a line is uniquely determined by its intersection with ∂X and its slope. More precisely, for $v, w \in \mathbf{R}$ we consider the halfline $L_{v,w} = \{(t, v + tw) | t > 0\}$. Then we can identify Ξ with $\mathbf{R} \times \mathbf{R}$ via $L_{v,w} \longleftrightarrow (v,w)$. Note that the affine action of G on X induces a transitive action of G on Ξ. It is given by

$$(\alpha,\beta,\gamma) \diamond (v,w) = (v + \gamma, \frac{w + \beta}{\alpha}).$$

This time we set $\xi_0 := (0,0) \in \Xi$ and note that the stabilizer ξ_0 is

$$H_\Xi = \{(\alpha,0,0) \in G | \alpha > 0\}.$$

Now we have the desired double fibration, which will give us the Radon transform

$$\begin{array}{ccc} & G & \\ \pi_X \swarrow & & \searrow \pi_\Xi \\ G/H_X = X & & \Xi = G/H_\Xi, \end{array}$$

where $\pi_X(\alpha,\beta,\gamma) = (\alpha, \beta + \gamma)$ and $\pi_\Xi(\alpha,\beta,\gamma) = (\gamma, \frac{\beta}{\alpha})$.

---

*Supported by a DFG Heisenberg-grant.

E. A. Tanner and R. Wilson (eds.), Noncompact Lie Groups and Some of Their Applications, 141–146.

2 The Radon transform

We use the following left invariant Haar measures on G, H_X and H_Ξ:

$$d\mu_G(\alpha,\beta,\gamma) = \frac{d\alpha}{\alpha}d\beta d\gamma, \quad d\mu_{H_X}(1,\beta,0) = d\beta, \quad d\mu_{H_\Xi}(\alpha,0,0) = \frac{d\alpha}{\alpha}$$

and define the Radon transform and its dual as

$$\mathcal{R}f(gH_\Xi) := \int_{H_\Xi} f(ghH_X)d\mu_{H_\Xi}(h) \quad \forall f \in C_c(X)$$

and

$$\check{\mathcal{R}}\phi(gH_X) := \Delta_G(g)\int_{H_X} \phi(ghH_\Xi)d\mu_{H_X}(h) \quad \forall \phi \in C_c(\Xi).$$

Explicitly we find

$$\mathcal{R}f(v,w) = \int_{\mathbf{R}_+^\times} f(a, v+aw)\frac{da}{a}$$

and

$$\check{\mathcal{R}}\phi(a,b) = \frac{1}{a}\int_{\mathbf{R}} \phi(v, \frac{b-v}{a})dv = \int_{\mathbf{R}} \phi(b-av, v)dv.$$

Note that $\mathcal{R}$ and $\check{\mathcal{R}}$ are dual to each other:

$$\int_{\mathbf{R}}\int_{\mathbf{R}_+^\times} f(a,b)\check{\mathcal{R}}\phi(a,b)\frac{da}{a}db = \int_{\mathbf{R}}\int_{\mathbf{R}} \mathcal{R}f(v,b)\phi(v,b)dvdb.$$

3 Plane waves and a Fourier transform

A *horocycle* in X is an orbit of a conjugate of H_Ξ. The set of horocylces coincides with Ξ. For each $v \in \partial X$ we define a pencil $\mathcal{P}_v$ of horocyles via

$$\xi \in \mathcal{P}_v \quad \text{if and only if} \quad \xi = (v,w).$$

For $x \in X$ and $\xi \in \mathcal{P}_v$ we define the *complex distance* $d_v(x,\xi)$ to be the (unique) element $(1,\beta,0) \in A := \{(1,\beta,0) \in G\}$ such that $(1,\beta,0)\odot x \in \xi$. Let $x = (a,b) \in X$ and $v \in \mathbf{R}$ then there exists a unique $\xi_{x,v} \in \mathcal{P}_v$ such that $x \in \xi_{x,v}$. It satisfies

$$d_v(x_0, \xi_{x,v}) = (1, v + \frac{b-v}{a}, 0).$$

We define our plane waves associated to a $\lambda \in \mathbf{a}_{\mathbf{C}}^*$, where $\mathbf{a}$ is the Lie algebra of A, which can be identified with $\mathbf{R}$, and v in $\mathbf{R}$:

$$e_{\lambda,v}(a,b) := e^{\langle \lambda, v + \frac{b-v}{a}\rangle}.$$

Proposition.

1. $e_{\lambda,v}$ *is a joint eigenfunction for all* $D \in \mathbf{D}(X)$, *the invariant differential operators on* X.

2. $e_{\lambda,v}$ *is constant on the line* ξ *for each* $\xi \in \mathcal{P}_v$.

3. $e_{\lambda,v}(gx) = e_{\lambda\Delta_G(g),g^{-1}\cdot v}(x)e_{\lambda,v}(gx_0)$ *for* $g \in G, x \in X$ *and* $\lambda \in \mathbf{a}^*_{\mathbf{C}}$, *where the action of* G *on* ∂X *is given by* $(\alpha,\beta,\gamma).v = v+\gamma$.

Proof. Consider the normal subgroup $P := \{(\alpha,\beta,0) \in G\}$ and note that G is the semidirect product of P and H_X. Then [1], Theorem II.4.9, tells us that $\mathbf{D}(X)$ is, as a vector space, isomorphic to $I(\mathbf{p})$, the space of H_X-invariants in the symmetric algebra $S(\mathbf{p})$ of the Lie algebra $\mathbf{p}$ of P. The group P is itself a semidirect product of the normal subgroup A with H_Ξ. Thus we may view the vector space $\mathbf{p}$ as a direct summand of $\mathbf{a}$ with the Lie algebra of H_Ξ. A simple calculation using the coadjoint action and the identification of the symmetric algebra with the polynomials on the dual shows that the H_X-invariants in $S(\mathbf{p})$ are just the elements of $S(\mathbf{a})$. The theorem quoted above now yields that $\mathbf{D}(X)$ is generated by $\frac{\partial}{\partial b}$. This proves the first claim. The second claim is proved by a straightforward verification. The cocycle condition from the last claim can also be verified by an elementary calculation.

It can be shown that the cocycle condition from the above proposition is closely related to unitary representations of G which are induced from characters of A. These representations are not irreducible, a fact that is reflected in some indeterminacy properties if the Fourier transform associated to our plane waves. We define the Fourier transform as

$$\mathcal{F}_X f(\lambda, gA) = \int_X f(x) e_{-i\lambda\Delta_G(g)^{-1},gP}(x)\,d\nu_X(x),$$

where $gP \in \partial X$ is the g-translate of $0 \in \partial X$. Identifying $G/A \cong H_X H_\Xi$ with $\mathbf{R}^\times_+ \times \mathbf{R}$ via

$$(\alpha, -\gamma\alpha, \gamma) = (1, -\gamma, \gamma)(\alpha, 0, 0) \leftrightarrow (\alpha, \gamma),$$

this reads

$$\mathcal{F}_X f(\lambda, \alpha, \gamma) = \int_{\mathbf{R}^\times_+}\int_{\mathbf{R}} f(a,b) e^{-i\alpha\langle\lambda, \gamma + \frac{b-\gamma}{a}\rangle}\,db\frac{da}{a}.$$

Note that $\mathcal{F}_X f(\lambda,\alpha,\gamma)$ does not depend explicitly on λ, γ and α but only on $\alpha\langle\lambda,\gamma\rangle$ and $\alpha\lambda$. Therefore we write $\mathcal{F}_X f(\lambda,\gamma) := \mathcal{F}_X(\lambda,1,\gamma)$ and introduce the map $\tilde{\mathcal{F}}_X f : \mathbf{R}\times\mathbf{R}^* \to \mathbf{C}$ defined by

$$\tilde{\mathcal{F}}_X f(r,\eta) := \int_{\mathbf{R}^\times_+}\int_{\mathbf{R}} f(\frac{1}{a}, -\frac{b}{a})\frac{1}{a}e^{+i(\langle\eta,b\rangle + ra)}\,db\,da.$$

We note $\mathcal{F}_X f^\flat(\lambda,\gamma) = e^{-i\langle\lambda,\gamma\rangle}\tilde{\mathcal{F}}_X f(\langle\lambda,\gamma\rangle,\lambda)$, where $f^\flat(a,b) := (1/a)f(a,b)$, and define $\tilde{f} \in C_c(\mathbf{R}\times\mathbf{R})$

$$\tilde{f}(t,v) = \begin{cases} \frac{1}{t}f(\frac{1}{t},-\frac{v}{t}) & t>0 \\ 0 & t \leq 0. \end{cases}$$

Then

$$(\frac{1}{2\pi})^2\tilde{\mathcal{F}}_X f(r,\eta) = (\mathcal{F}_{\mathbf{R}^2}^{-1}\tilde{f})(r,\eta),$$

where $\mathcal{F}_{\mathbf{R}^2}$ denotes the Fourier transform on $\mathbf{R}^2$.

4 Inversion and Plancherel formula

We can write the Fourier transform using the Radon transform

$$\tilde{\mathcal{F}}_X f(\langle\lambda,\gamma\rangle,\lambda) = \mathcal{F}_{\mathbf{R}}(\mathcal{R}f(\gamma,\cdot))(\lambda).$$

Define an operator $\tilde{A}$ on the Schwartz functions $S(\mathbf{R})$ via

$$2\pi\mathcal{F}_{\mathbf{R}}(\tilde{A}\psi)(\tau) = |\tau|\mathcal{F}_{\mathbf{R}}(\psi)(\tau)$$

and use to define an operator A on suitable functions in to variables by

$$(A\psi)(r,s) := \tilde{A}(\psi(r,\cdot))(s).$$

Hereafter, $\int$ denotes integration over $\mathbf{R}$, and $\int^+$ denotes integration over $\mathbf{R}_+$.

Proposition. *Let $f \in C_c^\infty(X)$. Then*

$$f(a,b) = a^{-1}\int(A\mathcal{R}f)(b+r,-\frac{r}{a})dr.$$

Proof. Note first that

$$\begin{aligned} f(a,b) &= a^{-1}\tilde{f}(a^{-1},-a^{-1}b) \\ &= a^{-1}\tfrac{1}{(2\pi)^2}\int\int(\mathcal{F}_{\mathbf{R}\times\mathbf{R}}\tilde{f})(\xi,r)e^{i(\frac{\xi}{a}-\frac{rb}{a})}d\xi dr \\ &= \tfrac{1}{(2\pi)^2}\int\int(\mathcal{F}_{\mathbf{R}\times\mathbf{R}}\tilde{f})(a\xi+br,r)e^{i\xi}d\xi dr \\ &= a\tfrac{1}{(2\pi)^2}\int\int|\xi|(\mathcal{F}_{\mathbf{R}\times\mathbf{R}}\tilde{f})((a+abr)\xi,ar\xi)e^{i\xi}d\xi dr, \end{aligned}$$

where $\mathcal{F}$ denotes the Fourier transform. We express the Fourier transform in terms of the Radon transform

$$\begin{aligned} \mathcal{F}_{\mathbf{R}\times\mathbf{R}}\tilde{f}((a+abr)\xi,ar\xi) &= \int\int^+ f(\tfrac{1}{t},b+\tfrac{1}{r}+\tfrac{x}{t})e^{iar\xi x}\tfrac{dt}{t}dx \\ &= \int\mathcal{R}f(b+\tfrac{1}{r},x)e^{iar\xi x}dx \\ &= \mathcal{F}_{\mathbf{R}}(\mathcal{R}f(b+\tfrac{1}{r},\cdot))(-ar\xi) \\ &= \tfrac{1}{a|r|}\mathcal{F}_{\mathbf{R}}(\mathcal{R}f(b+\tfrac{1}{r},-\tfrac{\cdot}{ar}))(\xi). \end{aligned}$$

This implies

$$\begin{aligned} f(a,b) &= \frac{1}{(2\pi)^2}\int\int|\xi|\mathcal{F}_{\mathbf{R}}(\mathcal{R}f(b+\tfrac{1}{r},-\tfrac{\cdot}{ar}))(\xi)e^{i\xi}d\xi\tfrac{dr}{|r|} \\ &= \frac{1}{2\pi}\int\int\mathcal{F}_{\mathbf{R}}(A(\mathcal{R}f)(b+\tfrac{1}{r},-\tfrac{\cdot}{ar}))(\xi)e^{i\xi}d\xi\tfrac{dr}{|r|} \\ &= \int\mathcal{F}_{\mathbf{R}}^{-1}\left(\mathcal{F}_{\mathbf{R}}(A(\mathcal{R}f)(b+\tfrac{1}{r},-\tfrac{\cdot}{ar}))\right)(1)\tfrac{dr}{|r|} \\ &= \int A(\mathcal{R}f)(b+\tfrac{1}{r},-\tfrac{\cdot}{ar})(1)\tfrac{dr}{|r|}. \end{aligned}$$

Note that the last integral in the above calculation is justified by this very calculation. It follows from the homogeneity properties of $\tilde{A}$ that

$$(\tilde{A}\mathcal{R}f(r,\frac{\cdot}{c}))(s)=\frac{1}{|c|}A\mathcal{R}f(r,\frac{s}{c}).$$

Thus

$$\begin{aligned} f(a,b) &= a^{-1}\int A\mathcal{R}f(b+\tfrac{1}{r},-\tfrac{1}{ar})\tfrac{dr}{r^2} \\ &= a^{-1}\int A\mathcal{R}f(b+r,-\tfrac{r}{a})dr \\ &= (\check{\mathcal{R}}\circ A\circ\mathcal{R}f)(a,b). \end{aligned}$$

The square root $\tilde{\Lambda}:=\tilde{A}^{\frac{1}{2}}$ of A is given by $\sqrt{2\pi}\mathcal{F}_{\mathbf{R}}(\tilde{\Lambda}\psi)(s)=|s|^{\frac{1}{2}}\mathcal{F}_{\mathbf{R}}(\psi)(s)$ and we set $\Lambda\phi(r,s):=\tilde{\Lambda}(\phi(r,\cdot))(s)$.

Theorem.

1. *$f(a,b)=\check{\mathcal{R}}\Lambda^2\mathcal{R}f(a,b)$ for $f\in C_c^\infty(X)$.*
2. *$\Lambda\circ\mathcal{R}$ can be extended to an isometry $L^2(\mathbf{R}_+\times\mathbf{R},\frac{da}{a}db)\to L^2(\mathbf{R}\times\mathbf{R},drds)$.*
3. *$\check{\mathcal{R}}\Lambda$ is the adjoint of $\Lambda\mathcal{R}$ in the L^2-sense.*

Proof. The first part is immediate with the above proposition. The second part is shown by the following calculation

$$\begin{aligned} \int\int\ &|\Lambda\mathcal{R}f(r,s)|^2dsdr = \int\int|\tilde{\Lambda}(\mathcal{R}f(r,\cdot))(s)|^2dsdr \\ &= \frac{1}{2\pi}\int\int|\mathcal{F}_{\mathbf{R}}(\tilde{\Lambda}(\mathcal{R}f(r,\cdot)))(\xi)|^2d\xi dr \\ &= \frac{1}{(2\pi)^2}\int\int|\xi||\mathcal{F}_{\mathbf{R}}(\mathcal{R}f(r,\cdot))(\xi)|^2d\xi dr \\ &= \frac{1}{(2\pi)^2}\int\int|\xi||\mathcal{F}_{\mathbf{R}}(\mathcal{R}f(r,\cdot))(r^{-1}\xi)|^2|r|^{-2}d\xi dr \\ &= \frac{1}{(2\pi)^2}\int\int|\mathcal{F}_{\mathbf{R}\times\mathbf{R}}^{-1}\tilde{f}(\xi,r^{-1}\xi)|^2|r|^{-2}|\xi|d\xi dr \\ &= \frac{1}{(2\pi)^2}\int\int|\mathcal{F}_{\mathbf{R}\times\mathbf{R}}^{-1}\tilde{f}(r\xi,\xi)|^2|\xi|d\xi dr \\ &= \int\int|\tilde{f}(t,s)|^2dtds \\ &= \int\int^+\frac{1}{t^2}|f(t^{-1},-\tfrac{s}{t})|^2dtds \\ &= \int\int^+|f(t,s)|^2\frac{dt}{t}ds. \end{aligned}$$

For the last claim we only have to recall from Section 1 that $\mathcal{R}$ and $\check{\mathcal{R}}$ transposes of each other.

References

[1] Helgason, S., *Groups and geometric analysis*, Acad. Press, Orlando, 1984

Department of Mathematics, University of Wisconsin
Madison, Wisconsin 53706.

Analytic torsion and automorphic forms.

Birgit Speh *

Abstract

In this note we prove a vanishing theorem for the analytic torsion of a locally symmetric space.

Introduction

For a compact manifold M and an orthogonal representation ρ of the fundamental group $\Pi_1(M)$ Ray and Singer defined the analytic torsion $\tau_\rho(M) \in (0, \infty)$ [R-S]. Using harmonic analysis of a linear combination of heat kernels Moscivici and Stanton prove

Theorem .1 *Let G be a semisimple Lie group with a maximal compact subgroup K. Suppose that Γ is a cocompact discrete torsion free subgroup and suppose that G has no factor locally isomorphic to $SL(3,\mathbf{R})$, or $SO(p,q)$, pq odd. For the trivial representation of Γ*

$$\tau_1(K\backslash G/\Gamma) = 1$$

In this note we present another proof of this theorem which doesn't rely on analysing a heat kernel. Our proof is modeled after the proof of the vanishing theorems for the cohomology of the locally symmetric space $K\backslash G/\Gamma$ [B-W],[Z]. In 2.1 we relate the torsion of compact locally symmetric space $K\backslash G/\Gamma$ to the "torsion" of unitary representations in the discrete spectrum

*Supported by NSF grant DMS 9104117

E. A. Tanner and R. Wilson (eds.), Noncompact Lie Groups and Some of Their Applications, 147–156.

of G/Γ. In 2.2 we prove a formula relating the logarithm of the torsion to special values of "Zeta functions of principal series representations." Then in section 3 we show that under the assumption that under assumptions of the theorem the torsion of all irreducible unitary representations is zero and hence the logarithm of the torsion is 0. In the appendix we explain how to extend these results to the torsion twisted by orthogonal representations ρ of the covering group Γ.

I would like to thank Bent Orsted for suggesting to take a closer look at this invariant and the University of Odense for its hospitality in the summer of 1992.

I Preliminaries

In this section we first recall the definition of analytic torsion and then discuss the complex of differential forms on a compact locally symmetric space.

1.1 Let D be a nonnegative elliptic operator on a compact manifold X. We define the Dirichlet series

$$\zeta_D(s) = \sum \lambda^{-s}$$

where we sum over all nonzero Eigenvalues of D. This Dirichlet series converges for Re(s) large enough. We can analytically continue this function to a meromorphic function in the complex plane. We define

$$det\ D = exp(-\zeta_D'(0)),$$

where ζ_D' denotes the derivative of the Zeta-function ζ_D.

Such operators arise in the following way. Consider the $\mathbf{R}$-valued differential forms on X. Since X is compact we can define an inner product on forms and apply Hodge theory. Taking the adjoint $d*$ of the differential D we can define the Hodge Laplacian Δ_j on j-forms. It is nonnegative and elliptic. The j-forms in $ker\ \Delta_j$ are called harmonic. They are a set of representatives for the cohomology $H^j(X)$. We define the *analytic torsion* by

$$\tau^2 = \frac{(det\Delta_1)(det\Delta_3)^3\dots}{(det\Delta_2)^2(det\Delta_4)^4\dots}$$

Cheeger and Müller showed that the analytic torsion is equivalent to the combinatorial torsion defined by Reidemeister, DeRham and Milnor [M],[C].

1.2 Let G be a semisimple connected Lie group with maximal compact subgroup K and let Γ be a discrete torsion free cocompact subgroup of G. The symmetric space $\tilde{X} = K\backslash G$ is simply connected and thus the universal covering space of $X = \tilde{X}/\Gamma$ with covering group Γ. Since a simply connected symmetric space $\tilde{X}$ is a product of symmetric spaces which are quotients of simple Lie groups we may assume from now on that G is simple and non-compact.

The Lie algebra $\mathbf{g}$ of G has the Cartan decompositon $\mathbf{g} = \mathbf{k} \oplus \mathbf{p}$ where $\mathbf{k}$ is the Lie algebra of K and $\mathbf{p} \cong T(X)_e$.

Let $\mathbf{A}(\tilde{X})$ be the complex of differential forms on $\tilde{X}$. The discrete group Γ acts on $\mathbf{A}(\tilde{X})$ and the subcomplex $\mathbf{A}(\tilde{X})^\Gamma$ is isomorphic to the complex $\mathbf{A}(X)$ of differential forms on X [B-W].

1.3 Following [B-W] we can define for every $(\mathbf{g},\mathbf{k})$–module M of the universal algebra $U(\mathbf{g})$ of $\mathbf{g}$ the complex

$$\mathbf{C}^*(\mathbf{g}, M) = Hom(\wedge^*\mathbf{g}, M)$$

and the relative Lie algebra complex

$$\mathbf{C}^*(\mathbf{g},\mathbf{k}, M) = Hom_{\mathbf{k}}(\wedge^*(\mathbf{k}\backslash\mathbf{g}), M) = Hom_{\mathbf{k}}(\wedge^*(\mathbf{p}), M)$$

where we consider $\mathbf{p}$ as module of K.

If M is unitary we can define an inner product on $Hom_{\mathbf{k}}(\wedge^*(\mathbf{p}), M)$ which allows us to define a Laplace operator

$$\Delta_M^j : Hom_{\mathbf{k}}(\wedge^j(\mathbf{p}), M) \to Hom_{\mathbf{k}}(\wedge^j(\mathbf{p}), M).$$

Kuga's lemma implies that for $\omega \in \wedge^j$ and $v \in M$.

$$\Delta_M(\omega \otimes v) = \omega \otimes (-Cv).$$

(see page 49 in [B-W]).

By 2.5 page 212 in [B-W] there exists an isomorphism of graded complexes $\mathbf{A}(X)$ and $\mathbf{C}^*(\mathbf{g},\mathbf{k}, C^\infty(G/\Gamma))$, where we regard the C^∞-functions on G/Γ by right invariant differentiation as a module for the envelloping algebra $U(\mathbf{g})$. The spectrum of the Laplace operator corresponds to the spectrum of the Casimir C on $L^2(G/\Gamma)$.

1.4 Let dx denote the Haar measure on G and also the associated measure on G/Γ. The Hilbert space $L^2(G/\Gamma)$ of square integrable functions with respect to dx is the completion of $C^\infty(G/\Gamma)$. It is a unitary G-module whose C^∞-vectors are $C^\infty(G/\Gamma)$. By a theorem of Gelfand and Piateckii-Shapiro $L^2(G/\Gamma)$ decomposes into a discrete direct sum of irreducible G-modules with finite multiplicities [G-G-P]. We can write

$$L^2(G/\Gamma) \cong \bigoplus m(\pi,\Gamma) H_\pi,$$

where we sum over all irreducible representations $\pi : G \to U(H_\pi)$ in the unitary dual $\hat{G}_u$ and $m(\pi,\Gamma) = dim Hom_G(\pi, L^2(G/\Gamma)$.

Lemma I.1 *Suppose that G/Γ is compact and let C be the Casimir in $U(\mathbf{g})$. For $\lambda \in \mathbf{R}$*

$$\dim\ ker(\Delta_j - \lambda) \cong \sum_{\substack{\pi\in\hat{G}\\ \pi(C)=\lambda}} m(\pi,\Gamma) \dim\ Hom_{\mathbf{k}}(\wedge^j(\mathbf{p}), \pi^\infty).$$

Proof: The operator Δ_j is elliptic and hence we can write $\mathbf{A}^j(G)$ in a unique way as direct sum of its Eigenspaces. Since by 1.3 the action of the Laplace operator corresponds to the action of the Casimir on $C^\infty(G/\Gamma)$ the decomposition claimed in the lemma is an Eigenspace decomposition. Q.E.D.

Thus we proved

Proposition I.2 *Suppose that $K\backslash G/\Gamma$ is a compact locally symmetric space and let $m(\lambda, j, \Gamma) = \sum_{\substack{\pi\in\hat{G}_u\\ \pi(C)=\lambda}} m(\pi,\Gamma) \dim\ Hom_{\mathbf{k}}(\wedge^j(\mathbf{p}), \Pi^\infty)$. Then*

$$\zeta_{\Delta_j}(s) = \sum_\lambda m(\lambda, j, \Gamma)\lambda^{-s}.$$

We conclude

Proposition I.3 *Suppose that X is a compact locally symmetric space. Then*

$$\log(\tau^2(X)) = \lim_{s\to 0} \sum_j (-1)^j j \sum_\lambda m(\lambda, j, \Gamma)\lambda^{-s}.$$

II Automorphic torsion

In this section we define the torsion of families of automorphic forms and relate it to the analytic torsion of the locally symmetric spaces. We use this to give a prove of the main theorem.

2.1 Let $P = MAN$ be a cuspidal parabolic subgroup, π a tempered representation of M and μ a character of A. Put $I(P, \pi, \mu) = ind_P^G \pi \otimes \mu \otimes 1$. If there is no confusion possible we suppress the index P.

For an admissible representation U we define

$$tor(U) = \sum_i (-1)^i i \dim \; Hom_K(\wedge^i \mathbf{p}, U).$$

Remark: The torsion $tor(I(\pi, \mu))$ is independent of the character μ since it depends only on the K–type structure of $I(\pi, \mu)$. We will denote $tor(I(\pi, \mu))$ by $tor(I(\pi))$ if no confusion is possible.

Lemma II.1 *Let U be an admissible representation of G and suppose that in the Grothendieck group $U = \sum m(U, \pi \otimes \mu) I(P, \pi, \mu)$. Then*

$$tor(U) = \sum m(U, \pi \otimes \mu) tor(I(P, \pi, \mu)).$$

Proof:

$$\begin{aligned}
\dim \; Hom_K(\wedge^j \mathbf{p}, U) &= \dim \; Hom_K(\wedge^j \mathbf{p}, U_{|K}) \\
&= \dim \; Hom_K(\wedge^j \mathbf{p}, \sum m(U, \pi \otimes \mu) I(P, \pi, \mu)_{|K}) \\
&= \sum m(U, \pi \otimes \mu) \dim \; Hom_K(\wedge^j \mathbf{p}, I(P, \pi, \mu)_{|K}) \\
&= \sum m(U, \pi \otimes \mu) \dim \; Hom_K(\wedge^j \mathbf{p}, I(P, \pi, \mu)).
\end{aligned}$$

Q.E.D.

Proposition II.2 *Suppose that X is a compact locally symmetric space. Then*

$$\sum_j (-1)^j j \cdot \zeta_{\delta_j}(s) = \sum_{U \; in \hat{G}'_u} tor(U) m(U, \Gamma) U(C)^{-s}$$

where $\hat{G}'_u$ denotes all irreducible representations with nontrivial Eigenvalue of the Casimir C.

Proof: This follows from I.2 since the series are absolutely convergent for $Re(s)$ large enough. Q.E.D.

2.2 For a fixed pair (P,π) we define the *ζ-function of the principal series* $I(P,\pi,\mu)$, $\mu \in \hat{A}$ by

$$\zeta_{P,\pi}(s) = tor(I(\pi))\sum_U m(U,\pi\otimes\mu)m(U,\Gamma)\cdot U(C)^{-s}$$

where we sum over all irreducible unitary representations U in $\hat{G}'_u$. This series converges absolutely for $\mathrm{Re}(s) >> 0$.

Theorem II.3 *Suppose that X is a compact locally symmetric space. Then*

$$\tau^2(X) = exp(\lim_{s\to 0}\sum_{P,\pi}\zeta'_{P,\pi}(s)).$$

Proof: By [V] there are only finite many pairs P,π so that

$$\dim\ \mathrm{Hom}_K(\Lambda^*\mathbf{p}, I(P,\pi,\mu)) \neq 0.$$

Thus we have a finite sum at the right hand side.
In the Grothendieck group every unitary representation U of G can be represented uniquely as a sum of principal series representations with coefficients $m(U,\pi\otimes\mu)$ [V]. Hence by II.2 and II.1

$$\sum_j(-1)^j j\cdot\zeta_{\delta_j}(s) = \sum_{P,\pi}\frac{d}{ds}\zeta_{P,\pi}(s)$$

for $\mathrm{Re}(s) >> 0$. Q.E.D.

2.3 In the next section we will prove

Proposition II.4 *If G does not have a cuspidal parabolic subgroup of split rank 1 and* $\dim K\backslash G$ *is odd , then every principal series has trivial torsion.*

Proof of the main theorem: The simple Liegroups which have cuspidal parabolic subgroups of real rank 1 and for which $\dim K\backslash G$ is odd are locally isomorphic to SL(3,$\mathbf{R}$) or S0(p,q) with $p\cdot q$ odd. Thus II.3 and II.4 imply the main theorem for odd dimensional locally symmetric spaces. On the other hand it is well known that even dimensional spaces have trivial analytic torsion [R-S]. Q.E.D.

III The torsion of principal series representations

This section is devoted to the proof of II.4.

3.1 The torsion of even dimensional spaces is trivial. Thus we assume in this section that G is locally isomorphic to SU*(2n), SL(n,**R**) $n \geq 5$ or $\mathrm{res}_{\mathbf{R}} G_{\mathbf{C}}$ for a complex simple group $G_{\mathbf{C}}$.

Let $H = TA$ be a Cartan subgroup of G where T is a compact torus and A is isomorphic to $\mathbf{R}^n$. We write a one dimensional representation of H is as a pair (σ, μ) where σ is the logarithm of a character of T and μ is a character of A. Let M be a the centralizer of A in G and $P = MAN$ a cuspidal parabolic subgroup. Let π be a tempered representation of M. As a representation of K the principal series representation $I(P, \pi, \mu)$ is isomorphic to $ind^K_{M\cap K}\pi_{|M\cap K}$. We consider $\mathbf{p}$ as a M module and deduce by Frobenius reciprocity

$$Hom_K(\wedge^i \mathbf{p}, I(P, \pi, \mu)_{|K}) \cong Hom_{M\cap K}(\wedge^i \mathbf{p}_{|M\cap K}, \pi_{|M\cap K}) \qquad (3.1)$$

Thus

Lemma III.1 *Let π be a irreducible representation of M. Then*

$$\dim\ Hom_K(\wedge^i \mathbf{p}, I(P, \pi, \mu)) = \sum_\sigma \dim Hom_{M\cap K}(\sigma, \wedge^i \mathbf{p}) \cdot \dim Hom_{M\cap K}(\sigma, \pi_{|M\cap K})$$

where we sum over all irreducible representations σ of $K \cap M$.

Since $\wedge^* \mathbf{p}$ is a direct sum of finitely many irreducible representations of $M \cap K$ we get

Lemma III.2 *The torsion of $I(P, \pi, \mu))$ is equal to*

$$\sum_\sigma (\sum_i (-1)^i i \cdot \dim Hom_M(\sigma, \wedge^i \mathbf{p})) \cdot \dim Hom_{M\cap K}(\sigma, \pi_{|m\cap K}).$$

3.2 Next we analyze the representations of $M \cap K$ on $\wedge^* \mathbf{p}$. For a representation ρ of $M \cap K$ on a vector space V we denote the isotypic component of type σ by V^σ.

Let $\mathbf{p}^\flat$ be the $M \cap K$ invariant complement to $\mathbf{a}$ in $\mathbf{p}$. All representations of $M \cap K$ in $\wedge^* \mathbf{p}$ are tensor products of the trivial representations on $\wedge^* \mathbf{a}$ and those in $\wedge^* \mathbf{p}^\flat$. Thus

Lemma III.3 *Let σ be an irreducible representation of $M \cap K$. There exist $\omega_j \in \wedge^* \mathbf{p}^\flat$, $j \in J$ so that as a graded vector space*

$$(\wedge^* \mathbf{p})^\sigma \cong \bigoplus_{j \in J} U(\mathbf{m} \cap \mathbf{k})\omega_j \otimes \wedge^* \mathbf{a}.$$

Corollary III.4 *Under the above assumptions*

$$\dim Hom_{M\cap K}(\sigma, \wedge^* \mathbf{p}) = |J| \dim(\wedge^* \mathbf{a}).$$

3.3 Recall the following 2 formulas for binomial coeffients

$$\sum_{i=0}^{l} (-1)^i \binom{l}{i} = 0 \quad for \quad l \leq 1 \tag{3.2}$$

$$\sum_{i=1}^{l} (-1)^i i \cdot \binom{l}{i} = 0 \quad for \quad l \leq 2 \tag{3.3}$$

Lemma III.5 *Suppose that dim $\mathbf{a} \geq 2$. Then*

$$\sum_{i=0}^{\dim \mathbf{p}} (-1)^i i \cdot \dim Hom_{M\cap K}(\sigma, \wedge^i \mathbf{p}) = 0.$$

Proof: By III.3 it suffices to prove the lemma for graded subspaces of the form $\omega \otimes \wedge^* \mathbf{a}$ where $\omega \subset \wedge^r \mathbf{p}^\flat$ is an irreducible representation of $M \cap K$. We have

$$\begin{aligned}
&\sum_{i=0}^{\dim \mathbf{a}} (-1)^{i+r} (i+r) dim\ Hom_{M\cap K}(\sigma, \omega \wedge (\wedge^i \mathbf{a})) \\
&= \sum_{i=0}^{n} (-1)^{i+r} (i+r) \binom{n}{i} \\
&= (-1)^r r \sum_{i=0}^{n} (-1)^i \binom{n}{i} + (-1)^r \sum_{i=0}^{n} (-1)^i i \binom{n}{i} = 0
\end{aligned}$$

by 3.2 and 3.3. Q.E.D.

Proof of II.4: This follows from III.2 and III.5.

IV Appendix

We sketch how our results can be extended to the torsion twisted with an orthogonal representation ρ of Γ.

Following [Fried] we can define a complex $\mathbf{A}(X,\rho)$ of twisted differential forms. We obtain an isomorphism between this complex and the $(\mathbf{g},\mathbf{k})$-cohomology with coeffients in

$$C^{\infty}(X,\rho) = (ind_{\Gamma}^{G}\rho)^{\infty})$$

In this case the result of Gelfand and Piatesky-Shapiro also applies and and we have

$$ind_{\Gamma}^{G}\rho \cong \bigoplus m(\pi,\Gamma,\rho)H_{\pi}.$$

Now the reduction to the the torsion of principal series representations is exactly as in the case of trivial representation of Γ.

References

[B-W] A.Borel and N.Wallach, *Continous cohomology, discrete Subgroups and representations of reductive groups*, Princeton University Press, 1980

[C] J.Cheeger, Analytic torsion and the heat equation. Ann.Math.109,259-322 (1979)

[F] D.Fried, Analytic torsion and closed geodesics on hyperbolic manifolds, Invent. Math. 84, 523-540, (1986)

[G-G-P] I.M.Gelfand,M.I.Graev and I.Pyateckii-Shapiro, *Representation theory and automorphic functions,* W.B.Saunders Co., Philidephia, 1969

[Mi] J.Millson, Closed geodesics and the η-invariant, Ann.Math.108. 1-39 (1978)

[M] W.Müller, Analytic torsion and and R-torsion of Riemanian manifolds, Adv.Math.288, 233-305 (1978)

[M-S] H.Moscovici and R.Stanton, Eta invariants of Dirac operators on locally symmetric manifolds, Invent.Math 95, 629-666,(1989)

[R-S] D.Ray and I.Singer: R–torsion and the Laplacian on Riemanian manifolds, Adv.Math.7 ,145-210 (1971)

[V] D.Vogan, *Representations of real reductive Lie groups*, Birkhauser 1981

[W] G.Warner, *Harmonic Analysis on Semisimple Lie Groups I*, Grundlehren der mathematischen Wissenschaften 188, Springer Verlag 1972

[Z] G.Zuckerman, Continuous Cohomology and unitary representations of real reductive Lie groups, Ann.Math.107, 495–516, (1978)

Birgit Speh
Department of Mathematics
Cornell University
Ithaca, N.Y. 14853

DIFFUSION ON COMPACT ULTRAMETRIC SPACES

A. Figà-Talamanca

ABSTRACT. We define a natural continuous diffusion process on an infinite, compact, metrically homogeneous, ultrametric space, and we compute the associated kernel. Spherical functions on the group of isometries are used as the main tool for the computation of the kernel.

1. Ultrametric Spaces and Trees.

An *ultrametric space* is a metric space in which the distance satisfies the *ultrametric inequality:*

$$d(x,y) \le \max[d(x,z),d(z,y)].$$

In other words a metric space is *ultrametric*, if the relation $d(x,y)\le r$ is an *equivalence relation* for every non negative real number $r \in \mathbb{R}$. This means that diameter and radius of a closed ball are one and the same number, and that every element belonging to the ball is also a center of the same ball.

Let X be an ultrametric space. We make the assumption that X is compact and *metrically homogeneous*. This means that the group of isometries of X acts *transitively*. In other words, if we denote by G the (compact) group of isometries of X, then given x, y $\in$ X, there exists g $\in$ G, such that gx=y.

It is not difficult to see that the assumptions imply that X is a complete metric space, that the distance d(x,y) attains a countable set of values $r_0 > r_1 > \dots r_n > \dots > 0$, (which is finite if X is finite, and otherwise has 0 as the only accumulation point) and finally that each closed ball $B(x,r_{j-1}) = \{y\colon d(x,y)\le r_{j-1}\}$ of radius r_{j-1} contains the same number of balls of radius r_j. We denote this number by ρ_j.

The group of isometries induces an invariant measure on X, which we denote by m. If the measure and the distance are normalized so that $m(X) = 1 = r_0$, m can be defined through the values assigned to the

E. A. Tanner and R. Wilson (eds.), Noncompact Lie Groups and Some of Their Applications, 157–167.

closed balls $B(x,r_j)$, as follows: $m(B(x,r_0)) = r_0 = 1$, and $m((B(x,r_j))\rho_j = m(B(x,r_{j-1}))$.

We associate to X a tree $\mathcal{T}$ in the following fashion. The *vertices* of the tree are the balls $B(x,r_j)$ and the edges are the two elements sets $\{B(x,r_j), B(x,r_{j+1})\}$. The tree has a root, namely the vertex $B(x,r_0) = X$. If X is infinite every other vertex of the tree $B(x,r_j)$ has $\rho_{j+1}+1$ adjacent vertices, namely $B(x,r_{j-1})$ and the ρ_{j+1} closed balls of radius r_{j+1} contained in $B(x,r_j)$. If X is finite and $r_n=0$, we also have an exception for the vertices $B(x,r_n)$ which have just one adjacent point namely $B(x,r_{n-1})$.

Observe that X can be identified with the set of *ends* of the tree $\mathcal{T}$. Only when X is finite the *ends* are actual vertices of the tree, that is balls of radius zero. If X is infinite the *ends* of $\mathcal{T}$ are all possible infinite chains starting from the root $B(x,r_0)$. In other words all possible maximal infinite nested sequences of balls, of radius (diameter) r_j. Each such sequence identifies one and only one point of X: the point belonging to the intersecion of all the balls of the sequence. For a more complete treatment of trees and their boundaries the reader may consult [1] or [5].

Observe that each isometry of X defines a transformation of the vertices of $\mathcal{T}$ (that is the balls of X) which maps edges into edges. In other words an isometry of X defines an automorphism of $\mathcal{T}$. Viceversa every automorphism of $\mathcal{T}$ induces an automorphism of its boundary which is readily seen to be an isometry of X.

2. Diffusion on a Finite Ultrametric Space.

Let X be a finite ultrametric space so that $r_0=1$, $r_n=0$, and $|X| = \rho_1\rho_2\cdots\rho_n$. The associated tree $\mathcal{T}$ is also finite in this case and X may be identified with the set of vertices of $\mathcal{T}$ which have only one neighbor, that is the balls of radius zero. A tree, or rather the set of vertices of a tree, is naturally a metric space, so that X, as a subset of the set of vertices of $\mathcal{T}$, inherits the metric of tree. It is easily seen that this metric is equivalent to the original metric and indeed defines the same set of closed balls, and has the same group of isometries. We shall work however with the distance defined above, keeping in mind however that a ball or radius r_j, as a subset of the set of ends of the tree has radius (or diameter) $2(n-j)$.The measure invariant under isometries is now the normalized counting measure: $m(\{x\}) = 1/|X|$.

We shall define a Markov chain on X which depends only on the distance between two points. The chain is defined considering the a *random walk* on the vertices of the tree $\mathcal{T}$ of which X is the set of *ends*.

Recall that the vertices of $\mathcal{T}$ are balls $B(x,r_j)$. We must define therefore the probability of going from the ball $B(x,r_j)$ to another ball. Let $0<p<1$. First of all we stipulate that one can go in one step only to an adjacent ball, that is a ball of radius r_{j+1}, contained in $B(x,r_j)$ (unless $j=n$) or to a ball of radius r_{j-1} containing $B(x,r_{j-1})$ (unless $j=0$). To define the walk we now stipulate that if j is neither 0 nor n, the probability of going from $B(x,r_j)$ to a bigger ball is p, and the probability of going to one of the ρ_{j+1} balls of radius r_{j+1} contained in $B(x,r_j)$ is $(1-p)/\rho_{j+1}$. To complete the definition let $1/\rho_1$ to be the probability of going from $B(x,r_0)$ to any of the ρ_1 balls of radius r_1, and let one to be the probability of going from a ball $B(x,r_n)$ of radius zero to the adjacent ball $B(x,r_{n-1})$.

We now define a Markov process $P(x,y)$ on the space X, or what is the same on the ends of the tree, which are the balls of radius $r_n=0$.

We let $P(x,y)$ to be the probability, that y is the first element of X reached, starting at x, by the random walk defined above. Observe that $\mathcal{T}$ is a finite tree therefore if we start with the random walk at a point x, sooner or later the random walk reaches another point of X. Thus $P(x,y)$ is well defined and we are now going to compute it.

Let $x=\xi_0, \xi_1, \ldots \xi_n$ be a sequence of vertices of $\mathcal{T}$ starting from the end $x_0 \in X$ and reaching the root of the tree ξ_n, so that ξ_j and ξ_{j+1} are adjacent. Let α_j denote the probability (with reference to the random walk defined above, starting at ξ_0) of *ever reaching* ξ_{j+1}, before touching again the set X, under the condition *that* ξ_j *be reached at least once.* Note that if ξ'_j is any other point adjacent to ξ_{j+1}, and *in the direction of the ends*, then α_j is also the probability of ever reaching ξ_{j+1}, under the condition that ξ'_j be reached at least once. Observe that by hypothesis, since x_0 is the starting point of the random walk, and ξ_1 is the only point adjacent to x_0, we have $\alpha_0=1$. Suppose now that we are at ξ_1. Of the points adjacent to ξ_1 one is ξ_2 and the other ρ_n points are in X. It follows that the only way to touch ξ_2 starting from ξ_1, and before touching again X, is to get there at the first step, and this happens with probability p. Let now $j>1$. Suppose we have reached ξ_j. In order to reach ξ_{j+1}, we must either reach it at the first step after reaching ξ_j, and this happens with probability p, or else reach it after having gone at the first step in the direction of the ends, having come back to ξ_j, and finally going from ξ_j to ξ_{j+1}. This implies that $\alpha_j = p + \alpha_{j-1}\alpha_j(1-p)$.

This defines the sequence α_j uniquely. If we let $q = (1-p)/p$, then one easily verifies that, for $p\neq 1/2$, $\alpha_j = (q^j-1)/(q^{j+1}-1)$, for $j = 1, \ldots n$, while for $p=1/2$ (and $q=1$), $\alpha_j = j/(j+1)$.

We compute now $P(x_0,x_0)$. Observe that if the point ξ_j is reached and the point ξ_{j+1} is not reached, then all elements of X which can be reached from ξ_j without passing by ξ_{j+1} have equal probability to be touched. This probability is nothing but the reciprocal of the number of points contained in a ball of radius r_{n-j}. Let m_{n-j} be this number, so that $m_0 = |X| = \rho_1 \ldots \rho_n$, $m_1 = \rho_2 \ldots \rho_n$, $\ldots$ $m_{n-1} = \rho_n$, and $m_n = 0$. Then

$$P(x_0,x_0) = m_{n-1}^{-1}(1-\alpha_1) + m_{n-2}^{-1}\alpha_1(1-\alpha_2) + \ldots + m_0^{-1}\alpha_1\alpha_2 \ldots \alpha_{n-1}.$$

If $x \in X$ and $d(x,x_0) = r_{n-1}$, then $P(x_0,x) = P(x_0,x_0)$. Suppose now x is a point at distance r_j, with $j<n-1$, from x_0. For x to be reached one must necessarily reach che point ξ_j. Therefore,

$$P(x_0,x) = m_{n-j-1}^{-1}\alpha_1\alpha_2 \ldots \alpha_j(1-\alpha_{j+1}) + \ldots + m_0^{-1}\alpha_1\alpha_2 \ldots \alpha_{n-1}.$$

In other words if $r_{n-1} < d(x,y) < 1$,

$$P(x,y) = m_{n-j-1}^{-1}\alpha_1\alpha_2 \ldots \alpha_j(1-\alpha_{j+1}) + \ldots + m_0^{-1}\alpha_1\alpha_2 \ldots \alpha_{n-1},$$

and if $d(x,y) = r_n = 1$,

$$P(x,y) = m_0^{-1}\alpha_1\alpha_2 \ldots \alpha_{n-1},$$

where $\alpha_j = (q^j-1)/(q^{j+1}-1)$, for $p\neq 2$, and $\alpha_j = j/(j+1)$ for $p=1/2$.

We shall now compute the eigenvalues of the kernel $P(x,y)$. We shall use the fact that P is invariant under isometries, that is $P(x,y) = P(gx,gy)$ for every isometry $g\in G$. In addition G acts doubly transitively on X, that is given two pairs (x,y) and (x',y') such that $d(x,y) = d(x',y')$ there exists $g\in G$ such that $gx=x'$ and $gy = y'$. This means that we can use the theory of Gelfand pairs and spherical functions to simultaneously diagonalize all G-invariant kernels on X. We refer to [2, Chapter 3F] and to [10] for an elementary concise accont of the theory of Gelfand pairs as applied to finite Markov chains. The general theory is due to I.M. Gelfand [6] and R. Godement [7]. Other useful references are [3], [8], [5] and [4].

We review here briefly the terminology with reference to our specific case. Let $x_0\in X$ be a fixed point of X. Let K be the subgroup of all isometries which do not move x_0, in other words $K = \{k \in G: kx_0 = x_0\}$. The space X may be identified with the quotient space G/K, that is the space of cosets gK, in such a way that the action of G on X is covariant with the left action of G on G/K. In particular functions on X may be identified with functions on G which are costant on the cosets of K, that is K - right -invariant functions: $f(gk) = f(g)$ for every $k \in K$. These functions form a convolution algebra, that is a subalgebra of the group algebra. We may also consider the K-bi-invariant function on G, satisfying $f(kgk') = f(g)$ that is functions which are constant on the double cosets KgK. These functions too form a convolution algebra and the double transitive action of G on X implies that the convolution algebra of K-bi-invariant functions is commutative. This, by definition, means that the pair (G,K) is a Gelfand pair. One can then find a set of simultaneous eigenfunctions for the convolution algebra of K-bi-Invariant functions. If these eigenfunctions are chosen to bi K- bi - invariant and normalized so that their value at the identity is one, they are the so-called *spherical functions.*

A G-invariant kernel such as $P(x,y)$ is associated to a K-bi-invariant probability measure μ on G. Let $\mu(x) = P(x_0,x)$. Then, for $k \in K$, $\mu(kx) = P(x_0,kx) = P(k^{-1}x_0,x) = P(x_0,x) = \mu(x)$. Therefore μ is constant on the orbits of K and may be identified with a bi-invariant probability measure on G. If $x=gx_0$ and $y=hy_0$, then $P(x,y) = P(gx_0,hx_0) = P(x_0,g^{-1}hx_0) = \mu(g^{-1}h)$. Since μ is bi-invariant it is diagonalized by the spherical functions. The same is true of the kernel P once the spherical functions are identified as functions on X. The spherical functions in this particular case were first computed by G.Letac [10]. See also [FT] for a different approach. As functions on X they can be defined as follows. The function $\varphi_0(x) \equiv 1$ is a spherical function and, for $j = 1,\dots n$,

$$\varphi_j(x) = \begin{cases} 1 \text{ if } d(x,x_0) < r_j \\ (1-\rho_j)^{-1} \text{ if } d(x,x_0) = r_j \\ 0 \text{ if } d(x,x_0) > r_j \end{cases}$$

The functions φ_0, φ_1, $\dots\varphi_n$ form an orthogonal system for the bi-invariant functions.

Observe that

$$\|\varphi_j\|^2 = (\varphi_j,\varphi_j) = |G|^{-1} \sum_{g\in G} \varphi_j(g)^2 = |X|^{-1} \sum_{x\in X} \varphi_j(x)^2 =$$

$$|X|^{-1}\Big[\rho_n\cdots\rho_{j-1} + (\rho_j-1)^{-2}(\rho_n\cdots\rho_j - \rho_n\cdots\rho_{j-1})\Big] =$$

$$|X|^{-1}\rho_n\cdots\rho_{j-1}(1+ (\rho_j-1)^{-1}) = (\rho_1\cdots\rho_{j-1})^{-1}(1-\rho_j)^{-1}.$$

The G-translates $\varphi_j(gx)$ of a spherical functions span an irreducible G-invariant subspace of the space of all functions on X. This subspace has dimension $d_j=(\rho_1 \dots \rho_{j-1})(1-\rho_j)$, which is exactly the inverse of $\|\varphi_j\|^2$.

We shall now compute the eigenvalues of P(x,y), which are the numbers $\lambda_j = \langle\mu,\varphi_j\rangle$. Observe first of all that $\lambda_0 = \langle\mu,\varphi_0\rangle = 1$, because μ is a probability measure. Next let x_j be a point at distance r_{n-1} from x_0 (that is distance 2j on the tree $\mathcal{T}$), and recall that $P(x_0,x_1)= P(x_0 . x_0)$. This means that $\langle\mu,\varphi_n\rangle = 0$.

If $1 \le j < n$, then

$$\langle\mu,\varphi_j\rangle = m_{n-1}P(x_0,x_1)+(m_{n-2}-m_{n-1})P(x_0,x_2)+\dots$$

$$+(m_j-m_{j+1})\ P(x_0,x_{n-j}) + (1-\rho_j)^{-1}(m_{j-1}-m_j)P(x_0,x_{n-j+1}) =$$

$$m_{n-1}(P(x_0,x_1)-P(x_0,x_2))+m_{n-2}(P(x_0,x_2)-P(x_0,x_3))+ \dots$$
$$m_{j+1}(P(x_0,x_{n-j-1})-P(x_0,x_{n-j}) + m_jP(x_0,x_{n-j})+$$

$$(1-\rho_j)^{-1}(m_{j-1}-m_j)P(x_0,x_{n-j+1}) =$$

$$\sum_{h=1}^{n-j} m_{n-h}\ (P(x_0,x_h)-P(x_0,x_{h+1})).$$

The last equality follows from the fact that $m_{j-1}= m_j\rho_j$. But

$$P(x_0,x_h)-P(x_0,x_{h+1}) = m_{n-h}^{-1}\ \alpha_1\ \dots\ \alpha_{h-1}(1-\alpha_h).$$

Therefore, for $j = 1, \dots n$,

$$\lambda_j= \langle\mu,\varphi_j\rangle= (1-\alpha_1)+\alpha_1(1-\alpha_2)+\ \alpha_1\dots\alpha_{n-j-2}(1-\alpha_{n-j})=$$

$$1-\alpha_1\dots\alpha_{n-j} = \begin{cases} 1- \dfrac{q-1}{q^{n-j+1}-1}, & \text{if } p\neq 2 \\ 1 - \dfrac{1}{n-j+1}, & \text{if } p=2 \end{cases}$$

Observe that the the eigenvalue λ_n is always zero. This is due to the fact that $P(x_0,x) = P(x_0,x_0)$ if $d(x,x_0) = r_{n-1}$.

We can now compute μ explicitly as a linear combination of the spherical functions. Keeping in mind that $d_j=(\rho_1 \dots \rho_{j-1})(1-\rho_j) = \|\varphi_j\|^{-2}$, one has:

$$\mu = |X|^{-1}(1 + \sum_{j=1}^{n} (\rho_1 \cdots \rho_{j-1})(1-\rho_j)\ \lambda_j\ \varphi_j. =$$

$$(\rho_1 \cdots \rho_n)^{-1}(1 + \sum_{j=1}^{n} (\rho_1 \cdots \rho_{j-1})(1-\rho_j)\ \lambda_j\ \varphi_j;$$

and for the convolution powers of μ,

$$\mu^k = (\rho_1 \cdots \rho_n)^{-1}(1 + \sum_{j=1}^{n} (\rho_1 \cdots \rho_{j-1})(1-\rho_j)\ \lambda_j^k\ \varphi_j$$

3. Diffusion on Compact Infinite Ultrametric Spaces.

From now on X is a compact *infinite* ultrametric space. This means that the set of values attained by the distance is an infinite decreasing sequence. We assume that $r_0 = 1$.

For every positive integer n, we consider the quotient space X_n obtained by collapsing to a point every ball $B(x,r_n)$ of radius r_n. Then X is a finite ultrametric space. If x_0 is a given point of X, the ball $B(x_0,r_n)$ identifies a point of X_n. We may therefore consider for each n the Markov process $P_n(x,y) = P(x,y)$ as defined in the previous section. To this process we associate a bi-invariant measure μ_n on the group G_n of isometries of X_n. But G_n is itself a quotient group of the compact group G of all isometries of X. Indeed $G_n = G/H_n$, where H_n is the subgroup of G which maps onto itself every ball of radius r_n. Therefore μ_n may be thought of as a measure defined on G. Thus the process $P_n(x,y) = P_n(gx_0,hx_0) = \mu_n(g^{-1}h)$ can be thought of as a process on X.

The theory of Gelfand pairs and spherical functions applies also in this context (cfr. [10], [11], [4]). In particular the convolution algebra $L^1(K\backslash G/K)$ is commutative and corresponds to the integrable functions on X which are constant on the orbits of K. In addition every spherical function defined with reference to X_n corresponds to a spherical function on X. Viceversa every spherical function on X has the property of being costant on the balls $B(x,r_n)$ for n sufficiently large. Indeed spherical functions on X are defined by the same conditions as spherical functions on X_n:

$$\varphi_j(x) = \begin{cases} 1 \text{ if } d(x,x_0) < r_j \\ (1-\rho_j)^{-1} \text{ if } d(x,x_0) = r_j, \\ 0 \text{ if } d(x,x_0) > r_j \end{cases}$$

the only difference being that the index j now runs through the positive integers.

This means that the process P_n and the measure μ_n have the eigenvalues:

$$\lambda_{nj} = \begin{cases} 1- \dfrac{q-1}{q^{n-j+1}-1}, \text{if } p\neq 2 \\ \\ 1 - \dfrac{1}{n-j+1}, \text{ if } p=2 \end{cases}$$

for $0<j<n$, and $\lambda_{nj} = 0$ for $j\geq n$, besides, of course the eigenvalue $\lambda_{n0} = 1$.

We shall now define a continuous process on X starting with the processes P_n. We first look at what happens as $n \rightarrow \infty$.

Suppose first that $p > 1/2$, then for $j\neq 0$, $\lim_n \lambda_{nj} = q < 1$. It follows that in this case the process P_n converges to a process $P(x,y)$ which may be thus defined: $P(x,x) = q$, and $P(x,y) = 1-q$, if $x\neq y$. This is clearly not an interesting case.

If $p\leq 1/2$ then $\lim_n \lambda_{nj} = 1$, for all j. This means that the process $P_n(x,y)$ approaches the identity. There is no diffusion. The same is true of any power $P^k_n(x,y)$, for fixed k, because $\lim_n \lambda^k_{nj} = 1$.

In order to define a continuous process on X we shall *scale* the steps taken with the process $P_n(x,y)$ so that an increasing number of steps can be taken in finite time. In other words, for each $t>0$, we shall define an increasing sequence $k(n)$ of positive integers so that the powers $P^{k(n)}_n(x,y)$ converge to a non trivial G-invariant kernel on X. This means that the convolution powers $\mu^{k(n)}_n$, of the bi-invariant measure μ_n associated to $P_n(x,y)$ converge to a bi-invariant measure μ_t. (Recall that the sequence $k(n)$ depends on t). The probability measure μ_t of course may be thought of as a bi-invariant measure on G, but also as a K-invariant measure on X, that is a measure defined on the algebra of sets which are unions of orbits of K.

For $t>0$ define $k(n)$ to be the first natural number such that

$$\lambda^{k(n)}_{n1} < e^{-t}.$$

This means that, for $p < 1/2$, $k(n)$ is the first integer such that

$$\left[1 - (q-1)(q^n-1)^{-1}\right]^{k(n)} < e^{-t}.$$

While if p = 1/2, k(n) is the first integer such that

$$\left[1 - \frac{1}{n}\right]^{k(n)} < e^{-t}.$$

The sequence $\mu_n^{k(n)}$ of K- bi-invariant probability measures on the group of isometries converges to a nontrivial K-bi-invariant probability measure μ_t. This is a nontrivial measure because all but the largest of its eigenvalues are less than or equal to e^{-t}, which is less than one.

Since the nontrivial orbits of K on G/K are of positive Haar measure, the measure μ_t is absolutely continuous with respect to the Haar measure, in other words μ_t defines a bi-invariant function $u_t(x)$ such that $u_t(x)dm = d\mu_t$.

The definition of k(n) implies that $u_{t+s}=u_t*u_s$. Finally for $t \to \infty$, $u_t(x)$ converges to the uniform distribution on X, and for $t \to 0$, u_t converges to the point measure at x_0. The function u_t depends only on the distance of x from x_0, therefore it defines a G-invariant kernel. We have thus defined the diffusion process on X.

We shall compute now explicitly the eigenvalues $\langle u_t, \varphi_j\rangle$ and the function u_t itself. We shall consider first the case $p < 1/2$. The case $p = 1/2$, a critical case, will be considered later. Let t=1, then the second largest eigenvalue of u_t is e^{-1}, and the sequence defined by (1) is simply $k(n) = (q^n-1)(q-1)^{-1}$. Therefore, always for t=1,

$$\langle u_t,\varphi_j\rangle = \lim_n\left[1-(q-1)(q^{n-j+1}-1)^{-1}\right]^{(q^n-1)/(q-1)} =$$

$$\lim_n\left[\left[1-(q-1)(q^{n-j+1}-1)^{-1}\right]^{(q^{n-j+1}-1)/(q-1)}\right]^{q^{j-1}} = e^{-q^{j-1}}.$$

If we replace the sequence k(n) defined for k=1, with the sequence [tk(n)] (the integral part of tk(n)), we obtain that for any t>0,

$$\langle u_t,\varphi_j\rangle = e^{-tq^{j-1}},$$

for j = 1, 2, ... This means that

$$u_t(x) = 1 + \sum_{j=1}^{\infty} \exp(-tq^{j-1})\ \varphi_j(x).$$

Recalling the definition of spherical functions, it follows that, for $d(x,x_0) = r_n > 0$,

$$u_t(x) = 1 + \sum_{j=1}^{n-1} \exp(-tq^{j-1}) - \exp(-tq^{n-1}).$$

While,

$$u_t(x_0) = 1 + \sum_{j=1}^{\infty} \exp(-tq^{j-1}).$$

A curious fact is that for large t, the sum appearing in the expression for $u_t(x_0)$ approximates the exponential integral [9, Chapter 3]:

$$-\mathrm{Ei}(-t) = \int_t^{\infty} \frac{e^{-y}}{y}\, dy.$$

If we think of $u_t(x) = u(t,x)$ as the ultrametric analogue of the fundamental solutions of the heat equation $du(t,x)/dt = Lu(x,t)$, we can use the equation above to define the "ultrametric" Laplace operator (which commutes with isometries of X) by means of its "symbol", or its "spherical transform". In other words we may define L by the equations

$$\langle Lf, \varphi_j \rangle = -q^{j-1} \langle f, \varphi_j \rangle, \text{ for } j \neq 0.$$

Of course L should be zero on the constants, that is $\langle Lf, \varphi_0 \rangle = 0$.

In order to give a more direct definition of the Laplace operator L one should consider the projections onto the irreducible G-invariant subspaces of $L^2(G/K)$, generated by left translates of the spherical functions. The description of these projections is not difficult. Recall that $X \cong G/K$ is endowed with a unique G - invariant measure of total mass one. Let $\mathcal{E}_j$ be the "conditional expectation" with respect to the partition of X consisting of the balls of radius r_j. In other words $\mathcal{E}_j f$ is the function which on each ball of radius r_j takes as value the m-average of f on that ball. Then for $f \in L^2(X)$,

$$f = \mathcal{E}_0 f + \sum_{j=1}^{\infty} [\mathcal{E}_j f - \mathcal{E}_{j-1} f].$$

It is not difficult to show that the irreducible G-invariant subspaces of $L^2(X)$ are the constants and the ranges of the operators $\mathcal{D}_j = \mathcal{E}_i - \mathcal{E}_{j-1}$. Indeed $\mathcal{D}_j$ commutes with the action of G, and $\mathcal{D}_j \varphi_j = \varphi_j$, while the dimension of the range of $\mathcal{D}_j$ is $d_j = (\rho_1 \ldots \rho_{j-1})(1-\rho_j)$ (one if $j=0$). This is exactly the dimension of the invariant subspace generated by φ_j.

We can now give the following definition of L which is equivalent to the definition given by means of the spherical transform:

$$-Lf = \sum_{j=1}^{n} q^{j-1}\mathcal{D}_j f.$$

We shall treat now briefly the critical case p = 1/2. In this case for each t > 0, the measures $\mu_n^{k(n)}$ converge to a measure which has eigenvalues $\langle\mu_t,\varphi_0\rangle = 1$, and $\langle\mu_t,\varphi_j\rangle = e^{-t}$, for each j > 0. In other words $\mu_t = e^{-t}\delta_{x_0} + (1-e^{-t})m$, where m is the G-invariant measure on X.

REFERENCES

[1] P. Cartier, Fonctions harmoniques sur un arbre, Symp. Math. 9 (1972), 203-270.

[2] P. Diaconis, Group Representations in Probability and Statistics, Institute of Mathematical Statistics, Lecture Notes - Monographs series, vol. 11, 1988.

[3] J. Faraut, Analyse harmonique sur les paires de Guelfand et les espaces hyperboliques, École d'Été d'Analyse Harmonique, Nancy 1980, Les cours de C.I.M.P.A., 1983.

[4] A. Figà-Talamanca. An application of Gelfand pairs to a problem of diffusion in compact ultrametric spaces, "Topics in probability and Lie groups: boundary theory", Montreal, September 1992 (to appear).

[5] A. Figà-Talamanca and C. Nebbia, Harmonic Analysis and Representation Theory for Groups Acting on Homogeneous Trees, London Math. Soc,, Lecture Notes Series 162, Cambridge University Press, Cambridge, 1991.

[6] I.M. Gelfand, Spherical functions on symmetric spaces, Dokl. Akad. Nauk SSSR 70 (1950) 5-8.

[7] R. Godement, A theory of spherical functions I, Trans. Amer. Math. Soc., 73 (1952) 496-556.

[8] S. Lang, SL(2,R), Addison-Wesley, Reading Mass. 1975.

[9] N.N. Lebedev, Special functions and their applications, Dover, New York, 1972

[10] G. Letac, Les fonctions sphériques d'une couple de Gelfand symmetrique et les chaînes de Markov. Advances Appl. Prob. 14, 272-294, (1981).

[11] C. Nebbia, Classification of all irreducible unitary representations of the stabilizer of the horocycles of a tree, Israel J. of Math., vol. 70, n.3 (1990), 343 - 351.

University of Rome "La Sapienza"
Department of Mathematics
piazzale A. Moro 2, 00185 Roma
Italy.

GENERALIZED SQUARE INTEGRABILITY AND COHERENT STATES

J.-P. ANTOINE

ABSTRACT. We present a method for the construction of coherent states, based on the notion of square integrability of a group representation on a homogeneous space. This generalized formalism allows to cover cases hitherto inaccessible, such as the Poincaré group.

1. Motivation

Coherent states (CS), originally introduced by Schrödinger [1] in the context of a harmonic oscillator, later popularized by Glauber [2] and Klauder for the description of coherent light, have found applications in almost all branches of quantum physics (see [3] for a review). To list a few : quantum optics, of course (lasers), but also nuclear, atomic or solid state physics, quantum electrodynamics (infrared problem), quantization/dequantization problem, path integrals, etc.

Can one obtain other states as useful as these so-called canonical CS ? The latter are usually defined as the states obtained from the ground state by displacement operators, $|z\rangle = D(z)|0\rangle$, where $D(z) = \exp(za^{\dagger} - \bar{z}a), z \in \mathbf{C}$ and $a, a^{\dagger}$ obey the familiar CCR $[a, a^{\dagger}] = I$. But they may also be taken as the orbit of the fixed vector $|0\rangle$ under a *square integrable* representation of the Weyl-Heisenberg group G_{WH}, the Lie algebra of which is precisely $\{a, a^{\dagger}, I\}$. In that sense, the construction of CS is a group representation problem. Now we are in a position to answer our question : for getting new types of CS, one has to look for other groups that have square integrable representations. This method was pioneered by Perelomov [4], but his method does not cover all the physically interesting cases, and it should be further extended. As can be expected, the key is to generalize the concept of square integrability, and this is what we shall do in the sequel. The discussion is based on joint work with S.T. Ali and J.-P. Gazeau [5]-[7]. First we briefly review the standard method.

2. Square Integrable Representations and Coherent States

Let G be a locally compact group, with left Haar measure dg, and U a strongly continuous, irreducible, unitary representation of G into a Hilbert space $\mathcal{H}$. Then one says that U is

E. A. Tanner and R. Wilson (eds.), Noncompact Lie Groups and Some of Their Applications, 169–180.

square integrable if there exists a vector $\eta \in \mathcal{H}$ (and in fact always a dense set of such vectors) for which:

$$c(\eta, \phi) = \int_G |\langle U(g)\eta|\phi\rangle|^2 \, dg < \infty, \ \forall \phi \in \mathcal{H} \tag{2.1}$$

(equivalently, U belongs to the discrete series). Given such an *admissible* vector η, normalized by $c(\eta,\eta) = 1$, its orbit under U, namely $\mathfrak{S} = \{\eta_g = U(g)\eta, g \in G\}$, is an overcomplete family of vectors, called *coherent states* associated to the representation U.

In many cases, the admissibility condition (2.1) is too strong, and one has to use the generalization due to Perelomov [4]. Let H_η denote the subgroup of G that leaves η invariant up to a phase: $U(h)\eta = \exp i\alpha(h)\eta, \forall h \in H_\eta$ (this is obviously motivated by Quantum Mechanics, where states are defined only up to a phase). Then the integrand in (2.1) does not depend on g, but only on the coset gH_η. Given an invariant measure ν on the coset space $X = G/H_\eta$, the whole construction of CS may be done under the weaker admissibility condition

$$c_X(\eta, \phi) = \int_X |\langle U(g)\eta|\phi\rangle|^2 \, d\nu(x) < \infty, \ \ \forall \phi \in \mathcal{H} \qquad (x \equiv gH_\eta). \tag{2.2}$$

In that case, we say that the representation U is *square integrable mod* H_η. The difference with the previous construction is that the CS (rays) are now indexed by points $x \in X$.

It is useful to rewrite the admissibility condition (2.2) in a slightly different but completely equivalent form, namely:

$$c_X(\eta, \phi) = \int_X |\langle U(\sigma(x))\eta|\phi\rangle|^2 \, d\nu(x) < \infty, \ \ \forall \phi \in \mathcal{H}, \tag{2.3}$$

where σ is an *arbitrary* section $\sigma : X \to G$ in the principal fibre bundle $\pi : G \to X$ (indeed it is readily seen that the integrand does not depend on the choice of the section).

If condition (2.2) or (2.3) is satisfied, for a given $\eta \in \mathcal{H}$, one defines a set of coherent states as the orbit of η under G, through the representation U and the section σ:

$$\mathfrak{S}_\sigma = \{\eta_x = U(\sigma(x))\eta, x \in X\} \tag{2.4}$$

(the vector η is normalized by $c_X(\eta,\eta) = 1$). Notice that, although the integral in (2.3) is independent of σ, the CS system (2.4) does depend on it.

These coherent states have the following properties, which generalize those of the canonical CS of Schrödinger:

(1) $\mathfrak{S}_\sigma$ is a total set : $\mathfrak{S}_\sigma^\perp = \{0\}$ (this is usually called overcompleteness).

(2) Define a linear map $W_\eta : \mathcal{H} \to L^2(X, d\nu)$ by

$$(W_\eta \phi)(x) = \langle \eta_x | \phi \rangle. \tag{2.5}$$

Then W_η is an isometry (i.e. $W_\eta^* W_\eta = I$) onto a closed subspace $\mathcal{H}_\eta$ of $L^2(X, d\nu)$. Equivalently the CS system $\mathfrak{S}_\sigma$ defines a *resolution of the identity*

$$\int_X |\eta_x\rangle\langle\eta_x| \, d\nu(x) = I. \tag{2.6}$$

(3) The projection operator $P_\eta = W_\eta W_\eta^*$ on $\mathcal{H}_\eta$ is an integral operator with kernel

$$K(x', x) = \langle \eta_{x'} | \eta_x \rangle. \tag{2.7}$$

In other words, $\mathcal{H}_\eta = P_\eta \mathcal{H}$ is a *reproducing kernel Hilbert space* of (continuous) functions:

$$\phi(x') = \int_X K(x', x)\phi(x)\, d\nu(x), \ \forall \phi \in \mathcal{H}_\eta. \tag{2.8}$$

(4) Being an isometry, the map W_η may be inverted on its range $\mathcal{H}_\eta$, and the inverse is simply the adjoint operator: $W_\eta^{-1} = W_\eta^*$ on $\mathcal{H}_\eta$. Thus one has an *inversion formula*:

$$W_\eta^{-1}\phi = \int_X \phi(x)\eta_x\, d\nu(x), \quad \phi \in \mathcal{H}_\eta. \tag{2.9}$$

This formalism applies to a large number of interesting cases, corresponding to familiar Lie groups [4]:

- G compact: any UIR is square integrable; for instance, $SU(2)$ yields the spin CS;
- G noncompact and semisimple: representations of the discrete series (if any) are square integrable; for instance, $SU(1,1)$ CS have been used for constructing path integrals;
- $G = G_{WH}$, the Weyl-Heisenberg group, yields the canonical CS; actually the corresponding map W_η is nothing but the Windowed or Short Time Fourier Transform familiar in signal analysis;
- for G the '$ax + b$' group, i.e. the connected affine group of the real line, one obtains *wavelet analysis* : W_η is the continuous wavelet transform and (2.9) is the reconstruction formula used for signal synthesis [8].

However the method does not always work. For instance, it is inapplicable in the case of the Euclidean group, the Galilei group or the Poincaré groups $\mathcal{P}_+^\uparrow(1,1)$ and $\mathcal{P}_+^\uparrow(1,3)$. We conclude that the Perelomov method is not sufficient and we are going to extend it.

3. A General Coherent State Formalism

In practice, the space $X = G/H$ is often given a priori, with $H \neq H_\eta$ for any η. For instance, in the context of (geometric) quantization, one takes for X a *phase space*, that is, a coadjoint orbit of G or, more generally, a symplectic homogeneous space for G.

For such a manifold $X = G/H$, the CS formalism has been shown to work if one uses a stronger notion of admissibility [9]. Define the vector η to be α-admissible if $c_X(\eta, \eta) < \infty$ and $H \subset H_\eta$, for some character α of G. This condition guarantees that the dependence on the choice of the section σ drops out in the expression of $c_X(\eta, \eta)$, and one may proceed as before. Using this more general definition, Healy and Schroeck have constructed CS systems for the Euclidean and the Galilei groups, but the Poincaré group remains inaccessible.

In the latter case, however, a detailed analysis gives a hint towards the general theory. Indeed, $\mathcal{P}_+^\uparrow(1,1)$ possesses a privileged homogeneous space $\Gamma = \mathcal{P}_+^\uparrow(1,1)/T$, where T is the subgroup of time translations. This Γ is the coadjoint orbit corresponding to the Wigner representation of mass $m > 0$, and thus a natural phase space. However the Wigner representation is not square integrable, and T is not the stability subgroup of any vector η, so that Perelomov's method does not apply (the same analysis may be done for $\mathcal{P}_+^\uparrow(1,3)$).

On the phase space Γ, consider standard coordinates $(\underline{q},\underline{p})$ and the $\mathcal{P}_+^\uparrow(1,1)$-invariant measure $d\underline{q}\,d\underline{p}$. Define the section $\sigma_o : \Gamma \to \mathcal{P}_+^\uparrow(1,1)$ by $\sigma_o(\underline{q},\underline{p}) = ((0,\underline{q}),\Lambda_p)$. Then there is a dense set of admissible vectors $\eta \in \mathcal{H}$, in the sense that the condition

$$\int_\Gamma |\langle U_W(\sigma_o(\underline{q},\underline{p}))\eta|\phi\rangle|^2 \, d\underline{q}\,d\underline{p} < \infty, \;\; \forall\phi \in \mathcal{H}, \tag{3.1}$$

is satisfied iff η is a finite energy state. For such an admissible vector η, the family $\mathfrak{S}_{\sigma_o} = \{\eta_{\underline{q},\underline{p}} = U_W(\sigma_o(\underline{q},\underline{p}))\eta,\; (\underline{q},\underline{p}) \in \Gamma\}$ has all the properties of a CS system. However, instead of the resolution of the identity (2.6), one gets:

$$\int_\Gamma |\eta_{\underline{q},\underline{p}}\rangle\langle\eta_{\underline{q},\underline{p}}| \, d\underline{q}\,d\underline{p} = A_{\sigma_o}, \tag{3.2}$$

where A_{σ_o} is a positive multiplication operator, bounded with bounded inverse, and $A_{\sigma_o} \neq \lambda I$ in general. The interesting fact is that this pattern extends to the general case, as we shall see now. We will come back to the example of the Poincaré group $\mathcal{P}_+^\uparrow(1,1)$ in Section 5 below.

Let $G, U, \mathcal{H}$ be as before, H a closed subgroup of G, $X = G/H$ with an invariant measure ν and $\sigma : X = G/H \to G$ a Borel section (in the principal bundle $G \to G/H$). Then we say that U is *square integrable mod* (H,σ) for the vector $\eta \in \mathcal{H}$ if the integral

$$\int_X U(\sigma(x))|\eta\rangle\langle\eta|U(\sigma(x))^* \, d\nu(x) \tag{3.3}$$

converges weakly to a bounded positive invertible operator A_σ on $\mathcal{H}$, i.e.

$$0 < \int_X |\langle U(\sigma(x))\eta|\phi\rangle|^2 \, d\nu(x) = \langle\phi|A_\sigma\phi\rangle \; < \infty, \forall\phi \in \mathcal{H}. \tag{3.4}$$

We also say that the vector η is *admissible* for (U,σ) or that the section σ is admissible for (U,η). As before one defines a set of coherent states based on X: $\mathfrak{S}_\sigma = \{\eta_{\sigma(x)} = U(\sigma(x))\eta, x \in X\}$. These states have the following properties [5]-[7]:

(1) The set $\mathfrak{S}_\sigma$ is total (overcomplete) : $\mathfrak{S}_\sigma^\perp = \{0\}$.

(2) Define the linear map $W_K : \mathcal{H} \to L^2(X,d\nu)$, by $(W_K\phi)(x) = \langle\eta_{\sigma(x)}|\phi\rangle$. Then the range $\mathcal{H}_K$ of W_K is complete with respect to the scalar product $\langle\Phi|\Psi\rangle_K \equiv \langle\Phi|W_K A_\sigma^{-1} W_K^{-1}\Psi\rangle$ and W_K is *unitary* from $\mathcal{H}$ onto $\mathcal{H}_K$. In other words, one has the resolution

$$\int_X |\eta_{\sigma(x)}\rangle\langle\eta_{\sigma(x)}| \, d\nu(x) = A_\sigma. \tag{3.5}$$

(3) In addition, the orthogonal projection from $L^2(X, d\nu)$ onto $\mathcal{H}_K$ is an integral operator K_σ. Thus $\mathcal{H}_K$ is a reproducing kernel Hilbert space of functions, and one has:

$$\Phi(x) = \int_X K_\sigma(x,y)\Phi(y)\, d\nu(y), \ \forall \Phi \in \mathcal{H}_K. \tag{3.6}$$

The kernel is given explicitly by $K_\sigma(x,y) = \langle \eta_{\sigma(x)} | A_\sigma^{-1}\, \eta_{\sigma(y)} \rangle$, if $\eta_{\sigma(y)} \in \mathcal{D}(A_\sigma^{-1}), \forall y \in X$; otherwise, Eq.(3.6) must be understood in a distributional sense [7].

(4) As before, the map W_K may be inverted on its range by the adjoint operator:

$$W_K^{-1}\phi = \int_X \phi(x)\, A_\sigma^{-1}\, \eta_x \, d\nu(x), \ \ \phi \in \mathcal{H}_K. \tag{3.7}$$

Depending on the properties of the operator A_σ, three cases have to be distinguished:

(i) A_σ^{-1} unbounded, the general case;

(ii) A_σ^{-1} bounded: then $\mathfrak{S}_\sigma$ (and by extension A_σ itself) is called a *frame*;

(iii) $A_\sigma = \lambda I$: then $\mathfrak{S}_\sigma$ is called a *tight frame*.

This terminology is borrowed from the theory of nonorthogonal expansions [10] and non-harmonic analysis [11]. The precise connection will be made clear in Sec. 4.4 below.

Actually, the relation (3.5) may be transformed into a genuine resolution of the identity if one introduces the vectors $\tilde{\eta}_x = T(x)\eta_{\sigma(x)}$, called *quasi-coherent* states, where $T(x)$ are suitable bounded operators (essentially $A_\sigma^{-1/2}$ acting 'fiberwise'):

$$\int_X |\tilde{\eta}_x\rangle\langle\tilde{\eta}_x| \, d\nu(x) = I. \tag{3.8}$$

Of course, the whole construction depends on the choice of a particular section σ, but this dependence is only apparent. Indeed:

- If the section σ is admissible for (U, η), so is the translated section σ_g, for all $g \in G$, where $\sigma_g(x) = g\sigma(g^{-1}x)$.

- If the two sections σ, σ' are both admissible for (U, η), then the corresponding CS systems $\mathfrak{S}_\sigma$, $\mathfrak{S}_{\sigma'}$ are in one to one correspondence, and thus physically equivalent. Indeed, there exist (nonunique) bounded operators $T_{\sigma\sigma'}(x), T_{\sigma'\sigma}(x)$ such that $\eta_{\sigma'(x)} = T_{\sigma'\sigma}(x)\eta_{\sigma(x)}$ and $\eta_{\sigma(x)} = T_{\sigma\sigma'}(x)\eta_{\sigma'(x)}$. The simplest choice are the rank one operators: $T_{\sigma'\sigma}(x) = |\eta_{\sigma'(x)}\rangle\langle\eta_{\sigma(x)}|$. In that case, we say that the two CS systems are *bundle equivalent*.

- Stronger equivalence relations may be considered [7]. For instance, the two CS systems will be called *kernel equivalent* if there exists a bounded operator $T \equiv T_{\sigma'\sigma}$ with bounded inverse such that $T_{\sigma'\sigma}(x) = T|\eta_{\sigma(x)}\rangle\langle\eta_{\sigma(x)}|, \forall x \in X$. Then one has $A_{\sigma'} = TA_\sigma T^*$ and $K_{\sigma'} = K_\sigma$ (hence the name).

This general formalism applies to a large number of situations. Besides standard and Perelomov CS, it covers also the Euclidean and the Galilei groups (here there exist several admissible sections, leading to tight frames, but they often require support restrictions on η). As for the Poincaré group $\mathcal{P}_+^{\uparrow}(1,1)$, it admits a whole class of admissible sections, without support restrictions, which yield both tight and nontight frames (see Section 5 below).

4. Further Generalizations

Although the theory just described works well in many interesting situations, it is not yet general enough for all physically meaningful applications. However, it may still be extended in several directions.

4.1. RANK n OR VECTOR COHERENT STATES

The projection operator $|\eta\rangle\langle\eta|$ in (3.3) may be replaced by an operator of rank n, with $n \geq 1$, of the form $F = \sum_{i=1}^{n} |\eta^i\rangle\langle\eta^i|$, where $\{\eta^i \in \mathcal{H}, i = 1,\ldots,n\}$ is a family of linearly independent vectors. Then the whole formalism outlined in Section 3 may be developed in essentially the same way. An example of application is the theory of vector CS developed in nuclear physics [12].

4.2. QUASI-INVARIANT MEASURES

It may be that the space $X = G/H$ carries no G-invariant measure, but only a *quasi-invariant* one, necessarily unique up to equivalence. Then two cases may arise.

(i) Covariant integrability condition: true coherent states

By this we mean an integrability condition of the form

$$\int_X U(\sigma(x))|\eta\rangle\langle\eta|U(\sigma(x))^* \,\lambda(\sigma(x),x)\, d\nu(x) = A_\sigma. \tag{4.1}$$

Here $\lambda(g,x)$ denotes the Radon-Nikodym derivative of the translated measure ν_g with respect to ν: $d\nu_g(x) = \lambda(g,x)\, d\nu(x)$. When the condition (4.1) holds, the family of vectors $\{\eta_{\sigma(x)} = \sqrt{\lambda(\sigma(x),x)}U(\sigma(x))\eta, x \in X\}$ are genuine *coherent states*, obtained by transporting a fixed vector η over X under the action of G, in a *covariant* way. This definition covers all cases treated previously in the literature [3,4].

(ii) Noncovariant integrability condition: quasi-coherent states

It may also happen that the integral (4.1) with the translated measure $\lambda(\sigma(x),x)\; d\nu(x)$ actually diverges, but that the corresponding one with another quasi-invariant measure μ, necessarily equivalent to ν, converges. In that case, one may still define a useful family of vectors, called *quasi-coherent* states, as $\{\tilde{\eta}_{\sigma(x)} = U(\sigma(x))\eta, x \in X,\}$, which enjoy all the nice properties of the coherent states (overcompleteness, resolution of a positive operator A_σ, reproducing kernel), but not covariance. In this case, there are no 'true' coherent states associated with the section σ and the given representation U.

4.3. QUASI-SECTIONS

Finally, it may happen that no section σ is admissible, but that the integrability condition (3.3) is satisfied with σ replaced by the combined map $\sigma_f \equiv \sigma \circ f : X = G/H \to G$, where $f : X \to X$ is a homeomorphism (reparametrization of the base space). This map, called a *quasi-section* of the principal bundle $\pi : G \to G/H$, satisfies the condition $\pi \circ \sigma_f = f$. Here again one may define, exactly as before, an overcomplete family of quasi-coherent states, with all the expected properties . A nice example [13], where this last construction is in fact necessary, is that of the CS associated to massless representations of $\mathcal{P}_+^\uparrow(1,1)$.

4.4. REPRODUCING TRIPLES

Actually the construction of CS just outlined is a particular case of a much more general set-up, where no reference is made to group theory. Indeed the only essential ingredient is a suitable generalization of the notion of resolution of the identity, which then leads to a reproducing kernel.

Let (X, ν) be a measure space, $\mathcal{H}$ a Hilbert space, A a bounded positive invertible operator on $\mathcal{H}, F : X \to \mathcal{B}(\mathcal{H})^+$ a measurable positive operator valued function. Then $\{\mathcal{H}, F, A\}$ is called a *reproducing triple* if the following relation holds, in the sense of a weak integral:

$$\int_X F(x)\, d\nu(x) = A. \tag{4.2}$$

Under these conditions, a full theory of CS based on X may be derived, along the same lines as above [6,7]. Again one may speak of a *frame* if A^{-1} is bounded, and of a *tight frame* if $A = \lambda I$. Two particular cases are then:

(i) the usual notion of frame [10], obtained when X is a discrete space and ν a counting measure;

(ii) the CS described above, which arise whenever X is a homogeneous space of a l.c. group G and $F(x)$ is obtained by transporting a fixed operator F covariantly under G, through some section $\sigma : X \to G$:

$$F(x) = U(\sigma(x)) F U(\sigma(x))^*, \; x \in X. \tag{4.3}$$

5. Example : the Poincaré group $\mathcal{P}_+^\uparrow(1,1)$

As an illustration of this theory, we shall now construct general CS systems for massive representations of the Poincaré group $\mathcal{P}_+^\uparrow(1,1)$ [5]-[7].

The group is parametrized in the standard way : $\mathcal{P}_+^\uparrow(1,1) = \mathbf{R}^2 \rtimes SO_o(1,1) \ni (a, \Lambda_p)$, with $a = (a_o, \underline{a}) \in \mathbf{R}^2$, and $\Lambda_p \in SO_o(1,1)$, a Lorentz boost indexed by $p = (p_o, \underline{p})$, where $p_o = \sqrt{\underline{p}^2 + 1}$. We consider the familiar Wigner representation $U_W \equiv U_W^{(m)}$ of mass $m > 0$:

$$(U_W(a, \Lambda_p)\phi)(k) = e^{ik\cdot a}\, \phi(\Lambda_p^{-1} k), \phi \in \mathcal{H}_m = L^2(\mathcal{V}_m, d\underline{k}/k_o), \tag{5.1}$$

where $\mathcal{V}_m = \{k = (k_o, \underline{k}), k^2 = k_o^2 - \underline{k}^2 = m^2, k_o > 0\}$ and $k \cdot a = k_o a_o - \underline{k} \cdot \underline{a}$. In this representation, the energy and momentum operators read, respectively:

$$P_o \psi(k) = k_o\, \phi(k), \quad \underline{P}\phi(k) = \underline{k}\, \phi(k). \tag{5.2}$$

A straightforward calculation shows that U_W is not square-integrable over $\mathcal{P}_+^\uparrow(1,1)$ (there is no discrete series !), so that we look for an appropriate quotient. As indicated in Section 2, the natural phase space, namely the coadjoint orbit of $\mathcal{P}_+^\uparrow(1,1)$ associated to U_W, is $\Gamma = \mathcal{P}_+^\uparrow(1,1)/T$, where T is the subgroup of time translations. In the corresponding bundle, we consider the following sections $\sigma : \Gamma \to \mathcal{P}_+^\uparrow(1,1)$:

- Galilean section $\sigma_o(\underline{q},\underline{p}) = ((0,\underline{q}),\Lambda_p)$, already introduced in Section 2.
- General section: $\sigma(\underline{q},\underline{p}) = \sigma_o(\underline{q},\underline{p})\,((f(\underline{q},\underline{p}),\underline{0}),I) \equiv (\hat{q},\Lambda_p)$; the expression multiplying $\sigma_o(\underline{q},\underline{p})$ is a general element of T.
- Affine sections: $f(\underline{q},\underline{p}) = \underline{q}\cdot\underline{\theta}(\underline{p})$ and $\hat{q}$ a space-like two-vector; equivalently, the coordinate $\hat{q}$ on the section $\sigma(\Gamma)$ obeys the following relation: $\hat{q}_o = \beta(\underline{p})\ \hat{\underline{q}}\cdot\underline{p}/|\underline{p}|$, where β is a momentum-dependent speed with $|\beta| < 1$.

Let σ be *any* affine section. Then a detailed analysis [6,7] yields the following results.

(1) A vector η is admissible mod(T,σ) iff it is of finite energy, that is, $\eta \in \mathcal{D}(P_o^{1/2})$.

(2) Any admissible vector η generates a resolution into a frame

$$\int_\Gamma |\eta_{\sigma(\underline{q},\underline{p})}\rangle\langle\eta_{\sigma(\underline{q},\underline{p})}|\; d\underline{q}\, d\underline{p} = A_\sigma^\eta, \text{ where } \eta_{\sigma(\underline{q},\underline{p})} \equiv U_W(\sigma(\underline{q},\underline{p}))\eta \tag{5.3}$$

with A_σ^η a positive bounded multiplication operator with bounded inverse: $(A_\sigma^\eta\psi)(k) = \mathcal{A}_\sigma^\eta(k)\psi(k)$, where

$$\mathcal{A}_\sigma^\eta(k) = \frac{2\pi}{m}\int_{\mathcal{V}_m^+} \frac{d\underline{p}}{p_o}\, |\eta(p)|^2\, \mathcal{A}_\sigma(k,p) \tag{5.4}$$

and the kernel $\mathcal{A}_\sigma(k,p)$ is strictly positive.

(3) The spectrum of the operator A_σ^η admits universal bounds, independent of σ:

$$\mathrm{Sp}\mathcal{A}_\sigma^\eta \subset \left[\frac{2\pi}{m}\langle P_o - |\underline{P}|\rangle_\eta,\ \frac{2\pi}{m}\langle P_o + |\underline{P}|\rangle_\eta\right], \tag{5.5}$$

where $\langle\cdot\rangle_\eta \equiv \langle\eta|\cdot\eta\rangle$ denotes a mean value in the state η.

In conclusion, for any affine section σ, every vector η admissible mod(T,σ) generates a family of coherent states indexed by points in phase space, which constitute a frame:

$$\mathfrak{S}_\sigma = \{\eta_{\sigma(\underline{q},\underline{p})} = U_W(\sigma(\underline{q},\underline{p}))\eta,\ (\underline{q},\underline{p}) \in \Gamma\}. \tag{5.6}$$

Furthermore, all these CS frames are bundle equivalent.

We conclude this section with some concrete examples.

(i) Admissible vectors η:

- Gaussian vector: $\eta_G(k) \sim e^{-k_o/U}$
- Binomial vector: $\eta_\alpha(k) \sim (1 + \frac{k_o - m}{U})^{-\alpha/2}$, $\alpha > 1/2$.

A remarkable property of these two vectors is that the corresponding CS go into CS under the group contraction that relates the different relativity groups [6,7]:

$$\text{Anti-deSitter} \xrightarrow{\kappa\to 0} \text{Poincaré} \xrightarrow{c\to\infty} \text{Galilei}.$$

(ii) Admissible sections σ:

- Galilean σ_o, $\theta(\underline{p}) = 0$: $A^{\eta}_{\sigma_o} = 2\pi m^{-1}\langle P_o\rangle_\eta \, I$, i.e. a tight frame, whenever $\langle \underline{P}\rangle_\eta = 0$.
- Lorentzian σ_L, $\theta(\underline{p}) = \underline{p}/m$: $A^{\eta}_{\sigma_L} = 2\pi m\langle P_o^{-1}\rangle_\eta \, I$, thus the frame is tight for any admissible vector η (this is the only section for which $\mathcal{A}_\sigma(k,p) = a_\sigma(p)$).
- Symmetric or self-dual σ_s, $\theta(\underline{p}) = \underline{p}/(p_o + m)$: $A^{\eta}_{\sigma_s} \neq \lambda I$, for any admissible η, i.e. the frame is *never* tight.

6. Orthogonality Relations, Wigner Transform

Suppose that U is a square integrable representation of G in the usual sense (2.1), that is, over G itself. Then, given any two admissible vectors η_1, η_2, the matrix elements of $U(g)$ obey the following orthogonality relations:

$$\int_G \overline{\langle U(g)\eta_2|\phi_2\rangle}\langle U(g)\eta_1|\phi_1\rangle \, dg = \langle C\eta_1|C\eta_2\rangle \, \langle \phi_2|\phi_1\rangle, \tag{6.1}$$

where C is a positive, closed, invertible operator, whose domain $\mathcal{D}(C)$ is exactly the subspace of admissible vectors [14]. Furthermore, $C = \lambda I$ iff G is unimodular.

Define the vectors $\psi_i = C\eta_i \in \operatorname{Ran} C = \mathcal{D}(C^{-1}), i = 1,2$ (the latter is a dense domain in $\mathcal{H}$). Then one has:

$$\langle U(g)C^{-1}\psi_i|\phi_i\rangle = \operatorname{Tr}(U(g)^* \rho_i C^{-1}) \equiv (\mathcal{W}\rho_i)(g), \; i = 1,2, \tag{6.2}$$

where we have set $\rho_i = |\phi_i\rangle \, \langle\psi_i| \in \mathcal{B}_2(\mathcal{H})$, the space of all Hilbert-Schmidt operators on $\mathcal{H}$. Thus the relation (6.2) defines a linear map $\mathcal{W}$ from $\mathcal{H} \otimes \mathcal{D}(C^{-1})$ into $L^2(G, dg)$, where the latter denotes the dense subspace of $\mathcal{B}_2(\mathcal{H})$ generated by vectors of the form $|\phi\rangle \, \langle\psi|$, $\phi \in \mathcal{H}, \psi \in \mathcal{D}(C^{-1})$. In this notation, the orthogonality relation (6.1) reads:

$$\int_G \overline{(\mathcal{W}\rho_2)(g)}(\mathcal{W}\rho_1)(g) \, dg = \operatorname{Tr} \rho_2{}^* \rho_1 = \langle \rho_2|\rho_1\rangle_{\mathcal{B}_2(\mathcal{H})}, \tag{6.3}$$

which means that $\mathcal{W}$ is isometric, and thus may be extended by continuity to an isometry (denoted by the same symbol) $\mathcal{W} : \mathcal{B}_2(\mathcal{H}) \to L^2(G, dg)$, called the *Wigner transform* . This map $\mathcal{W}$ induces a decomposition of Ran $\mathcal{W}$, which is a closed subspace of $L^2(G, dg)$, into a direct sum of *coherent sectors.* First the unitary irreducible representation U lifts to a unitary reducible representation U_L in $\mathcal{B}_2(\mathcal{H})$, defined as follows:

$$\mathsf{U}_L(g) = U(g) \vee I, \; g \in G, \tag{6.4}$$

where we have introduced the notation

$$(A \vee B)\rho = A\rho B^*, \rho \in \mathcal{B}_2(\mathcal{H}), A, B \in \mathcal{B}(\mathcal{H}). \tag{6.5}$$

Moreover the Wigner map $\mathcal{W}$ intertwines U_L with the left regular representation U_L on $L^2(G, dg)$:

$$\mathcal{W}\mathsf{U}_L(g) = U_L(g)\mathcal{W}, \; g \in G. \tag{6.6}$$

Next note that $\langle\psi_1|\psi_2\rangle = 0$ in $\mathcal{H}$ implies $\langle\rho_1|\rho_2\rangle = 0$ in $\mathcal{B}_2(\mathcal{H})$ by (6.3), and this in turn implies $\langle\mathcal{W}\rho_1|\mathcal{W}\rho_2\rangle = 0$ in $L^2(G, dg)$. Now let $\{\psi_j, j = 1, \ldots, \infty\}$ be an orthonormal basis of $\mathcal{H}$ contained in $\mathcal{D}(C^{-1})$. For each j, the set $\mathcal{H} \otimes \psi_j = \{|\phi\rangle\,\langle\psi_j|,\ \phi \in \mathcal{H}\}$ is a vector subspace of $\mathcal{B}_2(\mathcal{H})$, invariant under U_L. Taking the restrictions from $\mathcal{B}_2(\mathcal{H})$ to this subspace, we see that:

$$\begin{aligned} \mathsf{U}_L(g)\restriction \mathcal{H}\otimes\psi_j &\sim U(g) \ \text{(unitary equivalence)} \\ \mathcal{W}\restriction \mathcal{H}\otimes\psi_j &= W_{\eta_j}, \ \text{with } \psi_j = C\eta_j, \end{aligned}$$

where W_{η_j} is the isometric map defined in (2.5) (we are here in the case $X \equiv G$). It follows that the Wigner map $\mathcal{W}$ transforms the decomposition

$$\mathcal{B}_2(\mathcal{H}) = \bigoplus_{j=1}^{\infty} \mathcal{H}\otimes\psi_j \tag{6.7}$$

into the decomposition

$$\operatorname{Ran}\mathcal{W} = P_{\mathcal{W}} L^2(G, dg) = \bigoplus_{j=1}^{\infty} \mathcal{H}_{\eta_j}, \tag{6.8}$$

where $P_{\mathcal{W}} = \mathcal{W}\mathcal{W}^*$ is the projection operator onto $\operatorname{Ran}\mathcal{W}$ and $\mathcal{H}_{\eta_j}$ is the range of W_{η_j} in $\mathcal{H}\otimes\psi_j$. Eq.(6.8) is the announced decomposition of $\operatorname{Ran}\mathcal{W}$ into coherent sectors.

An interesting development of this result [15] follows from the observation that an entirely similar construction may be done on the right, starting from the representations $\mathsf{U}_R(g) = I \vee U(g)$ and $U_R(g)$, the right regular representation. Then, taking the two constructions together, one obtains the following *modular structure* on $\mathcal{B}_2(\mathcal{H})$. Consider the von Neumann algebras (bicommutants)

$$\mathcal{N}_L = \{\mathsf{U}_L(g), g \in G\}'', \ \ \mathcal{N}_R = \{\mathsf{U}_R(g), g \in G\}''. \tag{6.9}$$

Then $\mathcal{N}_L = \mathcal{N}_R'$ and there exists an anti-unitary map J such that $J\mathcal{N}_L J = \mathcal{N}_R$. This opens the possibility of using the powerful machinery of the Tomita-Takesaki theory [16] (modular operator, KMS states, ...). Furthermore, the Wigner map transports this modular structure to $P_{\mathcal{W}} L^2(G, dg)$. This construction allows to consider also noncoherent sectors, if one starts from an orthonormal basis of $\mathcal{H}$ not contained in $\mathcal{D}(C^{-1})$.

The whole construction just outlined rests on the assumption that the original representation U is square integrable over G. The crucial question now is, how does all this extend to the more general set-up described in this paper ?

(i) Square integrability on a homogeneous space X:

If η_1, η_2 are both α-admissible [9], then one obtains exactly the same results, orthogonality relations, Wigner transform, etc. In the general case of two admissible vectors η_1, η_2, one has to consider the integral:

$$\begin{aligned} I^{\eta_1\eta_2}_{\phi_1\phi_2} &= \int_X \overline{\langle U(\sigma(x))\eta_2|\phi_2\rangle}\langle U(\sigma(x))\eta_1|\phi_1\rangle\, d\nu(x) \\ &= \langle\phi_2|A^{\eta_1\eta_2}_{\sigma}\phi_1\rangle. \end{aligned} \tag{6.10}$$

Assume first that $A_\sigma^{\eta_1\eta_2} = \lambda(\eta_1, \eta_2)I$ (i.e. a tight frame). Then one gets

$$I_{\phi_1\phi_2}^{\eta_1\eta_2} = \lambda(\eta_1, \eta_2)\, \langle\phi_2|\phi_1\rangle, \tag{6.11}$$

and $\lambda(\eta_1, \eta_2)$ is a sesquilinear form. If that form is closed, one may write $\lambda(\eta_1, \eta_2) = \langle C\eta_1|C\eta_2\rangle$, as usual [14], and the whole development may be repeated. In the general case, however, the problem remains open.

(ii) The Poincaré case:

In the case of $\mathcal{P}_+^\uparrow(1,1)$, with an affine section σ, the operator $A_\sigma^{\eta_1\eta_2}$ reduces to the multiplication by the function

$$A_\sigma^{\eta_1\eta_2}(k) = \int_{V_m^+} \frac{d\underline{p}}{p_o}\, \overline{\eta_2(p)}\, \eta_1(p)\, \mathcal{A}_\sigma(k,p). \tag{6.12}$$

Then one gets an orthogonality relation:

$$\begin{aligned} I_{\phi_1\phi_2}^{\eta_1\eta_2} &= \int_\Gamma \overline{\mathrm{Tr}[U(\sigma(\underline{q},\underline{p}))^*\rho_2]}\, \mathrm{Tr}[U(\sigma(\underline{q},\underline{p}))^*\rho_1\, d\underline{q}d\underline{p} \\ &= \langle\rho_2|\mathcal{A}_\sigma\rho_1\rangle_{\mathcal{B}_2(\mathcal{H})}, \end{aligned} \tag{6.13}$$

where $\mathcal{A}_\sigma$ is a positive self-adjoint operator on $\mathcal{B}_2(\mathcal{H})$:

$$(\mathcal{A}_\sigma|\phi\rangle\langle\eta|)(k,p) = \mathcal{A}_\sigma(k,p)\phi(k)\overline{\eta(p)}. \tag{6.14}$$

If one takes the case of a tight frame, such as the Lorentzian section σ_L or the Galilean section σ_o with $\langle\underline{P}\rangle_\eta = 0$, then the usual orthogonality relations are valid and the whole construction follows. In particular, one gets a Wigner transform

$$(\mathcal{W}\rho)(\underline{q},\underline{p}) = \mathrm{Tr}[U(\sigma(\underline{q},\underline{p}))^* A_\sigma^{-1}\rho], \tag{6.15}$$

which is an isometry from $\mathcal{B}_2(\mathcal{H})$ into $L^2(\Gamma, d\underline{q}d\underline{p})$. If the frame is not tight, however, there seems to be no way of transforming (6.13) into a useful orthogonality relation.

7. Outcome

The general CS formalism outlined in the preceding sections opens interesting perspectives, both in the physical domain and for pure mathematics.

First, these more general CS realizations, based on phase space, may yield novel presentations of Quantum Mechanics, particularly suitable for the study of the quasi-classical limit. More generally, this approach may shed a new light on the general method of CS quantization, expressed in terms of frames [7].

On the other hand, this development may have interesting implications for the theory of quantum measurements. If one notices that the fiducial vector η plays here exactly the same role as the analyzing wavelet in wavelet analysis, it is tempting to consider it as a *quantum probe*. This point of view, which consists in applying in a quantum theory ideas and techniques borrowed from signal processing, a thoroughly classical theory, has been developed at length in a recent paper [7].

On the mathematical side too, several directions look promising for further research. For instance, the generalized orthogonality relations and the modular structure discussed in Section 6 should be further investigated. Also, the whole CS formalism may work for more general manifolds X, such as Kähler manifolds, which are good candidates for phase spaces. Finally, the analysis of the CS associated to the massless representations of $\mathcal{P}_{+}^{\uparrow}(1,1)$ should be extended to the corresponding conformal group, that is, essentially the Virasoro group.

References

[1] E. Schrödinger, Naturwiss. **14** (1926) 664

[2] R.J. Glauber, Phys. Rev. **130** (1963) 2529; ibid. **131** (1963) 2766

[3] J.R. Klauder and B.S. Skagerstam, *Coherent States - Applications in Physics and Mathematical Physics*, World Scientific, Singapore 1985

[4] A. Perelomov, *Generalized Coherent States and Their Applications*, Springer-Verlag, Berlin 1986

[5] S.T. Ali and J-P. Antoine, Ann. Inst. H.Poincaré **51** (1989) 23

[6] S.T. Ali, J-P. Antoine and J-P. Gazeau, Ann. Inst. H.Poincaré **52** (1990) 90; **55** (1991) 829, 857

[7] S.T. Ali, J-P. Antoine and J-P. Gazeau, Ann. Phys. (NY) **222** (1993) 1, 38

[8] J-M. Combes, A. Grossmann, P. Tchamitchian (eds.), *Wavelets: Time-Frequency Methods and Phase Space (Proc. Marseille 1987)* , Springer-Verlag, Berlin 1989

[9] D.M. Healy Jr. and F.E. Schroeck Jr., On informational completeness of covariant localization observables and Wigner coefficients, preprint, 1992

[10] I. Daubechies, A. Grossmann and Y. Meyer, J. Math. Phys. **27** (1986) 1271

[11] R.J. Duffin and A.C. Schaeffer, Trans. Amer. Math. Soc. **72** (1952) 341

[12] D.J. Rowe, G. Rosensteel and R. Gilmore, J. Math. Phys. **26** (1985) 2787; D.J. Rowe, G. Rosensteel and K.T. Hecht, J. Math. Phys. **29** (1988) 287

[13] J-P. Antoine and U. Moschella, J. Phys. A: Math. Gen. **26** (1993) 591

[14] A. Grossmann, J. Morlet, T. Paul, J. Math. Phys. **26** (1985) 2473; Ann. Inst. H. Poincaré **45** (1986) 293

[15] S.T. Ali, G.G. Emch and A. El Gradechi, in preparation

[16] O.Bratteli and D.Robinson, *Operator Algebras and Quantum Statistical Mechanics I, II*, Springer-Verlag, Berlin et al., 1979.

Institut de Physique Théorique
Université Catholique de Louvain
B - 1348 Louvain-la-Neuve, Belgium

Maximal Abelian Subgroups of SU(p,q) and Integrable Hamiltonian Systems

P. Winternitz

M.A. del Olmo

M.A. Rodríguez

Abstract *The classification of maximal abelian subgroups of the noncompact Lie group $SU(p,q)$ is reviewed and then applied to construct maximally superintegrable Hamiltonian systems on real hyperboloids. These Hamiltonian systems are obtained by projecting free motion on the homogeneous space $SU(p,q)/U(p-1,q)$ onto the space $O(p,q)/O(p-1,q)$. The projection is realized by introducing ignorable variables, corresponding to different maximal abelian subalgebras of $su(p,q)$. Classical and quantum mechanical systems are treated on the same footing.*

1.- INTRODUCTION

The purpose of this presentation is to show how the maximal abelian subgroups [1] of the noncompact pseudounitary Lie group $SU(p,q)$ can be used to generate completely integrable Hamiltonian systems, constrained to real $O(p,q)$ hyperboloids [2,3].

The obtained hamiltonians will have the form

$$H = \frac{1}{2} \sum_{\mu,\nu=1}^{n} g^{\mu\nu} p_\mu p_\nu + V(s) \tag{1.1a}$$

or

$$H = -\frac{1}{2} \sum_{\mu,\nu=1}^{n} \frac{1}{\sqrt{g_0}} \frac{\partial}{\partial s_\mu} \sqrt{g_0}\, g^{\mu\nu} \frac{\partial}{\partial s_\nu} + V(s), \tag{1.1b}$$

E. A. Tanner and R. Wilson (eds.), Noncompact Lie Groups and Some of Their Applications, 181–198.

with

$$g_{\mu\nu}s^{\mu}s^{\nu} = 1, \qquad g = \mathrm{diag}(I_p, -I_q), \qquad g_0 = \det g; \tag{1.2}$$

where $p \geq q \geq 0$, $p + q = n + 1$, $s \in \mathbb{R}^{n+1}$, in classical, or quantum mechanics, respectively. In view of the constraint (1.2), the momenta p_μ satisfy

$$p_\mu s^\mu = 0, \quad \text{or} \quad -is_\mu \frac{\partial \psi(s)}{\partial s_\mu} = 0, \tag{1.3a,b}$$

respectively, where $\psi(s) \in \mathbb{C}$ is the quantum mechanical wave function.

The classical system with Hamiltonian (1.1a) is completely integrable if it has n integrals of motion

$$\{Q_0 = H, Q_1(s,p), ..., Q_{n-1}(s,p)\} \tag{1.4}$$

that are well defined functions on phase space, are functionally independent, and are in involution (Poisson commute).

The systems presented below are actually "maximally superintegrable Hamiltonian systems" (MSIHS), in that they allow $2n+1$ functionally independent well defined integrals of motion, amongst which it is possible to choose 2 or more sets of n integrals in involution (all containing the Hamiltonian H).

Well known MSIHS are the Kepler (or Coulomb) system, and the harmonic oscillator. Systematic searches for MSIHS systems have been conducted e.g. in Euclidean spaces and have yielded several further systems, [4–6] amongst which only $V = \alpha r^{-1}$ and $V = \omega r^2$ are spherically symmetric.

Typical properties of MSIHS are that their finite classical trajectories are closed (periodic orbits) and that their quantum energy levels have higher degeneracies. The integrals of motion generate algebraic structures, either Lie algebras, or in some cases quadratic algebras [7]. From the practical point of view, MSIHS allow the separation of variables in more than one system of coordinates. They can be solved in an elementary manner in terms of known special functions, or doable integrals. They have physical applications in molecular and chemical physics, and also in nuclear and particle physics [8–11].

We shall obtain MSIHS of the type (1.1), ... , (1.4) in spaces of arbitrary dimension n by the simple device of projecting free motion on a homogeneous space of the group $SU(p,q)$ on to the hyperboloid (1.2). The space is a noncompact analogue of the complex projective space $\mathbb{C}P(n)$, namely

$$\mathbb{C}P(p,q) \sim SU(p,q)/[(SU(p-1,q) \times U(1))]. \tag{1.5}$$

This space is realized as a complex hyperboloid

$$g_{\mu\nu}\bar{y}^{\mu}y^{\nu} = 1, \quad y^{\mu} \in \mathbb{C}, \quad \mu = 0, 1, ..., n-1, \tag{1.6}$$

with the identification $y^\mu \sim y^\mu e^{i\phi}$ and the Hamiltonian

$$H = \frac{c}{2} g^{\mu\nu} P_\mu \bar{P}_\nu, \tag{1.7a}$$

or

$$H = -\frac{c}{2} \frac{1}{\sqrt{g_0}} \frac{\partial}{\partial y^\mu} \sqrt{g_0}\, g^{\mu\nu} \frac{\partial}{\partial \bar{y}^\nu} \quad (c = \text{const.}), \tag{1.7b}$$

respectively, and g as in eq. (1.2). The projection to the system (1.1) is realized as a phase space reduction by a maximal abelian subgroup of $SU(p,q)$. In practice this is equivalent to a coordinate transformation

$$(y^\mu, \bar{y}^\mu) \to (s^\mu, x^\mu), \qquad s, x \in \mathbb{R}^{n+1}, \tag{1.8}$$

where x^μ are ignorable variables.

Each different maximal abelian subalgebra (MASA) of the Lie algebra $su(p,q)$ provides a different system of ignorable variables and a different potential $V(s)$ in eq. (1.1). All the obtained systems on the real hyperboloid (1.2) are MSIHS because, as we shall see below, they inherit a large number of integrals of motion from the free system (1.1).

In Section 2 below we briefly review some results on the classification [1] of MASAs of $su(p,q)$. Section 3 is devoted to MSIHS obtained using the different Cartan subalgebras of $su(p,q)$. In Section 4 we restrict ourselves to the group $SU(2,2)$ and obtain further MSIHS, related to non-Cartan MASAs of $su(2,2)$.

2.- MAXIMAL ABELIAN SUBALGEBRAS OF SU(p,q).

Let L be a finite dimensional Lie algebra (in our case $su(p,q)$) and $G \sim \exp L$ the corresponding Lie group.

Definition. *A maximal abelian subalgebra M of L is Lie algebra satisfying*

$$\begin{aligned} &(i)\ M \subset L, \qquad [M, M] = 0, \\ &(ii)\ \mathit{cent}_L\ M = M. \end{aligned} \tag{2.1}$$

The centralizer of M in L, $\text{cent}_L M$ is defined as

$$\text{cent}_L M \sim \{x \in L \,|\, [x,y] = 0, \quad \forall y \in M\}. \tag{2.2}$$

A Cartan subalgebra is a selfnormalizing MASA:

$$\text{nor}_L M = M,$$

where the normalizer of M in L is defined by

$$\mathrm{nor}_L M \sim \{x \in L \,|\, [x,y] \in M, \quad \forall y \in M\}. \tag{2.3}$$

An important task in Lie algebra theory is that of classifying all MASA of a given Lie algebra L into conjugacy classes under the action of G. For the classical Lie algebras $gl(n,\mathbb{C})$ and $gl(n,\mathbb{R})$ this task is simply that of classifying maximal sets of commuting matrices. It is the natural extension of the classification of normal forms of elements of a Lie algebra, e.g. the Jordan canonical form for elements of $gl(n,\mathbb{C})$.

The Cartan subalgebras of all simple real Lie algebras were classified by Kostant [12] and Sugiura [13]. Kravchuk [14] provided the framework for classifying MASAs of another type, called maximal abelian nilpotent subalgebras (MANS) [15]. These consist entirely of nilpotent elements in L and are represented by nilpotent matrices in any finite dimensional representation of L. Recent articles have been devoted to a classification of MASAs of symplectic [16], pseudoorthogonal [17] and pseudounitary [1] Lie algebras.

Here we shall need some results on MASAs of $su(p,q)$ and their classification under $SU(p,q)$. We realize the corresponding Lie algebra and group by complex matrices

$$su(p,q) \sim \{X \in \mathbb{C}^{n\times n} | XK + KX^{+} = 0, \quad TrX = 0\}, \tag{2.4}$$

$$SU(p,q) \sim \{G \in \mathbb{C}^{n\times n} | GKG^{+} = K, \quad \det\ G = 1\}, \tag{2.5}$$

where

$$K = K^{+} \in \mathbb{C}^{n\times n}, \quad n = p + q, \quad \text{signature } K = (p,q). \tag{2.6}$$

A MASA of su(p,q) will be represented by a matrix set $\{M, K\}$ and two MASAs are equivalent if there exists a fixed matrix $H \in GL(n,\mathbb{C})$, such that

$$HX'H^{-1} = X, \quad HK'H^{+} = K, \tag{2.7}$$

for all $X' \subset M'$. Thus, we are not committing ourselves to a specific choice of the metric tensor K; only its signature is fixed.

Let us present some results, omitting all proofs.

Three types of MASAs of $su(p,q)$ exist:

1. Orthogonally decomposable MASAs (OD).
2. Orthogonally indecomposable, but decomposable MASAs (OID & D).
3. Orthogonally indecomposable and indecomposable MASAs (OID & ID).

Theorem 1. *All OD MASAs of $su(p,q)$ are obtained as follows: consider all partitions of (p,q) such that*

$$(p,q)=(p_1,q_1)+\cdots+(p_r,q_r),\quad p=\sum_{i=1}^{r}p_i,\ q=\sum_{i=1}^{r}q_i,\ r\geq 2,$$

$$p_1+q_1\geq p_2+q_2\geq\ldots\geq p_r+q_r\geq 1,\quad q_i=0 \Rightarrow p_i=1,\ p_i=0\Rightarrow q_i=1. \tag{2.8a}$$

For each partition form the matrix sets

$$M=\begin{pmatrix}M_1 & & \\ & \ddots & \\ & & M_r\end{pmatrix},\quad K=\begin{pmatrix}K_1 & & \\ & \ddots & \\ & & K_r\end{pmatrix}, \tag{2.8b}$$

such that M_i, K_i is an OID MASA of $u(p_i,q_i)$, i.e. we have $M_i=\mathbb{C}I_{p_i+q_i}\oplus$ OID MASA of $su(p_i,q_i)$. Then assure $TrX=0$ for all $X\in M$.

Theorem 2. *OID & D MASAs of $su(p,q)$ exist if and only if we have $p=q\geq 1$. They are represented by the matrix sets*

$$X=\begin{pmatrix}A & 0\\ 0 & -A^{+}\end{pmatrix},\quad K=\begin{pmatrix}O & I_p\\ I_p & O\end{pmatrix}, \tag{2.9}$$

where A is an indecomposable MASA of $gl(p,\mathbb{C})$, i.e. $A=\mathbb{C}I_p\oplus MANS$ of $sl(p,\mathbb{C})$.

For information on MANS of $sl(p,\mathbb{C})$ see ref. 14, 15, 18.

We see from Theorems 1 and 2 that the classification of MASAs of $su(p,q)$ (and all classical Lie algebras) reduces to some decomposition theorems and to a classification of MANS.

Theorem 3. *An OID & ID MASA of $su(p,q)$ is a MANS. It is characterized by a Kravchuk signature*

$$(\lambda,\ \mu,\ \lambda),\quad 2\lambda+\mu=p+q,\quad 1\leq\lambda\leq q,\quad \lambda,\ \mu\in Z^{>}. \tag{2.10a}$$

A MANS with Kravchuk signature (λ,μ,λ) can be brought to Kravchuk normal form and represented by the matrix sets

$$X=\begin{pmatrix}0 & A & Y\\ 0 & S & -HA^{+}\\ 0 & 0 & 0\end{pmatrix},\quad K=\begin{pmatrix} & & I_\lambda\\ & H & \\ I_\lambda & & \end{pmatrix}, \tag{2.10b}$$

$$H=H^{+}\in\mathbb{C}^{\mu\times\mu},\quad \mathrm{sgn}H=(p-\lambda,q-\lambda)$$

$$Y=-Y^{+}\in\mathbb{C}^{\lambda\times\lambda},\quad A\in\mathbb{C}^{\lambda\times\mu},\quad SH=-HS^{+}\in\mathbb{C}^{\mu\times\mu} \tag{2.10c}$$

The matrix S is nilpotent. Commutativity requires that we have

$$AHA'^{+} = A'HA^{+}, \quad AS' = A'S, \quad [S, S'] = 0, \tag{2.10d}$$

where $X \in M$ and $X' \in M$. The antihermitian matrix Y is free. The matrix S depends linearly on A.

For further information, see Ref. 1. We mention that two types of MANS of $su(p,q)$ exist:

(1) Free–rowed MANS: all entries in row 1 of A can be chosen real and otherwise free. Then all further rows of A and all entries in S depend linearly on row 1 of A.

(2) Non–free–rowed MANS. Certain constraints exist among the entries of row 1 of A. They cannot be chosen freely, but free entries may occur in other rows of A.

We note that the Cartan subalgebras of $su(p,q)$ for all cases but $su(1,1)$ are orthogonally decomposable. The decomposition pattern is [12, 13]

$$(p,q) = r(1,1) + (p-s)(1,0) + (q-r)(0,1), \quad 0 \le r \le q. \tag{2.11a}$$

The general Cartan subalgebra M_r has a maximal compact subalgebra of dimension $p-s$ and can be represented by the matrix sets

$$M_r = \begin{pmatrix} ic_0 & & & & & \\ & \ddots & & & & \\ & & ic_{n-2r} & & & \\ & & & Q_1 & & \\ & & & & \ddots & \\ & & & & & Q_r \end{pmatrix}, \quad K = \begin{pmatrix} I_{p-q} & \\ & I_q \times K_0 \end{pmatrix}, \tag{2.11b}$$

$$K_0 = \begin{pmatrix} 1 & 0 \\ 0 & -1 \end{pmatrix}, \quad Q_j = \begin{pmatrix} ia_j & ib_j \\ -ib_j & ia_j \end{pmatrix}, \quad a_j, b_j \in \mathbb{R},\ 1 \le j \le r \le q, \tag{2.11c}$$

$$\sum_{\mu=0}^{n-2r} c_\mu + 2\sum_{j=1}^{r} a_j = 0, \quad n = p+q-1.$$

To see how rich the structure of MASAs of $su(p,q)$ is, consider the case of $su(2,2)$. It has 12 different classes of MASAs. Of them, 3 are Cartan subalgebras M_0, M_1 and M_2 with 0, 1 and 2 linearly independent noncompact elements, respectively. Five more are orthogonally decomposable, corresponding to decompositions $(1,1)+(1,1)$ (twice), $(1,1)+(1,0)+(0,1)$, $(2,1)+(0,1)$ and $(1,2)+(1,0)$. One MASA is OID

& D, 3 more are MANS. Among the 12, 11 are of dimension 3 (the rank of $su(p,q)$), but one MANS is of dimension 4.

3.- PHASE SPACE REDUCTIONS BY CARTAN SUBGROUPS OF SU(p,q)

3.1.- Ignorable Variables and Reduction of the Hamiltonian.

Let us consider the free Hamiltonian (1.7) on the hyperboloid (1.6). The vector fields in the tangent space realizing the action of $u(p,q)$ on differentiable functions on $\mathbb{C}^{n+1}$ are given by

$$\hat{X} = -y^{\mu}(X)^{\nu}_{\mu}\,\partial y^{\nu} + c.c., \tag{3.1}$$

with $X \in u(p,q)$. Let us transform to a coordinate system, adapted to a chosen Cartan subalgebra. The result can be formulated as a theorem.

Theorem 4. *The vector fields $\hat{Y}_{\mu}$, $\mu = 0,1,...,n$, representing a Cartan subalgebra $M_r \subset u(p,q)$ can be simultaneously straightened out to $\hat{Y}_{\mu} = -\partial_{x^{\mu}}$ $(\mu = 0,1,...n)$. The corresponding coordinate transformation is*

$$y^{\mu} = B(x)^{\mu}_{\nu}s^{\nu}, \quad B(x) = \exp x^{\rho}Y_{\rho}, \tag{3.2}$$

with $s \in \mathbb{R}^{n+1}$ satisfying eq. (1.2). The vector fields $\hat{X}$ of eq. (3.1) are transformed to

$$\hat{X} = -\frac{1}{2}B^{\mu}_{\nu}s^{\nu}X^{\alpha}_{\mu}[(A^{-1})^{\beta}_{\alpha}\frac{\partial}{\partial x^{\beta}} + (B^{-1})^{\beta}_{\alpha}\frac{\partial}{\partial s^{\beta}}], \tag{3.3}$$

with

$$A^{\mu}_{\nu} = \frac{\partial y^{\mu}}{\partial x^{\nu}} = (Y_{\nu})^{\mu}_{\rho}y^{\rho}. \tag{3.4}$$

We shall not give a proof here, but mention the main ingredients:

(1) The algebra M_{μ} is abelian, dim $M_{\mu} = n + 1$.
(2) The algebra M_{μ} is spanned by $n+1$ vector fields that are linearly independent at a generic point of $\mathbb{C}P(p,q)$.
(3) All elements of M_r are represented by purely imaginary matrices in the defining representation of $u(p,q)$.

Thus, M_{μ} does not actually have to be a Cartan subalgebra. The above conditions are satisfied for every Cartan subalgebra, but also for some other MASAs. For instance, they hold for 11 of the 12 inequivalent MASAs of $u(2,2)$.

The transformation (3.2) takes the classical free Hamiltonian (1.7) into

$$H = \frac{1}{2}(p_s^T g p_s + p_x^T(A^+gA)^{-1}p_x), \tag{3.5}$$

where p_s and p_x are the momenta conjugate to the variables s and x, respectively. The matrix $(A^+gA)^{-1}$ depends on s, but not on the ignorable variables x. When we restrict to the real hyperboloid (1.2), i.e. set $x = 0$, $p_x = k$ =const., we obtain the reduced system (1.1) with potential

$$V(s,k) = \frac{1}{2}k^T(A^+gA)^{-1}k. \tag{3.6}$$

Different Cartan subalgebras (and other MASAs) provide different potentials.

It remains to show that the obtained systems are integrable, and what is more, superintegrable.

3.2.- Integrals of Motion.

We start from a specific Cartan subalgebra $M_r \sim < Y_0, Y_1, ..., Y_n > \subset u(p,q)$ and find all second order invariants of the corresponding Cartan subgroup $\exp M_r$ in the enveloping algebra of $su(p,q)$. A simple way of doing this is to construct the quadratic invariants of the coadjoint action of the maximal torus $\exp Y$ on the dual $su(p,q)^*$ and then to translate the results back into the enveloping algebra.

In order to present the results, we need some notations for the basis elements of $u(p,q)$ in the defining matrix representation with metric

$$g = K = \mathrm{diag}(I_{p-q}, K_0, ..., K_0), \quad K_0 = \mathrm{diag}(1,-1). \tag{3.7}$$

Let $E_{ik} \in \mathbb{R}^{(p+q)\times(p+q)}$ by the matrix with elements

$$(E_{ik})_{ab} = \delta_{ia}\delta_{kb}. \tag{3.8}$$

We introduce a canonical basis in which the compact elements are represented by the antihermitian matrices

$$\begin{aligned} &Y_0^c = iI_{n+1}, && Y_{\mu\mu}^c = i(E_{\mu\mu} - E_{\mu+1,\mu+1}), && \mu = 0,1,...n-1, \\ &X_{\mu\nu}^c = E_{\mu\nu} - E_{\nu\mu}, && Y_{\mu\nu}^c = i(E_{\mu\nu} + E_{\nu\mu}), && 0 \le \mu < \nu \le n. \end{aligned} \tag{3.9a}$$

Noncompact elements are represented by hermitian matrices

$$X_{\mu\nu}^{nc} = E_{\mu\nu} + E_{\nu\mu}, \quad Y_{\mu\nu}^{nc} = i(E_{\mu\nu} - E_{\nu\mu}), \quad 0 \le \mu < \nu \le n. \tag{3.10}$$

Thus, the matrices $X_{\mu\nu}$ are real, $Y_{\mu\nu}$ are imaginary and we can drop the superscripts since the values of μ and ν together with the metric K of eq.(3.7), determine whether the basis element is compact, or noncompact.

The results on invariants can be summed up as follows.

Theorem 5. *The elements of the Cartan subalgebra $M_\mu \subset u(p,q)$ commute with $n(n+1)/2$ linearly indepedent second order operators in the enveloping algebra of su(p,q). A basis for this set of invariants is provided by the operators*

$$
\begin{aligned}
R_{\mu\nu} &= X_{\mu\nu}^2 + Y_{\mu\nu}^2, \qquad (0 \le \mu < \nu \le n-2r),\\
S_k &= -(Y_{n-2r-1+2k,n-2r-1+2k})^2 + (X_{n-2r-1+2k,n-2r+2k})^2,\\
T_{\mu j} &= X_{\mu j}^2 + Y_{\mu j}^2 - X_{\mu j+1}^2 - Y_{\mu j+1}^2,\\
&\qquad (0 \le r \le n-2r, j = n-2r+1, n-2r+3, ..., n-1),\\
U_{\mu j} &= \{X_{\mu j}, X_{\mu j+1}\} + \{Y_{\mu j}, Y_{\mu j+1}\},\\
W_{1,jk} &= X_{jk}^2 + Y_{jk}^2 - X_{j,k+1}^2 - Y_{j,k+1}^2 - X_{j+1,k}^2 - Y_{j+1,k}^2 + X_{j+1,k+1}^2 + Y_{j+1,k+1}^2,\\
W_{2,jk} &= \{X_{jk}, X_{j,k+1}\} + \{Y_{jk}, Y_{j,k+1}\} - \{X_{j+1,k}, X_{j+1,k+1}\} - \{Y_{j+1,k}, Y_{j+1,k+1}\},\\
W_{3,jk} &= \{X_{jk}, X_{j+1,k}\} + \{Y_{jk}, Y_{j+1,k}\} - \{X_{j,k+1}, X_{j+1,k+1}\} - \{Y_{j,k+1}, Y_{j+1,k+1}\},\\
W_{4,jk} &= \{X_{jk}, X_{j+1,k+1}\} + \{Y_{jk}, Y_{j+1,k+1}\} + \{X_{j,k+1}, X_{j+1,k}\} + \{Y_{j,k+1}, Y_{j+1,k}\},\\
&\qquad (j = n-2r+1, n-2r+3, ..., n-3, \quad k = n-2r+3, n-2r+5, ..., n-1).
\end{aligned}
\tag{3.11}
$$

The curly brackets denote anticommutators. Further $n(n+1)/2$ trivial invariants exist. They lie in the enveloping algebra of the Cartan subalgebra itself, namely $T_{\mu\nu} = Y_\mu Y_\nu$, $(1 \le \mu \le \nu \le n)$.

Comments.

(1) A separate issue is that of choosing complete sets of $n-1$ commuting operators among those in eq. (3.11). This can be addressed in all generality, but we shall restrict ourselves to examples below. In each case more than one complete set exists.

(2) The results of Theorem 5 greatly simplify for the compact Cartan subalgebra M_0 and also for M_1 (one noncompact element). Indeed, for M_0 only the invariants $R_{\mu\nu}$, $(0 \le \mu < \nu \le n)$ exist and they are equivalent to Casimir operators of all $su(2)$ and $su(1,1)$ subalgebras. In this case $R_{\mu\nu}$ commutes with

$$R_{\alpha\beta}\ (\alpha \ne \mu, \beta \ne \nu), \quad R_{\mu\alpha} \pm R_{\nu\alpha}\ (r < \nu < \alpha), \quad R_{\beta\mu} \pm R_{\beta\nu}\ (\beta < \mu < \nu)$$

(the sign + applies if $R_{\mu\nu}$ is compact and - for $R_{\mu\nu}$ noncompact). For M_1 only the invariants $R_{\mu\nu}$ $(0 \le \mu < \nu \le n-2)$, S_1, $T_{\mu j}$ and $U_{\mu j}$, $(\mu = 0, 1, ...n-2;\ j = n-1)$ appear.

3.3.- Reduction to the Real Hyperboloid.

Consider the general Cartan subalgebra M_μ of eq. (2.11) and implement eq.

(3.2). We obtain the coordinate system

$$\begin{aligned}
y^0 &= e^{ic_0} s^0 \\
&\vdots \\
y^{n-2r} &= e^{ic_{n-2r}} s^{n-2r} \\
y^{n-2r+1} &= e^{ia_1}(s^{n-2r+1}\cosh b_1 + is^{n-2r+2}\sinh b_1) \\
y^{n-2r+2} &= e^{ia_1}(-is^{n-2r+1}\sinh b_1 + s^{n-2r+2}\cosh b_1) \\
&\vdots \\
y^{n-1} &= e^{ia_r}(s^{n-1}\cosh b_r + is^n \sinh b_r) \\
y^n &= e^{ia_r}(-is^{n-1}\sinh b_r + s^n\cosh b_r),
\end{aligned} \tag{3.12}$$

with $0 \le c_\mu < 2\pi$, $0 \le a_k < 2\pi$, $s^\mu \ge 0$ and

$$(s^0)^2 + (s^1)^2 + \cdots + (s^{p-q-1})^2 + [(s^{p-q})^2 - (s^{p-q+1})^2] + \cdots + [(s^{n-1})^2 - (s^n)^2] = 1. \tag{3.13}$$

The reduced potential (3.6) in this case is

$$V_r = \frac{c}{4}\left(\sum_{\mu=0}^{p-q-1}\frac{k_\mu^2}{(s^\mu)^2} + \sum_{\mu=p-q}^{n-2r}\frac{(-1)^{p-q-\mu}k_\mu^2}{(s^\mu)^2} + \right.$$
$$\left.\sum_{\mu=n-2r+1}^{n}\frac{((s^\mu)^2 - (s^{\mu+1})^2)(k_\mu^2 - k_{\mu+1}^2) + 4s^\mu s^{\mu+1}k_\mu k_{\mu+1}}{((s^\mu)^2 + (s^{\mu+1})^2)^2}\right). \tag{3.14}$$

The potentials V_r are all singular along the hypersurfaces $s^\mu = 0$, $0 \le \mu \le n-2r$, and the surfaces $s^\mu = s^{\mu+1} = 0$, $n-2r+1 \le \mu \le n$.

Let us now restrict the vector fields (3.3) to the real hyperboloid (3.13). We use the form of the matrices B and A inherent in the coordinates (3.12). Then we set the ignorable variables x_μ equal to $x_\mu = 0$ and the conjugate momenta $p_x = k$ =const. The vector fields $\hat{X}_{\mu\nu}$ corresponding to the real matrices $X_{\mu\nu}$ in (3.9) and (3.10) reduce to O(p,q) generators

$$\hat{S}^c_{\mu\nu} = s_\mu p_\nu - s_\nu p_\mu, \qquad \hat{S}^{nc}_{\mu\nu} = -(s_\mu p_\nu + s_\nu p_\mu), \quad 0 \le \mu \le \nu \le n, \tag{3.15}$$

where as usual, p_μ is either the classical momentum, conjugate to s^μ, or the quantum mechanical operator $p_\mu = -i\frac{\partial}{\partial s^\mu}$. The vector fields $\hat{Y}_{\mu\nu}$ corresponding to the imaginary matrices $Y_{\mu\nu}$ of eq. (3.9) and (3.10) reduce to functions on the reduced configuration space, i.e. functions of s. The general reduction formulas are quite complicated [2]. We shall present examples below in section 4 for the case $p = q = 2$.

The reduction of the vector fields $\hat{X}_{\mu\nu}$ and $\hat{Y}_{\mu\nu}$ of course also provides a reduction of the invariants (3.11) of the corresponding Cartan subalgebra.

For instance we have

$$\begin{aligned}
&\hat{R}^{c}_{\mu\nu} \to (S^{c}_{\mu\nu})^2 + (k_\mu \frac{s_\nu}{s_\mu} + k_\nu \frac{s_\mu}{s_\nu})^2, \\
&\hat{R}^{nc}_{\mu\nu} \to (S^{nc}_{\mu\nu})^2 + (k_\mu \frac{s_\nu}{s_\mu} - k_\nu \frac{s_\mu}{s_\nu})^2, \\
&\hat{S}_k \to (S^{nc}_{n-2r-1+2k,n-2r+k})^2 + \\
&+ \frac{k_{n-2r-1+2k}[(s_{n-2r-1+2k})^2 - (s_{n-2r+2k})^2] + 2k_{n-2r+2k}\, s_{n-2r+2k}\, s_{n-2r-1+2k}}{(s_{n-2r-1+2k})^2 + (s_{n-2r+2k})^2}.
\end{aligned} \tag{3.16}$$

4.- THE GROUP SU (2,2) AND PHASE SPACE REDUCTIONS.

4.1.- General Comments.

In this section we restrict ourselves to the case $p = q = 2$. Since $SU(2,2)$ is locally isomorphic to the conformal group Conf (3,1) of 3+1 dimensional Minkowski space, this example is of particular physical interest. This restriction to a six-dimensional space makes it possible to present formulas more explicitly and, more important, to consider other MASAs (not only the Cartan ones).

As stated above, su(2,2) has 12 inequivalent MASAs [1]. The 4 dimensional one does not satisfy the conditions of Theorem 4 and cannot be used for phase space reductions.

All the other ones do. We shall consider several of them as examples and in particular show that different commuting sets of reduced operators can be chosen in each case. The sets will have the form

$$\{H, Q_1, Q_2\}, \tag{4.1}$$

where H is the reduced Hamiltonian. The invariants Q_1 and Q_2 are chosen among the invariants obtained by reducing the SU(2,2) invariants of the form of eq. (3.11), or similar invariants corresponding to MASAs of su(2,2) that are not Cartan subalgebras.

The reason we restricted ourselves to second order operators in the Lie algebra of su(p,q) is that we wish to solve the reduced Hamilton–Jacobi and Schrödinger equations by separation of variables. There exists an intimate relation between separation of variables for these equations and the existence of complete sets of commuting second order operators [19, ... , 24].

In the case under consideration we obtain, for each MASA of su(2,2), several different operators (Q_1, Q_2), satisfying $[H, Q_1] = [H, Q_2] = [Q_1, Q_2] = 0$ and this leads to separation of variables on the $O(2,2)$ hyperboloid in several coordinate systems. We shall make use of the known results [22] on the separation of variables in free Hamilton–Jacobi (HJ) and Laplace–Beltrami (LB) equations on the $O(2,2)$ hyperboloid.

For the $o(2,2)$ algebra we shall use the basis:

$$\begin{aligned} K_{01} &= s_0\partial_{s_1} + s_1\partial_{s_0}, & K_{23} &= s_2\partial_{s_3} + s_3\partial_{s_2}, \\ K_{03} &= s_0\partial_{s_3} + s_3\partial_{s_0}, & K_{02} &= s_0\partial_{s_2} - s_2\partial_{s_0}, \\ K_{12} &= s_1\partial_{s_2} + s_2\partial_{s_1}, & K_{13} &= s_1\partial_{s_3} - s_3\partial_{s_1}. \end{aligned} \tag{4.2}$$

We recall that separable coordinates for free HJ or LB equations on 3 dimensional Riemannian (or pseudo–Riemannian) spaces of constant nonzero curvature can be of several types [23], which for $O(2,2)$ hyperboloids reduce to the following:

Type 1. Two ignorable variables.

Then Q_1 and Q_2 are squares of elements of the Lie algebra.

Type 2. One ignorable variable.

2a. *Subgroup type.*

The operator Q_1 is the square of an element of the Lie algebra, Q_2 is the quadratic Casimir operator of a subgroup H of $O(2,2)$. This corresponds to group reductions

$$\begin{aligned} &O(2,2) \supset O(2,1) \supset G_0, \qquad G_0 \sim O(2),\ O(1,1), \quad \text{or} \quad T(1), \\ &O(2,2) \supset E(1,1) \supset O(1,1). \end{aligned}$$

Here $E(1,1)$ is the pseudoeuclidean group in $1+1$ dimensions and $T(1)$ corresponds to a nilpotent element of H.

2b. *Generic type:*

Q_1 as above, Q_2 generic (not a pure square, not a Casimir operator).

Type 3. No ignorable variables

3a. *Subgroup type*

Q_1 is the Casimir operator of a subgroup $H \subset O(2,2)$, Q_2 is a second order element in the enveloping algebra of the Lie algebra of H. This corresponds to the reduction

$$O(2,2) \supset H \supset \Gamma,$$

where $H \sim O(2,1)$, or $E(1,1)$ and Γ is a discrete subgroup of H.

3b. *Generic type.*

Q_1 and Q_2 are not Casimir operators, nor pure squares.

Let us now turn to individual MASAs of su(2,2) and the corresponding reductions.

4.2.- The Compact Cartan Subalgebra

The coordinates (3.12) reduce to

$$Y_\mu = s_\mu e^{ix_\mu}, \qquad \mu = 0,1,2,3, \tag{4.3}$$

and the reduced Hamiltonian is

$$H_0 = \frac{1}{2}\left[p_0^2 - p_1^2 + p_2^2 - p_3^2 + \frac{k_0^2}{s_0^2} - \frac{k_1^2}{s_1^2} + \frac{k_2^2}{s_2^2} - \frac{k_3^2}{s_3^2}\right]. \tag{4.4}$$

The 6 integrals of motion are:

$$\begin{aligned}
T_1 &= K_{01}^2 + (k_0\frac{s_1}{s_0} - k_1\frac{s_0}{s_1})^2, \quad T_2 = L_{02}^2 + (k_0\frac{s_2}{s_0} + k_2\frac{s_0}{s_2})^2,\\
T_3 &= K_{03}^2 + (k_0\frac{s_3}{s_0} - k_3\frac{s_0}{s_3})^2, \quad T_4 = K_{12}^2 + (k_1\frac{s_2}{s_1} - k_2\frac{s_1}{s_2})^2,\\
T_5 &= L_{13}^2 + (k_1\frac{s_3}{s_1} + k_3\frac{s_1}{s_3})^2, \quad T_6 = K_{23}^2 + (k_2\frac{s_3}{s_2} - k_3\frac{s_2}{s_3})^2.
\end{aligned} \tag{4.5}$$

The Hamiltonian (4.4) is equals to

$$H_0 = \frac{1}{2}(-T_1 + T_2 - T_3 - T_4 + T_5 - T_6) + \sum_{\mu<\nu}(k_\mu - k_\nu)^2.$$

A classification of commuting pairs of integrals and the corresponding coordinate systems is as follows:

<u>Type 1</u>.

1. – (T_2, T_5) $\quad O(2,2) \supset O(2) \times O(2)$,
2. – (T_1, T_6) $\quad O(2,2) \supset O(1,1) \times O(1,1)$.

<u>Type 2a</u>.

3. – $(T_1 + T_4 - T_2, T_2)$ $\quad O(2,2) \supset O(2,1) \supset O(2)$
4. – $(T_1 + T_4 - T_2, T_1)$ $\quad O(2,2) \supset O(2,1) \supset O(1,1)$.

<u>Type 2b</u>.

5. – $(T_2, T_1 + T_4 + a(T_3 + T_6) + bT_5)$,
6. – $(T_1, T_4 - T_2 + a(T_3 - T_5) + bT_6)$, $\quad a, b \in \mathbb{R}$.

Type 3a.

$$7. - \quad (T_1 + T_4 - T_2,\, T_1 + aT_4), \qquad a \neq 0.$$

Type 3b.

$$8. - \quad (T_2 + aT_1 + bT_3,\, T_5 + \frac{a-b}{b}T_1 + \frac{a+b}{(1+a)b}T_4), \quad b \neq 0,\, a \neq -1.$$

The superintegrability of the Hamiltonian (4.4) is manifiest: any two of the above pairs of operators guarantee it.

4.3.- The Orthogonally Indecomposable, but Decomposable MASA (OID& D).

This MASA can be represented by the matrices.

$$X = \begin{pmatrix} 0 & ia & ib & ic \\ -ia & 0 & ic & -ib \\ 0 & 0 & 0 & -ia \\ 0 & 0 & ia & 0 \end{pmatrix}, \quad K = \begin{pmatrix} 0 & 0 & 1 & 0 \\ 0 & 0 & 0 & 1 \\ 1 & 0 & 0 & 0 \\ 0 & 1 & 0 & 0 \end{pmatrix}. \tag{4.6}$$

The coordinates on the $SU(2,2)$ hyperboloid are

$$\begin{aligned} y^0 &= e^{ix^0}([s^0 + i(s^2x^2 + s^3x^3)]\cosh x^1 + i[s^1 + i(s^2x^3 - s^3x^2)]\sinh x^1), \\ y^1 &= e^{ix^0}(-i[s^0 + i(s^2x^2 + s^3x^2)]\sinh x^1 + i[s^1 + i(s^2x^3 - s^3x^2)]\cosh x^1), \\ y^2 &= e^{ix^0}(s^2\cosh x^1 - is^3\sinh x^1), \\ y^3 &= e^{ix^0}(is^2\sinh x^1 + s^3\cosh x^1). \end{aligned} \tag{4.7}$$

The Hamiltonian, in the metric (4.6) is

$$\begin{aligned} H_8 =&(p_0p_2 + p_1p_3) + \frac{(k_0k_2 + k_1k_3)(s_2^2 - s_3^2) + 2(k_0k_3 - k_1k_2)s_2s_3}{(s_2^2 + s_3^2)^2} \\ &+ \frac{1}{(s_2^2 + s_3^2)^3}\{(k_2^2 - k_3^2)[(s_1s_3(3s_2^2 - s_3^2) - s_0s_2(s_2^2 - 3s_3^2)] \\ &- 2k_2k_3[s_1s_2(s_2^2 - 3s_3^2) + s_0s_3(3s_2^2 - s_3^2)]\}. \end{aligned} \tag{4.8}$$

An interesting point is that this Hamiltonian, contrary to that of eq. (4.4), is nonsingular: the point $s_2 = s_3 = 0$ does not lie on the hyperboloid, which in this metric is given as $2(s_0s_2 + s_1s_3) = 1$.

The 6 invariants are

$$
\begin{aligned}
T_1 &= \frac{1}{4}(L_{02} - K_{03} - L_{13})^2 + f_1(s),\\
T_2 &= (K_{01} + K_{23})^2 - (L_{02} + L_{13})^2 + f_2(s),\\
T_3 &= K_{01}^2 + K_{23}^2 + 2K_{03}^2 + K_{12}^2 - L_{02}^2 - L_{13}^2 - \{K_{01},\ K_{23}\} + f_3(s),\\
T_4 &= \frac{1}{2}\{K_{01} - K_{23},\ L_{02} - K_{03} + K_{12} - L_{13}\} + f_4(s),\\
T_5 &= \frac{1}{2}\{K_{03} + K_{12},\ L_{02} - K_{03} + K_{12} - L_{13}\} + f_5(s),\\
T_6 &= \{K_{01} + K_{23},\ L_{02} + L_{13}\} + f_6(s).
\end{aligned}
\tag{4.9}
$$

(we do not specify the functions $f_i(s)$ here, though that can easily be done).

Commuting pairs in this case are:

Type 1:

1.– (T_2, T_6), $O(2,2) \supset O(1,1) \times O(2)$,
2.– (T_1, T_4), $O(2,2) \supset O(1,1) \times E(1)$.

Type 2b:

3.– $(T_1, T_4 + aT_5)$, $a \neq 0$.

Type 4b:

4.– $(-(a^2 + b^2)T_1 + T_2 + aT_4 + bT_5,\ -bT_4 + aT_5 + T_6)$, $(a,\ b) \neq (0,\ 0)$.

The pairs (T_2, T_6) and (T_1, T_4) correspond to separation of variables in nonorthogonal coordinates on the $O(2,2)$ hyperboloid.

4.4.- A Maximal Abelian Nilpotent Subalgebra (MANS).

Let us consider the MANS of $su(2,2)$ represented by the matrix set

$$
X = \begin{pmatrix} 0 & ia & ib & ic \\ 0 & 0 & ia & ib \\ 0 & 0 & 0 & ia \\ 0 & 0 & 0 & 0 \end{pmatrix}, \quad K = \begin{pmatrix} & & & 1 \\ & & 1 & \\ & 1 & & \\ 1 & & & \end{pmatrix}, \tag{4.10}
$$

The coordinates on the SU(2,2) hyperboloid are

$$
\begin{aligned}
y^0 &= e^{ix^0}(s^0 + is^1x^1 + (ix^2 - \frac{(x^1)^2}{2})s^2 + [i(x^3 - \frac{(x^1)^2}{6}) - x^1x^2]s^3),\\
y^1 &= e^{ix^0}(s^1 + ix^1s^2 + (ix^2 - \frac{(x^1)^2}{2})s^3),\\
y^2 &= e^{ix^0}(s^2 + ix^1s^3),\\
y^3 &= e^{ix^0}s^3.
\end{aligned}
\tag{4.11}
$$

There are again 6 second order invariants, including the Hamiltonian, namely

$$\begin{aligned}
T_1 &= \frac{1}{4}(-K_{01} + L_{13} + L_{02} + K_{23})^2 + f_1(s),\\
T_2 &= \frac{1}{2}(K_{01} - L_{13} + L_{02} + K_{23})^2 + f_2(s),\\
T_3 &= 2(K_{03}^2 + K_{12}^2 + K_{01}^2 + K_{23}^2 - L_{13}^2 - L_{02}^2) + f_3(s),\\
T_4 &= \frac{1}{2}((L_{02} + K_{23})^2 - (K_{01} - L_{13})^2) + f_4(s),\\
T_5 &= \frac{1}{2}\{K_{03} - K_{12}, -K_{01} + L_{13} + L_{02} + K_{23}\} + f_5(s),\\
T_6 &= \frac{1}{2}((K_{23} + L_{13})^2 - (K_{01} - L_{02})^2\\
&\quad + \{K_{03} + K_{12},\ K_{01} - L_{13} + L_{02} + K_{23}\}) + f_6(s).
\end{aligned} \tag{4.12}$$

Commuting pairs are:

Type 1:

$$1. - \quad (T_1, T_4), \qquad O(2,2) \supset E(1) \times E(1).$$

Type 2b:

$$2. - \quad (T_1, T_5 + aT_4), \qquad a \neq 0.$$

Type 3a:

$$3. - \quad (T_4, T_2). \qquad O(2,2) \supset E(1,1) \supset \Gamma.$$

Type 3b:

$$4. - \quad (2abT_1 + aT_2 + bT_4 + T_6,\ 2bT_1 + T_2 + 4aT_4 + 2T_5).$$

5.- CONCLUSIONS.

The results presented above can be viewed as the intersection of three research programs:

(i) A classification of all MASAs of the classical Lie algebras.

(ii) A systematic search for integrable and maximally superintegrable Hamiltonian systems on various homogeneous spaces.

(iii) A group theoretical approach to the separation of variables in partial differential equations.

Applications of the obtained integrable systems can be found in several directions. The Pöschl–Teller potentials [8, 9], succesfully applied e.g. in studies of heavy ion scattering are special cases of those presented in this article. We mention that integrable potentials in spaces with indefinite metric were studied by Kibler [25].

A quite different application is to the study of infinite dimensional completely integrable Hamiltonian systems, i.e. soliton equations [26]. Indeed, interesting classes of solutions of such equations (multi soliton solutions, quasiperiodic solutions) are obtained from solutions of associated finite dimensional systems [27–29]. The problem then is to determine the soliton equations, related to the finite dimensional ones, obtained above.

Acknowledgements

The research of the first two authors is supported by CICYT de España (grants AEN90–0027 and PB92–0255). That of the third by grants from NSERC of Canada and FCAR du Québec. All three authors benefitted from a NATO collaborative research grant. This manuscript was prepeared while P.W. was visiting the Universidad de Valladolid, whose hospitality is acknowledged.

References

[1]. del Olmo, M.A., Rodríguez, M.A., Winternitz, P. and Zassenhaus, H., Lin. Alg. Appl. **135**, 79 (1990).

[2]. del Olmo, M.A., Rodríguez, M.A. and Winternitz, P., J. Math. Phys. **34**, ..., (1993).

[3]. del Olmo, M.A., Rodríguez, M.A., Winternitz, P., Preprint, Valladolid, (1993).

[4]. Fris, I., Mandrosov, V., Smorodinsky, Y. A., Uhlir, M. and Winternitz, P., Phys. Lett. **16**, 354 (1965).

[5]. Makarov, A.A., Smorodinsky, Y.A., Valiev, Kh. and Winternitz, P., Nuovo Cim. **A 52**, 1061 (1967).

[6]. Evans, N.W., Phys. Rev. **41 A**, 5666 (1990); Phys. Lett. **147 A**, 483 (1990); J. Math. Phys. **32**, 3369 (1991).

[7]. Letourneau, P. and Vinet, L., in *Symmetries in Science* VII, B. Gruber ed., Pergamon Press, to appear.

[8]. Alhassid, Y., Gursey, F. and Iachello, F., Phys. Rev. Lett. **50**, 873 (1983); Chem. Phys. Lett. **99**, 27 (1983); Ann. Phys (NY). **148**, 346 (1983).

[9]. Frank, A. and Wolf, K.B., Phys. Rev. Lett. **52**, 1737 (1984).

[10]. Kibler, M., Lamot, G.H. and Winternitz, P., Int. J. Quantum Chem. **43**, 625 (1992).

[11]. Wehrhahn, R.F., Smirnov, Yu F., and Shirokov, A.M., J. Math. Phys. **33**, 2384 (1992).

[12]. Kostant, B., Proc. Nat. Acad. Sci. USA **41**, 967 (1955).

[13]. Sugiura, M., J. Math. Soc. Jap. **11**, 374 (1959).

[14]. Kravchuk, M.F., Rend. Circ. Mat. Palermo **51**, 126 (1927).

[15]. Suprunenko, D.A. and Tyshkevich, B.I. *Commutative Matrices*, (Academic, New York, 1968).

[16]. Patera, J., Winternitz, P. and Zassenhaus, H., J. Math. Phys. **24**, 1973 (1983).

[17]. Hussin, V., Winternitz, P. and Zassenhaus, H., Lin. Alg. Appl, **141**, 183 (1990); **173**, 125 (1992).

[18]. Winternitz, P. and Zassenhaus, H., Preprint CRM 1199 (1984).

[19]. Winternitz, P. and Fris, I., Sov. J. Nucl. Phys. **1**, 636 (1965).

[20]. Winternitz, P., Lukac, I. and Smorodinsky, Ya. A., Sov. J. Nucl. Phys. **7**, 139 (1968).

[21]. Miller, W. Jr., *Symmetry and Separation of Variables*, (Addison Wesley, New York, 1977).

[22]. Kalnins, E.G. and Miller, W. Jr., Proc. Roy. Soc. Edinb. **79 A**, 227 (1977).

[23]. Miller, W. Jr., Patera, J. and Winternitz, P., J. Math. Phys. **22**, 251 (1981).

[24]. Kalnins, E.G., *Separation of Variables for Riemannian Spaces of Constant Curvature*, Pitman, New York, (1986).

[25]. Kibler, M., in *Group Theoretical Methods in Physics*, Lecture Notes in Physics 318, p. 238 (1988).

[26]. Ablowitz, M.J., Clarkson, P.A., *Solitons, Nonlinear Evolution Equations and Inverse Scattering*, Cambridge University Press, Cambridge (1991).

[27]. Previato, E., Physica **D18**, 312 (1986)., Duke Math. J. **52**, 329 (1985).

[28]. Adams, M.R., Harnad, J. and Previato, E., Commun. Math. Phys. **117**, 451 (1988).

[29]. Adams, M.R., Harnad, J. and Hurtubise, J., Commun. Math. Phys. **134**, 555 (1990).

Centre de recherches mathématiques, Université de Montréal
CP 6128-A, Montréal, H3C 3J7, Québec (Canada).

Dept. Física Teórica, Facultad de Ciencias,
Universidad de Valladolid, E-47011 Valladolid, (Spain).

Dept. Física Teórica, Facultad de Físicas,
Universidad Complutense, E-28040 Madrid (Spain).

PATH INTEGRALS AND LIE GROUPS

Akira Inomata

Georg Junker

ABSTRACT. The roles of Lie groups in Feynman's path integrals in non-relativistic quantum mechanics are discussed. Dynamical as well as geometrical symmetries are found useful for path integral quantization. Two examples having the symmetry of a non-compact Lie group are considered. The first is the free quantum motion of a particle on a space of constant negative curvature. The system has a group $SO(d,1)$ associated with the geometrical structure, to which the technique of harmonic analysis on a homogeneous space is applied. As an example of a system having a non-compact dynamical symmetry, the d-dimensional harmonic oscillator is chosen, which has the non-compact dynamical group $SU(1,1)$ besides its geometrical symmetry $SO(d)$. The radial path integral is seen as a convolution of the matrix functions of a compact group element of $SU(1,1)$ on the continuous basis.

1. Prologue

Needless to say, Lie groups have a special importance in quantum mechanics. The Lie group often details the kinematical symmetry of a quantum system. The analysis of angular momentum on the basis of the rotation group $SO(3)$ or the unitary group $SU(2)$ is a classic example. The Lie group sometimes reveals itself in the structure of quantum dynamics. The well-known accidental degeneracy of the Coulomb problem has been ascribed to its dynamical symmetry $SO(4)$ which is not shared by other spherically symmetric systems. In the applications to non-relativistic quantum mechanics, it is usually associated with Schrödinger's differential equation or involved in making a algebraic framework of a quantum system. Rather independent of physics, there are extensive studies of differential equations and special functions from the Lie group representation aspect [1; 2]. The spectrum-generating algebraic approach has placed many of the standard exactly soluble problems in quantum mechanics within the $su(1,1)$ algebraic scheme and opened the field of infinite component theories for composite systems [3]. However, little is known about the roles

E. A. Tanner and R. Wilson (eds.),
Noncompact Lie Groups and Some of Their Applications, 199–224.

of Lie groups in Feynman's path integral. As the kernel of the heat equation is easily obtainable from the Fourier analysis, the path integral representing the kernel of the Schrödinger equation may be studied from a more general harmonic analysis point of view. In recent years, there have been considerable developments in the study of the Lie group theoretical approach to Feynman's path integral [4-10]. In this report, we would like to highlight the main ideas and results of the study.

Any standard text book of quantum mechanics contains the treatment of central potential problems in polar coordinates, whereas nearly all of the books dealing with Feynman's path integral ignore the polar coordinate treatment. This is simply a reflection of the situations that path integration is difficult in non-cartesian variables and hence that Feynman's path integral can be explicitly calculated only for the harmonic oscillator or more general quadratic systems. The polar coordinate formulation was already initiated, to the authors' knowledge, by Ozaki [11] as early as 1955, and independently, by Edwards and Gulyaev [12] in 1964. They separated the angular part and the radial part of the path integral for a spherically symmetric system in polar coordinates and calculated the propagator for a free particle. The angular path integration for a free particle or a particle in a central potential is in essence an application of harmonic analysis based on the unitary representation of $SO(3)$ – an elementary application of the Peter-Weyl theorem.

However, the method of explicit calculation of the radial path integral [13] was not known until 1967, without which the polar coordinate path integral has of little interest. The energy spectrum of a central potential problem arises only from the radial path integral. Harmonic analysis of the radial path integral is not as simple as that of the angular part. For a free particle, extension of $SO(3)$ to the Euclidean group $E(3)$ is sufficient for covering both the radial part as well as the angular part [14]. In the presence of a central potential, the translational symmetry is broken, so that the Euclidean group becomes inapplicable. It turns out that the radial path integral is most conveniently analyzed on the basis of the dynamical group $SU(1,1)$ [15].

In 1968, in order to demonstrate that the sum over classical paths is exact, Schulman [16] calculated semiclassically the path integral for the symmetric top on the manifold of $SU(2)$. In relation with Schulman's observation, Dowker [17] showed that the sum over classical paths is exact on the manifold of a class of Lie groups. Since explicit evaluation of Feynman's path integral was difficult, attention was mainly focused on semiclassical approximation and examination of its exactness. What Schulman and Dowker have studied are harmonic expansions of the propagators in terms of spherical functions. The semiclassical approximation is not generally exact even if space is symmetric.

It is a strange fact that Feynman's path integral cannot provide the solution for the hydrogen atom problem. Since no closed form expression for the propagator of the hydrogen atom is available, there have been a number of attempts to construct one. Historically, Feynman's path integral was once expected as a possible means to find the Coulomb propagator. Soon it was recognized that the power of the path integral method was very limited in deriving exact results. None of the integral representations of the Coulomb propagator so far available was originated from Feynman's path integral. In the summers of 1976-78, the Kustaanheimo-Stiefel transformation became a topic of discussions in the seminars of the University of Munich. Barut and Wilson [18] were mainly investigating the KS mapping as a realization of the subgroup of a wider dynamical group $SO(4,2)$. One of the present authors examined its use in the path integral for the hydrogen atom, and realized that the path integral given in the KS coordinates does not help in constructing a closed form expression for the Coulomb propagator. The discussions of Munich were carried over to Trieste in 1978, involving Barut, Duru, Wilson and others [19]. Barut was more optimistic about the use of the KS transformation in path integration. In 1979, Duru and Kleinert [20], formally applying the Kustaanheimo-Stiefel transformation of space and time to the Hamiltonian path integral for the hydrogen atom, succeeded to derive an integral representation for the Coulomb propagator. Giving up the hope for expressing the Coulomb propagator in closed form, the authors of ref. [22] applied the KS transformation to the Lagrangian path integral, carrying out explicitly Feynman's prescription for the time-sliced path integral calculation, to arrive at the exact expression for the energy-dependent Green function of the hydrogen atom, previously obtained by Hostler [23] from Schrödinger's equation.

The Kustaanheimo-Stiefel transformation was originally introduced for regularization of the classical Kepler orbit [21]. It consists of a space transformation $\mathbb{R}^3 \to \mathbb{R}^4$ and a position-dependent time transformation. In particular, the time transformation is integrable only along the classical orbit. In path integration, there is no unique orbit, and such a formal calculation often involves ambiguity [24]. The significance of these results is that Feynman's path integral, if slightly modified, can produce solutions for systems other than quadratic systems. Furthermore, the successful use of the Kustaanheimo-Stiefel map that takes the geometrical symmetry group $SO(3)$ to the dynamical symmetry group $SO(4)$ suggests that the dynamical symmetry could play a role in path integral calculation. The fact of matter is that the path integral for the Coulomb problem can be solved in terms of polar variables without the help of the KS coordinates [25]. Yet, the local time transformation of Kustaanheimo and Stiefel was found essential for solving the Coulomb path integral problem.

Noticing that the Pöschl-Teller oscillator has the dynamical symmetry of $SU(2)$ [26], the one-dimensional path integral for this system is extended to the path integral on $SU(2)$ with the aid of the asymptotic form of an integral representation for the modified Bessel function [4; 6]. This nontrivial dimensional extension of the path integral has enabled us to carry out path integration in the way angular path integration was completed. Naturally, a particle bound in the modified Pöschl-Teller potential, having the dynamical symmetry of $SU(1,1)$ [27], can be path-integrated in the $SU(1,1)$ manifold. In fact, in combination with a local time transformation of the Kustaanheimo-Stiefel type, a number of systems have been solved in the manifolds of $SU(2)$ and $SU(1,1)$. Of course, $SU(2)$ and $SU(1,1)$ can be identified with $S^3 = SO(4)/SO(3)$ and $\Lambda^3 = SO(3,1)/SO(2,1)$, respectively. We shall call those systems path-integrable on S^d or Λ^d the systems of the *hypergeometric class.* The stochastic processes corresponding to Feynman's path integrals of the hypergeometric class are called the *Legendre processes* [28]. Although a system of $SU(2)$ can be converted to a system of $SU(1,1)$ by analytic continuation, the role of the non-compact group $SU(1,1)$ in path integration is not completely expressed by analogy from the case of $SU(2)$. More subtle treatments are needed for non-compact systems. On these backgrounds, a general formulation of the path integration on a group manifold, compact or non-compact, has been developed [7; 10].

As has been mentioned above, the radial path integral can be handled in a way very similar to the spectrum generating algebra of $SU(1,1)$. Although this method yields solutions only for an isotropic harmonic oscillator in an inverse-square potential, the path integral for a number of systems can be transformed into such a standard harmonic oscillator form if the local time transformation technique is used [15]. We shall refer to the systems soluble by radial path integration as those of the *confluent hypergeometric class.* Again, the stochastic counterparts are called the *Bessel processes* [29]. For the hypergeometric class, the spherical functions involved are the matrix elements on the discrete basis. In the case of the confluent hypergeometric class, the matrix elements adopted are on a continuous basis [30]. The matrix elements of $SU(1,1)$ on continuous bases have been discussed by Barut and Fronsdal [31], Barut and Phillips [32], Lindblad and Nagel [33], and Mukunda and Radhakrishnan [34].

Section 2 briefly reviews the relations of Feynman's path integral to the evolution operator, the propagator, the resolvent, the energy-dependent Green function and the promotor. In Section 3, path integrals on homogeneous spaces are discussed. In Sections 4 and 5, two specific path integrals are solved with the help of harmonic analysis. Section 4 deals with a free particle on a space of constant negative curvature and Section 5 analyses the dynamical group of the path integral for the harmonic oscillator in d-dimensions. A brief review of the rudiments of harmonic analysis is given in Appendix.

2. Feynman's Path Integral

Let us start with the Hamiltonian of the standard form,

$$H = H_0 + V(\mathbf{x}) = \mathbf{p}^2/(2M) + V(\mathbf{x}), \tag{1}$$

which acts on the Hilbert space $\mathcal{H} = \mathcal{L}^2(\mathbf{R}^d)$. The operators $\mathbf{p}$ and $\mathbf{x}$ represent the momentum and position of a particle of mass M moving in the d-dimensional Euclidean space $\mathbf{R}^d$ under the influence of a scalar potential V. They obey Heisenberg's commutation relations $x_i p_k - p_k x_i = \mathrm{i}\hbar\delta_{ik}$ $(i,k = 1,2,\ldots,d)$ where x_i and p_k are cartesian components of $\mathbf{x}$ and $\mathbf{p}$. As the Schrödinger operator acting on the Hilbert space $\mathcal{H} = \mathcal{L}^2(\mathbf{R}^d)$, the Hamiltonian (1) is self-adjoint with $D(H) = C_0^\infty(\mathbf{R}^d)$, but not necessarily bounded. The time evolution of the system from the initial time t_0 to the final time t is described by the unitary evolution operator $U(t,t_0) = \exp[-(\mathrm{i}/\hbar)(t-t_0)H]$. As long as the Hamiltonian (1) is self-adjoint, the time evolution forms a one-parameter group. Under the causal restriction $t \geq t_1 \geq t_0$, the evolution operator satisfies the initial condition, $U(t_0,t_0) = 1$, and the composition rule, $U(t,t_1)U(t_1,t_0) = U(t,t_0)$.

The propagator is a matrix element of the evolution operator in the position representation defined only for $t'' > t'$:

$$K(\mathbf{x}'', \mathbf{x}'; t'' - t') = \langle \mathbf{x}'' \mid \exp\{-(\mathrm{i}/\hbar)(t'' - t')H\} \mid \mathbf{x}' \rangle, \tag{2}$$

where $\mathbf{x}' = \mathbf{x}(t')$ and $\mathbf{x}'' = \mathbf{x}(t'')$. The initial condition of the evolution operator provides us the normalization condition,

$$\lim_{t \to t'} K(\mathbf{x}, \mathbf{x}'; t - t') = \delta(\mathbf{x} - \mathbf{x}'). \tag{3}$$

From the composition rule readily follows the semi-group property,

$$K(\mathbf{x}'', \mathbf{x}'; t'' - t') = \int_{\mathbf{R}^d} \mathrm{d}\mathbf{x}\, K(\mathbf{x}'', \mathbf{x}; t'' - t)\, K(\mathbf{x}, \mathbf{x}'; t - t'), \tag{4}$$

where $\mathrm{d}\mathbf{x}$ is the translation-invariant Lebesgue measure in $\mathbf{R}^d$. The property (4) leads to

$$K(\mathbf{x}'', \mathbf{x}'; t'' - t') = \int_{\mathbf{R}^d} \mathrm{d}\mathbf{x}_1 \cdots \int_{\mathbf{R}^d} \mathrm{d}\mathbf{x}_{N-1}\, K(\mathbf{x}_N, \mathbf{x}_{N-1}; \tau_N) \cdots K(\mathbf{x}_1, \mathbf{x}_0; \tau_1), \tag{5}$$

where $\mathbf{x}_j = \mathbf{x}(t_j)$, $\tau_j = t_j - t_{j-1} > 0$, $t' = t_0$ and $t'' = t_N$. For convenience, as Feynman did, we adopt now on the isometric subdivision of the time interval $\tau = t'' - t'$ as $\tau_j = \tau/N = \epsilon$ for all j.

In the celebrated 1948 paper, Feynman [35] asserted that the (infinitesimally) short-time propagator can be given by

$$\tilde{K}(\mathbf{x}_j, \mathbf{x}_{j-1}; \epsilon) = \left[\frac{M}{2\pi\mathrm{i}\hbar\epsilon}\right]^{d/2} \exp\left[\frac{\mathrm{i}}{\hbar} S_\epsilon(\mathbf{x}_j, \mathbf{x}_{j-1})\right], \tag{6}$$

with the short-time action,

$$S_\epsilon(\mathbf{x}_j, \mathbf{x}_{j-1}) = \frac{M}{2\epsilon}(\mathbf{x}_j - \mathbf{x}_{j-1})^2 - \frac{1}{2}\epsilon\Big(V(\mathbf{x}_j) + V(\mathbf{x}_{j-1})\Big). \tag{7}$$

Feynman's assertion implies that the propagator can be calculated with (6) by the infinite convolution formula,

$$K(\mathbf{x}'', \mathbf{x}'; \tau) = \lim_{N \to \infty} \int_{\mathbf{R}^d} \prod_{j=1}^{N-1} \mathrm{d}\mathbf{x}_j \prod_{j=1}^{N} \tilde{K}(\mathbf{x}_j, \mathbf{x}_{j-1}; \epsilon). \tag{8}$$

Feynman's formula for the propagator was proven stochastically by Kac [36] for an imaginary time $\beta = \mathrm{i}\tau/\hbar > 0$. For real time, Nelson [37] has a proof for potentials of the Kato class, and Faris [38; 39] has a proof for the Rollnik class. Both are based on Trotter's product formula,

$$\mathrm{e}^{(\mathrm{i}/\hbar)\tau H} = \lim_{\varepsilon \downarrow 0} \lim_{N \to \infty} \left(\mathrm{e}^{(\mathrm{i}/\hbar)\tau H_0/N(1-\mathrm{i}\varepsilon)}\, \mathrm{e}^{(\mathrm{i}/\hbar)\tau V/N}\right)^N. \tag{9}$$

In practice, Feynman's path integral can be evaluated only for quadratic systems. To expand the scope of Feynman's path integral, we also pay attention to the resolvent of the Hamiltonian,

$$G = \frac{1}{E-H} = \frac{1}{\mathrm{i}\hbar}\int_0^\infty \mathrm{d}\tau \, \exp\{(\mathrm{i}/\hbar)\tau(E-H)\}, \qquad \mathrm{Im}\, E > 0. \tag{10}$$

The matrix element of the resolvent in the position representation, which is often referred to as the energy-dependent Green function, is given by

$$G(\mathbf{x}'', \mathbf{x}'; E) = \langle \mathbf{x}'' | (E-H)^{-1} | \mathbf{x}' \rangle = \frac{1}{\mathrm{i}\hbar}\int_0^\infty \mathrm{d}\tau \, P(\mathbf{x}'', \mathbf{x}'; \tau), \tag{11}$$

where

$$P(\mathbf{x}'', \mathbf{x}'; \tau) = \langle \mathbf{x}'' | \exp\{(\mathrm{i}/\hbar)\tau(E-H)\} | \mathbf{x}' \rangle. \tag{12}$$

The last entity $P(\mathbf{x}'', \mathbf{x}'; \tau)$, which we call the *promotor*, is also expressible as a path integral in Feynman's form. The Green function contains the same quantum-mechanical information as that the propagator has. Therefore, the path integral may be calculated for the promotor rather than the propagator. Once the Green function is found with the help of the promotor, it can also be converted into the propagator by a Fourier transformation.

It is true that the path integral structure of the promotor is identical with that of the propagator except for the additional energy term. When Feynman's path integral is difficult to evaluate, the path integral for the promotor is equally difficult to calculate. Nevertheless, the promotor has a unique advantage. Suppose it is transformed into

$$\tilde{P}(\mathbf{x}'', \mathbf{x}'; \sigma) = \langle \mathbf{x}'' | \exp\{(\mathrm{i}/\hbar)\sigma f(\mathbf{x})(E-H)g(\mathbf{x})\} | \mathbf{x}' \rangle \tag{13}$$

with $\sigma = \tau/[f(\mathbf{x}')g(\mathbf{x}'')]$ and positive-definite q-number functions $f(\mathbf{x})$ and $g(\mathbf{x})$. Then we can show that the Green function (11) can also be evaluated by

$$G(\mathbf{x}'', \mathbf{x}'; E) = \frac{1}{\mathrm{i}\hbar}\int_0^\infty \mathrm{d}\sigma \, \tilde{P}(\mathbf{x}'', \mathbf{x}'; \sigma)\, (\mathrm{d}\tau/\mathrm{d}\sigma). \tag{14}$$

The result remains the same. This is a remarkable property. The above time transformation of the promotor is sometimes called the *local time rescaling trick* [15]. This added flexibility in the path integral has contributed much to the development of path integral calculus in the last decade. In this regard, when we talk about Feynman's path integral, we may include the path integral for the promotor as well as the original path integral for the propagator. All the arguments given now on are equally applicable to the propagator and the promotor.

Suppose Feynman's assertion is fully justified for a certain class of potentials. However, we have to note that the background space of the time-sliced path integral (11) is $\mathbf{R}^d$. In fact, Feynman recognized that the cartesian coordinate system played a special role in defining Feynman's path integral, and suggested that only the short-time propagator found in cartesian variables may be expressed by a coordinate transformation in any choice of coordinate variables [35]. The special role of the cartesian coordinates is not unique in Feynman's path integral. We can find a similar situation in the canonical quantization procedure. Recall that Heisenberg's commutation relations are applicable only to cartesian variables. In setting up Schrödinger's equation from the Hamiltonian (1), we have to replace $\mathbf{p}$ by $-i\hbar\,\nabla$ in cartesian variables. Then, for a spherically symmetric system, for instance, we transform the Laplace-Beltrami operator ∇^2 from cartesian variables to polar coordinates. The connection of these two similar situations must be of a profound significance, though we do not discuss it any further.

By expressing the short-time propagator in polar coordinates and expanding in terms of Legendre functions (the zonal spherical functions on S^2), Feynman's path integral in three dimensions has been separated into the radial path integral and the angular part [12; 13] as ($r' = |\mathbf{x}'|$, $r'' = |\mathbf{x}''|$)

$$K(\mathbf{x}'', \mathbf{x}'; \tau) = \frac{1}{4\pi} \sum_{l=0}^{\infty} (2l+1)\, K_l(r'', r'; \tau)\, P_l(\mathbf{x}' \cdot \mathbf{x}''/r'r''). \tag{15}$$

Similarly, it is possible to express Feynman's path integral in any desired coordinate system as long as the background space remains flat. However, there is no established formulation of a path integral in a general curved space. In this article, our interest is not in formulating a path integral in a general curved space. We are rather interested in the questions as to how we can take advantage of a symmetry of the physical system when we carry out path integration and whether the path integral can be extended to a homogeneous space (or more restrictively a symmetric space) and solved in much the same fashion that the polar coordinate path integral is treated.

3. Path Integral on a Homogeneous Space

Let a group G be a transformation group on a space $\mathcal{M}$. If G acts transitively on $\mathcal{M}$, then $\mathcal{M}$ is a homogeneous space with respect to G. If H is the isotropy group of G at a point q_a of $\mathcal{M}$, then $\mathcal{M} = G/H$. For details, see Appendix.

In an effort to understand Feynman's path integral on a homogeneous space, we assert that the propagator on a space $\mathcal{M}$ equipped with a measure $\mathrm{d}q$ can be given by the following multi-convolution, ($q'' = q_N$, $q' = q_0$)

$$K(q'', q'; \tau) = \lim_{N\to\infty} \int_{\mathcal{M}} \mathrm{d}q_1 \cdots \int_{\mathcal{M}} \mathrm{d}q_{N-1}\, \tilde{K}(q_N, q_{N-1}; \epsilon) \cdots \tilde{K}(q_1, q_0; \epsilon), \tag{16}$$

in analogy to the time-sliced path integral (8). The finite-time propagator in $\mathcal{M}$ is defined in the Hilbert space $\mathcal{H} = \mathcal{L}^2(\mathcal{M})$. Here we assume that it has the following properties,

$$\lim_{t\to t'} K(q, q'; t - t') = \delta(q - q'), \tag{17}$$

and

$$\int_{\mathcal{M}} dq\, K(q'', q; t'' - t)\, K(q, q'; t - t') = K(q'', q'; t'' - t). \tag{18}$$

Note that the finite-time propagator $K(q'', q'; \tau)$ approaches the short-time propagator $\tilde{K}(q, q'; \epsilon)$ as τ tends to ϵ, but the converse is not necessarily true. The functional form of $K(q'', q'; \tau)$ is generally different from $\tilde{K}(q, q'; \epsilon)$. The latter can be an approximation of the former. However, the short-time propagator has to obey the normalization condition (17). It must also satisfy the semi-group property (18) when the exponential contributions of $\mathcal{O}(\epsilon^2)$ are ignored. Keeping this subtle difference in mind, we use now on the same notation for the short-time propagator and the finite-time propagator by dropping the tilde from the short-time propagator.

Then we utilize the techniques in harmonic analysis to evaluate the path integral on a homogeneous space. The transformation group of the homogeneous space may be directly related to a geometrical symmetry of the system in question. Or it may be of a dynamical origin.

In order to convert a path integral defined on the homogeneous space $\mathcal{M}$ into one on the group manifold G, we restrict the short-time propagator $K(q, q'; \epsilon)$ to be symmetric under the interchange of two end points, $q \rightleftharpoons q'$, and invariant under the action of $g \in G$. Namely,

$$K(q, q'; \epsilon) = K(q', q; \epsilon) = K(gq', gq; \epsilon), \quad \text{for all} \ \ g \in G. \tag{19}$$

Then, we introduce for the fixed q_a the following function,

$$u_\epsilon(g) = K(q_a, gq_a; \epsilon). \tag{20}$$

It is obvious from the properties (19) that

$$K(q, q'; \epsilon) = u_\epsilon(g^{-1}g') = u_\epsilon(g'^{-1}g) \ \text{ for } \ q = gq_a\,, \quad q' = g'q_a. \tag{21}$$

Furthermore, it is easily to verify that

$$u_\epsilon(g) = u_\epsilon(h_1^{-1}gh_2) \ \text{ for all } \ h_1, h_2 \in H. \tag{22}$$

Hence, the function $u_\epsilon(g)$ is a zonal function [40; 41].

Consequently, we can express Feynman's path integral (16) as the limit of a multi-convolution (see Appendix),

$$K(q'', q'; \tau) = \lim_{N\to\infty} \int_G dg_1 \cdots \int_G dg_{N-1}\, u_\epsilon(g_0^{-1}g_1) \cdots u_\epsilon(g_{N-1}^{-1}g_N) \tag{23}$$

or

$$K(q, q_0; \tau) = \lim_{N\to\infty} \prod_{j=1}^{N} * \, u_\epsilon(g_{j-1}^{-1}g_j) \tag{24}$$

where $q_j = g_j q_a$, $j = 0, \ldots, N$. Thus, the path integral in a homogeneous space is reduced to a convolution in a group manifold G.

Since the short-time propagator $u_\epsilon(g)$ is a spherical function which becomes constant on the two-sided cosets HgH, it can be expanded in terms of the zonal spherical functions $D^l_{00}(g)$. At this point, however, we make an assumption that the group G can be given as a direct product $A \otimes B$ of two unimodular groups A and B, and the isotropy group H of G at q_a is a subgroup of A. Let $a \in A$, $b \in B$ and $g = ab \in G$. Then, the zonal spherical function $D^l_{00}(g)$ of G may be decomposed in terms of the zonal spherical functions $D^l_{00}(a)$. As a result, the short-time propagator can be expressed as

$$u_\epsilon(g) = \sum_{l \in \Lambda} d_l \, \lambda_l(b;\epsilon) \, D^l_{00}(a) \tag{25}$$

where Λ stands for the set of all spherical representations (see Appendix) and the expansion coefficients are

$$\lambda_l(b;\epsilon) = \int_A \mathrm{d}a \, u_\epsilon(ab) \, D^l_{00}(a^{-1}). \tag{26}$$

Inserting the series expansion (25) into (23) and noticing that $\mathrm{d}g = \mathrm{d}a\mathrm{d}b$, we calculate the convolution. Use of the orthogonality relation of the zonal spherical functions (see Appendix) immediately leads to

$$K(q'', q'; \tau) = \sum_{l \in \Lambda} d_l \, K_l(b_0^{-1} b_N; \tau) \, D^l_{00}(a_0^{-1} a_N) \tag{27}$$

where

$$K_l(b_0^{-1} b_N; \tau) = \lim_{N \to \infty} \prod_{j=1}^{N} * \lambda_l(b_{j-1}^{-1} b_j; \epsilon). \tag{28}$$

The last convolution will be calculated later for a specific example.

If, in particular, the subgroup B consists of the unit element e alone, that is, if $g = a$, we have

$$K_l(e;\tau) = \lim_{N \to \infty} [\lambda_l(\epsilon)]^N. \tag{29}$$

Since the normalization condition (17) must be satisfied, we obtain

$$\lim_{\epsilon \to 0} \lambda_l(\epsilon) = 1 \quad \text{for all } l \in \Lambda. \tag{30}$$

This limiting condition allows us to calculate the remaining limit in (29). To be more explicit,

$$\lim_{N \to \infty} [\lambda_l(\tau/N)]^N = \lim_{N \to \infty} \left[1 + (\tau/N)\dot{\lambda}_l(0)\right]^N = \exp\left\{\tau \dot{\lambda}_l(0)\right\} \tag{31}$$

where $\dot{\lambda}_l(\epsilon) = \mathrm{d}\lambda_l(\epsilon)/\mathrm{d}\epsilon$. The resulting propagator reads [10]

$$K(q'', q'; \tau) = \sum_{l \in \Lambda} d_l \, \exp\left\{\dot{\lambda}_l(0)\tau\right\} \, D^l_{00}(g_0^{-1} g_N). \tag{32}$$

In this particular case, since the Hilbert space can be decomposed as $\mathcal{H} = \bigoplus_{l\in\Lambda} \mathcal{H}^l$, we can make the spectral decomposition of the Hamiltonian,

$$H = \sum_{l\in\Lambda} \sum_{m} E_l \, |l,m\rangle\langle l,m| \tag{33}$$

where E_l are the eigenvalues of H and $\{|l,m\rangle\}$ is a complete orthonormal basis in $\mathcal{H}^l$. Therefore, we have

$$K(q'',q';\tau) = \sum_{l,m} \exp\{-(\mathrm{i}/\hbar)E_l\tau\}\langle q''|l,m\rangle\langle l,m|q'\rangle. \tag{34}$$

Comparing this with the result (32) we can identify the spectrum as well as the corresponding eigenstates of H:

$$E_l = \mathrm{i}\hbar\dot{\lambda}_l(0) \quad , \quad \langle q|l,m\rangle = \sqrt{d_l}\; D^l_{m0}(g). \tag{35}$$

Although this result is rather obvious, it provides a simple prescription to obtain the correct Hamiltonian associated with a given semiclassical short-time propagator. This will be illustrated in the example discussed in the following section.

4. Quantum Mechanics on a Space of Constant Negative Curvature

Recent interest in quantum mechanics of classically chaotic systems has revived the study of quantum motion on spaces of constant negative curvature. For example, the classical free motion on a two-dimensional compact space of constant negative curvature exhibits chaotic behavior [42]. Naturally, it is important to investigate the quantum aspect of such a motion.

In what follows, we shall study quantum-mechanically a particle of mass M on a d-dimensional non-compact space of constant negative curvature, which is an integrable system.

The line element of a space having curvature $K = -1/R^2$ is given by [43]

$$\mathrm{d}s^2 = (1 + r^2/R^2)^{-1}\mathrm{d}r^2 + r^2\mathrm{d}\Omega^2. \tag{36}$$

Here, $R\,(>0)$ is the "radius" of curvature, and $\mathrm{d}\Omega^2$ represents the $(d-1)$-dimensional angular part of the line element, which is identical with that of a flat Euclidean space $\mathbf{R}^d$. Introducing a new variable θ by setting $\sinh\theta = r/R$ $(\theta \geq 0)$, the line element can be put into the form

$$\mathrm{d}s^2 = R^2\mathrm{d}\theta^2 + R^2\sinh^2\theta\mathrm{d}\Omega^2. \tag{37}$$

This geometry can be embedded in a $(d+1)$-dimensional pseudo-Euclidean space as follows:

$$\mathbf{x} = R\sinh\theta\,\mathbf{e}(\Omega), \quad x_{d+1} = R\cosh\theta \quad \Rightarrow \quad \mathrm{d}s^2 = \mathrm{d}\mathbf{x}^2 - \mathrm{d}x_{d+1}^2. \tag{38}$$

The vector $\mathbf{e}(\Omega)$ is a unit vector in the Euclidean subspace $\mathbf{R}^d$ having polar coordinates Ω. Remember that Ω stands short for the $d-1$ angular coordinates. By this embedding the space $\mathcal{M}$ can be identified with the upper sheet of a time-like subspace of constant "radius" R of a pseudo-Euclidean space with metric $(+1, \ldots, +1, -1)$.

The transformation group acting transitively on $\mathcal{M}$ is $SO(d,1)$. The isotropy group about the origin $\theta = 0$ is the compact subgroup $SO(d)$. This means that the space $\mathcal{M}$ may be identified with the quotient space $SO(d,1)/SO(d)$. The variables, θ and Ω, indicating the position of the particle on $\mathcal{M}$, may be identified with the group parameters. Namely, a group element of $SO(d,1)$ which transforms the origin $q_a = (0,0)$ into a point $q = (\theta, \Omega)$ by $q = gq_a$ is parameterized as $g = g(\theta, \Omega)$. The invariant Lebesgue measure on $\mathcal{M}$ is related to the Haar measure of $SO(d,1)$ (as defined in Ref. [2] p.509):

$$\int_{\mathcal{M}} \mathrm{d}\theta \mathrm{d}\Omega \sinh\theta\, f\Big(q(\theta,\Omega)\Big) = \frac{2\pi^{(d+1)/2}}{\Gamma\left(\frac{d+1}{2}\right)} \int_{SO(d,1)} \mathrm{d}g\, f(gq_a). \tag{39}$$

This relation is indeed compatible with (A.2) of the Appendix. Note that the Haar measure is only unique up to a multiple constant.

Having set up the geometry of the configuration space, we have to construct the short-time propagator for the particle in motion. The Lagrangian for the classical free motion in the above geometry is given by

$$L = \frac{m}{2}\left(\frac{\mathrm{d}s}{\mathrm{d}t}\right)^2. \tag{40}$$

As Feynman pointed out already in his 1948 paper, we must recognize the special role of cartesian coordinates. In the present problem, there is no cartesian coordinate system. The coordinates closest to the cartesian ones are those given in (37). In analogy with the Ozaki-Edward-Gulyaev prescription [11; 12] in formulating the polar coordinate path integral, we choose [10]:

$$S_\epsilon(q,q') = \frac{MR^2}{\epsilon}(\cosh\Theta - 1) + \frac{\hbar^2\epsilon}{8MR^2} \tag{41}$$

where Θ is the angle of the hyperbolic rotation $g = g(\Theta)$ transforming q' into q, $q = gq'$. When analytical continuation in Θ is made, this short-time action goes over to the counterpart for the compact group $SO(d+1)$. However, there is ambiguity in the choice of the additive constant. This is a common problem for both the compact and non-compact cases. As far as a free motion is concerned, there is no decisive criterion, physical or mathematical, for the selection of the constant term.

With the short-time action chosen above, we have the short-time propagator,

$$K(q,q';\epsilon) = \left(\frac{MR^2}{2\pi\mathrm{i}\hbar\epsilon}\right)^{d/2} \exp\{(\mathrm{i}/\hbar)S_\epsilon(q,q')\}. \tag{42}$$

The prefactor has been chosen in order to comply with the normalization condition (17).

In passing from the path integral on the homogeneous space $\mathcal{M}$ to the one on the group manifold $G = SO(d,1)$, we have to bear the relation (39) in mind. Then, we apply the Fourier transformation to the short-time propagator. Since there are no bound states for the free motion, the energy spectrum expected are continuous. Therefore, we use the irreducible unitary representations of $SO(d,1)$ belonging to the fundamental series characterized by a complex label $l = l(\rho) = -(d-1)/2 + \mathrm{i}\rho$, $(\rho \geq 0)$. The integral decomposition of the short-time propagator is

$$\frac{2\pi^{(d+1)/2}}{\Gamma(\frac{d+1}{2})} K(q,q';\epsilon) = \int_0^\infty \mathrm{d}\rho\, d_l\, \lambda_l(\epsilon)\, D^l_{00}(g'^{-1}g). \tag{43}$$

In this representation, the zonal functions are explicitly given by Gegenbauer functions [2]

$$D^l_{00}(g) = \frac{\Gamma(d-1)\Gamma(l+1)}{\Gamma(l+d-1)} C_l^{(d-1)/2}(\cosh\Theta) \tag{44}$$

and are related to those of $SO(d+1)$ by analytical continuation in Θ and l. The constant d_l is given by [7]

$$d_l = 2\frac{|\Gamma(\frac{d-1}{2} + \mathrm{i}\rho)|^2}{\Gamma(d)|\Gamma(\mathrm{i}\rho)|^2}. \tag{45}$$

The Fourier coefficients calculated with (26) take the form [7],

$$\lambda_l(\epsilon) = \left(\frac{2MR^2}{\pi \mathrm{i}\hbar\epsilon}\right)^{1/2} \exp\left\{\frac{\mathrm{i}\hbar\epsilon}{8MR^2}\right\} K_{\mathrm{i}\rho}(MR^2/\mathrm{i}\hbar\epsilon) \tag{46}$$

where $K_\nu(z)$ is the modified Bessel function of the third kind. Use of the limiting relation,

$$\lim_{N\to\infty}\left[\sqrt{\frac{2Nz}{\pi}}\mathrm{e}^{1/(8Nz)}\mathrm{e}^{Nz}K_{\mathrm{i}\rho}(Nz)\right]^N = \exp\{-\rho^2/2z\} \tag{47}$$

which is obtained from the asymptotic form of the Bessel function

$$K_{\mathrm{i}\rho}(Nz) \sim \sqrt{\pi Nz/2}\,\mathrm{e}^{-Nz}\left(1 - \frac{\rho^2 + 1/4}{2Nz}\right), \quad N \to \infty,$$

enables us to put the propagator into the form [7]

$$K(q'',q';t) = \frac{\Gamma(\frac{d+1}{2})}{2\pi^{(d+1)/2}} \int_0^\infty \mathrm{d}\rho\, d_l \exp\left\{-\frac{\mathrm{i}}{\hbar}\frac{\hbar^2\rho^2}{2MR^2}t\right\} D^l_{00}(g_0^{-1}g_N). \tag{48}$$

Obviously the spectrum of the Hamiltonian associated with (48) is given by $E_\rho = \hbar^2\rho^2/2MR^2$ $(\rho \geq 0)$. This may be compared with the spectrum of the Laplace-Beltrami operator ∇^2 on $\mathcal{M}$ [2] which has the spectrum $l(l+d-1)$. The Hamiltonian which corresponds to our choice (41) of the short-time action is

$$H = -\frac{\hbar^2}{2M}\nabla^2 + \frac{(d-1)^2}{8MR^2}\hbar^2. \tag{49}$$

There have also been other suggestions for short-time actions in the literature [7; 44]. However, the present choice (41) has the consequence that for its corresponding Hamilton operator (49) Huygens' principle is valid [45].

Although the propagator cannot be given in a simple form, its Laplace transform, that is, the energy-dependent Green function, can be given in closed form [46]. The propagator (48) may be written, if d is odd, in the form [10]

$$K(q'',q';\tau) = R\left(\frac{M}{2\pi i\hbar\tau}\right)^{1/2}\left(\frac{-1}{2\pi\sinh\Theta}\frac{\partial}{\partial\Theta}\right)^{(d-1)/2}\exp\{(i/\hbar)S_{cl}(\Theta,\tau)\} \tag{50}$$

where $S_{cl}(\Theta,\tau) = (MR^2\Theta^2/2\tau)$ is the classical action and Θ is the hyperbolic angle between the two positions q' and q''. For d even, it takes a similar but different form,

$$\begin{aligned} K(q'',q';\tau) = {} & \sqrt{2}R^2\left(\frac{M}{2\pi i\hbar\tau}\right)^{3/2}\left(\frac{-1}{2\pi\sinh\Theta}\frac{\partial}{\partial\Theta}\right)^{(d-1)/2} \\ & \times\int_{\Theta}^{\infty} dz\,\frac{z\exp\{(iMR^2/\hbar 2\tau)z^2\}}{\sqrt{\cosh z - \cosh\Theta}}. \end{aligned} \tag{51}$$

5. The Harmonic Oscillator

The harmonic oscillator is so ordinary that it can hardly be an attractive object. Historically, it has been the only object that Feynman's path integral can handle. As is mentioned earlier, the hydrogen atom is too difficult to path-integrate. However, thanks to several techniques developed in the past ten years, it has become possible to reduce many non-quadratic path integrals including the one for the hydrogen atom into the path integral for the harmonic oscillator. By the harmonic oscillator, here, we mean an isotropic harmonic oscillator in higher dimensions. A path integral of the confluent hypergeometric class is reducible to the path integral for the radial harmonic oscillator in an inverse-square potential. Performing path integration in polar coordinates even for the harmonic oscillator is not trivial. In fact, the path integration in polar coordinates was done already in 1967 by brute force [13]. It was only after the work of Böhm and Junker [7] when the study of the link between the radial propagator and the matrix element of the $SU(1,1)$ began. Because of the spherical symmetry, it is expected that the propagator is decomposable in terms of the spherical functions. However, it is somewhat surprising that there is a beautiful underlying structure in the radial path integral that is connected to the non-compact group $SU(1,1)$.

5.1. The Dynamical Group

First we examine the group structure of an isotropic harmonic oscillator in $\mathbf{R}^d$, having mass M and spring constant $M\omega^2$. The Hamiltonian is given by

$$H = \frac{1}{2M}\mathbf{p}^2 + \frac{1}{2}M\omega^2\mathbf{x}^2. \tag{52}$$

The dynamical group of the three-dimensional isotropic harmonic oscillator has been identified with the symplectic group $Sp(6)$ by Moshinski and Quesne [47]. A subgroup $Sp(2)\otimes SO(3)$ of $Sp(6)$ has been chosen as the basis for the separation in spherical polar coordinates. The group $SO(3)$ is associated with the rotational symmetry of the oscillator, and the symplectic group $Sp(2)$, which is isomorphic with $SU(1,1)$, is the spectrum-generating group. Generalizing this to the d-dimensional case, we use the group $SU(1,1)\otimes SO(d)$ to separate the system in polar coordinates.

Let us realize the algebra of $SU(1,1)$ by

$$\begin{aligned} J_1 &= -\frac{1}{4M\hbar\omega}\left(\mathbf{p}^2 - M^2\omega^2\mathbf{x}^2\right), \quad J_2 = -\frac{1}{4\hbar}\left(\mathbf{x}\cdot\mathbf{p}+\mathbf{p}\cdot\mathbf{x}\right), \\ J_3 &= \frac{1}{4M\hbar\omega}\left(\mathbf{p}^2 + M^2\omega^2\mathbf{x}^2\right). \end{aligned} \tag{53}$$

Namely, these operators satisfy

$$[J_1, J_2] = -\mathrm{i}J_3, \quad [J_2, J_3] = \mathrm{i}J_1, \quad [J_3, J_1] = \mathrm{i}J_2. \tag{54}$$

The Casimir operator is $\mathbf{J}^2 = -J_1^2 - J_2^2 + J_3^2$. On the standard orthonormal basis $|J,\mu\rangle$,

$$\mathbf{J}^2|J,\mu\rangle = J(J+1)|J,\mu\rangle \quad , \quad J_3|J,\mu\rangle = \mu|J,\mu\rangle. \tag{55}$$

The unitary irreducible representations of $SU(1,1)$ are labeled by the number J. There are two discrete series denoted by D_J^+ and D_J^- and two continuous series denoted by C_J and E_J [48].

In the present realization, the Hamiltonian is proportional to J_3 as

$$H = 2\hbar\omega J_3. \tag{56}$$

Since J_3 is positive-definite by construction, the eigenvalues of J_3 must be positive. Therefore, the set of the operators (53) provides a realization of the representation D_J^+ of $SU(1,1)$ for which $\mu = -J + n \quad (n \in \mathbb{N}_0)$ [31].

It can be verified that the Casimir operator $\mathbf{J}^2$ calculated with (53) is related to the Casimir invariant $\mathbf{L}^2$ of $SO(d)$ as

$$\mathbf{J}^2 = \frac{1}{4\hbar^2}\mathbf{L}^2 + \frac{1}{16}\,d(d-4). \tag{57}$$

Here $\mathbf{L}$ is the angular momentum in $\mathbb{R}^d$ whose components are $L_{ik} = x_i p_k - x_k p_i$ $(i,k = 1,2,\ldots,d)$. The Casimir invariant of $SO(d)$,

$$\mathbf{L}^2 = \frac{1}{2}\sum_{i,k=1}^{d} L_{ik}L_{ik}, \tag{58}$$

has the spectrum $\hbar^2 l(l+d-2)$ $(l \in \mathbb{N}_0)$. Therefore, we have

$$J(J+1) = \frac{1}{4}l(l+d-2) + \frac{1}{16}d(d-4). \tag{59}$$

Thus, the subspace of the Hilbert space $\mathcal{H} = \mathcal{L}^2(\mathbb{R}^d)$ with a fixed angular momentum l carries also the irreducible representation space of $D^+_{J(l)}$ with

$$J(l) = -\frac{1}{2}l - \frac{1}{4}d. \tag{60}$$

The Hilbert space $\mathcal{H}$ may be decomposed into an orthogonal sum of subspaces $\mathcal{H}^l$ with $l \in \mathbb{N}_0$. Each subspace is a product of the $SU(1,1)$ representation D_J^+ and the $SO(d)$ representation denoted by D^l. Namely,

$$\mathcal{L}^2(\mathbf{R}^d) = \mathcal{L}^2(\mathbf{R}^+) \otimes \mathcal{L}^2(S^{d-1}) = \bigoplus_{l=0}^{\infty} \left(D_{J(l)}^+ \otimes D^l \right). \tag{61}$$

The basis in $\mathcal{H} = \mathcal{L}^2(\mathbf{R}^d)$ is given as a tensor product of the discrete basis $\{|J(l), \mu\rangle\}$ in $D_{J(l)}^+$ and the basis $\{|l, m\rangle\}$ in D^l, that is,

$$|\mu, l, m\rangle = |J(l), \mu\rangle \otimes |l, m\rangle. \tag{62}$$

Obviously, these are the eigenstates of the Hamiltonian (56):

$$H|\mu, l, m\rangle = (2\hbar\omega J_3 \otimes \mathbf{1})\, |\mu, l, m\rangle = 2\hbar\omega\mu|\mu, l, m\rangle. \tag{63}$$

Using $\mu = -J(l) + \nu = \nu + l/2 + d/4 = n/2 + d/4$, we obtain the expected result,

$$E_n = \hbar\omega(n + d/2) \ , \quad n = 2\nu + l \in \mathbb{N}_0 \ . \tag{64}$$

Here we have to note that the evolution operator $\exp\{-(\mathrm{i}/\hbar)H\tau\}$ which is now identified with $\exp\{-2\mathrm{i}\omega\tau J_3\}$ is an element of $SU(1,1)$. The third generator J_3 of $SU(1,1)$ acts as the generator of the one-parameter group of time evolution. The group $SU(1,1)$ is literally the dynamical group of the harmonic oscillator.

The orthonormal set $\{|J, \mu\rangle\}$ used above forms a discrete basis, which diagonalizes the compact operator J_3. Although the discrete basis is widely used, the use of a continuous basis which diagonalizes a non-compact operator is often overlooked. There have been extensive studies of continuous bases [32; 33; 34]. For instance, $K_+ = J_1 + J_3$ is a non-compact operator. The basis which makes K_+ diagonal is a continuous basis:

$$K_+|J, \eta\rangle = \eta|J, \eta\rangle \tag{65}$$

where the eigenvalue η is a continuous variable. In the present realization, it happens to be

$$K_+ = \frac{M\omega}{2\hbar}\mathbf{x}^2. \tag{66}$$

Obviously, the eigenvalue of K_+ is

$$\eta = \alpha r^2 \tag{67}$$

where $\alpha = M\omega/(2\hbar)$ and $r = |\mathbf{x}|$. The corresponding eigenstates in $\mathcal{H}$ are given by

$$|\eta, l, m\rangle = |J(l), \eta\rangle \otimes |l, m\rangle. \tag{68}$$

In parallel to Wigner's d-function, $d^l_{mn}(\theta) = \langle l, m|\mathrm{e}^{-\mathrm{i}\theta J_2}|l, m\rangle$ defined for the compact group $SU(2)$, Bargmann defined the functions, $b^l_{mn}(\theta) = \langle l, m|\mathrm{e}^{-\mathrm{i}\theta J_2}|l, m\rangle$ on the discrete basis of $SU(1,1)$. In the case of $SU(2)$, all the group generators are compact, whereas J_3 of $SU(1,1)$ is the only compact operator and J_2 is non-compact. The function defined by Bargmann is a matrix element of a non-compact member of the $SU(1,1)$ group on the discrete basis. It is certainly interesting to define a function which is a matrix element of a member of the maximal compact subgroup of $SU(1,1)$ on the continuous basis that diagonalizes K_+. Namely,

$$v^J_{\eta\eta'}(\theta) = \langle J, \eta| \exp\{-\mathrm{i}\theta J_3\} |J, \eta'\rangle \tag{69}$$

where $0 < \theta < 2\pi$. In the unitary irreducible representation of the discrete series D^+_J, it has been shown by Lindblad and Nagel that the v-function takes the following explicit form [33; 15],

$$v^J_{\eta\eta'}(2\varphi) = \frac{1}{\mathrm{i}\sin\varphi} \exp\{\mathrm{i}(\eta + \eta')\cot\varphi\}\, \mathrm{I}_{-2J-1}\left(\frac{2\sqrt{\eta\eta'}}{\mathrm{i}\sin\varphi}\right), \tag{70}$$

where $\mathrm{I}_\nu(z)$ is the modified Bessel function of the first kind and $0 < \varphi < \pi$. This function turns out to be the core function of the radial propagator of the harmonic oscillator.

5.2. The propagator

According to the relation (56), the propagator as a matrix element of the evolution operator can be put in the form,

$$K(\mathbf{x}'', \mathbf{x}'; \tau) = \langle \mathbf{x}'' \mid \exp\{-2\mathrm{i}\omega\tau J_3\} \mid \mathbf{x}'\rangle. \tag{71}$$

Using the basis (68) which diagonalizes the non-compact operator K_+, we can express (71) as

$$K(\mathbf{x}'', \mathbf{x}'; \tau) = \sum_{l,m} \int_{\mathbf{R}^+} \mathrm{d}\eta'' \int_{\mathbf{R}^+} \mathrm{d}\eta' \, \langle \mathbf{x}''|\eta'', l, m\rangle\langle J(l), \eta''|\mathrm{e}^{-2\mathrm{i}\omega\tau J_3}|J(l), \eta'\rangle\langle \eta', l, m|\mathbf{x}'\rangle. \tag{72}$$

Let $\mathbf{x} = r\mathbf{u}$ where $r = |\mathbf{x}|$ and $\mathbf{u} = \mathbf{x}/r$, and let also $\mathbf{u} = h\mathbf{u}_a$ where $h \in SO(d)$ and $\mathbf{u}_a$ is a fixed point on S^{d-1}. Then we can write the position states as $|\mathbf{x}'\rangle = |r', h'\mathbf{u}_a\rangle$ and $|\mathbf{x}''\rangle = |r'', h''\mathbf{u}_a\rangle$, where $h', h'' \in SO(d)$. The propagator (72) can now be expressed in the form,

$$K(\mathbf{x}'', \mathbf{x}'; \tau) = \frac{\Gamma(d/2)}{2\pi^{d/2}} \sum_{l=0}^{\infty} K_l(r'', r'; \tau)\, d_l\, D^l_{00}(h''^{-1}h') \tag{73}$$

with

$$K_l(r'', r'; \tau) = \int_{\mathbf{R}^+} \mathrm{d}\eta'' \int_{\mathbf{R}^+} \mathrm{d}\eta' \, \langle r'' \mid \eta''\rangle\langle J(l), \eta'' \mid \mathrm{e}^{-2\mathrm{i}\omega\tau J_3} \mid J(l), \eta'\rangle\langle \eta' \mid r'\rangle. \tag{74}$$

The dimension of the representation D^l is given by

$$d_l = (2l + d - 2)\frac{(l + d - 3)!}{l!\,(d-2)!}, \tag{75}$$

and the zonal spherical functions are expressed in terms of Gegenbauer polynomials

$$D^l_{00}(h''^{-1}h') = \frac{l!\,\Gamma(d-2)}{\Gamma(d+l-2)}\, C_l^{(d-2)/2}(\mathbf{u}''\cdot\mathbf{u}'). \tag{76}$$

Note that the area of $S^{d-1} = SO(d)/SO(d-1)$ is $2\pi^{d/2}/\Gamma(d/2)$.

Since $\mathrm{d}\eta = 2\alpha r\,\mathrm{d}r$, and since the completeness relations of the sets, $\{|r\rangle\}$ and $\{|J(l),\eta\rangle\}$, are given by

$$\int_{\mathbf{R}^+} \mathrm{d}r\, r^{d-1}\, |r\rangle\langle r| = 1, \qquad \int_{\mathbf{R}^+} \mathrm{d}\eta\, |J(l),\eta\rangle\langle J(l),\eta| = 1.$$

we have

$$\langle r \mid J(l),\eta\rangle = \sqrt{2\alpha}\, r^{-(d-2)/2}\,\delta(\eta - \alpha r^2), \qquad \alpha = M\omega/(2\hbar). \tag{77}$$

Making use of this relation in (74), we can immediately obtain the radial propagator in the form,

$$\begin{aligned} K_l(r'',r';\tau) &= \frac{M\omega}{\hbar}(r'r'')^{-(d-2)/2}\,\langle J(l),\eta''|\exp\{-2\mathrm{i}\omega\tau J_3\}|J(l),\eta'\rangle \\ &= \frac{M\omega}{\hbar}(r'r'')^{-(d-2)/2}\, v^{J(l)}_{\eta''\eta'}(2\omega\tau) \end{aligned} \tag{78}$$

where $\eta' = (M\omega/2\hbar)r'^2$ and $\eta'' = (M\omega/2\hbar)r''^2$.

5.3. Path Integration for the Propagator

Next, let us make an explicit calculation of Feynman's path integral for the oscillator:

$$K(\mathbf{x}'',\mathbf{x}';\tau) = \lim_{N\to\infty}\int_{\mathbf{R}^d}\prod_{j=1}^{N-1}\mathrm{d}\mathbf{x}_j\prod_{j=1}^{N}K(\mathbf{x}_j,\mathbf{x}_{j-1};\epsilon) = \lim_{N\to\infty}\prod_{j=1}^{N} * K(\mathbf{x}_j,\mathbf{x}_{j-1};\epsilon). \tag{79}$$

The Lagrangian corresponding to the Hamiltonian (52) is

$$L = \frac{1}{2}M\dot{\mathbf{x}}^2 - \frac{1}{2}M\omega^2\,\mathbf{x}^2. \tag{80}$$

The short-time propagator of the harmonic oscillator in $\mathbf{R}^d$ is given by

$$K(\mathbf{x},\mathbf{x}';\epsilon) = \left(\frac{M}{2\pi\mathrm{i}\hbar\epsilon}\right)^{d/2}\exp\left\{\frac{\mathrm{i}}{\hbar}\left[\frac{M}{2\epsilon}(\mathbf{x}-\mathbf{x}')^2 - \frac{M}{4}\omega^2\epsilon(\mathbf{x}^2+\mathbf{x}'^2)\right]\right\}. \tag{81}$$

The prefactor of the exponential function is chosen so as to meet the normalization condition (3). Since the kinematic factor $\exp[(\mathrm{i}M/2\hbar\epsilon)(\mathbf{x}-\mathbf{x}')^2]$ is invariant under the group of motion in $\mathbf{R}^d$, it can be expanded in terms of the zonal spherical functions of the rotation group $SO(d)$ or the zonal spherical functions of the Euclidean group $E(d)$. If we put it into the form,

$$\exp\left[\frac{\mathrm{i}M}{2\hbar\epsilon}(\mathbf{x}-\mathbf{x}')^2\right] = \exp\left[\frac{\mathrm{i}M}{2\hbar\epsilon}\left(r^2+r'^2\right)\right]\exp\left[-\frac{\mathrm{i}M}{2\hbar\epsilon}\mathbf{x}\cdot\mathbf{x}'\right], \tag{82}$$

the second factor is a zonal function with respect to $SO(d)$. The expansion of the second factor in terms of the zonal spherical function $D^l_{00}(h)$ of $SO(d)$ turns out to be the well-known Gegenbauer expansion,

$$\exp(z\,\mathbf{u}\cdot\mathbf{u}') = (2/z)^\lambda\Gamma(\lambda)\sum_{l=0}^{\infty}(l+\lambda)\,\mathrm{I}_{l+\lambda}(z)\,C_l^\lambda(\mathbf{u}\cdot\mathbf{u}'),$$

with $z = (M/2\mathrm{i}\hbar\epsilon)$ and $\lambda = (d-2)/2$. Consequently, the short-time propagator can be expanded as

$$K(\mathbf{x},\mathbf{x}';\epsilon) = \frac{\Gamma(d/2)}{2\pi^{d/2}}\sum_{l=0}^{\infty}K_l(r,r';\epsilon)\,d_l\,D^l_{00}(h^{-1}h') \tag{83}$$

with the short-time radial propagator,

$$\begin{aligned}K_l(r,r';\epsilon) = {} & -2\mathrm{i}(\alpha/\epsilon)\,(rr')^{-(d-2)/2}\exp\left[\mathrm{i}(\alpha/\omega\epsilon)\left(1-\frac{1}{2}\omega^2\epsilon^2\right)(r^2+r'^2)\right] \\ & \times\mathrm{I}_{l+(d-2)/2}\left(\frac{\alpha}{\mathrm{i}\omega\epsilon}rr'\right).\end{aligned} \tag{84}$$

For convenience, let us define the following function,

$$v_\lambda(\eta,\eta';\varphi) = -\mathrm{i}\csc\varphi\,\exp[\mathrm{i}(\eta+\eta')\cot\varphi]\,\mathrm{I}_\lambda\left(-2\mathrm{i}\sqrt{\eta\eta'}\csc\varphi\right). \tag{85}$$

If we use Weber's integral formula,

$$\int_0^\infty dr\,r\,\exp(\mathrm{i}\beta r^2)\,\mathrm{I}_\lambda(-iar)\,\mathrm{I}_\lambda(-ibr) = \frac{\mathrm{i}}{2\beta}\exp\left[-\frac{\mathrm{i}}{4\beta}(a^2+b^2)\right]\mathrm{I}_\lambda\left(-\frac{ab}{2\beta}\right), \tag{86}$$

valid for $\mathrm{Re}\,\beta > 0$ and $\mathrm{Re}\,\lambda > -1$, then we can verify the recurrence convolution,

$$v_\lambda * v_\lambda(\eta'',\eta';\varphi''+\varphi) = \int_0^\infty d\eta\,v_\lambda(\eta'',\eta;\varphi'')\,v_\lambda(\eta,\eta';\varphi) = v_\lambda(\eta'',\eta';\varphi''+\varphi). \tag{87}$$

This convolution formula can be extended to the multi-convolution formula,

$$v_\lambda(\eta_N,\eta_0;\varphi) = \prod_{j=1}^{N} * v_\lambda(\eta_j,\eta_{j-1};\varphi_j), \tag{88}$$

where

$$\varphi = \sum_{j=1}^{N} \varphi_j. \tag{89}$$

Now we can utilize this convolution formula to calculate the path integral as the multi-convolution of the short-time propagator. The multi-convolution of the N short-time propagators is given by

$$K(\mathbf{x}_N, \mathbf{x}_0; N\epsilon) = \prod_{j=1}^{N} *K(\mathbf{x}_j, \mathbf{x}_{j-1}; \epsilon) = \frac{\Gamma(d/2)}{2\pi^{d/2}} \sum_{l=0}^{\infty} K_l(r_N, r_0; N\epsilon)\, d_l\, D^l_{00}(h_N^{-1} h_0), \tag{90}$$

with

$$K_l(r_N, r_0; N\epsilon) = \prod_{j=1}^{N} *K_l(r_j, r_{j-1}). \tag{91}$$

If we let

$$\sin\varphi_j = \omega\epsilon, \qquad \cos\varphi_j = 1 - \frac{1}{2}\omega^2\epsilon^2 + \mathcal{O}(\epsilon^4), \tag{92}$$

the short-time radial function may be approximated as

$$K_l(r, r'; \epsilon) = 2\alpha(rr')^{-(d-2)/2}\, v_{l+(d-2)/2}(\alpha r^2, \alpha r'^2; \varphi_j). \tag{93}$$

Therefore, the multi-convolution formula of the v-function helps to evaluate the multi-convolution of the radial propagator. In the limit $N \to \infty$ (i.e., $\epsilon \to 0$),

$$\varphi = \lim_{N\to\infty} N \arcsin(\omega\epsilon) = \omega\tau.$$

As a result, we can obtain the radial propagator for a finite time interval τ,

$$K_l(r'', r'; \tau) = 2\alpha(r''r')^{-(d-2)/2}\, v_{l+(d-2)/2}(\alpha r''^2, \alpha r'^2; \omega\tau), \tag{94}$$

or

$$\begin{aligned} K_l(r'', r'; \tau) = & -2\mathrm{i}\alpha(r''r')^{-(d-2)/2} \csc(\omega\tau) \exp\left[\frac{\mathrm{i}M\omega}{2\hbar}(r'^2 + r''^2)\cot(\omega\tau)\right] \\ & \times \mathrm{I}_{l+(d-2)/2}\left(\frac{M\omega}{\mathrm{i}\hbar} r'r'' \csc(\omega\tau)\right). \end{aligned} \tag{95}$$

Comparison of (78) and (95) leads us to the relation,

$$\langle J, \eta | \mathrm{e}^{-2\mathrm{i}\varphi J_3} | J, \eta' \rangle = -\mathrm{i}\csc\varphi\, \exp[\mathrm{i}(\eta + \eta')\cot\varphi]\, \mathrm{I}_{-2J-1}\left(-2\mathrm{i}\sqrt{\eta\eta'}\csc\varphi\right) \tag{96}$$

which coincides with the result (70) obtained by Lindblad and Nagel [33]. In this way, we can also determine the matrix element on the continuous basis by path integration.

Shapiro and Vilenkin [49] expressed Weber's integral formula (86) in terms of the matrix element in question. Here, conversely, we have used Weber's formula to determine the matrix element.

6. Epilogue

Fourier analysis is well-known as a tool for finding the heat kernel. Then it is not surprising that harmonic analysis works well in calculating various kernels of the Schödinger equation. The path integral itself is known as a tool of finding the Schrödinger kernel or the Feynman kernel or the propagator in non-relativistic quantum mechanics. What is surprising is that path integrals and harmonic analysis can mix very well. In the preceding sections, we have demonstrated how the techniques of harmonic analysis can be utilized in path integration in nontrivial ways.

We have picked up only two examples, but the same or similar techniques used in those examples have been applied to various other problems. As generalized Fourier expansions may be applied to (*i*) functions on the homogeneous space $\mathcal{M} = G/H$, (*ii*) zonal functions defined on spheres $S = HgH$, and (*iii*) central functions on the group manifold G, so can the propagators be treated differently, depending on their structure. In the first example, we have expanded the propagator in terms of zonal functions. However, in dealing with a particle on a three-dimensional sphere, we may apply the character expansion. This is because S^3 can be identified with the group manifold of $SU(2)$. In this case, the group $SU(2)$ is directly related to the rotational symmetry of the background space. The one-dimensional Pöschl-Teller oscillator, for instance, has no spherical symmetry, but can be solved on the dynamical group manifold $SU(2)$. Even though the physical natures of the group $SU(2)$ for these two cases are different, the structures of the two path integrals are basically the same. Both belong to the hypergeometric class. Path integrals of the hypergeometric class solved by harmonic analysis include the free motion on a space of constant positive curvature $S^d = SO(d+1)/SO(d)$, the free motion on the Euclidean space $\mathbf{R}^d$ by means of the zonal function of the Euclidean group, and the rotations about the origin $\mathbf{R}^d = E^d/SO(d)$. For details, see refs. [10; 14; 50].

For the second example, we have used the matrix elements on the continuous basis. The radial path integral belongs to the confluent hypergeometric class. A number of problems have also been solved with the help of the radial path integral of this type. For a review of the path integrals of $SU(2)$ and $SU(1,1)$, we refer to ref. [15].

In concluding, we may say that these group theoretical methods together with other techniques (see also [15]) have made Feynman's path integral as accessible as Schrödinger's equation.

Acknowledgement

We would like to thank Raj Wilson for his invitation to this stimulating conference from which we greatly benefited.

Appendix: Rudiments of Harmonic Analysis on Homogeneous Spaces

This Appendix reviews the rudiments of harmonic analysis on homogeneous spaces in a way pertinent to path integral calculus. For details, see refs. [2] and [51].

Transformation Group and Homogeneous Space:

First, we consider a transformation group G of a space $\mathcal{M}$. A transformation g of $\mathcal{M}$ is a one-to-one map of $\mathcal{M}$ onto itself. If q is a point of $\mathcal{M}$, then the transform of q by g, denoted by $q' = gq$, also belongs to $\mathcal{M}$. A set G of such transformations is a transformation group of $\mathcal{M}$ if the inverse g^{-1} transforms q' back into q by $q = g^{-1}q'$ and if $(g_1g_2)q = g_1(g_2q)$ for any two transformations, g_1 and g_2, of G. We assume that G acts transitively on $\mathcal{M}$. In other words, for every pair (q', q) of $\mathcal{M}$, there is a transformation $g \in G$ such that $q' = gq$. We also assume that G is effective on $\mathcal{M}$, that is, for every group element g not equal to the identity element there exists a $q \in \mathcal{M}$ such that $gq \neq q$. Then, $\mathcal{M}$ is a homogeneous space with respect to the transitive group G. A subgroup H of G whose action leaves a point q_a of $\mathcal{M}$ fixed is called the *isotropy group* of G at q_a. Since any point q of $\mathcal{M}$ can be reached from the fixed point q_a by a group action $g \in G$, we have $q = gq_a = ghq_a$ where $h \in H$. Thus, there is one-to-one correspondence between the homogeneous space $\mathcal{M}$ and the coset space G/H. Hence $\mathcal{M}$ is identified with G/H, that is, $\mathcal{M} = G/H$.

Suppose a fixed point q of $\mathcal{M}$ is taken to a point $q' = hq$ of $\mathcal{M}$ under the action h of the isotropy group H at q_a. The set $S = \{q' \in hq;\ h \in H\}$ is called *the sphere* centered at q_a, passing through q. Since q corresponds to the left coset gH, the sphere S can be viewed as the two-sided coset HgH with g fixed.

Next, we assume that the transformation group G is locally compact and unimodular. Then, G has a unique (up to a multiple constant) invariant Haar measure for all integrable functions $f:\ G \mapsto \mathbb{C}$, denoted by $\mathrm{d}g$:

$$\int_G \mathrm{d}g\, f(g) = \int_G \mathrm{d}g\, f(g'g) = \int_G \mathrm{d}g\, f(gg') = \int_G \mathrm{d}g\, f(g^{-1}) \tag{A.1}$$

where $g' \in G$. The Haar measure $\mathrm{d}g$ induces a G-invariant measure $\mathrm{d}q$ on $\mathcal{M}$ [51]:

$$\int_{\mathcal{M}} \mathrm{d}q\, f(q) = \int_G \mathrm{d}g\, f(gq_a) \tag{A.2}$$

where q_a is a fixed point in $\mathcal{M}$ and $f:\ \mathcal{M} \mapsto \mathbb{C}$ is an integrable function on $\mathcal{M}$. The G-invariance of the measure on $\mathcal{M}$,

$$\int_{\mathcal{M}} \mathrm{d}q\, f(gq) = \int_{\mathcal{M}} \mathrm{d}q\, f(q), \tag{A.3}$$

is a direct consequence of the invariance of the Haar measure.

Spherical Representations:

Let us denote by D^l a unitary irreducible representation of the group G in an invariant subspace $\mathcal{H}^l \subset \mathcal{L}^2(\mathcal{M})$. The representation $D^l(g)$ is called a *spherical* representation (or a representation of *class 1*) of G if $\mathcal{H}^l$ has non-null vectors $|\psi_\alpha\rangle$ which are invariant under transformations of the subgroup H [2], $D^l(h)|\psi_\alpha\rangle = |\psi_\alpha\rangle$ for all $h \in H$. Let the subgroup H be *massive* [2], that is, let there be only one such vector $|\psi_0\rangle$ in $\mathcal{H}^l$. Then a function $\psi^l(g)$ defined on G by

$$\psi^l(g) = \langle\psi \mid D^l(g) \mid \psi_0\rangle \tag{A.4}$$

is called a *spherical function* of $D^l(g)$. It is obvious that $\psi^l(gh) = \psi^l(g)$. Thus, a spherical function $\psi^l(g)$ of $D^l(g)$ is constant on a left coset gH. Since $\mathcal{M}$ is identified with the space of the left cosets gH, $\psi^l(g)$ can be regarded as a function on the homogeneous space $\mathcal{M} = G/H$.

Let $\{|e_m\rangle\}$ with $\langle e_m \mid e_n\rangle = \delta_{mn}$ $(m, n = 0, 1, 2, \ldots, d_l - 1)$ be an orthonormal basis with $|e_0\rangle = |\psi_0\rangle$ in the d_l-dimensional subspace $\mathcal{H}^l$. The matrix elements of $D^l(g)$ for $g \in G$ on this basis are given by

$$D^l_{mn}(g) = \langle e_m \mid D^l(g) \mid e_n\rangle. \tag{A.5}$$

The unitary property implies $D^{l\,*}_{nm}(g) = D^l_{mn}(g^{-1})$. It is also easy to show that

$$D^l_{mn}(g_1 g_2) = \sum_k D^l_{mk}(g_1) D^l_{kn}(g_2). \tag{A.6}$$

Certainly the matrix elements

$$D^l_{m0}(g) = \langle e_m \mid D^l(g) \mid e_0\rangle \tag{A.7}$$

are spherical functions of $D^l(g)$ on G, which are more specifically called *associated spherical functions*. They satisfy $D^l_{m0}(gh) = D^l_{m0}(g)$ for any $h \in H$. A special case of associated spherical functions is

$$D^l_{00}(g) = \langle e_0 \mid D^l(g) \mid e_0\rangle, \tag{A.8}$$

which is called a *zonal spherical function*. Naturally, the zonal spherical function is constant on the two-sided coset HgH, that is, for any $h, h' \in H$,

$$D^l_{00}(h^{-1} g h') = D^l_{00}(g). \tag{A.9}$$

For instance, if $G = SO(3)$ and if $H = SO(2)$ about the north pole $q_a = (0, 0, 1)$ of $\mathcal{M} = S^2$, then the well-known spherical harmonics $Y_l^m(\theta, \phi)$ are associated spherical functions of irreducible unitary representations of $SO(3)$, and the Legendre polynomials $P_l(\cos\theta)$ are the zonal spherical functions. The latter has a constant value along a circle (i.e., $\theta =$ constant) on S^2.

Let us add a few more examples. The zonal spherical functions on an d-dimensional Euclidean space E^d, the unit sphere $S^{d-1} = SO(d)/SO(d-1)$ in d dimensions, and the $(d-1)$-dimensional Lobačevskiĭ space are given, respectively, by

$$D_{00}^{k}(r) = 2^{\nu}\Gamma(\nu+1)\,(kr)^{-\nu}\,J_{\nu}(kr), \qquad (0 < k, r < \infty),$$
$$D_{00}^{\ell}(\theta) = \frac{\Gamma(\ell+1)\Gamma(2\nu)}{\Gamma(\ell+2\nu)} C_{\ell}^{\nu}(\cos\theta), \qquad (\ell = 0, 1, 2, ...;\ 0 \le \theta \le \pi),$$
$$D_{00}^{\rho}(t) = 2^{\nu-1/2}\Gamma(\nu+\tfrac{1}{2})(\sinh t)^{-\nu+1/2}\, P_{-\frac{1}{2}+i\rho}^{\frac{1}{2}-\nu}(\cosh t), \qquad (0 < \rho, t < \infty),$$

where $2\nu = d-2$. In the above, $J_\nu(z), C_l^\nu(z)$, and $P_\mu^\nu(z)$ are the Bessel function, the Gegenbauer polynomials, and the associated Legendre function, respectively.

The character $\chi^l(g)$ of a finite dimensional irreducible unitary representation D^l is defined by

$$\chi^l(g) = \mathrm{Tr} D^l(g) = \sum_m D_{mm}^l(g). \tag{A.10}$$

In particular, $\chi^l(e) = d_l = \dim \mathcal{H}^l$ where e is the unit element of G. For an infinite dimensional representation, the character cannot be defined in the way it is defined for the finite dimensional case. The character of an infinite dimensional representation is defined as a distribution.

Fourier Expansions:

In the case of compact groups, the harmonic analysis in the Hilbert space $\mathcal{H} = \mathcal{L}^2(G)$ is based on the Peter-Weyl theorem. The theorem tells us that the matrix elements, $\sqrt{d_l}\, D_{mn}^l(g)$, form a complete orthonormal set on $\mathcal{H}$ in regard to a normalized invariant Haar measure dg, satisfying the orthogonality relation,

$$\int_G \mathrm{d}g\, D_{mn}^l(g)\, D_{m'n'}^{l'\,*}(g) = d_l^{-1}\,\delta_{ll'}\delta_{mm'}\delta_{nn'}. \tag{A.11}$$

An arbitrary function $f(g) \in \mathcal{H}$ may be expanded in the form,

$$f(g) = \sum_{l,m,n} d_l\, \hat{f}_{mn}^l\, D_{mn}^l(g), \tag{A.12}$$

where

$$\hat{f}_{mn}^l = \int_G \mathrm{d}g\, f(g)\, D_{mn}^{l\,*}(g). \tag{A.13}$$

For a non-compact group G, the spectra of some invariant operators of G are continuous, and the eigenstates should be understood as distributions. Therefore, we have to deal with the so-called Gel'fand triplet $\Phi \subset \mathcal{H} \subset \Phi'$. Here, Φ is a certain nuclear space of smooth functions in $\mathcal{H}$, and Φ' is the dual space of Φ. The orthogonality relation (A.11), which holds only in special cases, should be broadly interpreted. For instance, the Kronecker delta $\delta_{ll'}$ may be replaced by the delta function $\delta(l-l')$, and the "dimension" d_l may be defined by the relation (A.11) itself. The Fourier expansion formula (A.12) may be expressed in the form,

$$f(g) = \sum_{l\in\Lambda} d_l\, \mathrm{Tr}\Big(\hat{f}^l D^l(g)\Big), \tag{A.14}$$

where Λ is the set of all inequivalent unitary irreducible representations to which D^l belongs. The summation in the above expression must be replaced by an appropriate Lebesgue-Stieltjes integral when l is continuous.

The zonal spherical functions (A.9) satisfy the relations,

$$\int_G dg\, D^l_{00}(g) D^{l'}_{00}(g^{-1}) = \frac{1}{d_l}\,\delta_{ll'} \tag{A.15}$$

and

$$\int_G dg_j\, D^l_{00}(g_{j-1}^{-1} g_j) D^{l'}_{00}(g_j^{-1} g_{j+1}) = \frac{1}{d_l}\,\delta_{ll'}\, D^l_{00}(g_{j-1}^{-1} g_{j+1}), \tag{A.16}$$

both of which readily follow from the orthogonality relation (A.11). Any function $f(g)$ constant on two-sided cosets HgH, that is, satisfying $f(h^{-1}gh) = f(g)$ for $h \in H$ and $g \in G$, can be expanded in terms of the zonal spherical function $D^l_{00}(g)$ as

$$f(g) = \sum_{l \in \Lambda} d_l\, \hat{f}^l\, D^l_{00}(g), \tag{A.17}$$

where the "Fourier" coefficients are given by

$$\hat{f}^l = \int_G dg\, f(g)\, D^l_{00}(g^{-1}). \tag{A.18}$$

It is evident from the orthogonality relation (A.11) that the character function (A.10) satisfies

$$\int_G dg\, \chi^{l\,*}(g)\, \chi^{l'}(g) = \delta_{ll'}, \tag{A.19}$$

and

$$\int_G dg_2\, \chi^l(g_3 g_2^{-1}) \chi^{l'}(g_2 g_1^{-1}) = d_l^{-1}\, \delta_{ll'}\, \chi^l(g_3 g_1^{-1}). \tag{A.20}$$

A function $f(g)$ on G, satisfying $f(g_1^{-1} g_2 g_1) = f(g_2)$ for any $g_1, g_2 \in G$, is called a *central function*. Evidently, the character $\chi^l(g)$ is a central function on G. Any central function on G can be decomposed by means of the character function:

$$f(g) = \sum_{l \in \Lambda} d_l\, \hat{f}^l\, \chi^l(g), \tag{A.21}$$

with

$$\hat{f}^l = d_l^{-1} \int_G dg\, f(g)\, \chi^l(g^{-1}). \tag{A.22}$$

For a compact group, the presence of the dimensional constant d_l in the series expansion seems redundant. However, if G is non-compact, and if l takes a continuous value, then the replacement of the sum by an integral necessarily brings the factor d_l as part of the Jacobian. Therefore, for a unified treatment of the discrete and continuous cases, it is convenient to keep d_l explicitly in (A.21) and (A.22).

Convolutions:

The convolution $f_1 * f_2(g)$ of two integrable functions $f_1(g)$ and $f_2(g)$ on a locally compact unimodular group G is defined by

$$f_1 * f_2(g) = \int_G dg_1\, f_1(gg_1^{-1}) f_2(g_1), \tag{A.23}$$

which is also integrable on G. The operation of convolution is associative;

$$f_1 * (f_2 * f_3) = (f_1 * f_2) * f_3. \tag{A.24}$$

It is sometime convenient to use the following short hand notation for a multi-convolution of N functions,

$$(\cdots((f_1 * f_2) * f_3) * \cdots) * f_N = \prod_{j=1}^{N} * f_j. \tag{A.25}$$

It is noteworthy that if two square integrable functions $f_1(g)$ and $f_2(g)$ are expanded as

$$f_1(g) = \sum_l d_l \sum_{m,n} a^l_{mn}\, D^l_{mn}(g), \tag{A.26}$$

$$f_2(g) = \sum_l d_l \sum_{m,n} b^l_{mn}\, D^l_{mn}(g), \tag{A.27}$$

then the Fourier transform c^l_{mn} of their convolution $f_1 * f_2(g)$ is given by

$$c^l_{mn} = \sum_k a^l_{mk} b^l_{kn}. \tag{A.28}$$

References

[1] W. Miller, *Lie Theory and Special Functions*, (Academic Press, New York, 1968).
[2] N.Ya. Vilenkin, *Special Functions and the Theory of Group Representations*, (Amer. Math. Soc., Providence, RI, 1968).
[3] A.O. Barut, *Dynamical Groups and Generalized Symmetries in Quantum Theory*, (University of Canterbury Publ., Christchurch, New Zealand, 1971).
[4] A. Inomata and M.A. Kayed, Phys. Lett. A **108** (1985) 9.
[5] G. Junker and A. Inomata, in M.C. Gutzwiller, A. Inomata, J.R. Klauder and L. Streit (eds.), *Path Integrals from meV to MeV*, (World Scientific, Singapore, 1986) p. 315.
[6] A. Inomata and R. Wilson, in A.O. Barut and H.D. Doebner (eds.), *Conformal Groups and Related Symmetries*, (Springer, Berlin, 1986) p.42.
[7] M. Böhm and G. Junker, J. Math. Phys. **28** (1987) 1978.
[8] A.O. Barut, A. Inomata and G. Junker, J. Phys. A **20** (1987) 6271.
[9] A.O. Barut, A. Inomata and G. Junker, J. Phys. A **23** (1990) 1179.
[10] G. Junker, in V. Sayakanit, W. Sritrakool, J. Berananda, M.C. Gutzwiller, A. Inomata, S. Lundqvist, J.R. Klauder and L.S. Schulman (eds.), *Path Integrals from meV to MeV*, (World Scientific, Singapore, 1989) p.217.
[11] S. Ozaki, Lecture notes at Kyushu University (1955) unpublished.
[12] S.F. Edwards and Y.V. Gulyaev, Proc. Roy. Soc. London A **279** (1964) 229.
[13] A. Inomata, Benét Laboratories Technical Report WVT- 6718, (1967); D. Peak and A. Inomata, J. Math. Phys. **10** (1969) 1422.

[14] M. Böhm and G. Junker, J. Math. Phys. **30** (1989) 1195.
[15] A. Inomata, H. Kuratsuji and C.C. Gerry, *Path Integrals and Coherent States of $SU(2)$ and $SU(1,1)$*, (World Scientific, Singapore, 1992).
[16] L.S. Schulman, Phys. Rev. **176** (1968) 1558.
[17] J.S. Dowker, Ann. Phys. (NY) **62** (1971) 361.
[18] A.O. Barut, C.K.E. Schneider and R. Wilson, J. Phys. **20**, (1979) 2244.
[19] Private communication with Raj Wilson.
[20] I.H. Duru and H. Kleinert, Phys. Lett. B **84** (1979) 185.
[21] P. Kustaanheimo and E. Stiefel, J. Reine Angew. Math. **218** (1965) 207.
[22] R. Ho and A. Inomata, Phys. Rev. Lett. **48** (1982) 231.
[23] L.H. Hostler, J. Math. Phys. **5** (1964) 591.
[24] A. Inomata, in M.C. Gutzwiller, A. Inomata, J.R. Klauder and L. Streit (eds.), *Path Integrals from meV to MeV*, (World Scientific, Singapore, 1986) p. 433, and A. Inomata, in V. Sayakanit, W. Sritrakool, J. Berananda, M.C. Gutzwiller, A. Inomata, S. Lundqvist, J.R. Klauder and L.S. Schulman (eds.), *Path Integrals from meV to MeV*, (World Scientific, Singapore, 1989) p.112.
[25] A. Inomata, Phys. Lett. A **101** (1984) 253.
[26] A.O. Barut, A. Inomata and R. Wilson, J. Phys. A **20** (1987) 4075.
[27] A.O. Barut, A. Inomata and R. Wilson, J. Phys. A **20** (1987) 4083.
[28] See, e.g., W. Fischer, H. Leschke and P. Müller, J. Phys. A **25** (1992) 3835.
[29] See, e.g., W. Fischer, H. Leschke and P. Müller, Ann. Phys. (NY) (1993) in press.
[30] A. Inomata and G. Junker, in H.D. Doebner, W. Scherer and F. Schroeck (eds.), *Classical and Quantum Systems – Foundations and Symmetries*, (World Scientific, Singapore, 1993) p.334.
[31] A.O. Barut and C. Fronsdal, Proc. Roy. Soc. (Lond.) A **287** (1966) 532.
[32] A.O. Barut and E.C. Phillips, Comm. Math. Phys. **8** (1968) 52.
[33] G. Lindblad and B. Nagel, Ann. Inst. Henri Poincaré A **13** (1970) 27.
[34] N. Mukunda and B. Radhakrishnan, J. Math. Phys. **14** (1973) 254.
[35] R.P. Feynman, Rev. Mod. Phys. **20** (1948) 367.
[36] M. Kac, *Probability and Related Topics in the Physical Science*, (Interscience, New York, 1959).
[37] E. Nelson, J. Math. Phys. **5** (1964) 332.
[38] W. Faris, Pac. J. Math. **22** (1967) 47.
[39] B. Simon, *Quantum Mechanics for Hamiltonians Defined as Quadratic Forms* (Princeton University Press, Princeton, 1971).
[40] I.M. Gel'fand, Dokl. Akad. Nauk SSSR **70** (1950) 769.
[41] F.A. Berezin and I.M. Gel'fand, Trans. Amer. Math. Soc. Series 2 **21** (1962) 193.
[42] M.C. Gutzwiller, Physica Scripta T **9** (1985) 184.
[43] C.W. Misner, K.S. Thorne and J.A. Wheeler, *Gravitation*, (Freeman, San Fransisco, 1970).
[44] C. Grosche and F. Steiner, Ann. Phys. (NY) **182** (1987) 120.
[45] P.D. Lax and R.S. Phillips, Comm. Pure Appl. Math. **31** (1978) 415.
[46] Grosche and Steiner [44] were the first to realize that the Green function can be given in closed form, but their result was incorrect. The correct form of the Green function is given in ref. [10].
[47] M. Moshinsky and C. Quesne, J. Math. Phys. **12** (1971) 1780.
[48] V. Bargmann, Ann. Math. **48** (1947) 568.
[49] R.L. Shapiro and N.Ya. Vilenkin, in N.A. Markov, V.I. Man'ko and V.V. Dodonov (eds.), *Group Theoretical Methods in Physics* Vol. I, (VNU Science Press, Utrecht, 1986) p.551.
[50] G. Junker, J. Phys. A **23** (1989) L881.
[51] A.O. Barut and R. Raczka, *Theory of Group Representations and Applications*, 2nd edition (Polish Scientific Publ., Warszawa, 1980).

Department of Physics, State University of New York at Albany,
Albany, N.Y. 12222, USA

Institut für Theoretische Physik I, Universität Erlangen–Nürnberg,
Staudtstr. 7, D-91058 Erlangen, Germany.

Representations of Diffeomorphism Groups and the Infinite Symmetric Group

Takeshi HIRAI

ABSTRACT. Unitary representations of diffeomorphism groups of manifolds are studied in connection with those of the infinite symmetric group. First we study quasi-invariant measures on the spaces of ordered configurations. Then, using them, we construct quite a big family of representations by a measurable version of the method of associated vector bundles. Irreducibility and equivalence relations are studied.

Introduction. Let M be a connected, paracompact $C^{(r)}$-manifold $(1 \leq r \leq \infty)$ and $\mathit{Diff}(M)$ be the group of all the $C^{(r)}$-diffeomorphisms on M. In this paper we take a subgroup

$$G = \mathit{Diff}_0(M) = \{g \in \mathit{Diff}(M); \quad \mathrm{supp}(g) \quad \mathrm{compact}\},$$

and study representations of G. Here $\mathrm{supp}(g) = \mathrm{Cl}\{p \in M;\ gp \neq p\}$ and $\mathrm{Cl}\{\cdot\}$ denotes the closure. Introduce in G a topology of compact uniform convergence of every differentials of $M \ni p \mapsto gp \in M$, then G becomes a topological group, not locally compact. We construct irreducible unitary representations (= IURs) of G, associated with any IURs of the infinite symmetric group $\mathcal{S}_\infty$ of all the finite permutations on $\mathbf{N}$. Denote also by $\tilde{\mathcal{S}}_\infty$ the group of all permutations on $\mathbf{N}$: $\tilde{\mathcal{S}}_\infty \supset \mathcal{S}_\infty$. Put

$$X = X_M = \prod_{i \in \mathbf{N}} M_i, \quad M_i = M \quad (i \in \mathbf{N}).$$

Then G (resp. $\tilde{\mathcal{S}}_\infty$) acts on X from the left (resp. right): for $x = (x_i)_{i \in \mathbf{N}} \in X, g \in G$, and $\tau \in \tilde{\mathcal{S}}_\infty$,

$$gx = (gx_i)_{i \in \mathbf{N}}, \quad x\tau = (x_{\tau(i)})_{i \in \mathbf{N}}.$$

We call a point $x = (x_i)_{i \in \mathbf{N}}$ of X an *ordered* configuration of points in M if $x_i \neq x_j (i \neq j)$ and the series (x_i) has no accumulation points in M, and let $\tilde{X}$ be the

E. A. Tanner and R. Wilson (eds.),
Noncompact Lie Groups and Some of Their Applications, 225–237.

set of all ordered configurations. Then $\tilde{X}$ is stable under $G \times \tilde{\mathcal{S}}_\infty$. The space $\Gamma = \Gamma_M$ of (non-ordered) configarations is isomorphic to the quotient space $\tilde{X}/\tilde{\mathcal{S}}_\infty$.

In the work [7], I.M. Gelfand and others utilized the principal fibre bundle $\tilde{X} \to \Gamma = \tilde{X}/\tilde{\mathcal{S}}_\infty$, and IURs of $\tilde{\mathcal{S}}_\infty$ and G-quasi-invariant measures on Γ such as Poisson measures. Here we wish to utilize another principal fibre bundle $\tilde{X} \to \Omega = \tilde{X}/\mathcal{S}_\infty$ and IURs of $\mathcal{S}_\infty$. However this time the fibre bundle is very pathological in topology and so we should content ourselves to do with measurable structures. Furthermore we construct G-quasi-invariant measures first on X and then prove that they are carried by the subset $\tilde{X}$. Using these measures on $\tilde{X}$, we can apply a measurable version of the method of associated vector bundles to construct IURs of G.

Remark. From another point of view, our results show a strong analogy with the classical Weyl's case of $G = GL(n, \mathbf{C})$ and the finite symmetric groups $\mathcal{S}_N$, $N = 1, 2, \cdots$, studied in his book [8]. The group G acts naturally on $V = \mathbf{C}^n$. Take N-times tensor product $\otimes^N V$ of the G-module V, then the algebra of intertwining operators on $\otimes^N V$ are canonically isomorphic to the group algebra of $\mathcal{S}_N$, and irreducible components are given by Young tableaux of size N. Thus, finite-dimensional (holomorphic) irreducible representations of G are essentially classified by Young diagrams.

In our present case, take $G = \mathit{Diff}_0(M)$ and $V = L^2(M, \mu_0)$ for a certain measure μ_0 on M and consider the infinite tensor product of the G-module V (with respect to a reference vector). Then, under a certain condition, this infinite tensor product can be decomposed into the IURs which we will construct in this paper.

Acknowledgements. The author would like to thank Professor A. Hora and Professor A. Shimomura for their kind discussions on the theory of measures.

1. Quasi-Invariant Measures

1.1. Let $\mathcal{M}_M$ be the σ-algebra of all Borel sets of M. A measure defined on $(M, \mathcal{M}_M)$ is called an L-measure if it is locally equivalent to Lebesgue measures in local coordinates. Denote by $\mathcal{M}_X$ the σ-algebra of all measurable subsets of $X = \prod_{i \in \mathbf{N}} M_i$, $M_i = M$ $(i \in \mathbf{N})$ with respect to the product measurable structure of $(M, \mathcal{M}_M)$.

Now we take a system $\mu = (\mu_i)_{i \in \mathbf{N}}$ of measures μ_i of M satisfying

(M1) each μ_i is an L-measure on M;

(M2) $\forall K \subset M$ compact, $\exists c_K > 0$ such that for any mesurable decomposition $K = \coprod_{i \in \mathbf{N}} B_i$, we have $\sum_{i \in \mathbf{N}} \mu_i(B_i) \leq c_K$.

Note. Define a mesure $\sup_{i \in \mathbf{N}} \mu_i$ on M as follows:

for $B \in \mathcal{M}_M$, we take any measurable decomposition $B = \coprod_{i \in \mathbf{N}} B_i$, and put

$$\left(\sup_{i\in\mathbf{N}} \mu_i\right)(B) := \sup\{\textstyle\sum_{i\in\mathbf{N}} \mu_i(B_i);\quad B = \coprod_{i\in\mathbf{N}} B_i\}.$$

Then the condition (M2) is equivalent to the following

$$\text{(M2}'\text{)} \qquad \left(\sup_{i\in\mathbf{N}} \mu_i\right)(K) < \infty \quad (\forall K \subset M \quad \text{compact}).$$

1.2. Measurable Structures.

A product type subset $E = \prod_{i\in\mathbf{N}} E_i$ of $X = \prod_{i\in\mathbf{N}} M_i, M_i = M$, is called *$\mu$-unital* if

(U1) $E_i \in \mathcal{M}_M$ are mutually disjoint, and $\mu_i(E_i) > 0$ $(\forall i \in \mathbf{N})$;

(U2) $\sum_{i\in\mathbf{N}} \mid 1 - \mu_i(E_i) \mid < \infty$.

Definition. Two unital sets $E = \prod_{i\in\mathbf{N}} E_i$, $E' = \prod_{i\in\mathbf{N}} E'_i$ are called mutually *μ-cofinal* (Notation: $E \overset{\mu}{\sim} E'$) if

(CF) $\sum_{i\in\mathbf{N}} \mu_i(E_i \ominus E'_i) < \infty$

with $E_i \ominus E'_i = (E_i \setminus E'_i) \cup (E'_i \setminus E_i)$. Further E and E' are *strongly μ-cofinal* (Notation: $E \overset{\mu}{\approx} E'$) if

(SCF) $\mu_i(E_i \ominus E'_i) = 0 \quad (i \gg 0)$.

Now fix a μ-unital set E, and put

$$\mathcal{E}(E) = \mathcal{E}(\mu, E) := \{E';\quad \mu\text{-unital}, E' \overset{\mu}{\sim} E\}.$$

It generates a σ-ring denoted by $\mathcal{M}(E) = \mathcal{M}(\mu, E)$. Here we call a family $\mathcal{A}$ of subsets of X a σ-ring after Halmos [1] if it satisfies

(i) $A_n \in \mathcal{A}$ $(n = 1, 2, \cdots)$ $\Longrightarrow$ $\bigcup_{n=1}^{\infty} A_n \in \mathcal{A}$,

(ii) $A, B \in \mathcal{A}$ $\Longrightarrow$ $A \setminus B \in \mathcal{A}$.

Lemma 1.1. *Let $A \subset X$. Then $A \in \mathcal{M}(E)$ if and only if $A \in \mathcal{M}_X$ and is covered by a countably infinite number of elements of $\mathcal{E}(E)$.*

Lemma 1.2. *For $E' = \prod_{i\in\mathbf{N}} E'_i, g \in G$, and $\sigma \in \mathcal{S}_\infty$, put*

$$gE' = \textstyle\prod_{i\in\mathbf{N}} gE'_i, \quad E'\sigma = \prod_{i\in\mathbf{N}} E'_{\sigma(i)}.$$

Then $gE' \overset{\mu}{\sim} E'$, $E'\sigma \overset{\mu}{\approx} E'$. Hence $\mathcal{E}(E)$ is $G \times \mathcal{S}_\infty$-invariant and so is $\mathcal{M}(E)$.

SKETCH OF PROOF. Let us prove only $gE' \overset{\mu}{\sim} E'$. Put $K = \mathrm{supp}(g)$. Then

$$gE'_i = gE'_i \cap K + gE'_i \setminus K = gE'_i \cap K + E'_i \setminus K,$$

and so

$$\sum_{i\in\mathbf{N}} \mu_i(gE'_i \ominus E'_i) \le \sum_i \mu_i(gE'_i \cap K) + \sum_i \mu_i(E'_i \cap K) \le 2c_K \quad \text{(by (M2))}.$$

1.3. Construction of Measures.

Let $\mu = (\mu_i)_{i\in\mathbf{N}}$ be a system of measures on M satisfying the conditions (M1)–(M2), and let E be a μ-unital subset of X. We define a kind of product measure on $(X, \mathcal{M}(E))$, associated to E, as follows. First, take any μ-unital set $E' = \prod_{i\in\mathbf{N}} E'_i \in \mathcal{E}(E)$, then we can apply Kolmogorov's theorem of product of probability measures, on the system $\mu_i|E'_i$, $i \in \mathbf{N}$, and get a unique product measure

$$\nu_\mu^{E'} = \prod_{i\in\mathbf{N}} (\mu|E'_i) \quad \text{on} \quad (E', \mathcal{M}(E)|E')$$

with $\mathcal{M}(E)|E' := \{A \cap E'; A \in \mathcal{M}(E)\}$. Now, for any $B \in \mathcal{M}(E)$, there exists a covering $E^{(k)} \in \mathcal{E}(E), k \in \mathbf{N}$, of B by Lemma 1.1: $B \subset \bigcup_{k\in\mathbf{N}} E^{(k)}$. Take a mesurable decomposition $B = \coprod_{k\in\mathbf{N}} B_k$, $B_k \subset E^{(k)}$, and put

$$\nu_{\mu,E}(B) = \textstyle\sum_{k\in\mathbf{N}} \nu_\mu^{E^{(k)}}(B_k).$$

Then we see that this gives actually a measure on $\mathcal{M}(E)$.

Proposition 1.3.

(i) $E \overset{\mu}{\sim} F \implies \mathcal{M}(E) = \mathcal{M}(F)$ *and* $\nu_{\mu,E} = \nu_{\mu,F}$.

(ii) $E \not\sim F \implies \nu_{\mu,E}(B) = \nu_{\mu,F}(B) = 0$ *for* $\forall B \in \mathcal{M}(E) \bigcap \mathcal{M}(F)$.

Thus, associated to each equivalence class $[E] := \{F;\ F \overset{\mu}{\sim} E\}$, we get a measure $\nu_{\mu,E}$ on $(X,\ \mathcal{M}(E))$, and they are mutually disjoint.

Note. The measure space $(X,\ \mathcal{M}(E),\ \nu_{\mu,E})$ is σ-finite in the sense that for any $B \in \mathcal{M}(E)$, there exist $B_n \in \mathcal{M}(E), n \in \mathbf{N}$, such that

$$B = \textstyle\coprod_{n\in\mathbf{N}} B_n, \quad \nu_{\mu,E}(B_n) < \infty \quad (\forall n \in \mathbf{N}).$$

Proposition 1.4. *The measure $\nu_{\mu,E}$ is supported by $\tilde{X}$ in the sense that for any $B \in \mathcal{M}(E)$,*

$$B \cap \tilde{X} \in \mathcal{M}(E) \quad \text{and} \quad \nu_{\mu,E}(B) = \nu_{\mu,E}(B \cap \tilde{X}).$$

1.4. Quasi-Invariance.

For $\sigma \in \mathcal{S}_\infty$, put $\operatorname{supp}(\sigma) := \{i \in \mathbf{N};\ \sigma(i) \neq i\}$.

Theorem 1.5. *The measure $\nu_{\mu,E}$ is $G \times \mathcal{S}_\infty$-quasi-invariant. More exactly, for any $x = (x_i)_{i \in \mathbf{N}} \in X, \sigma \in \mathcal{S}_\infty$, and $g \in G$,*

$$\frac{d\nu_{\mu,E}(x\sigma)}{d\nu_{\mu,E}(x)} = \prod_{i \in \operatorname{supp}(\sigma)} \frac{d\mu_{\sigma^{-1}(i)}(x_i)}{d\mu_i(x_i)},$$

$$\frac{d\nu_{\mu,E}(gx)}{d\nu_{\mu,E}(x)} = \prod_{i \in \mathbf{N}} \frac{d\mu_i(gx_i)}{d\mu_i(x_i)},$$

on every $B \in \mathcal{M}(E)$.

Sketch of Proof. This is proved by Kakutani's theorem [6] and the condition (M2) on $\mu = (\mu_i)_{i \in \mathbf{N}}$.

We see that the measure $\nu_{\mu,E}$ is $\mathcal{S}_\infty$-invariant if and only if there exists a measure μ_0 on M such that $\mu_i = \mu_0$ ($\forall i \in \mathbf{N}$).

2. Construction of Unitary Representations of $G = \mathit{Diff}_0(M)$

2.1. Let us take $\Sigma = (\mu, E; \Pi)$ with

$\mu = (\mu_i)_{i \in \mathbf{N}}$: satisfying (M1)–(M2),

$E = \prod_{i \in \mathbf{N}} E_i$: a μ-unital set, and

Π : an IUR of $\mathcal{S}_\infty$ on a Hilbert space $V(\Pi)$.

Here we utilize the measurable principal fibre bundle $\tilde{X} \to \Omega = \tilde{X}/\mathcal{S}_\infty$ and its associated vector bundles.

For a $V(\Pi)$-valued function f on $\tilde{X}$, we put formally as follows: for $\sigma \in \mathcal{S}_\infty$, and $g \in G$,

$$(2.1) \quad R_\Sigma(\sigma)f(x) = \sqrt{\frac{d\nu_{\mu,E}(x\sigma)}{d\nu_{\mu,E}(x)}}\ \Pi(\sigma)(f(x\sigma)),$$

$$(2.2) \quad T_\Sigma(g)f(x) = \sqrt{\frac{d\nu_{\mu,E}(g^{-1}x)}{d\nu_{\mu,E}(x)}}\ f(g^{-1}x) \quad (x \in X).$$

Then the actions of $\mathcal{S}_\infty$ and G commute with each other.

2.2. First Realization.

Take an $E' \in \mathcal{E}(E)$, and denote the Hilbert space of $V(\Pi)$-valued L^2-functions on the measure space $(E', \mathcal{M}(E)|E', \nu_{\mu,E}|E')$ by

$$\begin{aligned} H^{\Sigma}_{E'} &= L^2(E', \mathcal{M}(E)|E', \nu_{\mu,E}|E'; V(\Pi)) \\ &\cong L^2(E', \mathcal{M}(E)|E', \nu_{\mu,E}|E') \otimes V(\Pi). \end{aligned}$$

For $E^{(1)}, E^{(2)} \in \mathcal{E}(E)$, we patch together $H^{\Sigma}_{E^{(1)}}$ and $H^{\Sigma}_{E^{(2)}}$ so that, for $\varphi_j \in H^{\Sigma}_{E^{(j)}}$ $(j = 1, 2)$,

$$< \varphi_1, \varphi_2 >= \sum_{\sigma \in \mathcal{S}_\infty} \int_{E^{(1)} \cap E^{(2)}\sigma} < \varphi_1(x), R_\Sigma(\sigma^{-1})\varphi_2(x) >_{V(\Pi)} d\nu_{\mu,E}(x).$$

Then we get a Hilbert space

$$H^{\Sigma} = \bigvee_{E' \in \mathcal{E}(E)} H^{\Sigma}_{E'} \quad \text{(spanned)},$$

which turn out to be separable. The global structure of M reflects to how to patch together $H^{\Sigma}_{E'}$'s.

Theorem 2.1. *On the Hilbert space H^{Σ}, $T_\Sigma(g)$ in (2.2) gives a continuous unitary representation of G.*

2.3. Second Realization.

Take a $V(\Pi)$-valued measurable function f on $\tilde{X}$ such that

$$\begin{aligned} R_\Sigma(\sigma)f &= f \quad (\sigma \in \mathcal{S}_\infty), \\ \mathrm{supp}(f) &:= \{x \in \tilde{X};\ f(x) \neq 0\} \in \mathcal{M}(E). \end{aligned}$$

Then $\mathrm{supp}(f)$ is $\mathcal{S}_\infty$-invariant and there exists an $F \in \mathcal{M}(E)$, a section, such that

$$\mathrm{supp}(f) = F \cdot \mathcal{S}_\infty, \quad F \cap F\sigma = \emptyset \quad (\forall \sigma \in \mathcal{S}_\infty, \neq 1).$$

We put

$$\begin{aligned} \|f\|^2 &:= \int_F \|f(x)\|^2_{V(\Pi)}\, d\nu_{\mu,E}(x), \\ \mathcal{H}^{\Sigma} &:= \{f;\ \|f\| < \infty\}/\{f;\ \|f\| = 0\}. \end{aligned}$$

Theorem 2.2.
(i) *On the Hilbert space $\mathcal{H}^{\Sigma}$, $T_\Sigma(g)$ $(g \in G)$, give a continuous unitary representation of G.*

(ii) *The map* $H^{\Sigma} \ni \varphi \longmapsto f \in \mathcal{H}^{\Sigma}$ *with*

$$f = \sum_{\sigma \in \mathcal{S}_\infty} R_{\Sigma}(\sigma)\varphi$$

gives a natural equivalence of unitary G-modules.

3. Normalization of Parameter $\Sigma = (\mu, E; \Pi)$

3.1. Normalization of $E = \prod_{i \in \mathbf{N}} E_i$.

The μ-unital set E in the parameter Σ of the unitary representation T_Σ can be replaced by any $F \in \mathcal{M}(E)$. Hence we may assume additional conditions on E as follows, replacing it if necessary.

(U3) Each E_i is open, and in case $\dim M \geq 2$, is connected, simply-connected.

(U4) Assume $\dim M \geq 2$. For any i, j $(i \neq j)$, $\mathrm{Cl}(E_i)$ and $\mathrm{Cl}(E_j)$ can be connected by a path outside of $\mathrm{Cl}(\cup_{k \neq i,j} E_k)$.

To give some consequences of these conditions, we introduce several notations. For an open set U of M and a μ-unital set E satisfying (U1)–(U4), put

$$G(U) := \text{the connected component of the identity in } \mathit{Diff}_0(U),$$
$$G(E) := \textstyle\prod'_{i \in \mathbf{N}} G(E_i) \qquad \text{(restricted direct product)},$$
$$G((E)) := \{\, g \in G;\ gE_i = E_{\sigma(i)} \quad (\exists \sigma \in \mathcal{S}_\infty, \forall i \in \mathbf{N})\}.$$

Lemma 3.1. *Assume* $\dim M \geq 2$. *Then, for any* $E' = \prod_{i \in \mathbf{N}} E'_i \in \mathcal{E}(E)$ *which satisfies* (U1)–(U4), *there holds that, for any* $\sigma \in \mathcal{S}_\infty$, *there exists an element* $g_\sigma \in G((E))$ *such that*

$$g_\sigma E'_i = E'_{\sigma(i)} \quad (i \in \mathbf{N}) \quad \text{or} \quad g_\sigma E' = E'\sigma.$$

Lemma 3.2. *Let the assumptions be as in Lemma* 3.1.
(i) *The quasi-regular representation* $V_{\mu,E'}$ *of* $G(E')$ *on* $L^2(E', \mathcal{M}(E)|E', \nu_{\mu,E}|E')$ *is irreducible.*
(ii) *The representation* $T_\Sigma|G((E'))$ *on* $H^{\Sigma}_{E'}$ *is irreducible.*

This lemma is fundamental to prove the irreducibility of the representation T_Σ of G on H^{Σ}.

3.2. Normalization of $\mu = (\mu_i)_{i \in \mathbf{N}}$.

Let μ and $\mu' = (\mu'_i)_{i \in \mathbf{N}}$ both satisfy the conditions (M1)–(M2).

Definition. μ and μ' are said to be mutually *strongly equivalent* on M (Notation: $\mu \overset{st}{\sim} \mu'$) if

$$\left(\sup_{i\in\mathbf{N}} | \mu_i - {\mu'}_i |\right)(M) < \infty.$$

By certain variants of Kakutani's theorem, we get

Lemma 3.3. *Assume $\mu \overset{st}{\sim} \mu'$.*
(i) *For an $F = \prod_{i\in\mathbf{N}} F_i \subset X$ with F_i mutually disjoint,*

$$\mu\text{-unital} \iff \mu'\text{-unital}.$$

(ii) *For such an F,*

$$\prod_{i\in\mathbf{N}} (\mu_i|F_i) \sim \prod_{i\in\mathbf{N}} ({\mu'}_i|F_i) \qquad (\textit{equivalent}) \quad \textit{on } F.$$

Corollary 3.4. *Assume $\mu \overset{st}{\sim} \mu'$. Then $E = \prod_{i\in\mathbf{N}} E_i$ is μ-unital if and only if it is μ'-unital, and for $\Sigma = (\mu, E, \Pi)$ and $\Sigma' = (\mu', E, \Pi)$*

$$T_\Sigma \cong T_{\Sigma'} \qquad (\textit{unitary equivalent})$$

as representations of G.

Thus, as a parameter $\mu = (\mu_i)_{i\in\mathbf{N}}$ in Σ, we may take such a one that satisfies the following condition together with (M1)–(M2):

(M3) Each μ_i has a $C^{(r)}$-class density with respect to local coordinates.

4. Irreducibility of the representation T_Σ

Theorem 4.1. *The unitary representation T_Σ of G on H^Σ is always irreducible if* $\dim M \geq 2$.

Sketch of Proof. The points of the proof are the following.

(1) The representation space H^Σ is spanned by $H^\Sigma_{E'}$'s with $E' \in \mathcal{E}(E)$ such that $E' \overset{\mu}{\approx} E$, and satisfies (U1)–(U4).
(2) The restriction $T_\Sigma|G((E'))$ on $H^\Sigma_{E'}$ is irreducible (by Lemma 3.2).
(3) The IUR of $G((E'))$ on $H^\Sigma_{E'}$ does not appear in the orthogonal complement $(H^\Sigma_{E'})^\perp$ in H^Σ.

5. Equivalence and Inequivalence

5.1. Natural Equivalence.

Let $\Sigma = (\mu, E; \Pi)$ be the parameter of our unitary representation T_Σ. For an element $a \in \tilde{S}_\infty$, put

$$\mu^a := (\mu_{a(i)})_{i \in \mathbf{N}}, \qquad Ea := \prod_{i \in \mathbf{N}} E_{a(i)},$$
$$\Pi^a(\sigma) := \Pi(a\sigma a^{-1}) \qquad (\sigma \in S_\infty), \quad \text{and}$$
$$\Sigma^a := (\mu^a,\ Ea;\ \Pi^a).$$

Then we have the following natural equivalence.

Lemma 5.1. *For any $a \in \tilde{S}_\infty$, we have* $T_\Sigma \cong T_{\Sigma^a}$.

5.2. A Sufficient Condition.

We have also a good sufficient condition for unitary equivalence between two IURs T_Σ and $T_{\Sigma'}$ with $\Sigma' = (\mu', E'; \Pi')$, so that we can conjecture a necessary and sufficient condition for unitary equivalence. To state them, we introduce some definitions.

Definition. We say that μ and μ' are mutually *comparable* on a Borel set B of M if, for any disjoint Borel subsets D_i $(i \in \mathbf{N})$ of B, we have

$$\sum_{i \in \mathbf{N}} \mu_i(D_i) < \infty \quad \Longleftrightarrow \quad \sum_{i \in \mathbf{N}} \mu'(D_i) < \infty.$$

We say that (μ, E) is *equivalent* to (μ', E') (Notation: $(\mu, E) \sim (\mu', E')$) if there exists an $F = \prod_{i \in \mathbf{N}} F_i \subset X$ with $F_i \in \mathcal{M}_M$ mutually disjoint such that

$$F \overset{\mu}{\sim} E, \quad F \overset{\mu'}{\sim} E'; \qquad \prod_{i \in \mathbf{N}} (\mu_i | F_i) \sim \prod_{i \in \mathbf{N}} (\mu' | F_i) \quad \text{on } F$$

and that μ and μ' are comparable outside of $\mathrm{supp}(F) := \cup_{i \in \mathbf{N}} F_i$.

Using these notions we can state the following result.

Lemma 5.2. *For two parameters Σ and Σ' of the unitary representations, assume*

$$(\mu, E) \sim (\mu', E'), \qquad \Pi \cong \Pi'.$$

Then, $T_\Sigma \cong T_{\Sigma'}$ naturally.

Conjecture. Lemmas 5.1 and 5.2 generate the equivalence relations among T_Σ's.

5.3. The Case of $\mathcal{S}_\infty$-invariant Measures.

This case is very important to treat the general case and we get in [3] the following result for this fundamental case. Recall that, for the measure space $(X, \mathcal{M}(E), \nu_{\mu,E})$, the measure $\nu_{\mu,E}$ is $\mathcal{S}_\infty$-invariant if and only if $\mu_i = \mu_0$ ($\forall i \in \mathbf{N}$) for some L-measure μ_0 on M.

Theorem 5.3. *For two parameters Σ and Σ', assume $\mu_i = \mu'_i = \mu_0$ ($\forall i \in \mathbf{N}$) for some μ_0. Then, $T_\Sigma \cong T_{\Sigma'}$ if and only if*

$$E' \overset{\mu}{\sim} Ea, \quad \Pi \cong \Pi^a \quad \textit{for some} \quad a \in \tilde{\mathcal{S}}_\infty.$$

Sketch of Proof. We reduce the essential part of the proof to a problem on a "probability matrix" of infinite size, stated below.

Suppose we are given an infinite matrix $C = (c_{ij})_{i,j \in \mathbf{N}}$ with $c_{ij} \geq 0$ $(i, j \in \mathbf{N})$ such that

$$d_i := \sum_{j \in \mathbf{N}} c_{ij} > 0, \quad 0 < \prod_{i \in \mathbf{N}} d_i < \infty \text{ (unconditionally)},$$

$$e_j := \sum_{i \in \mathbf{N}} c_{ij} > 0, \quad 0 < \prod_{j \in \mathbf{N}} e_j < \infty \text{ (unconditionally)}.$$

Expand the product $e_1 e_2 \dots e_N$ in terms of

$$c_{i_1 1} c_{i_2 2} \dots c_{i_N N}$$

and pick up all the terms with indices $(i_1, i_2, \dots, i_N)$ of all different integers. Their sum is denoted by $p_N(C)$. Also expand the product $d_1 d_2 \dots d_N$ in terms of $c_{1 j_1} c_{2 j_2} \dots c_{N j_N}$ and define the sum $q_N(C)$ similarly. Then the problem is the following.

Problem. Assume $\prod_{i \in \mathbf{N}} c_{i,b(i)} = 0$ for any permutation $b \in \tilde{\mathcal{S}}_\infty$. Then,

$$\text{either} \quad p_N(C) \to 0 \ (N \to \infty) \quad \text{or} \quad q_N(C) \to 0 \ (N \to \infty)\,?$$

The answer is affirmative and it proves the theorem.

5.4. Inequivalence.

As mentioned in the introduction, I.M. Gelfand and others have first constructed IURs of G in [7]. In connection to these representations, we have

Theorem 5.4. *None of the IURs T_Σ is equivalent to the IURs constructed in [7] by means of Poisson measures on $\Gamma = \tilde{X}/\tilde{\mathcal{S}}_\infty$.*

Recently H. Shimomura clarified completely the equivalence relation and the mutual disjointness among Poisson measures on the configuration space Γ.

6. Application of Explicit Form of IURs of $\mathcal{S}_\infty$

Until now, we do not use the explicit form of IURs Π of $\mathcal{S}_\infty$. Here we recall their construction in [2] and explain briefly its application. Firstly we can generalize our construction of representations T_Σ of G, as explained below. And secondly we can realize more visibly the representation space H^Σ.

6.1. IURs of the Infinite Symmetric Group $\mathcal{S}_\infty$.

Let us recall how to construct IURs of $\mathcal{S}_\infty$ in [2]. It is by a standard induction method from subgroups of "wreath product" type of $\mathcal{S}_\infty$.

(1) Take any finite group T and its faithful permutation representation into a symmetric group, say, $\mathcal{S}_N$ of order N. Or equivalently, we may consider from the beginning that $T \hookrightarrow \mathcal{S}_N$. Let A be a set and $\mathcal{S}_A$ the symmetric group consisting of all finite permutations on A. Let

$$D_A(T) = \prod\nolimits'_{\alpha\in A} T_\alpha, \quad T_\alpha = T \ (\alpha \in A)$$

be the restricted direct product of copies of T. Then the group $\mathcal{S}_A$ acts naturally on it by permuting coordinates. Then the wreath product of T with $\mathcal{S}_A$ is defined as the semidirect product

$$\mathcal{S}_A(T) := \mathcal{S}_A \ltimes D_A(T).$$

where $D_A(T)$ is normal in $\mathcal{S}_A(T)$. According to $T \hookrightarrow \mathcal{S}_N$, we have a natural embedding $\mathcal{S}_A(T) \hookrightarrow \mathcal{S}_A(\mathcal{S}_N)$.

(2) Take an IUR ρ of T on a Hilbert space $V(\rho)$. Then, choosing a reference vector $v = (v_\alpha)_{\alpha\in A}$, $v_\alpha \in V_\alpha = V(\rho)$, $\|v_\alpha\| = 1$, we have a tensor product representation

$$\otimes^v_{\alpha\in A}\,\rho_\alpha \quad \text{with} \quad \rho_\alpha = \rho \ (\alpha \in A)$$

of the group $D_A(T)$ on the space $\otimes^v_{\alpha\in A} V_\alpha$. This representation can be extended to that of the wreath product $\mathcal{S}_A(T)$ by making the group $\mathcal{S}_A$ act as permutations of coordinates (cf.[2]). We denote it by $\pi(\rho)$.

(3) We imbed $\mathcal{S}_A(T)$ into $\mathcal{S}_\infty$ through a so-called "saturated" imbedding $\iota : \mathcal{S}_A(\mathcal{S}_N) \hookrightarrow \mathcal{S}_\infty$. The latter is defined by a kind of infinite Young tableau (of rectangular form of size $|A| \times N$) given as follows. For $\alpha \in A$, let $I_\alpha = (i_{\alpha,1}, i_{\alpha,2}, \cdots, i_{\alpha,N})$ be an ordered set of N different integers. It defines an imbedding of $\mathcal{S}_\alpha := \mathcal{S}_N$ into $\mathcal{S}_\infty = \mathcal{S}_{\mathbf{N}}$ through the correspondece $j \to i_{\alpha,j}$, $1 \le j \le N$. Denote

by $\overline{I}_\alpha$ the underlying set of integers $\{i_{\alpha,1}, i_{\alpha,2}, \cdots, i_{\alpha,N}\}$. If the sets of N integers $\overline{I}_\alpha\ (\alpha \in A)$ gives a partition of **N**, we say that the imbedding into $\mathcal{S}_\infty$ of $D_A(\mathcal{S}_N) = \prod'_{\alpha\in A}\mathcal{S}_\alpha$ and accordingly that of the wreath product $\mathcal{S}_A(\mathcal{S}_N) = \mathcal{S}_A \ltimes D_A(\mathcal{S}_N)$ are *saturated*. Thus we get a saturated imbedding ι of $\mathcal{S}_A(T)$ into $\mathcal{S}_\infty$ as

$$\iota : \mathcal{S}_A(T) \hookrightarrow \mathcal{S}_A(\mathcal{S}_N) \hookrightarrow \mathcal{S}_\infty.$$

The image $H = \iota(\mathcal{S}_A(T))$ has naturally an IUR $\pi = \pi(\rho) \circ \iota^{-1}$. We see in [2] that the induced representation $\Pi = \mathrm{Ind}(\pi;\ H \uparrow \mathcal{S}_\infty)$ is irreducible.

More generally we define wreath product type subgroups H of $\mathcal{S}_\infty$ through "saturated" imbeddings of restricted direct products of some (maybe infinite) number of groups like $\mathcal{S}_A(T)$ and a finite symmetric group. Taking an IUR like $\pi(\rho)$ for each component $\mathcal{S}_A(T)$ and an IUR for the component of finite symmetric group, we define an IUR π of a wreath product type subgroup H, and then get an IUR Π of $\mathcal{S}_\infty$ by inducing up π from H (cf. [2]). Here we omit the details and discuss only about the above simple case of IUR Π.

6.2. A Generalization of the Method of Constructing IURs of G.

Let an integer N and an IUR Π of $\mathcal{S}_\infty$ be as in (1)–(3) above. From the explicit construction of this Π, we see that, to construct representations of G, instead of starting with $\Sigma = (\mu, E\,; \Pi)$ with

$$\mu = (\mu_i)_{i\in\mathbf{N}},\ \ \mu_i \text{ an } L\text{-measure on } M,$$

$$E = \textstyle\prod_{i\in\mathbf{N}} E_i,\ \ E_i \subset M, \text{ a } \mu\text{-unital set,}$$

we can start with $\Sigma^{(N)} = (\mu^{(N)}, E^{(N)}; \Pi)$ with

$$\mu^{(N)} = (\mu_j^{(N)})_{j\in\mathbf{N}},\ \mu_j^{(N)} \text{ an } L\text{-measure on } Y = M^N$$

$$E^{(N)} = \textstyle\prod_{j\in\mathbf{N}} E_j^{(N)},\ \ E_j^{(N)} \subset Y_j = Y \text{ a Borel subset.}$$

This means in particular that, it is not necessary to assume the disjointness condition on E_i's in (M1), as subsets of M, but is sufficient to assume the disjointness of $E_j^{(N)}$'s, as subsets of $Y = M^N = M \times M \times \cdots \times M$ (N-times), while we should put some assumption on the shape of each $E_j^{(N)} \subset M^N$.

The space $X = \prod_{i\in\mathbf{N}} M_i,\ M_i = M$, is understood, this time, as the direct product space of copies of $Y = M^N$ as follows. For $\alpha \in A$, put

$$Y_\alpha = \textstyle\prod_{i\in I_\alpha} M_i := M_{i_{\alpha,1}} \times M_{i_{\alpha,2}} \times \cdots \times M_{i_{\alpha,N}} \cong M^N,$$

then X is isomorphic to $\prod_{\alpha\in A} Y_\alpha$, and hence, under an bijection between A and **N**, it is also isomorphic to the above product $\prod_{j\in\mathbf{N}} Y_j,\ Y_j = Y$.

Applying Proposition 1.4 to the present case of $\mu^{(N)}, E^{(N)}$ and Y, we see that the "product" measure $\nu_{\mu^{(N)},E^{(N)}}$ on $\prod_{j\in\mathbf{N}} Y_j \cong X$ and $\mathcal{M}(E^{(N)})$ is carried by the subset consisting of all ordered configurations of points in Y. In this connection, it is worthwhile to give here the following remark.

Remark. Let $(\mu_i)_{i\in\mathbf{N}}$ be, as in **1.1**, a system of L-measures on M which satisfies the conditions (M1)–(M2). Let $E = \prod_{i\in\mathbf{N}} E_i \in X$, $E_i \subset \mathcal{M}_M$, satisfy the condition (U2) and the one (U1c) below, instead of (U1).

(U1c) For any compact subset $K \subset M$, we have $\sum_{i\in\mathbf{N}} \mu_i(E_i \cap K) < \infty$, and $\mu_i(E_i) > 0$ $(\forall i \in \mathbf{N})$.

Then, again we can define a product measure $\nu_{\mu,E}$ on $(X, \mathcal{M}(E))$. Furthermore the point is the following.

Lemma 6.1. *The measure $\nu_{\mu,E}$ is carried by the space $\tilde{X} \subset X$ of all ordered configurations of points in M.*

References

[1] P. R. Halmos, *Measure Theory*, Springer-Verlag, 1988.

[2] T. Hirai, *Construction of irreducible unitary representations of the infinite symmetric group S_∞*, J. Math. Kyoto Univ., **31** (1991), 495–541.

[3] T. Hirai, *Irreducible unitary representations of the group of diffeomorphisms of a non-compact manifold*, to appear in J. Math. Kyoto Univ.

[4] R. S. Ismagilov, *Unitary representations of the group of diffeomorphisms of the space* $\mathbf{R}^n$, $n \geq 2$, Funct. Anal. Appl., **9** (1975), 144–145 (= Funct. Anal., **9**(1975), 154-155).

[5] R. S. Ismagilov, *Imbedding of a group of measure-preserving diffeomorphisms into a semidirect product and its unitary representations*, Mat. Sb., **113** (1980), 81–97 (= Math. USSR Sb., **41**(1982), 67–81).

[6] S. Kakutani, *On equivalence of infinite product measures*, Ann. Math., **49** (1948), 214–224.

[7] A. M. Vershik, I. M. Gelfand and M. I. Graev, *Representations of the group of diffeomorphisms*, Usp. Mat. Nauk, **30** (1975), 3–50 (= Russ. Math. Surv., **30**(1975), 1–50).

[8] H. Weyl, *The Classical Groups, Their Invariants and Representations*, 2nd ed., Princeton University Press, Princeton, 1946.

Department of Mathematics
Faculty of Science
Kyoto University

CHARACTERS OF LIE GROUPS

M. ANOUSSIS

ABSTRACT. - We consider a connected Lie group with co-compact radical G. Let g be the Lie algebra of G. Let l be in the dual of g. Under the assumption that g(l) is commutative and reductive in g,we construct an application $\varphi \to F_{l,\varphi}$ from D(G) to the space of C^∞functions on an open dense subset of G(l). If G is compact, $F_{l,\varphi}$ is - up to a scalar - the invariant integral of φ relative to the Cartan subgroup G(l) of G. Using this, we obtain a formula for the trace of the operator T(l, G)(φ), where T(l, G) is the unitary representation of G associated to l.

1. Introduction

We consider a connected, simply connected unimodular Lie group G with co-compact radical. We denote by **g** the Lie algebra of G. Let l be an admissible linear form on **g**. We assume that l has a good polarisation and that **g**(l) is reductive in **g**. Let T(l, G) be the equivalence class of unitary irreducible representations of G associated with l. In what follows a formula for the character of T(l, G) is obtained. First we construct an application $\varphi \to F_{l,\varphi}$ from D(G) into $C^\infty(G(l)')$ where G(l)' is a dense open subset of G(l). If G is compact,G(l) is a Cartan subgroup of G and $F_{l,\varphi}$ is up to a scalar the invariant integral of φ with respect to G(l).
The character formula for T(l, G) in theorem 4 (see below) uses explicitely $F_{l,\varphi}$. A special case of this formula was considered by Duflo in [5, Ch. IX, prop 2.4.1.] One can obtain a character formula for T(l, G) by using orbital integral in the dual of **g**. [8]. This formula holds for functions φ in D(G) supported in a neighbourhood of the neutral element of G. Our formula is valid for every φ in D(G). If G is compact, this formula is a consequence of the Weyl character formula. It follows from [1,2,6] that theorem 4 applies to the unitary representations of G which are square integrable modulo the center Z(G) of G.

2. Notations

1. If V is a vector space we denote by V' the dual of V. If V is a real vector space we denote by V^C the complexified vector space of V. Let $x \to \bar{x}$ be the involution of V^C defined by the real form V. For λ in $(V^C)'$ we define: $\bar{\lambda} \in (V^C)'$ by $\bar{\lambda}(x)=\overline{\lambda(\bar{x})}$
2. Let G be a Lie group with Lie algebra **g**. The group G acts of **g'** via the co-adjoint reresentation. If l is in **g'** we denote by G(l) the stabiliser of l and **g**(l) the Lie algebra of the group G(l).
3. If X is a C^∞ manifold, C^∞ (X) is the space of C^∞ real valued functions on X and D(X) is the space of compactly supported C^∞ real valued functions on X.
4. Let **g** be a Lie algebra and l a linear form on **g**. Let W be a polarisation in $\mathbf{g}^C$ with respect to l.We say that W is a good polarisation if W is solvable and satisfies the Pukanszky condition.

3. The invariant integral on G

Let G be a connected, simply connected unimodular Lie group G with co - compact radical. We fix a Haar mesure dg on G. Let g be the Lie algebra of G. We consider a linear form l on **g** which is admissible and has a good polarisation. We assume that **g**(l) is rductive in **g**. Then, the group G(l) is abelian and connected. If λ is in $(g^C)'$, $\lambda \neq 0$ we set:

E. A. Tanner and R. Wilson (eds.), Noncompact Lie Groups and Some of Their Applications, 239–242.

$(g^C)_\lambda=\{X \in g^C: [A, X] = \lambda(A)X$, for every A in $g(l)^C\}$.
We denote by P(l) the set of λ's in $(g(l)^C)'$ which are different from 0 and for which $(g^C)_\lambda$ is non trivial. If λ is in P(l) we say that λ is a weight for g(l).
The set of weights P(l) is invariant under the involution $\lambda \to \bar{\lambda}$ of $g(l)^C$. If λ is in P(l) then -λ is also in P(l).
We have

$g^C = h^C \oplus \sum_{\lambda \in P(l)} (g^c)_\lambda$ where h is the centraliser of g(l) in g.

Lemma 1 [3, Prop. 2.1, and Lemma 3.1.]

Let H_l be the hermitian form on g^C defined by: $H_l(X,Y) = il([X, Y])$. Then the restriction H_l^λ of H_l to $(g^C)_\lambda$ is non-degenarate for every λ in P(l) such that $\lambda=-\bar{\lambda}$. Moreover if (p_λ,n_λ) is the signature of H_l^λ and W is a positive polarisation in g^C with respect to l we have:

$$\dim((g^C)_\lambda \cap W) = p_\lambda.$$

□

Let F be a subset of P(l). We set: $-F = \{ \lambda \in P(l): -\lambda \in F\}$
$F_r = \{ \lambda \in F: \lambda = \bar{\lambda} \}$
$F_i = \{ \lambda \in F: \lambda = -\bar{\lambda} \}$
$F_c = F - (F_i \cup F_r)$

Definition
Let F be a subset of P(l). We call F a system of positive weights if :
i) $F \cap -F = \varnothing$, $F \cup -F = P(l)$.
ii) $\lambda \in F_c \Rightarrow \bar{\lambda} \in F_c$

We set $d_\lambda = \dim (g^C)_\lambda$, for λ in P(l).

Lemma 2 [4, Lemma 4.1.1.]
Let F be a system of positive weights.

We put $\rho_F = \frac{1}{2}\sum_{\lambda \in F} d_\lambda \cdot \lambda$.
Then, there exists a character of G(l) with differential ρ_F. □

For λ in P(l), ξ_λ denotes the character of G(l) of differential λ. Let F be a system of positive weights.Then ξ_{ρ_F} denotes the character of G(l) which has differential ρ_F.

We set for x in G(l): $C_{r,F}(x) = \operatorname{sgn} \prod_{\lambda \in F_r} (1-\xi_{-\lambda}(x))^{d_\lambda}$ and

$i(F) = \sum_{\lambda \in F_i} n_\lambda$ where $n_\lambda = d_\lambda - p_\lambda$.
We have the following:

Proposition 3 [4, Lemma 4.1.2.]

The function $r_{l,F}$ on G(l) defined by: $r_{l,F}(x) = (-1)^{i(F)}\, C_{r,F}(x)\xi_{\rho_F}(x)\prod_{\lambda\in F}(1-\xi_{-\lambda}(x))^{d_\lambda}$
does not depend on the choice of the system of positive weights F. □

We will denote by r_l the function defined in proposition 3. If x is in G(l), we call x regular if $r_l(x)\neq 0$. The set of regular points will be denoted by G(l)'.
Let H be the centraliser of G(l) in G. If x is a regular point of G(l) and φ is in D(G) the function $g \to \varphi(gxg^{-1})$ on G is compactly supported modulo H [4, Lemma 4.3.2.]. The group H is unimodular [4, Prop. 2.2.]. It follows that there exists a G - invariant measure on G/H. We denote by $d\dot{g}$ this measure, normalised as in [4,§4.] . This normalisation depends only on l and dg.

Definition

Let x be in G(l)' and φ in D(G). We set: $F_{l,\varphi}(x) = r_l(x)\int_{G/H}\varphi(gxg^{-1})d\dot{g}$

The function $F_{l,\varphi}(x)$ is in $C^\infty(G(l)')$.[4,§4.3]. We say that $F_{l,\varphi}$ is the invariant integral of φ with respect to l.

Remarks

1. Let g_1 be in G and $\varphi_1(g) = \varphi(g_1 g g_1^{-1})$. Then $F_{l,\varphi} = F_{l,\varphi_1}$
2. If G is compact, G(l) is a Cartan subgroup of G, P(l) is a root system and F is a system of positive roots. Then, $F_{l,\varphi}$ is up to a scalar the invariant integral of φ with respect to the Cartan subgroup G(l) of G, as defined in [7].
3. There exist functions φ in D(G) such that the function $F_{l,\varphi}$ is not integrable with respect to the Haar measure on G(l).

4. The character formula

We denote by T(l, G) the equivalence class of unitary irreducible representations associated to G as in [8.§3.2]. Let T be in T(l, G) and χ_l the character of G(l) with differental il. We denote by dz (resp. $d\dot{x}$) the Haar measusre on the center Z(G) of G (resp. on $G(l)/Z(G)$) normalised as in [4.§5.] This normalisation depends only on l and dg. Then we have the following.

Theorem 4

Let φ in D(G). Then the operator $T(\varphi) = \int_G \varphi(g)\,T(g)\,dg$ is a trace class operator. Its trace is given by the formula:

$$Tr(T(\varphi)) = \int_{G(l)/Z(G)}\int_{Z(G)} F_{l,\varphi}(xz)\chi_l(xz)dz d\dot{x}$$

the successive integral being convergent. □

Remarks

1. If G is compact the formula of the theorem is a consequence of the Weyl character formula.
2. The character formula given in [8,Th.7.3] is valid for functions φ in D(U) where U is a neighbourhood of the neutral element in G. The formula given in Theorem 4 above is valid for every φ in D(G).
3. It follows from theorem 4 that the support of the distribution $\varphi \rightarrow Tr(T(\varphi))$ is contained in the closure of the set of conjugates of the elements of G(l).
4. It follows from [1,2,6] that every unitary irreducible representation of G which is square integrable modulo Z(G) is associated with a linear form l on **g** which verifies our assumptions. Thus, our formula applies to this class of representations.

5. Questions

Does theorem 4 remain valid if we relax the assumption that **g**(l) is reductive in **g**? More specifically, could one prove it for every unitary irreducible representation of a connected nilpotent Lie group? What is the analogue of the application $\varphi \rightarrow F_{l,\varphi}$ in this case?

6. References

1. Anh, N.H., (1976) "Lie groups with square integrable representations", Ann. of Math. t. 104, 431 - 458
2. Anh, N. H., (1980) "Classification of connected unimodular Lie groups with discrete series", Ann. Inst. Fourier, t. 30,1, 159 - 192.
3. Anoussis, M. (1991) "Sur les caracteres des groupes de Lie resolubles", Ann Inst. Fourier. t. 41,1 27 - 48.
4. Anoussis, M. (1992), "Integrales invariantes et formules de caracteres pour un groupe de Lie connexe a radical co-compact". Bull. Soc. math. France, t. 120, 347 - 370
5. Bernat, P. et al. (1972) "Representations des groupes de Lie resolubles" Dunod, Paris.
6. Charbonnel, J. Y. (1977) "La formule de Plancherel pour un groupe de Lie resoluble connexe" Lecture Notes in Math., t. 587, 32 - 76.
7. Harish - Chandra, (1957) "A formula for semi-simple Lie groups" Amer. J. Math t. 79, 733 - 760.
8. Khalgui, M. S. (1981) "Sur les caracters des groupes de Lie a radical co-compact" Bull. Soc. math. France, t. 79, 733 - 760.

Department of Mathematics
University of the Aegean
Karlovassi 83 200
Samos - Greece

WEYL GROUP ACTIONS ON LAGRANGIAN CYCLES AND ROSSMANN'S FORMULA

W. SCHMID AND K. VILONEN

1. Introduction.

Let $G_{\mathbb{R}}$ be a connected Lie group, $\mathfrak{g}_{\mathbb{R}}$ its Lie algebra, and $\mathfrak{g}_{\mathbb{R}}^*$ the vector space dual of $\mathfrak{g}_{\mathbb{R}}$. Each coadjoint orbit, i.e., $G_{\mathbb{R}}$-orbit in $\mathfrak{g}_{\mathbb{R}}^*$, has an intrinsically defined $G_{\mathbb{R}}$-invariant symplectic structure, and thus carries a distinguished $G_{\mathbb{R}}$-invariant measure. Kirillov's character formula - in those cases when it applies - expresses the irreducible unitary characters of $G_{\mathbb{R}}$ as Fourier transforms of the distinguished measures on coadjoint orbits, which are then lifted from the Lie algebra to the group via the exponential map. Kirillov [Ki] originally established the formula for nilpotent groups, and Auslander-Kostant [AK] for a large class of solvable groups.

Matters are considerably more complicated in the case of a semisimple Lie group $G_{\mathbb{R}}$. As was shown by Rossmann [R1], Kirillov's formula is then still valid for characters of irreducible tempered representations. Characters of non-tempered irreducible unitary representations, on the other hand, usually do not arise as Fourier transforms of invariant measures on coadjoint orbits. Rossmann [R2] proposes a different type of integral formula: irreducible characters of $G_{\mathbb{R}}$ - even of non-unitary representations - can be expressed as Fourier transforms of certain cycles in coadjoint orbits of the complexified group, which we denote by G. To make this formula useful in practice, it is necessary, of course, to identify the cycles. Rossmann [R3] has done so when $G_{\mathbb{R}}$ is itself a complex group; the general case has remained open until now. Our main result, theorem 3.1, establishes an explicit Rossmann type integral formula for any semisimple group $G_{\mathbb{R}}$.

One of the main ingredients of our proof is of independent interest. Let W denote the Weyl group of G and X the flag variety. Rossmann [R3] constructs a homotopy action of W on the cotangent bundle T^*X (see also [KL]), and this homotopy action induces a W-action on the group of Lagrangian cycles in T^*X. The group of Lagrangian cycles is naturally isomorphic to to K-group of the derived category $D_c^b(X)$. Via intertwining functors [BB] W acts on the K-group of $D_c^b(X)$, and hence on the group of Lagrangian cycles. Our theorem 2.3 asserts that these two W-actions on the group of Lagrangian cycles coincide.

W.Schmid was partially supported by NSF
K.Vilonen was partially supported by NSA, NSF, and the AMS Centennial Fellowship

E. A. Tanner and R. Wilson (eds.), Noncompact Lie Groups and Some of Their Applications, 243–250.

Applications and complete proofs of the results announced in this note will appear elsewhere.

2. Weyl group actions on Lagrangian cycles.

Let G be a semi-simple complex algebraic group and fix $U_{\mathbb{R}} \subset G$, a maximal compact subgroup. We further denote by $\mathcal{B} \to X$ the bundle of Borel subgroups of G. Then $\mathcal{B}/[\mathcal{B},\mathcal{B}]$ splits G-equivariantly into a product $X \times H$. Here H, the universal Cartan, is a complex torus. We make the convention that the root spaces in $[\mathcal{B},\mathcal{B}]$ are considered negative. Any concrete Cartan together with an ordering of its root system can then be identified canonically with H. As a homogenous space, $X \cong G/B$, where B is a Borel subgroup of G. Let $\mathfrak{g}$, $\mathfrak{u}_{\mathbb{R}}$ and $\mathfrak{h}$ denote the Lie algebras of G, $U_{\mathbb{R}}$ and H respectively. We denote by $\tilde{\mathfrak{g}} \to X$ the bundle of Borel subalgebras of $\mathfrak{g}$ and by W the Weyl group, which operates naturally as a group of automorphisms on $\mathfrak{h}$.

Let us denote $\tilde{\mathfrak{n}} = [\tilde{\mathfrak{g}}, \tilde{\mathfrak{g}}]$ and consider the following vector bundles on X:

$$\tilde{\mathfrak{n}} \subset \tilde{\mathfrak{g}} \subset \mathfrak{g} \times X. \tag{2.1}$$

At any Borel subalgebra, viewed as a point of X, the tangent space of X is naturally isomorphic to $\mathfrak{g}/\mathfrak{b}$. Globally this gives an exact sequence of vector bundles

$$0 \to \mathfrak{h} \times X \to (\mathfrak{g} \times X)/\tilde{\mathfrak{n}} \to TX \to 0, \tag{2.2}$$

and by dualising we get

$$0 \to T^*X \to (\mathfrak{g} \times X/\tilde{\mathfrak{n}})^* \to \mathfrak{h}^* \times X \to 0. \tag{2.3}$$

The second of these maps and various natural projections give the correspondence

$$\mathfrak{g}^* \xleftarrow{p} (\mathfrak{g} \times X/\tilde{\mathfrak{n}})^* \xrightarrow{q} \mathfrak{h}^*. \tag{2.4}$$

If $\lambda \in \mathfrak{h}^*$ is regular,

$$\Omega_\lambda = pq^{-1}\lambda \tag{2.5}$$

defines a regular, semisimple coadjoint orbit, which depends only on the W-orbit of λ. In this way the regular W-orbits in $\mathfrak{h}^*$ parametrize precisely the regular coadjoint orbits in $\mathfrak{g}^*$.

The group $U_{\mathbb{R}} \subset G$ acting on X chooses for us a concrete Cartan $\mathfrak{t}_x$ for every point $x \in X$, and the root system of this Cartan has a natural order. Diagonalising $\mathfrak{g}$ under the adjoint action of $\mathfrak{t}_x$ gives us real algebraic $U_{\mathbb{R}}$-invariant complements for $\tilde{\mathfrak{n}} \subset \tilde{\mathfrak{g}}$ and for $\tilde{\mathfrak{g}} \subset \mathfrak{g} \times X$, thus splitting both (2.2) and (2.3) $U_{\mathbb{R}}$-invariantly.

This gives us a $U_{\mathbb{R}}$-invariant real algebraic isomorphism

$$(\mathfrak{g} \times X/\tilde{\mathfrak{n}})^* \cong (\mathfrak{h}^* \times X) \oplus T^*X. \tag{2.6}$$

For $\lambda \in \mathfrak{h}^*$, the section $\{\lambda\} \times X \subset \mathfrak{h}^* \times X$, followed by (2.6) and p, as in (2.4), defines a $U_{\mathbb{R}}$-invariant, real algebraic map

$$\mu_\lambda : T^*X \to \mathfrak{g}^*.$$

If $\lambda \in \mathfrak{h}^*$ is regular, μ_λ takes values in Ω_λ, and μ_λ becomes an isomorphism

$$\mu_\lambda : T^*X \xrightarrow{\sim} \Omega_\lambda,$$

which, following Rossmann [R3], we call the twisted moment map. At the opposite extreme, for $\lambda = 0$,

$$\mu =_{\text{def}} \mu_0 : T^*X \to \mathcal{N}^*$$

is the usual moment map; here $\mathcal{N}^* \subset \mathfrak{g}^*$ denotes the nilpotent cone. In this case μ no longer depends on the choice of the compact real form $U_{\mathbb{R}} \subset G$, and μ is, in fact, G-invariant and complex algebraic.

As is explained in [R3], the compositions $\mu_{w\lambda}^{-1} \circ \mu_\lambda$ for regular λ define a proper homotopy action of W on T^*X, which is independent of the choice of regular λ. As Rossman shows, this action coincides with the one defined in [KL].

Let $\mathcal{O} \subset \mathcal{N}^*$ be a nilpotent, coadjoint orbit. Its inverse image $\mu^{-1}(\mathcal{O})$ under the moment map is a G-invariant algebraic subvariety. The spaces T^*X, $\mathcal{O}$, and Ω_λ, for $\lambda \in \mathfrak{h}^*$ regular, come equipped with nondegenerate, holomorphic two forms σ, $\sigma_{\mathcal{O}}$ and σ_λ, which give each of these spaces the structure of a complex symplectic manifold. All three symplectic structures are G-invariant.

In order to relate the above symplectic structures, we define a $U_{\mathbb{R}}$-invariant two form τ_λ on X by the formula

$$\tau_\lambda(u, v) = \lambda([u, v]),$$

where $u, v \in \mathfrak{u}_{\mathbb{R}}$ are vector fields defined by the derivative of the $U_{\mathbb{R}}$-action. Here the $U_{\mathbb{R}}$-invariant splitting of (2.1) is used to view λ as $\mathfrak{u}_{\mathbb{R}} \cap \mathfrak{b}$-invariant linear functional on the tangent space of X at any Borel subalgebra $\mathfrak{b}$. When $\lambda \in \mathfrak{h}^*$ happens to be integral, τ_λ is the curvature form of the (essentially unique) $U_{\mathbb{R}}$-invariant metric of the G-invariant algebraic line bundle $\mathcal{L}_\lambda \to X$ parametrized by $\lambda \in \mathfrak{h}^*$ [GS].

Let $\pi : T^*X \to X$ be the natural projection.

Proposition 2.1. *The symplectic structures defined above are related as follows:*

(1) $\mu_\lambda^* \sigma_\lambda = \sigma + \pi^* \tau_\lambda$;
(2) $\mu^* \sigma_{\mathcal{O}} = \sigma|\mu^{-1}(\mathcal{O})$ *at the smooth points of* $\mu^{-1}(\mathcal{O})$;
(3) $\mu_\lambda^* \sigma_\lambda|\mu^{-1}(\mathcal{O}) = \mu^* \sigma_{\mathcal{O}} + \pi^* \tau_\lambda|\mu^{-1}(\mathcal{O})$ *at the smooth points of* $\mu^{-1}(\mathcal{O})$.

Part (3) of this proposition is proved in [R3]. Part (2) follows from the fact that if $\mathfrak{n}$ is a nilradical of a Borel, then $\mathfrak{n} \cap \mathcal{O}$ is Lagrangian in $\mathcal{O}$; see [Gi], [J]. Part (1) follows from (2) and (3).

Consider T^*X as a real algebraic, symplectic manifold. The real symplectic form $\sigma_{\mathbb{R}}$ is given by $\sigma_{\mathbb{R}} = 2Re(\sigma)$. Consider the group $\mathcal{L}$ of real semi-algebraic Lagrangian $\mathbb{R}^+$-invariant integral cycles on T^*X. With n denoting the real dimension of X, this group can formally be defined, as in [KSa], as

$$\mathcal{L} = \varinjlim_{\Lambda} \mathrm{H}^n_\Lambda(\pi^{-1}(or_X)), \tag{2.7}$$

where Λ runs through the isotropic real semi-algebraic $\mathbb{R}^+$-invariant subvarieties of T^*X and or_X denotes the orientation sheaf of X. If one wants to consider the cycles in $\mathcal{L}$ as cycles in the usual sense, i.e., as coming equipped with an orientation, it is enough to choose an orientation for X. The point of definition (2.7) is that it makes sense intrinsically, independently of the choice of orientation of X, even when X cannot be oriented. In our situation we shall give X its natural orientation as a complex manifold, thus making each cycle in the group (2.7) an oriented cycle. In the language of [KSa], we view the Lagrangian cycle $[X] \in \mathcal{L}$ as the usual orientation class of X. We will use this convention for the rest of this paper.

Lemma 2.2. *Given a cycle $C \in \mathcal{L}$, there exists, for each G-orbit $\mathcal{O} \subset \mathcal{N}^*$, a Lagrangian, semialgebraic subset $S_\mathcal{O} \subset \mathcal{O}$ such that*

1. *the boundary $\partial S_\mathcal{O} \subset \mathcal{N}^*$ of $S_\mathcal{O}$ satisfies $dim_\mathbb{R} \partial S_\mathcal{O} \leq dim_\mathbb{R} S_\mathcal{O} - 2$.*
2. *$S = \bigcup S_\mathcal{O}$ is closed in $\mathcal{N}^*$*
3. *$\mu^{-1}(S)$ is Lagrangian in T^*X and contains the support of C.*

To see this, we consider a G-orbit $\mathcal{O}$ and a Lagrangian submanifold $M \subset T^*X$ which is contained in $\mu^{-1}(\mathcal{O})$. By proposition 2.1.(2), M is locally of the form $\mu^{-1}(M_\mathcal{O})$, where $M_\mathcal{O}$ is a Lagrangian submanifold of $\mathcal{O}$. Applying this statement inductively with respect to the orbit stratifiaction of $\mathcal{N}^*$, we get the lemma.

Let us consider a cycle $C \in \mathcal{L}$, and let S be as in lemma 2.2. As shown by Rossmann, the proper homotopy action of W on T^*X induces an action on the Borel-Moore homology groups $\mathrm{H}_*^{inf}(\mu^{-1}(S), \mathbb{Z})$, i.e., on the homology groups where cycles are allowed to have infinite support. By lemma 2.2, $\mathrm{H}_*^{inf}(\mu^{-1}(S), \mathbb{Z})$ injects into $\mathcal{L}$, and $C \in \mathcal{L}$ determines an element in $\mathrm{H}_*^{inf}(\mu^{-1}(S), \mathbb{Z})$ for our particular choice of S. This, in effect, determines an action of W on $\mathcal{L}$.

Remark. In view of 2.1 (2), the diffeomorphisms $\mu_{w\lambda}^{-1} \circ \mu_\lambda : T^*X \to T^*X$, which define Rossmann's proper homotopy action, are symplectic for each regular λ. While we do not use this fact explicitly, it explains the existence of the W-action on $\mathcal{L}$.

We can define another action of W on $\mathcal{L}$ as follows. Let us consider $\mathrm{D}_c^b(X)$, the bounded derived category of complexes of $\mathbb{C}$-sheaves whose cohomology is constructible in the semi-algebraic sense. We have the map $\mathrm{CC} : \mathrm{D}_c^b(X) \to \mathcal{L}$, the characteristic cycle map, as defined in [KSa], which becomes a bijection on the level of K-groups. The intertwining functors of Beilinson-Bernstein [BB] give a W-action on the K-group of $\mathrm{D}_c^b(X)$, and therefore induce a W-action on $\mathcal{L}$.

Theorem 2.3. *The two Weyl group actions on the Lagrangian cycles $\mathcal{L}$ defined above coincide.*

In the special case when the Lagrangian cycles are assumed to be supported on the conormal bundles of orbits under the action of a Borel $B \subset G$, theorem 2.3 is a result of Kashiwara and Tanisaki [KT].

Remark. The geometric action of W which is defined via the twisted moment maps is of course defined on all semi-algebraic Lagrangian cycles on T^*X. The action of W which is defined via the intertwining functors gives us an action only on the $\mathbb{R}^+$-invariant Lagrangian cycles.

A crucial ingredient in the proof of theorem 2.3 is the following result.

Theorem 2.4. *Let X be a real analytic manifold and let U be an open subanalytic subset with $j : U \hookrightarrow X$ denoting the inclusion. Let f be a smooth function on X which vanishes precisely on the boundary $\bar{U} - U$ of U. If $\mathcal{F} \in \mathrm{D}_c^b(X)$ then*

$$\mathrm{CC}(Rj_*\mathcal{F}) = \lim_{s \to 0+} \left(\mathrm{CC}(\mathcal{F}) + s\frac{df}{f} \right).$$

When X is complex analytic and U is a complement of a divisor, this theorem is a result of Ginsburg [Gi2], though our proof is quite different from his.

3. The main theorem.

We retain the notation of section 2. We consider a real form $G_{\mathbb{R}}$ of G which, for simplicity, we assume to be linear, and denote by $\mathfrak{g}_{\mathbb{R}}$ the Lie algebra of $G_{\mathbb{R}}$. We now choose $U_{\mathbb{R}}$ so that the complex conjugations defining $G_{\mathbb{R}}$ and $U_{\mathbb{R}}$ commute; this can always be done [He]. Recall that the enhanced flag variety $\hat{X}$ is the quotient $G/[B, B]$ (see [BB]). Then H acts on $\hat{X}$ by right translation, $h : g[B, B] \mapsto gh^{-1}[B, B]$, and $X \cong \hat{X}/H$.

Let $\lambda \in \mathfrak{h}^*$ be fixed, and let $D_{G_{\mathbb{R}}}(X)_\lambda$ denote the $G_{\mathbb{R}}$-equivariant derived category of H-monodromic sheaves with twist λ. These are sheaves on $\hat{X}$ which, when pulled back to $H \times \hat{X}$ by the action map above, have monodromy e^λ on the H-factor. For $\lambda = 0$, for example, this is precisely the $G_{\mathbb{R}}$-equivariant derived category of sheaves on X. Note that the definition of $D_{G_{\mathbb{R}}}(X)_\lambda$ only depends on the value of λ modulo the weight lattice.

Let $\rho \in \mathfrak{h}$ be half the sum of positive roots. We write $\mathcal{O}_X(\lambda)$ for the sheaf of H-monodromic holomorphic functions with twist $\lambda + \rho$. Note that this does not agree with the notation in [SV], where the ρ-shift is not built in to the notation. $\mathcal{O}_X(\lambda)$ again is a sheaf on $\hat{X}$. When $\lambda = -\rho$, it is the inverse image (as a sheaf of $\mathbb{C}$-vector spaces) of $\mathcal{O}_X$ under $\hat{X} \to X$. More generally, when λ is an integral weight and $\mathcal{L}_\lambda \to X$ is the G-equivariant line bundle modeled on the character e^λ, then $\mathcal{O}_X(\lambda)$ is the sheaf pullback to $\hat{X}$ of $\mathcal{O}_X(\mathcal{L}_{\lambda+\rho})$.

We define a homomorphism

$$D_{G_{\mathbb{R}}}(X)_{-\lambda} \to \{\text{virtual } G_{\mathbb{R}}\text{-representations with infinitesimal character } \chi_\lambda\} \tag{3.1}$$

by the formula

$$\mathcal{F} \mapsto \sum (-1)^p \operatorname{Ext}^p(\mathbb{D}\mathcal{F}, \mathcal{O}_X(\lambda)), \tag{3.2}$$

where $\mathbb{D}$ denotes the Verdier dual. The insertion of the Verdier dual serves the purpose of making the correspondence (3.1) covariant. For the fact that this formula makes sense see [KSd]. On the level of K-groups, which suffices for our purposes, it was done in [SW]. These matters are discussed in greater detail in [SV].

Consider the moment map $\mu : T^*X \to \mathcal{N}^*$. Let $\Lambda_{\mathbb{R}} = \mu^{-1}(\mathfrak{g}_{\mathbb{R}}^{\perp})$, where $\mathfrak{g}_{\mathbb{R}}^{\perp}$ denotes the annihilator of $\mathfrak{g}_{\mathbb{R}}$ in $\mathfrak{g}^*$. Then $\Lambda_{\mathbb{R}}$ is the union of the conormal bundles of the various $G_{\mathbb{R}}$-orbits on X. Consider now the group $\mathcal{L}_{\Lambda_{\mathbb{R}}}$ of Lagrangian cycles supported on $\Lambda_{\mathbb{R}}$. It is a finitely generated free abelian group with one generator for each $G_{\mathbb{R}}$-orbit on X. Note, however, that the conormal bundles of the $G_{\mathbb{R}}$-orbits do, in general, have boundaries; for a discussion, see [R2].

We introduce a map

$$\mathrm{CC} : D_{G_{\mathbb{R}}}(X)_\lambda \to \mathcal{L}_{\Lambda_{\mathbb{R}}} \tag{3.3}$$

as follows. First associate to $\mathcal{F} \in D_{G_{\mathbb{R}}}(X)_\lambda$ its underlying object $\hat{\mathcal{F}} \in \mathrm{D}^b_c(\hat{X})$. Because $\hat{\mathcal{F}}$ is H-monodromic, its characteristic cycle descends from $T^*\hat{X}$ to T^*X. Because of $G_{\mathbb{R}}$-equivariance this characteristic cycle, now viewed on T^*X, is supported on $\Lambda_{\mathbb{R}}$. This gives us the map (3.3).

With $\lambda \in \mathfrak{h}^*$ fixed as before, we associate to each Lagrangian cycle in $\mathcal{L}_{\Lambda_{\mathbb{R}}}$ an invariant eigendistribution on the Lie algebra $\mathfrak{g}_{\mathbb{R}}$, i.e., we construct a map

$$\mathcal{L}_{\Lambda_{\mathbb{R}}} \to \{\text{invariant eigendistributions on } \mathfrak{g}_{\mathbb{R}} \text{ with infinitesimal character } \chi_\lambda\}. \tag{3.4}$$

This is done by a process due to Rossmann [R3]. Let $\phi(x) \in C_c^\infty(\mathfrak{g}_{\mathbb{R}})$. We normalise the Fourier transform $\hat{\phi}$, viewed as holomorphic function on $\mathfrak{g}^*$, as

$$\hat{\phi}(\xi) = \int e^{\xi(x)}\phi(x)dx, \tag{3.5}$$

where dx is a Euclidean measure on $\mathfrak{g}_{\mathbb{R}}$. We then get the map (3.4), for regular $\lambda \in \mathfrak{h}^*$, by associating to the Lagrangian cycle $C \in \mathcal{L}_{\Lambda_{\mathbb{R}}}$ the invariant eigendistribution

$$\phi(x) \mapsto \frac{1}{(-2\pi i)^n n!}\int_{\mu_\lambda(C)} \hat{\phi}\sigma_\lambda^n. \tag{3.6}$$

The Paley-Wiener theorem assures that this integral converges for every $C \in \mathcal{L}_{\Lambda_{\mathbb{R}}}$.

To extend the defintion of the map (3.4) to all $\lambda \in \mathfrak{h}^*$, we rewrite formula (3.6), using proposition 2.1 (1), as follows:

$$\phi(x) \mapsto \frac{1}{(-2\pi i)^n n!}\int_C (\hat{\phi} \circ \mu_\lambda)(\sigma + \pi^*\tau_\lambda)^n. \tag{3.7}$$

This formula has a good meaning also for non-regular λ.

The virtual representations in (3.1-2) have characters in the sense of Harish-Chandra. By definition, the characters are invariant eigendistributions on the group $G_{\mathbb{R}}$. Via the exponential map, invariant eigendistributions on $G_{\mathbb{R}}$ descend to invariant eigendistributions on the Lie algebra $\mathfrak{g}_{\mathbb{R}}$; this transfer process involves a twist by $\sqrt{\det(\exp_*)}$ [HC]. We can thus speak of the character of a virtual representation on the Lie algebra. By construction, it depends on the choice of measure on $\mathfrak{g}_{\mathbb{R}}$, but is otherwise canonical.

Consider the following diagram

$$\begin{array}{ccc} D_{G_\mathbb{R}}(X)_{-\lambda} & \xrightarrow{(3.1)} & \{\text{virtual } G_\mathbb{R}\text{-representations with infinitesimal character } \chi_\lambda\} \\ {\scriptstyle CC}\downarrow & & \downarrow{\scriptstyle Char} \\ \mathcal{L}_{\Lambda_\mathbb{R}} & \xrightarrow{(3.4)} & \{\text{invariant eigendistributions on } \mathfrak{g}_\mathbb{R} \text{ with infinitesimal character } \chi_\lambda\}; \end{array}$$

here *Char* is the map associating the character on $\mathfrak{g}_\mathbb{R}$ to each virtual representation, with the same choice of Euclidean measure on $\mathfrak{g}_\mathbb{R}$ as in (3.5).

Theorem 3.1. *The diagram above is commutative.*

Rossmann [R2] shows that each virtual character can be expressed in the form (3.4), (3.6) for some unspecified cycle $C \in \mathcal{L}_{\Lambda_\mathbb{R}}$. Our theorem 3.1 pins down what the cycle C is. When $G_\mathbb{R}$ is itself a complex group, Rossmann [R3] had already done so[1], in a way which - not entirely obviously - reduces to a special instance of Theorem 3.1.

The proof of this theorem follows the same outline as the proof in [SV], i.e. it can be reduced to the following steps:

(1) The theorem holds for discrete series representations.
(2) The theorem is compatible with parabolic induction.
(3) (3.1) is compatible with coherent continuation.
(4) (3.4) is compatible with intertwining functors.

Checking (3) is not difficult, most of (2) is done in [R2], and (4) is a consequence of theorem 2.3. Using theorem 2.4 we deduce (1) from Rossmann's description of the discrete series characters [R1].

The character of a representation π, viewed as an invariant eigendistribution on $\mathfrak{g}_\mathbb{R}$, has an asymtotic expansion around 0, whose leading term is a linear combination of Fourier transforms of nilpotent coadjoint orbits (Barbasch-Vogan [BV]). As a corollary of theorem 3.1, this leading term can be read off from the characteristic variety of the sheaf $\mathcal{F} \in D_{G_\mathbb{R}}(X)_{-\lambda}$ which corresponds to π via (3.1). In the case of a complex group $G_\mathbb{R}$, this is Rossmann's argument [R3].

REFERENCES

[AK] L.Auslander and B.Kostant, *Polarisation and unitary representations of solvable Lie groups*, Inventiones Math. **14**, 255–354.

[BB] A.Beilinson and J.Bernstein, Proof of Jantzen's conjecture, preprint.

[BV] D.Barbasch and D.Vogan, *The Local Structure of Characters*, Jour. Func. Anal. **37** (1980), 27–55.

[Gi1] V.Ginsburg, *𝔊-modules, Springer Represenations and bivariant Chern classes*, Advances Math. **61** (1986), 1–48.

[Gi2] V.Ginsburg, *Characteristic Varieties and vanishing cycles*, Inventiones Math. **84** (1986), 327–402.

[GS] P.A.Griffiths and W.Schmid, *Locally Homogenous Complex Manifolds*, Acta Math. **123** (1969), 253–302.

[1] As J.-T. Chang has pointed out, Rossmann's argument applies slightly more generally to groups with a single conjugacy class of Cartans

[He] S.Helgason, *Differential Geometry, Lie Groups, and Symmetric Spaces*, Academic Press, 1978.

[HC] Harish-Chandra, *Invariant Eigendistributions on a Semisimple Lie Group*, Trans. Amer. Math. Soc. **119** (1965), 457–508.

[J] A.Joseph, *On the Variety of a Highest Weight Module*, Jour. Algebra **88** (1984), 238–278.

[KL] D.Kazhdan and G.Lusztig, *A Topological Approach to Springer's Representations*, Advances Math. **38** (1980), 222–228.

[KSa] M.Kashiwara and P.Schapira, *Sheaves on manifolds*, Springer, 1990.

[KSd] M.Kashiwara and W.Schmid, Equivariant Derived Category and Representations of Semisimple Lie Groups, to appear.

[KT] M.Kashiwara and T.Tanisaki, *The Characteristic Cycles of Holonomic Systems on a Flag Manifold*, Inventiones Math. **77** (1984), 185–198.

[Ki] A.Kirillov, *Unitary representations of nilpotent Lie groups (in Russian)*, Uspehi Mat. Nauk. **17** (1962), 57–110.

[R1] W. Rossmann, *Kirillov's Character Formula for Reductive Lie Groups*, Inventiones Math. **48** (1978), 207–220.

[R2] W.Rossmann, Characters as contour integrals, in Springer Lecture Notes in Mathematics, Vol. 1077, pp. 375-388, 1984.

[R3] W.Rossmann, *Invariant Eigendistributions on a Semisimple Lie Algebra and Homology Classes on the Conormal Variety I, II*, Jour. Func. Anal. **96** (1991), 130–193.

[SV] W.Schmid and K.Vilonen, *Character, Fixed Points and Osborne's Conjecture*, Contemp. Math **145** (1993).

[SW] W.Schmid and J.Wolf, *Geometric quantization and derived functor modules for semisimple Lie groups*, Jour. Funct. Anal. **90** (1990), 457–508.

DEPARTMENT OF MATHEMATICS, HARVARD UNIVERSITY, CAMBRIDGE, MA 02138, USA

DEPARTMENT OF MATHEMATICS, BRANDEIS UNIVERSITY, WALTHAM, MA 02254, USA

Taylor Formula, Tensor Products, and Unitarizability

Eugene Angelopoulos

Abstract

A method is described which yields necessary and sufficient conditions of unitarizability for representations of a real reductive Lie algebra, in the form of algebraic inequalities. There are some matrix rings with entries in the enveloping algebra which are closely related to the reduction of tensor products of representations. Combined use, for both the Lie algebra and its maximal compact subalgebra, of the properties of these rings, in particular the Taylor expansion of matrix polynomials, yields formulas which provide the desired conditions.

1 Introduction

The classification of the unitary representations of real semisimple Lie algebra is still an open problem, though great advances have been made in this direction, especially by Vogan [5] and Barbash [2]. In particular, there are five large classes of families, namely $su(p,q), so(p,q), sp(p,q), sp(N,\mathbf{R}), so^*(N)$ for which the unitary dual is not yet determined.

The scope of this talk is to present an approach of the unitarizability problem which reduces the question to what we believe is its hard core: to express a necessary and sufficient condition for a representation to be unitary, in the form of a minimal set of inequalities to be satisfied by some algebraic expressions of its parameters - and, if possible, to solve them.

This approach has been inspired by the algebraic methods used by Bargman [3] and Naimark [4] to determine the unitary dual of $so(2,1)$ and $so(3,1)$ respectively; when dealing with larger Lie algebras like $sl(3,\mathbf{R})$ or $so(p,2)$, these methods were insufficient to find the unitary dual, mostly because of difficulties arising from nontrivial multiplicity of isotypic components of the maximal compact subalgebra. Insufficient but not useless, since the problem was solved with the help of some additional calculatory devices [1a]; since then, a more elaborate formalism has been developed out of them [1b] in order to treat more general situations.

E. A. Tanner and R. Wilson (eds.), Noncompact Lie Groups and Some of Their Applications, 251–263.

This formalism is essentially based on the properties of the reduction of the tensor product of finite-dimensional representations. Indeed, if $\mathfrak{k}_0$ is the maximal compact subalgebra of a Lie algebra $\mathfrak{g}_0$ and $\mathscr{M}$ a unitarizable $\mathfrak{g}_0$-module, the noncompact generators act on a $\mathfrak{k}_0$-submodule W of $\mathscr{M}$ by sending it into some $\mathfrak{k}_0$-submodule W' isomorphic to a tensor product $V \otimes W$. If $V \otimes W$ is reduced to a sum of $\mathfrak{k}$-types W_u then the hermitian square b^*b of a shift operator b, sending W into W_u, and W_u back to W, must have positive spectrum. To study this spectrum in terms of the parameters characterizing $\mathscr{M}$ and W, one has to transfer these structures in the enveloping algebras $\mathscr{U}$ of $\mathfrak{k}_0$ and $\mathscr{U}(\mathfrak{g}_0)$ of $\mathfrak{g}_0$. It turns out that the ring of self-intertwining operators of $V \otimes W$, for fixed V and variable W, is the homomorphic image of a ring of matrices with entries in $\mathscr{U}$. However, this archetype ring glues together the isotypic components W_u; but they can be split apart inside a Galois extension by the Weyl group of $\mathfrak{k}_0$ of the center $\mathscr{Z}$ of $\mathscr{U}$ (roughly speaking, this means adding to $\mathscr{Z}$ the coordinates of the dominant weight). The Taylor expansion of polynomials plays a great role in this construction: in particular, it is closely connected with the projections of $V \otimes W$ on each isotypic components.

The above formalism is valid for every reductive Lie algebra, hence for both $\mathfrak{g}_0$ and $\mathfrak{k}_0$. Combined use of the Taylor expansion yields algebraic relations (see (17) and following, below) involving squared shift operators and parameters characterizing W and $\mathscr{M}$. These relations, used along with some convenient ordering of the $\mathfrak{k}_0$-types, yield some equalities and inequalities, and it results that, for given $\mathfrak{k}_0$-content of $\mathscr{M}$, unitarizability is granted if and only if the central character satisfies a set of diophantine inequalities - a great part of which can be expressed by means of linear inequalities: this depends on the particular structure of $\mathfrak{g}_0$. For families with real rank not greater than two, as well as for those solved by Vogan and Barbash, the nonlinear part is either inexistent or can be solved with no much trouble: and the results obtained are in accordance with theirs.

This will be exposed in more detail in what follows.

2 Unitarizable Representations

Let $\mathfrak{g}_0 = \mathfrak{k}_0 \oplus \mathfrak{p}_0$ be a reductive real Lie algebra, $\mathfrak{k}_0$ being its maximal compact subalgebra, and $\mathfrak{g} = \mathfrak{k} \oplus \mathfrak{p}$ the corresponding complexified ones. We shall denote by $\mathscr{U}(\mathfrak{g})$ the enveloping algebra of $\mathfrak{g}$ and by $\mathscr{Z}(\mathfrak{g})$ its center, by $\mathscr{U}$ and $\mathscr{Z}$ the corresponding ones of $\mathfrak{k}$. Let $\mathscr{M}$ be a simple Harish-Chandra $\mathfrak{g}$-module, that is $\mathscr{M}$ is the direct sum of finite-dimensional $\mathfrak{k}$-types with finite multiplicities. To simplify the formulas, we shall often identify $\mathscr{U}(\mathfrak{g})$ to the quotient $\mathscr{U}(\mathfrak{g})/\mathscr{I}$, where $\mathscr{I}$ is the bilateral ideal of elements of $\mathscr{U}(\mathfrak{g})$ vanishing on the whole of $\mathscr{M}$. Notice that $\mathscr{Z}(\mathfrak{g})$ is scalarly represented on $\mathscr{M}$, that is, $\mathscr{Z}(\mathfrak{g})/\mathscr{Z}(\mathfrak{g}) \cap \mathscr{I} = \mathbf{C}$, and all algebraic equations can be solved.

Assuming that there is a $\mathfrak{g}_0$-invariant sesquilinear form $(\varphi|\varphi)$ on $\mathscr{M}$ (it is then essentially unique), we shall adopt the following definition:

Definition: *$\mathscr{M}$ is unitarizable if and only if its sesquilinear form is positive definite.*

This form defines an involutory semilinear antiautomorphism $b \to b^*$ on the enveloping algebra $\mathscr{U}(\mathfrak{g})$, such that

$$(b^*\varphi|\varphi') = (\varphi|b\varphi'). \tag{1}$$

On the other hand, since $\mathscr{M}$ is simple, one may identify it with $\mathscr{U}(\mathfrak{g}) \cdot \varphi$, for some $\varphi \in \mathscr{M}$. Taking φ inside a $\mathfrak{k}$-isotypic component W, one obviously gets:

Proposition 1: *$\mathscr{M}$ is unitarizable if and only if every element which has the form b^*b and every positive linear combination of such elements has positive spectrum when restricted to the stabilizer of W.*

It is a tremendous task to check the positivity of all such elements, and that is why we shall refer to this statement as the *Ionesco condition* [6]. Using the splitting of $\mathscr{U}(\mathfrak{g})$ into $\mathfrak{k}$-isotypic components under the bracket representation, and the existence of multiplicity-free components in $\mathscr{M}$, one can ameliorate the above statement to:

Proposition 1a: *$\mathscr{M}$ is unitarizable if and only if every positive linear combination of elements b^*b commuting with $\mathfrak{k}$ is represented by a positive number when restricted to the multiplicity-free type W (and $b^*bW = 0$ implies $bW = 0$).*

However, even this simplified version is quite unpractical. To make it operative, one should confine the elements to check into a finite set. To do so, one may first choose $b \in \mathscr{U}$ to be a *shift operator*, sending W into another component W' and then prove that taking W' in some finite set "near" W is sufficient.

This leads to the determination of elements of $\mathscr{U}$ which act as shift operators. For, say, $\mathfrak{g}_0 = so(3,1)$ and $\mathfrak{k}_0 = so(3)$, such elements behave as ordinary 3-vectors under the action of $so(3)$, and the Ionesco condition reduces to the checking of their ordinary inner product. This is equal to a polynimial expression involving the parameters which characterize W and $\mathscr{M}$ (the angular momentum and complex angular momentum in physicists' language): Naimark has calculated these expressions and solved the corresponding inequialities, exhibiting thus the unitary dual of the Lorentz group.

In the general case, one has to find where $b \cdot \varphi$ lies, for $\varphi \in W$ and $b \in \mathscr{U}(\mathfrak{g})$. Since for every $X \in \mathfrak{k}$ one has

$$X \cdot (b\varphi) = [X, b] \cdot \varphi + b \cdot X\varphi,$$

this problem is equivalent to the reduction into $\mathfrak{k}$-isotypic components of $\mathscr{U}(\mathfrak{g}) \otimes W$, $\mathscr{U}(\mathfrak{g})$ being considered as a $\mathfrak{k}$-module under the bracket representation. Since $\mathscr{U}(\mathfrak{g})$ is itself the direct sum of isotypic components, one may restrict b to such a component V, so that $b \cdot \varphi$ lies in $V \otimes W$. However, $V \otimes W$ is not simple $\mathfrak{k}$-module in general; so that one has to connect $\mathscr{U}(\mathfrak{g})$ with the reduction of $V \otimes W$ into a direct sum of $\mathfrak{k}$-modules.

3 Reduction of the Tensor Product of $\mathfrak{k}$-modules

Let V, W be $\mathfrak{k}$-modules, π and ρ the corresponding representations; assume V finite dimensional and semisimple; let $\left\{e^A\right\}$ be a basis of V, $\{e_A\}$ its dual basis, spanning the dual space $\widetilde{V}$ on which $\mathfrak{k}$ acts through the contragradient representation $\check{\pi}$; and $\left\{E^A_B\right\}$ be the corresponding basis of the ring $\mathscr{L}(V)$ of linear self-mappings of V, which is itself a $\mathfrak{k}$-module under the action of $\pi \otimes \check{\pi}$, the adjoint of π.

Consider the $\mathfrak{k}$-module $V \otimes W$, and assume it has the reduction

$$V \otimes W = \oplus_u W_u \ ; \ W_u = M_u \otimes \overline{W}_u, \tag{2}$$

where W_u is an isotypic component of type $\overline{W}_u$ and multiplicity space M_u (if $\dim M_u = 1$ then W_u and $\overline{W}_u$ are isomorphic). It is clear that this reduction is known if and only if the ring of self-intertwinning operators of $V \otimes W$, denoted by $\mathrm{End}_{\mathfrak{k}}(V \otimes W)$, is known (it is isomorphic to $\oplus_u \mathscr{L}(M_u)$). This ring is connected to the space $\mathrm{Hom}_{\mathfrak{k}}(\mathscr{L}(V), \mathscr{L}(W))$ of intertwinning operators between the $\mathfrak{k}$-modules $\mathscr{L}(V)$ and $\mathscr{L}(W)$ by:

Proposition 2: *There is an invertible linear mapping* $\mathfrak{j}$ *from* $\mathrm{Hom}_{\mathfrak{k}}(\mathscr{L}(V), \mathscr{L}(W))$ *to* $\mathrm{End}_{\mathfrak{k}}(V \otimes W)$ *defined by:*

$$[\mathfrak{j}T]\left(e^A \otimes \varphi\right) = e^B \otimes \left[T(E^A_B)\right] \varphi. \tag{3}$$

This result is immediate to establish; finite dimension of W is not necessary for it; however, we shall assume W finite-dimensional hereafter.

By transfer of structure from $\mathrm{End}_{\mathfrak{k}}(V \otimes W)$, one can define a ring structure on the space $\mathrm{Hom}_{\mathfrak{k}}(\mathscr{L}(V), \mathscr{L}(W))$, denoted by $(\pi)_{\mathscr{L}(W)}$, that is [7]:

$$\{T * Y\}\left(E^B_A\right) = T(E^C_A) Y(E^B_C). \tag{4}$$

Such a ring will be referred to as a *ctp* (= contracted tensor product) ring: it involves multiplication of matrices with entries in $\mathscr{L}(W)$. However, $\mathscr{L}(W)$ plays no role in (4): any ring $\mathscr{A}$ may take its place, provided $\mathfrak{k}$ acts on $\mathscr{A}$ by derivations.

Since the bracket representation of $\mathfrak{k}$ on its enveloping algebra $\mathscr{U}$ is a representation by derivation, a *ctp* ring structure $(\pi)_{\mathscr{U}}$ can be defined by (4) on the space $\mathrm{Hom}_{\mathfrak{k}}(\mathscr{L}(V),\mathscr{U})$ as well: here T denotes a mapping, and $T_A^B = T(E_A^B) = T(e_A \otimes e^B)$ an element of $\mathscr{U}$, such that

$$\left[X, T_A^B\right] = T(e_A \otimes (\pi_X e^B) + (\check{\pi}_X e_A) \otimes e^B) \ ; \ X \in \mathfrak{k}. \tag{5}$$

Now, since ρ is a representation, it sends $\mathscr{U}$ into $\mathscr{L}(W)$, hence it defines a mapping τ_ρ from $(\pi)_{\mathscr{U}}$ into $(\pi)_{\mathscr{L}(W)}$ by $\tau_\rho \circ T(E) = \rho(T(E))$; if W is simple then ρ and τ_ρ are surjective by Burnside's lemma, so that $(\pi)_{\mathscr{U}}$ has the following universal property:

Proposition 3: *If (π, V) is fixed and (ρ, W) varies among simple finite-dimensional $\mathfrak{k}$-modules, the ring* $\mathrm{End}_{\mathfrak{k}}(V \otimes W)$ *is the homomorphic image of $(\pi)_{\mathscr{U}}$ by $\mathbf{j} \circ \tau_\rho$.*

4 The ring $(\pi)_{\mathscr{U}}$

One easily establishes that the ring $(\pi)_{\mathscr{U}}$ has finite dimension over the center $\mathscr{Z}$ of $\mathscr{U}$, and that it is semisimple. Moreover, semisimple ideals of $(\pi)_{\mathscr{U}}$ are sent onto semisimple ones of $\mathrm{End}_{\mathfrak{k}}(V \otimes W)$ by $\mathbf{j} \circ \tau_\rho$. This is not true, however, for simple ideals; in particular, the ideals of $\mathrm{End}_{\mathfrak{k}}(V \otimes W)$ which are isomorphic to some $\mathscr{L}(M_u)$ (that is those which restrict to a simple isotypic component) are not, in general, images of an ideal of $(\pi)_{\mathscr{U}}$.

This situation can be amended as follows: every element Y of $(\pi)_{\mathscr{U}}$ (thought of as a matrix) has a characteristic polynomial C_Y, the degree of which cannot exceed the dimension of $(\pi)_{\mathscr{U}}$ over $\mathscr{Z}$. C_Y is a polynomial with coefficients in $\mathscr{Z}$; in general it has no solutions in $\mathscr{Z}$, but it can be fully factorized into $\deg C_Y$ linear factors in a Galois extension $\overline{\mathscr{Z}}$ of $\mathscr{Z}$ (which is still entire and factorial). One then has:

Proposition 4: *Let Y be an element which spans algebraically the center of $(\pi)_{\mathscr{U}}$. Let $\overline{\mathscr{Z}}$ be the Galois extension defined by its characteristic polynomial C_Y, called hereafter the* **Weyl extension** *of $\mathscr{Z}$, and let $\overline{\mathscr{U}} = \mathscr{U} \otimes_{\mathscr{Z}} \overline{\mathscr{Z}}$. If W is simple, then every simple ideal of the ring $(\pi)_{\overline{\mathscr{U}}} = (\pi)_{\mathscr{U}} \otimes_{\mathscr{Z}} \overline{\mathscr{Z}}$ is sent on a simple ideal of* $\mathrm{End}_{\mathfrak{k}}(V \otimes W)$ *by $\mathbf{j} \circ \tau_\rho$ and every simple ideal of* $\mathrm{End}_{\mathfrak{k}}(V \otimes W)$ *is the image of a simple ideal of $(\pi)_{\mathscr{U}}$.*

When $\mathfrak{k}$ is a classical Lie algebra, it has been shown [1b] that the Weyl extension $\overline{\mathscr{Z}}$ is independent of the choice of V, that the spanning element Y can be taken of degree one (that is, Y_B^A belongs to $\mathfrak{k}$), and that $\overline{\mathscr{Z}}$ is isomorphic to $\mathbf{C}[z'_1, \cdots, z'_r]$, the ring of polynomials in r indeterminates with complex coefficients.

The proof uses the commutation relations of $\mathfrak{k}$, and the fact that the central character determines the type for finite $\mathfrak{k}$-modules. Since the center $\mathscr{Z}$ is the subring of $\mathbf{C}[z'_1, \cdots, z'_r]$, which is invariant by the Galois group of the extension (in fact the Weyl group of $\mathfrak{k}$), one can think of the generators z'_i as of variables which parametrize the center $\mathscr{Z}$: only their symmetric functions are needed to determine a $\mathfrak{k}$-type, but the symmetry must be broken to effect the reduction of tensor products. And the roots z_u of C_Y are affine functions of them.

Notice that the usual labelling of representations by extremal weights introduces the symmetry breaking at an earlier stage (by the choice of the weight vector); indeed, generators z'_i are affinely related to the coordinates of the dominant weight, like the roots z_u; this is a translation of the Harish-Chandra isomorphism, and it can be used to prove the last properties enounced above for every reductive Lie algebra.

5 Elementary idempotents and Taylor Polynomials

We shall now focus attention on simple ideals S_u of $(\pi)_{\overline{\mathscr{U}}}$, since they are closely related to the shift operators from which we started: indeed, every S_u is sent on some $\mathscr{L}(M_u)$; that is, if $T \in S_u$ and $\varphi \in W$ then $e^A \otimes \rho(T_A^B)\varphi$ belongs to the isotypic component W_u of $V \otimes W$.

Assuming that T is central and generates the simple ideal S, one sees that its characteristic polynomial must be of the form $t(t-\zeta)$, with $\zeta \in \overline{\mathscr{Z}}$, so that $\zeta^{-1}T$ is an idempotent element $\mathbf{p}$ spanning S. Notice that ζ is not invertible in general, so that $\mathbf{p}$ belongs to the local ring $(\pi)_{\overline{\mathscr{U}}} \otimes_{\overline{\mathscr{Z}}} \overline{\mathscr{Z}}[\zeta^{-1}]$ and not to $(\pi)_{\overline{\mathscr{U}}}$ itself.

One can be more precise about that, Indeed, let Y be a spanning central element with characteristic polynomial $C_Y = C$. Denoting by δ (the Kronecker symbol) the identity element of $(\pi)_{\overline{\mathscr{U}}}$, and taking t in $\mathbf{C}$ (or even in $\overline{\mathscr{Z}}$), one may write the Taylor expansion of $C(t)$ as:

$$\delta C(t) = C(t \cdot \delta) = C(t \cdot \delta) - C(Y) = (t \cdot \delta - Y) * T(t) \tag{6}$$

T being a $(\pi)_{\overline{\mathscr{U}}}$-valued polynomial function of t, of degree deg$C$-1: it will be called hereafter the *Taylor polynomial* of Y.

From (6) one easily gets:

$$(t-t')T(t) * T(t') = C(t) \cdot T(t') - C(t') \cdot T(t). \tag{7}$$

By derivation and equalling t and t' it follows:

$$T(t) * T(t) = (\frac{dC(t)}{dt}) \cdot T(t) - C(t) \cdot (\frac{dT(t)}{dt}) \tag{8}$$

so that if $t = z$ is a root of C, one gets, denoting by dC the derived polynomial of C:

$$T(z) * (T(z) - dC(z) \cdot \delta) = 0 \tag{9}$$

so that $T(z)$ is an absorbing element of the center, and proportional to an elementary idempotent.

Notice that C has no double roots (this would contradict semisimplicity), hence $dC(z) \neq 0$, and the discriminant Δ of C is nonzero: every elementary idempotent

$$\mathbf{p}_u = \frac{T(z_u)}{dC(z_u)}$$

belongs to the local ring $(\pi)_{\overline{\mathscr{U}}} \otimes_{\overline{\mathscr{Z}}} \overline{\mathscr{Z}}[\Delta^{-1}]$.

When $(\pi)_{\overline{\mathscr{U}}}$ is sent to $\mathscr{L}(W)$ through τ_ρ, it may happen that $\rho(dC(z)) = 0$, Then, by (9), $T(z)$ is nilpotent: because of semisimplicity of $V \otimes W$, $T(z)$ must be zero and the idempotent $\mathbf{p}$ can still be defined as the limit of $\frac{T(t)}{dC(t)}$ for $t \to z$ (for infinite-dimensional ρ, semisimplicity is no more granted, and $T * T = 0 \neq T$ yields indecomposable non-simple components of $V \otimes W$).

Moreover, one has the following formulas:

$$\begin{aligned} \delta &= \sum_u \mathbf{p}_u, \\ Y &= \sum_u z_u \mathbf{p}_u, \\ T(t) &= C(t) \cdot \sum_u (t - z_u)^{-1} \mathbf{p}_u, \end{aligned} \tag{10}$$

the summation running over the spectrum $\{z_u\}$ of Y: these formulas describe the partition of unity into elementary idempotents.

Without entering into much details, we may say that Y can be normalized so that $\rho(z_u) - \rho(z_v)$ is an integer; it follows that the set of roots can be ordered (this breaks the symmetry of the Weyl group), so that the uth component W_u of $V \otimes W$ can be "followed" when W varies inside the finite dual. In particular, the multiplicity of each W_u cannot exceed the rank of the corresponding $\mathbf{p}_u$ (considered as a matrix); and the ratio $\dim W_u / \dim W$ equals the trace $\rho((\mathbf{p}_u)^A_A)$ of $\tau_\rho \circ \mathbf{p}_u$ (the sum of all such traces being $\delta^A_A = \dim V = \dim(V \otimes W)/\dim W$).

Notice that the multiplicity of W_u may be less than its maximal value, and even zero (in this case $\tau_\rho \circ \mathbf{p}_u = 0$: if W is one-dimensional this happens for every u but one).

6 The squared shift operators

Assume now that $\mathfrak{k}$ is the maximal compact subalgebra of $\mathfrak{g}$ (in fact, for many purposes it is sufficient to take $\mathfrak{k}$ to be reductive in $\mathfrak{g}$). Since $\mathscr{U}(\mathfrak{g})$ is a $\mathfrak{k}$-module, the involution $b \to b^*$ defines, on every $\mathfrak{k}$-submodules V of $\mathscr{U}(\mathfrak{g})$, a mapping $F \to F^\dagger$ from $\mathrm{Hom}_{\mathfrak{k}}(V, \mathscr{U}(\mathfrak{g}))$ to $\mathrm{hom}_{\mathfrak{k}}(\tilde{V}, \mathscr{U}(\mathfrak{g}))$ which satisfies:

$$(F^A\varphi|\varphi') = (\varphi|F^\dagger_B\varphi')\, g^{BA} \tag{11}$$

g being the $\mathfrak{k}$-invariant tensor in V.

If φ is taken in W, then $F^A\varphi$ is in $V \otimes W$: by what precedes, $F^B\rho(\mathbf{p}^A_{uB}) \cdot \varphi$ is in the component W_u and one has

$$g_{AB}(F^C \cdot \rho(\mathbf{p}^A_{uC}) \cdot \varphi \,|\, F^D \cdot \rho(\mathbf{p}^B_{uD}) \cdot \varphi) = (\varphi \,|\, F^\dagger_B \cdot F^D \cdot \rho(\mathbf{p}^B_{uD}) \cdot \varphi). \tag{12}$$

The above formula relates the restrictions to W_u and to W of the sesquilinear form of $\mathscr{M}$. After taking some precautions (imbed $\overline{\mathscr{Z}}$ in $\mathscr{U}(\mathfrak{g})$ and show how to handle vanishing denominators) we may say that the squared shift operators b^*b of the Ionesco condition are precisely the elements of the form $F^\dagger_B \cdot F^D \cdot \rho(\mathbf{p}^B_{uD})$.

Still, it is not easy to express the spectrum of all such elements. Nevertheless, if one wants to express it algebraically, one must use parameters which describe the action of $\mathfrak{k}$ on W and the one of $\mathfrak{g}$ on $\mathscr{M}$: hence both central chatacters must be taken into account. We have just seen how $\mathscr{Z}$ is parameterized by the roots of C: the same properties hold for the reductive Lie algebra $\mathfrak{g}$.

7 Combined use of *ctp* rings of $\mathfrak{g}$ and $\mathfrak{k}$

Let V be a finite-dimensional $\mathfrak{g}$-module and π the corresponding representation; V is also a $\mathfrak{k}$-module by restriction. To simplify formulas, we shall assume it $\mathfrak{g}$-irreducible (This does not imply $\mathfrak{k}$-irreducible). One may define a *ctp* ring structure $(\pi)_{\mathscr{U}(\mathfrak{g})}$ on $\mathrm{Hom}(\mathscr{L}(V), \mathscr{U}(\mathfrak{g}))$, as it was done for $\mathfrak{k}$, and define inside it characteristic and Taylor polynomials, as well as the Weyl extension. To avoid confusion we shall use script characters $(\mathscr{C}, \mathscr{T}, \mathscr{X}, \cdots)$ for objects related to $\mathfrak{g}$ and ordinary ones $(C, T, X, \cdots)$ for objects related to $\mathfrak{k}$.

Let $\mathscr{X} \in \mathrm{Hom}_{\mathfrak{g}}(\mathscr{L}(V), \mathfrak{g})$ and let $\mathscr{C}$ and $\mathscr{T}$ be its characteristic and Taylor polynomials satisfying, for $t \in \mathbf{C}$:

$$(t\delta - \mathscr{X}) * \mathscr{T}(t) = \mathscr{T}(t) * (t\delta - \mathscr{X}) = \delta \cdot \mathscr{C}(t). \tag{13}$$

The element $\mathscr{X}$ is central (it is closely related to the variation of the spectrum of Casimir operators of $\mathfrak{g}$ on $V \otimes W$) and it spans the center of $(\pi)_{\mathscr{U}(\mathfrak{g})}$ so that its spectrum determines the Weyl extension of $\mathscr{U}(\mathfrak{g})$. Considered as a $\mathfrak{k}$-homomorphism, $\mathscr{X}$ splits into two parts:

$$\mathscr{X} = X + L \ , \ X \in \mathrm{Hom}_{\mathfrak{k}}(\mathscr{L}(V), \mathfrak{k}) \ , \ L \in \mathrm{Hom}_{\mathfrak{k}}(\mathscr{L}(V), \mathfrak{p}) \tag{14}$$

and X is a semisimple element of the ring $(\pi)_{\overline{\mathscr{U}}}$, the spectrum of which is known for every ρ (though X may be non-central if V is $\mathfrak{k}$-reducible).

One can show that the mapping $\mathscr{X} \to \mathscr{X}^\dagger$ is given by:

$$\mathscr{X}^\dagger = X - L \tag{15}$$

and that the set of roots $\{\xi_\alpha\}$ of the characteristic polynomial $\mathscr{C}^\dagger$ of $\mathscr{X}^\dagger$ must be represented by the complex conjugate set to the one of $\mathscr{C}$ when $\mathfrak{g}$ acts on $\mathscr{M}$, in order to preserve the $\mathfrak{g}_0$-invariance of the sesquilinear form; the Taylor formula can then be written as

$$(\bar{t}\delta - \mathscr{X}^\dagger) * \mathscr{T}(t)^\dagger = \delta \cdot \mathscr{C}(t)^\dagger. \tag{16}$$

By combining equations (13) to (16) one gets the following formula, the implications of which are quite important:

$$\mathscr{T}(t)^\dagger * ((\bar{t} + t')\delta - 2X) * \mathscr{T}(t') = \mathscr{C}(t')\mathscr{T}(t)^\dagger + \mathscr{C}(t)^\dagger \mathscr{T}(t'). \tag{17}$$

Consider now the action of elements $\mathscr{T}(t)^A_B$ on W: for $\varphi \in W$ it is clear that $\mathscr{T}(t)^A_B\varphi$ lies in $(\tilde{V} \otimes V) \otimes W$ and that the reduction of this tensor product can be related to that of $V \otimes W$ and $\tilde{V} \otimes W'$ through the corresponding *ctp* rings. One gets

$$\mathscr{L}(V) \otimes W = \oplus_u(\oplus_v W^v_u), \tag{18}$$

where summations take place over the spectrum of X in $(\pi)_{\overline{\mathscr{U}}}$ and the component W^v_u is spanned by elements of the form

$$(\mathbf{p}_v * \mathscr{T}(t) * \mathbf{p}_u)^A_B\varphi = \mathbf{p}_v{}^C_B \mathscr{T}(t)^D_C \mathbf{p}_u{}^A_D\varphi \tag{19}$$

the idempotents $\mathbf{p}_u$ (not necessarily central) being obtained by the diagonalization of the tensor X, and satisfying:

$$(X * \mathbf{p}_v * \mathscr{T}(t) * \mathbf{p}_u)^A_B\varphi = (\mathbf{p}_v \mathscr{T}(t) * \mathbf{p}_u)^A_B\varphi \cdot m^v_u, \tag{20}$$

where the quantities m^v_u are real-valued affine expressions of the generators of $\overline{\mathscr{Z}}$ (the coefficients of these expressions depend on V). Notice that $\mathscr{L}(V)$ contains a trivial submodule, so that $\mathscr{L}(V) \otimes W$ must contain submodules isomorphic to W: indeed for every choice of $\mathbf{p}_u$, *the component* W^u_u (no summation on u) *is isomorphic to* W.

Taking in (17) $t = t' = \xi$ to be a root of $\mathscr{C}$ and combining with (20) yields, for every choice of u:

$$\sum_v (\Re e(\xi) - m_u^v) \sum_A \sum_B \|(\mathbf{p}_v * \mathscr{T}(\xi) * \mathbf{p}_u)_A^B \varphi\|^2 = 0. \tag{21}$$

Hence the following necessary condition for unitarizability:

Theorem 1: *Assume $\mathscr{M}$ unitarizable. Let ξ be a root of $\mathscr{C}$ and let W be a $\mathfrak{k}$-isotypic component of $\mathscr{M}$. For every component W_u of $V \otimes W$ one must have either*

$$\inf_v (m_u^v) \leq \Re e(\xi) \leq \sup_v (m_u^v) \tag{22}$$

or else all elements $\mathscr{T}(\xi)_C^D \mathbf{p}_{uD}^A$ vanish on W. Moreover, if ξ fails to satisfy (22) for every u, then elements $\mathscr{T}(\xi)_C^D$ vanish on the entire $\mathfrak{k}$-module $\mathscr{M}$.

The "moreover" part of this statement comes from the fact that, for $b \in \mathscr{U}(\mathfrak{g})$, and using the commutation relations of $\mathfrak{g}$, $\mathscr{T}(\xi)_C^D \cdot b$ can be rewritten as a linear combination of elements $b' \cdot \mathscr{T}(\xi)_{C'}^{D'}$ so that if all elements $\mathscr{T}(\xi)_C^D$ vanish on W they vanish everywhere.

One may choose W so to maximize the lower bound of (22) and/or minimize the upper one, finding thus a smallest interval into which $\Re e(\xi)$ must belong - otherwise the corresponding shift operator vanishes on W. The exceptions to the inequalities (22) may undergo a further algebraic treatment, too technical to be exposed here; one should notice however, that these inequalities may not be skipped by every root ξ: indeed since δ is in the span of the $\mathscr{T}(\xi)$'s (and of some of their derivatives if double roots occur - but this does no great harm), this would imply $\mathbf{p}_B^A \varphi = 0$, which is impossible.

8 Ordering of $\mathfrak{k}$-types

The set of isotypic components of $\mathscr{M}$ can be assimilated to a lattice of points parametrized by the set $\{\rho(z_i')\}$. We shall suppose that there is some partial order relation in this lattice (it might be the one induced by a positive quadratic form on $\{\rho(z_i')\}$, but there are other orders which yield the same results). If one succeeds in determining a finite subset F (the smaller, the better) of this lattice, such that positivity of the sesquilinear form on points of F implies positivity everywhere, then the unitarizability criteria are much easier to check.

On this purpose, let us introduce:

Definition: *Given V, a $\mathfrak{k}$-type will be called V-***smooth** *if there is a component W_u in $V \otimes W$ such that*

(a) *$m_u^v = \sup_w(m_u^w)$ implies $W_u^v = W_u^u$ isomorphic to W, and*

(b) *$m_u^v < \sup_w(m_u^w)$ implies $W_u^v < W$.*

Fixing $t = t'$ with $\Re e(t) = \sup_w(m_u^w)$ in (17), dividing both sides of (17) by $\mathscr{C}(t) \cdot \mathscr{C}(t)^\dagger$, contracting with $\mathbf{p}_u{}_B^A$, and integrating over the imaginary part of t, yields the following result:

Theorem 2: *the necessary condition of Theorem 1 is satisfied, and if W is V-smooth, then either* **(a)** *or* **(b)** *is true:*

(a) *For every $v \neq u$ there is a mapping $(\varphi, \lambda) \to [\varphi_u^v(\lambda)]_B^A$ from $W \times \mathbf{R}$ to W_u^v such that*

$$||\varphi||^2 = \sum_{v \neq u} \int_{-\infty}^{+\infty} \sum_A \sum_B ||\,[\varphi_u^v(\lambda)]_B^A\,||^2 \, d\lambda. \tag{23}$$

(b) *All components W_u^v vanish (u being fixed by the smoothness of V) except $W_u^u = W$, and there is a root ξ of $\mathscr{C}$ with real part $\Re e(\xi) = m_u^u$.*

The importance of these two theorems, both directly derived from (17), is that they drastically restrict the checking domain of the Ionesco condition to squared shift operators sending W into $\mathfrak{k}$-types which, for every choice of V, either satisfy condition (b) of Theorem 2, or are non-V-smooth.

9 Application to classical real forms

It is beyond the limits of this intervention to give a precise description of the results obtained for the various real forms; we shall be limited to rough indications only.

For the $\mathfrak{gl}(N, \mathbf{C})$, $\mathfrak{gl}(N, \mathbf{R})$, and $\mathfrak{u}^*(N)$ families, the checking is limited to the minimal $\mathfrak{k}$-type and its immediate neighbourhood. Some of the generators of the Weyl extension of $\mathscr{Z}(\mathfrak{g})$ have their real part fixed by the minimal $\mathfrak{k}$-type; for the remaining ones the real parts are confined into some intervals determined by inequalities (22); one may also have some additional constraints, relating affinely several generators.

For $\mathfrak{so}(N, \mathbf{C})$ and $\mathfrak{sp}(N, \mathbf{C})$ the situation is more complicated, because there are more squared shift operators to check than before (they correspond to minimal $\mathfrak{k}$-types with trivial restriction to a subalgebra of the form $\mathfrak{so}(N-2k)$ or $\mathfrak{sp}(N-2k)$). Besides linear constraints of the above type, some polynomials of higher degree must be positive.

For the remaining families only partial results are available. The basic difference with the preceding families is that there is no more a natural ordering among the generators of $\overline{\mathscr{Z}}(\mathfrak{k})$; one has to deal either with two naturally ordered sets, or with one set of numbers differing by integers, extending to both sides of the real line and ordered by their absolute value. Consequently, there is a quite large set of non-smooth $\mathfrak{k}$-types, for which higher degree polynomial inequalities are needed, the exact expression of which is not yet known.

To conclude this intervention, a remarkable feature of this approach should be pointed out: no use at all is made of Cartan subalgebras and their properties. Indeed, it only uses direct considerations on the centers $\mathscr{Z}$ and $\mathscr{Z}(\mathfrak{g})$, followed by breaking the Weyl symmetry as long as constraints are produced, and keeping it when there is no need of breaking it. In particular, it gives no direct information on inducing characters for a representation; but this is of little harm to the problem of unitarizability, since the real problem when one starts with induced representations is to find when non-unitary representations induce unitary (or partly unitary) ones. And it is always possible, by using classical tools which involve again a breaking of the Weyl symmetry, given a $\mathfrak{g}$-module $\mathscr{M}$ isomorphic to a subquotient of $\mathscr{U}(\mathfrak{g}) \otimes W$, ($W$ being, preferably but not necessarily, a lowest $\mathfrak{k}$-type) to relate it to induced representations. And if unitarizability is the goal, direct access to algebraic expressions of squared shift operators is by no means to neglect.

BIBLIOGRAPHY

1. E. Angelopoulos, (a) Comm. Math. Phys., **89** (1983), 41- and references therein.
   (b) *Tensor calculus in enveloping algebras and parametrization of their center for classical Lie algebras*, preprint, Universite de Dijon (1990); See also C, R. Acad. Sci. Paris, **314** (1992), 705-708.

2. D. Barbash, Invent. Math., **96** (1989), 103-176.

3. V. Bargmann, *Irreducible unitary representations of the Lorentz group*, Ann. Math., **48** (1947), 568-640.

4. M. A. Naimark, *Linear representations of the Lorentz group*, Amer. Math. Soc. Transl., **6** (1957), 379-458.

5. D. Vogan, Invent. Math., **83** (1986), 449-505.

6. In Ionesco's comedy *la Lecon*, the pupil, who could not understand how to multiply integer numbers, has avoided the difficulty by learning by heart the results of all possible multiplications...

7. The Einstein summation convention on repeated indices is systematically adopted in this text.

Department of Mathematics, National Technical University,
Zografou Campus, 15773 Athens, Greece.
and
Physique Mathematique, Universite de Dijon,
6 bd Gabriel, 2100 Dijon, France.

A CONNECTION BETWEEN LIE ALGEBRA ROOTS AND WEIGHTS AND THE FOCK SPACE CONSTRUCTION

George W. Mackey

1. Introduction.

The bulk of my talk today will be presented in the spirit of pure mathematics – as a systematic account of the properties of certain Lie algebras of linear operators. The study of these Lie algebras was inspired by the study of the physics literature and to a considerable extent can be regarded as a unification, systematization, and reorganization of algebraic techniques which have already been discovered and found useful by physicists working on a variety of physical problems. I hope in doing this I have turned up some useful novelties. However it is difficult to be certain about this because of the very nature of the physical literature. Physicists are much more inclined to learn methods of attacking problems than to formulate and apply general theorems. Indeed, I believe that the physicists in solving their problems have in effect discovered a number of theorems interesting to but unknown to mathematicians and hidden away by not being explicitly formulated.

2. Roots and weights. Let $\mathcal{L}$ be any Lie algebra over the complex numbers and let $\mathcal{A}$ be any commutative subalgebra. If $V, x \to V_x$ is any representation of $\mathcal{L}$ by linear operators in a vector space $\mathcal{H}(V)$ one defines a vector φ in $\mathcal{H}(V)$ to be a *weight vector* for V (with respect to $\mathcal{A}$). If ϕ is an eigenvector of V_a for all a in $\mathcal{A}$, the associated eigenvalue will depend upon a. If we denote it by $\ell(a)$, then ℓ is clearly a linear functional on the vector space $\mathcal{A}$. This linear functional is called the *weight* of V (with respect to $\mathcal{A}$) corresponding to the weight vector φ. Conversely if ℓ is any linear functional on $\mathcal{A}$ then ℓ is said to be a weight of V (with respect to $\mathcal{A}$) if the set $\mathcal{H}_\ell$ of all φ in $\mathcal{H}(V)$ with $V_a(\varphi) = \ell(a)\varphi$ for *all* a in $\mathcal{A}$ does not reduce to the zero element. The set $\mathcal{H}_\ell$ is a vector subspace of $\mathcal{H}(L)$ called the *weight space* of ℓ.

It is easy to see that the weight spaces $\mathcal{H}_\ell$ for distinct weights ℓ are all linearly

E. A. Tanner and R. Wilson (eds.), Noncompact Lie Groups and Some of Their Applications, 265–284.

independent. If their linear span is the whole of $\mathcal{H}(L)$ then $\mathcal{H}(L)$ is the direct sum of the $\mathcal{H}_\ell$ and one says that V admits a *weight space decomposition* (with respect to $\mathcal{A}$).

Now consider the special case in which V is the so-called adjoint representation of $\mathcal{L}$. This means that $\mathcal{H}(V) = \mathcal{L}$ and $L_a(x) = ax - xa = [a, x]$ for all $a \in \mathcal{A}$ and all $x \in \mathcal{L}$. In that case the nonzero weights of V (with respect to $\mathcal{A}$) are called *roots*, the weight vectors are called root vectors and the weight spaces are called root spaces.

There is a simple but important connection between roots and weights which we now recall. Let V be a representation of a Lie algebra $\mathcal{L}$ and let $\mathcal{A}$ be a commutative subalgebra. Let ℓ be a weight of V (with respect to $\mathcal{A}$) and let w be a root of $\mathcal{L}$ (with respect to $\mathcal{A}$). Let φ be a weight vector of V and let x be a root vector (both with respect to $\mathcal{A}$). Then $V_x(\varphi)$ is either zero or is a weight vector whose weight is $\ell + w$.

In the important special case in which $\mathcal{L}$ is finite dimensional and semi simple it is known that there exists a choice of $\mathcal{A}$ which is unique to within automorphisms of $\mathcal{L}$ and has the following properties:

(a) The adjoint representation of $\mathcal{L}$ (with respect to $\mathcal{A}$) has a weight space decomposition.

(b) In this decomposition the 0 weight space is $\mathcal{A}$ itself and all other weight spaces (i.e., all root spaces) are one dimensional.

(c) Every finite dimensional reprsentation of $\mathcal{L}$ has a weight space decomposition with respect to $\mathcal{A}$.

(d) For every root α, $-\alpha$ is also a root and it is possible to choose one member out of each pair $\alpha, -\alpha$ in such a way that if α_1 and α_2 are chosen, then $\alpha_1 + \alpha_2$ is either chosen or not a root. Fixing such a choice, one speaks of the chosen members as the positive roots.

The reader unfamiliar with roots and weights is advised to examine the case in which $\mathcal{L}$ is the Lie algebra of all $n \times n$ complex matrices of trace zero and $\mathcal{A}$ is the subalgebra of diagonal matrices. He should have no difficulty in verifying assertions (a), (b) and (d). The systematic use of roots and weights is the key to the structure and representation theory of the finite dimensional semi simple Lie algebras.

3. The concept of a δ algebra and the canonically associated Lie algebra of operators. Let $\delta = \pm 1$ and let X be a vector space over the complex numbers. Let V be a vector space of linear operators on X. We define V to be a *δ algebra* if $AB + \delta BA$ is a constant times the identity for all A and B in V. Of course the constant depends upon A and B and we define $\lambda(A, B)$ by the equation $AB + \delta BA = \lambda(A, B)I$. Evidently, λ is a bilinear form which is symmetric or anti symmetric according as $\delta = 1$ or -1. We distinguish these two cases by saying that the δ algebra is fermionic or bosonic according as $\delta = 1$ or -1. This terminology is suggested by the physical applications but that need not concern us now. Much of the theory can be developed without knowing whether our δ algebra is fermionic or bosonic.

Now let V be any δ algebra and let V^2 denote the linear span of all products AB where A and B are arbitrary elements of V.

THEOREM 1. *For any δ algebra V, the vector space V^2 is closed under the usual commutator bracket and hence is a Lie algebra. Moreover the commutator bracket of any element of V with any element of V^2 is an element of V and the resulting mapping of the Lie algebra V^2 into linear operators in V is a representation of V^2.*

The truth of this rather unexpected result is an immediate consequence of two simple lemmas which have some independent interest and will find other uses below.

LEMMA 1. *If A, B and C are arbitrary elements of V then $(AB)C - C(AB) = \lambda(B,C)A - \delta\lambda(A,C)B$.*

PROOF:

$$\begin{aligned}(AB)C &= A(BC) = A(BC + \delta CB - \delta CB)\\ &= A(\lambda(B,C)I - \delta CB) = \lambda(B,C)A - \delta ACB = \lambda(B,C)A - \delta(AC)B\\ &= \lambda(B,C)A - \delta(AC + \delta CA - \delta CA)B = \lambda(B,C)A - \delta(\lambda(A,C)I - \delta CA)B\\ &= \lambda(B,C)A - \delta\lambda(A,C)B + \delta^2 CAB. \quad \text{Hence } (AB)C - C(AB)\\ &= \lambda(B,C)A - \delta\lambda(A,C)B \quad \text{as was to be proved.}\end{aligned}$$

LEMMA 2. *If A, B, C, and D are arbitrary elements of V then*

$$(AB)(CD) - (CD)(AB) = \lambda(B,D)CA - \delta\lambda(A,D)CB + \lambda(B,C)AD - \delta(A,C)BD.$$

PROOF: $(AB)(CD) = ((AB)C)D$ and by Lemma 1 applied to $(AB)C$ this is equal to $(C(AB) + \lambda(B,C)A - \delta\lambda(A,C)B)D = (C(AB)D + \lambda(B,C)AD - \delta\lambda(A,C)BD)$. But we may apply Lemma 1 again, this time to $(AB)D$, and conclude that $C(AB)D = C(D(AB) + \lambda(B,D)A - \delta\lambda(A,D)B = (CD)(AB) + \lambda(B,D)CA - \delta\lambda(A,D)CB$. Thus $(AB)(CD) = (CD)(A,B) + \lambda(B,D)CA - \delta\lambda(A,D)CB + \lambda(B,C)AD - \delta\lambda(A,C)BD$ and the announced result follows.

PROOF OF THEOREM 1: The fact that V^2 is a Lie algebra is an immediate consequence of Lemma 2. Moreover the second statement of Theorem 1 as an immediate consequence of Lemma 1. To prove the final statement note that it follows from Lemmas 1 and 2 that the linear union of V and V^2 is a Lie algebra and that the mapping in question is just the restriction to the invariant subspace V of the restriction of the adjoint representation of $V + V^2$ to the subalgebra V^2.

4. Sums of orthogonal δ algebras and δ super algebras. Let V_1 and V_2 be two δ algebras in the same vector space X with $\delta = \delta_1$ and δ_2, respectively. If $\delta_1\delta_2 = 1$; that is, if $\delta_1 = \delta_2$ we shall say that V_1 and V_2 are orthogonal if $AB + \delta BA = 0$ whenever A is in V_1 and B is in V_2. If $\delta_1\delta_2 = -1$, that is, if $\delta_1 \neq \delta_2$ we shall say that V_1 and V_2 are orthogonal if $AB - BA = 0$ whemever A is in V_1 and B is in V_2.

THEOREM 2. *Let V_1 and V_2 be orthogonal δ algebras with the same δ in the same vector space X. Then $V_1 + V_2$, the linear span of V_1 and V_2, is again a δ algebra.*

PROOF: Let $A, A' \in V_1$ and $B, B' \in V_2$ be otherwise arbitrary. Then $(A+A')(B+B')+\delta(B+B')(A'_A) = (AB+\delta BA)+(A'B'+\delta B'A')+(A'B+\delta BA')+(AB'+\delta B'A)$. But the first two terms in this four fold sum are constant multiples of the identity by the definition of δ algebra and the other two are zero by the definition of orthogonality. Hence $V_1 + V_2$ is a δ-algebra.

It is clear that Theorem 2 remains true for any linear span of δ algebras with the same δ no matter how many terms there are. If we take a linear span of many mutually orthogonal δ algebras where δ is sometimes 1 and sometimes -1 we may write it as a sum of the form $V_1 + V_2$ where V_1 is a linear span of orthogonal fermionic δ algebras and hence is a fermionic δ algebra and where V_2 is a linear span of bosonic δ algebras and hence is a bosonic δ algebra and where V_1 and V_2 are orthogonal. However it is *not* true that the linear sum of a fermionic δ algebra and a bosonic δ algebra is a δ algebra even if they are orthogonal and it is *not* true that the square of such a linear sum is a Lie algebra. On the other hand, such linear sums and their squares have interesting properties which we now propose to examine.

First of all, it will be useful to generalize the concept V^2 of the square of a vector space V of linear operators by making the following definition. If V_1 and V_2 are two vector spaces of linear operators in the same vector space X then we define V_1V_2 as the linear span of all products AB where A is in V_1 and B is in V_2. Clearly $V_1V_2 = V_2V_1$ whenever the elements of V_1 and V_2 commute and $VV = V^2$. Equally clearly $(V_1 + V_2)^2 = V_1^2 \dot{+} V_2^2 \dot{+} V_1V_2 \dot{+} V_2V_1$ and when the elements of V_1 and V_2 commute this reduces to $V_1^2 + V_2^2 + V_1V_2$.

THEOREM 3. *Let X be a vector space and let $\mathcal{F}$ and $\mathcal{B}$ respectively be fermionic and bosonic δ algebras in X. Suppose that they are orthogonal. Then $\mathcal{FB} = \mathcal{BF}$ and $(\mathcal{F} + \mathcal{B})^2 = \mathcal{F}^2 + \mathcal{B}^2 + \mathcal{FB}$ and by Theorem 1, $\mathcal{F}^2$ and $\mathcal{B}^2$ are Lie algebras. Since the elements of $\mathcal{F}^2$ commute with the elements of $\mathcal{B}^2$, $\mathcal{F}^2 + \mathcal{B}^2$ is also a Lie algebra. The Lie algebra $\mathcal{F}^2 + \mathcal{B}^2$ and the vector space $\mathcal{FB}$ are related to one another as follows:*

(a) The commutator of an element of the Lie algebra $\mathcal{F}^2 + \mathcal{B}^2$ with an element of the vector space $\mathcal{FB}$ as an element of $\mathcal{FB}$ and under the action so defined $\mathcal{FB}$ is a representation space for the Lie algebra $\mathcal{F}^2 + \mathcal{B}^2$.

(b) *The anti commutator $C_1C_2 + C_2C_1$ of any two elements C_1, C_2 of $\mathcal{FB}$ is an element of the Lie algebra $\mathcal{F}^2\dot{+}\mathcal{B}^2$.*

PROOF: All statements except those under (a) and (b) are either obvious or are immediate consequences of the foregoing. To prove (a) let F_1, F_2 and F_3 be arbitrary members of $\mathcal{F}$ and let B_1, B_2 and B_3 be arbitrary members of B. By linearity it will suffice to prove that $F_1F_2(F_3B_3) - (F_3B_3)F_1F_2 \in \mathcal{B}$. But since B_3 commutes with F_1, F_2 and F_3 we may write this as $((F_1F_2)F_3 - F_3(F_1F_2))B_3$ and by Lemma 1 $(F_1F_2)F_3 - F_3(F_1F_2)$ is in $\mathcal{F}$. Thus the first statement under (a) is proved and the rest is straightforward.

To prove (b) let F_1 and F_2 be arbitrary members of $\mathcal{F}$ and let B_1 and B_2 be arbitrary members of $\mathcal{B}$. Then F_1B_1 and F_2B_2 are members of $\mathcal{FB}$ and their anti commutator

$$\begin{aligned}
F_1B_1F_2B_2 + F_2B_2F_1B_1 &= F_1F_2B_1B_2 + F_2F_1B_2B_1 \\
&= F_1F_2B_1B_2 + (\lambda(F_1F_2)I - F_1F_2)(B_1B_2 - \lambda(B_1, B_2)I) \\
&= F_1F_2B_1B_2 - F_1F_2B_1B_2 - \lambda(F_1F_2)\lambda(B_1B_2)I + \lambda(F_1F_2)B_1B_2 + \lambda(B_1, B_2)F_1F_2 \\
&= \lambda(F_1F_2)B_1B_2 + \lambda(B_1, B_2)F_1F_2 - \lambda(F_1F_2)\lambda(B_1, B_2)I.
\end{aligned}$$

Now since $F_1F_2 \in \mathcal{F}^2$ and $B_1, B_2 \in \mathcal{B}^2$ and $I \in \mathcal{F}^2 + \mathcal{B}^2$ unless $\lambda(F_1, F_2) = \lambda(B_1, B_2) = 0$ it follows that $(F_1B_1)(F_2B_2) + (F_2B_2)(F_1B_1)$ is in $\mathcal{F}^2 + B^2$ and hence by linearity that (b) holds and the proof of Theorem 3 is complete.

Now consider the non degenerate case in which $(\mathcal{F}^2 + \mathcal{B}^2) \cap \mathcal{FB} = 0$. Then every element in $(\mathcal{F} + \mathcal{B})^2$ may be written uniquely in the form $S + L$, where $L \in \mathcal{F}^2 + \mathcal{B}^2$ and $S \in \mathcal{FB}$. Moreover if $S_1 + L_1$ and $S_2 + L_2$ are two such elements we may *define* their "product" by the formula

$$(S_1+L_1)\circ(S_2+L_2) = ((S_1L_2-L_2S_1)+(L_1S_2-S_2L_1))+((S_1S_2+S_2S_1)+(L_1L_2-L_2L_1))$$

since the first two commutators are in $\mathcal{FB}$ and the anti commutator and the third commutator are in $\mathcal{F} + \mathcal{B}^2$. This definition converts $(\mathcal{F} + \mathcal{B})^2$ into a distributive algebra which in general is neither a Lie algebra nor an associative algebra. It is an example of what was once called a graded Lie algebra but is now (less misleadingly) called a Lie superalgebra (a graded Lie algebra is *not* a special kind of Lie algebra). While Lie superalgebras are not Lie algebras they share many of the properties of Lie algebras. In particular one can speak of roots and weights and work out an analogous representation theory using roots and weights.

The general definition is as follows. Let $\mathcal{L}$ be any Lie algebra of operators in a vector space X and let W be any vector space of linear operators X which has the following properties with respect to $\mathcal{L}$.

(1) If $L \in \mathcal{L}$ and $A \in W$ then $LA - AL \in W$.
(2) If $A \in W, B \in W$, then $AB + BA \in \mathcal{L}$.
(3) $\mathcal{L} \cap W = 0$.

Then there is a unique distributive multiplication law $\circ$ in $W\dot{+}\mathcal{L}$ such that $A\circ B = AB-BA$ whenever at least one of A and B is in $\mathcal{L}$ and $A\circ B = AB+BA$ whenever both A and B are in W. The vector space $W+\mathcal{L}$ equipped with the multiplicative law $\circ$ is then called a ***Lie superalgebra***. The elements in $\mathcal{L}$ are called the ***even*** elements and those in W are called the ***odd*** elements. Beginning in 1974 Lie superalgebras began to play an important role in physics, generalizing the role played earlier by ordinary Lie algebras, and making it possible to have symmetries which connected bosons with fermions.

For reasons which should now be obvious we shall call the linear span of a fermionic δ algebra and a bosonic δ algebra a δ superalgebra.

5. Canonical bases. The simplest non trivial case of a δ algebra occurs when V is two dimensional. The bilinear form λ such that $AB+\delta BA = \lambda(A,B)I$ defines a linear map T^b of V into its dual V^* which will either be identically zero, have a one dimensional range or be an isomorphism of V onto V^*. In the last case we shall say that V is non singular. In the non singular case it is a well known and easy theorem of linear algebra that V admits a basis A, B such that $\lambda(A,B)=1$, $\lambda(A,A)=\lambda(B,B)=0$. Such a basis will be called canonical. More generally, a δ algebra V will be said to admit a canonical basis if it may be written as a direct sum of (possibly infinitely many) mutually orthogonal δ algebras each of which is two dimensional and non singular. If V is written as the direct sum of two dimensional δ algebras V_α and A_α, B_α is a canonical basis for V_α, the basis for V defined by all the A_α and all the B_α will be said to be a ***canonical basis*** for V. Equivalently, a canonical basis for V is a family A_α, B_α of pairs of operators such that $A_\alpha B_\beta + \delta B_\beta A_\alpha = \delta_\alpha^\beta$, $A_\alpha A_\beta + \delta A_\beta A_\alpha = B_\alpha B_\beta + \delta B_\beta A_\alpha = 0$ for all α and β.

6. Canonical bases and roots and weights for the Lie algebra V^2. Let V be a δ algebra and suppose that it admits a canonical basis $A_1, B_1, A_2, B_2, \cdots$ where the sequence may be finite or infinite. (Actually the set of A's and B's can even be non countably infinite but we shall not bother with this refinement here.) The main purpose of this section is to show that using this basis one can find an explicit commutative subalgebra $\mathcal{A}$ of the Lie algebra V^2 with respect to which the natural representation of V^2 has a weight space decomposition. Both decompositions are into one dimensional components and can be given explicitly.

LEMMA 3. *For any i and j the operators B_jA_j and B_iA_i commute with one another.*

PROOF: Apply Lemma 2 (Section 3) to $(B_jA_j)(B_iA_i)-(B_iA_i)(B_jA_j)$. Using the definition of a canonical basis it follows at once that this commutator is zero.

COROLLARY. *The linear span $\mathcal{A}$ of all the B_jA_j is a commutative subalgebra of V^2.*

LEMMA 4. *Let $A_1, B_1, A_2, B_2, \cdots$ be as above. Then for all i, j and k we have*
(a) $(B_kA_k)(B_iA_j) - (B_iA_j)(B_kA_k) = \lambda_{ijk}B_iA_j$, where

$$\begin{aligned} \lambda_{ijk} &= 0 \quad \text{if } i \neq k \quad \text{and } j \neq k \text{ or if } i = j \\ &= 1 \quad \text{if } i = k \quad \text{and } i \neq j \\ &= -1 \quad \text{if } i \neq j \quad \text{and } j = k \end{aligned}$$

(b) $(B_kA_k)(A_iA_j) - (A_iA_j)(B_kA_k) = \lambda^A_{ijk}A_iA_j$, where

$$\begin{aligned} \lambda^A_{ijk} &= 0 \quad \text{if } i \neq k,\ j \neq k \text{ or if } i = j \neq k \\ &= -2 \quad \text{if } i = j = k \\ &= -1 \quad \text{if } i = k \text{ or } j = k \quad \text{and } i \neq j \end{aligned}$$

(c) $(B_kA_k)(B_iB_j) - (B_iB_j)(B_kA_k) = \lambda^B_{ijk}B_iB_j$, where

$$\begin{aligned} \lambda^B_{ijk} &= 0 \quad \text{if } i \neq k \quad \text{and } j \neq k \text{ or } i = k = j \\ &= 2 \quad \text{if } i = j = k \\ &= 1 \quad \text{if } i = k \text{ or } j = k \text{ and } i \neq j. \end{aligned}$$

PROOF: Apply Lemma 2 and the definition of canonical basis just as in the proof of Lemma 3.

THEOREM 4. *Let $A_1, B_1, A_2, B_2 \cdots$ be a canonical basis for the δ algebra V and let $\mathcal{A}$ be the commutative subalgebra of V^2 consisting of all linear combinations of the binary products B_jA_j. Then $\mathcal{A}$ is a "Cartan subalgebra" for V^2 in the sense that V^2 admits a root space decomposition with respect to $\mathcal{A}$. Indeed, when $\delta = -1$ each B_iA_j with $i \neq j$ is a root vector and so is each A_iA_j and each B_iB_j. When $\delta = 1$, the same is true except that the prohibition $i \neq j$ also applies to A_iA_j and B_iB_j. Moreover, if we consider only those products in which $i \leq j$ ($i < j$ when $\delta = 1$), the root vectors we obtain are linearly independent and together with the B_jA_j form a basis for V^2. In particular, all root spaces are one dimensional. the weights associated to the roots of the form, B_iA_j, A_iA_j and B_iB_j, respectively, are the unique linear functionals ℓ_{ij}, ℓ^A_{ij} and ℓ^B_{ij} on $\mathcal{A}$ such that $\ell_{ij}(B_kA_k) = \lambda_{ijk}$, $\ell^A_{ij}(B_kA_k) = \lambda^A_{ijk}$ and $\ell^B_{ij}(B_kA_k) = \lambda^B_{ijk}$. Here $\lambda_{ijk}, \lambda^A_{ijk}$ and λ^B_{ijk} are as defined in Lemma 4.*

PROOF: The truth of the theorem is a more or less immediate consequence of Lemma 4.

Having associated a "Cartan subalgebra" with every canonical basis for V^2 and found the associated root space decomposition, let us look at the natural representation of V^2 on V.

LEMMA 5. *Let V be any δ algebra and let $n = 1,2,3,\cdots$. Then if $A, B, C_1, C_2, \cdots, C_n$ are arbitrary elements of V we have*

$$(AB)(C_1C_2\cdots C_n) - (C_1C_2\cdots C_n)(AB) = \Sigma_1 - \delta\Sigma_2,$$

where Σ_1 is the sum of all n terms of the form $\lambda(B, C_j)C_1C_2\cdots C_{j-1}AC_{j+1}C_n)$ and Σ_2 is the sum of all n terms of the form $\lambda(A, C_j)C_1C_2\cdots C_{j-1}BC_{j+1}\cdots C_n$.

PROOF: $(AB)(C_1C_2\cdots C_n) = (ABC_1)(C_2\cdots C_n)$. Using Lemma 1 on (ABC_1) this becomes

$$\begin{aligned}&(C_1AB + \lambda(B, C_1)A - \delta\lambda(A, C_1)B)(C_2\cdots C_n) = \\ &\quad C_1AB(C_2\cdots C_n) + \lambda(B, C_1)AC_2\cdots C_n - \delta\lambda(A, C_1)BC_2\cdots C_n.\end{aligned}$$

Now write $C_1ABC_2\cdots C_n = C_1(ABC_2)C_3\cdots C_n$ and apply Lemma 1 to ABC_2. The result is to replace
$C_1ABC_2\cdots C_n$ by $C_1C_2ABC_3\cdots C_n$ and add on the two terms $\lambda(B, C_2)C_1AC_3\cdots C_n$- $\delta\lambda(A, C_2)C_1BC_3\cdots C_n$. We can repeat this maneuver until we reach $C_1C_2\cdots C_nAB$ moving AB a step to the right each time and adding two terms – one of the Σ_1 form and one of the Σ_2 form. The truth of the lemma follows at once.

COROLLARY. *If $A = V^2$ and $B \in V^n$ then $AB - BA \in V^n$.*

THEOREM 5. *Let V be a δ algebra and let $n = 1,2,3\cdots$. Let $A_1, B_1, A_2, B_2, A_3, B_3, \cdots$ be a canonical basis for V. Let $C_1C_2\cdots C_n$ be a product of elements of V such that each C_j is a member of this canonical basis. Then for all k, $B_kA_k(C_1C_2\cdots C_n) - (C_1C_2\cdots C_n)B_kA_k$ is equal to a constant multiple of $C_1C_2\cdots C_n$ so that $C_1C_2\cdots C_n$ is a weight vector for the representation of V^2 defined by V^n with respect to the "Cartan subalgebra" spanned by the B_kA_k.*

The proof is a straightforward calculation using Lemma 5 and the definition of a canonical basis.

COROLLARY. *Choosing a maximal set of linearly independent members of the products $C_1\cdots C_n$ where each C_j is an A_k or a B_k, we obtain a weight space decomposition of the representation of V^2 defined by V^n.*

The reader should have no difficulty in computing the weights explicitly.

7. Fock space and the weight vectors of the representation of V^2 on X. Given a δ algebra V on X, by definition it is a vector space of linear operators on this same vector space and hence defines a representation of V^2 on the space X. Let V admit a canonical basis $A_1, B_1, A_2, B_2\cdots$. The question arises as to whether one can use the A_i and B_j to construct a weight space decomposition of the representation of V^2 on X as was done almost trivially for the representations on $V, V^2, V^3, V^4 \cdots$. The answer is that a weight space decomposition need not exist. Indeed it is possible to find δ algebras with canonical bases for which no non zero weight vectors exist at all in X. Moreover, even when a weight space decomposition does exist its construction from the A_i and B_j is less obvious and more complicated.

Let us define a member φ of X to be a *vacuum vector* with respect to the canonical basis $A_1, B_1, A_2, B_2, \cdots$ if $A_j(\varphi) = 0$ for all j. It is obvious that the set of all vacuum vectors φ for a given canonical basis as a vector space W which we shall call the *vacuum space*. It may reduce to $\{0\}$. In any case we define $\widetilde{W}$ to be the smallest subspace of X which contains W and is carried into itself by all A_2 and all B. $\widetilde{W}$ itself is then the space of a representation of V^2 which could be trivial but could also be non trivial and could even be the whole of X. When $\widetilde{W}$ is neither trivial nor all of X it defines a sort of splitting of the representation of V^2 defined by X into two parts; namely the subrepresentation whose space is $\widetilde{W}$ and the associated quotient representation whose space consists of classes of elements of X where two elements in X are put in the same class when their difference is in $\widetilde{W}$. We shall show below that the quotient representation has no non zero vacuum vectors. Thus the splitting of the representation of V^2 in X is into two pieces one of which has no non zero vacuum vectors and the other of which is generated by its vacuum vectors. The case in which $\widetilde{W} = X$, so that the quotient with no non trivial vacuum is trivial, is of importance in physics and can be given a very complete analysis.

LEMMA 6. *Let A and B be operators in the vector space X such that $AB - BA = 1$. Then for all positive integers k, $AB^k - B^kA = kB^{k-1}$.*

PROOF: $AB^k = (AB)B^{k-1} = (BA+1)B^{k-1} = BAB^{k-1} + B^{k-1} = B(AB^{k-1}) + B^{k-1} = B(BAB^{k-2} + B^{k-2}) + B^{k-1} = B^2AB^{k-2} + 2B^{k-1}$. Continuing in this way one moves A past another B each time and adds on another B^{k-1}. Finally one gets $B^kA + kB^{k-1}$ as announced.

LEMMA 7. *Let the δ algebra V in X have a canonical basis $A_1, B_1, A_2, B_2, \cdots$ and let there exist a non zero vector φ in X which is a vacuum vector for this basis. Then for every finite sequence $j_1, j_2 \cdots j_n$ of positive integers with $j_1 \leq j_2 \cdots \leq j_n$ the vector $A_k(B_{j_1}B_{j_2}\cdots B_{j_n}(\varphi))$ is zero unless k is equal to some j_t. If k is equal to j_t for s different values of t, then $A_k(B_{j_1}B_{j_2}\cdots B_{j_n}(\varphi))$ is equal to $s(-\delta)^{t-1}B_{j_1}B_{j_2}\cdots B_{j_{t-1}}B_{j_{t+1}}\cdots B_{j_n}(\varphi)$ unless $\delta = 1$ and $s > 1$ in which case it is zero.*

PROOF: Since $A_kB_j = -\delta B_jA_k$ when $j \neq k$ and $A_k(\varphi) = 0$ the first statement is obvious. If t is the least positive integer such that $k = j_t$ and there are s values of t for which $k = j_t$ then

$$A_k(B_{j_1}B_{j_2}\cdots B_{j_n}(\varphi)) = (-\delta)^{t-1}(A_kB_k^s)B_{j_{t+s}}B_{j_{t+s+1}}\cdots(\varphi)$$

and by Lemma 6 either $\delta = -1$ and $A_k B^s = B_k^s A_k + s B_k^{s-1}$ or $\delta = 1$ and $s = 1$ so $A_k B_k^s = -B_k^s A_k + 1$ or $\delta = 1$ and $s > 1$ so $A_k B_k^s = 0$. Correspondingly

$$A_k(B_{j_1} B_{j_2} \cdots B_{j_r}(\varphi)) = (-\delta)^{t-1} s(B_{j_1} B_{j_2} \cdots B_{j_{k-1}} B_{j_{k+1}} \cdots B_{j_r} \varphi)$$

in the first two cases and zero in the last.

THEOREM 6. *Let the δ algebra V in X have a canonical basis $A_1, B_1, A_2, B_2, \cdots$ which admits a vacuum vector φ in X. Let $\mathcal{A}$ be the commutative subalgebra of V^2 consisting of all finite linear combinations of the products $B_k A_k$. Then*

(1) For every finite sequence $j_1, j_2 \cdots j_n$ of positive integers the element $B_{j_1} B_{j_2} \cdots B_{j_r}(\varphi)$ is a (possibly zero) weight vector for the representation of V^2 in X with respect to the "Cartan subalgebra" $\mathcal{A}$.

(2) Let X_φ denote the linear span of all weight vectors of the form $B_{j_1} B_{j_2} \cdots B_{j_r}(\varphi)$. Then X_φ is carried into itself by all operators in V and hence by all operators in V^2 so that it defines a subrepresentation of our representation of V^2 in X.

(3) If $\delta = -1$ then $B_{j_1} B_{j_2} \cdots B_{j_n}(\varphi)$ is never zero and since the B_j's commute we may assume that $j_1 \leq j_2 \cdots \leq j_n$. Restricting the sequences $j_1, j_2 \cdots j_n$ in this way the weight vectors $B_{j_1} B_{j_2} \cdots B_{j_n}(\varphi)$ are linearly independent and hence constitutes a basis for X_φ.

(4) If $\delta = 1$, $B_{j_1} B_{j_2} \cdots B_{j_n}(\varphi) = 0$ if and only if the sequence $j_1, j_2, \cdots, j_n$ has at least two equal members. Since the B_j's anti commute, we need only consider the case when $j_1 < j_2 \cdots < j_n$ and then the vectors $B_{j_1} B_{j_2} \cdots B_{j_n}(\varphi)$ are linearly independent and constitute a basis for X_φ.

(5) If $\delta = -1$, the vectors $B_{j_1} B_{j_2} \cdots B_{j_n}(\varphi)$ where $j_1 \leq j_2 \leq j_3 \cdots \leq j_n$ define a weight space decomposition of the representation of V^2 on X_φ and if $\delta = 1$ the vectors $B_{j_1} B_{j_2} \cdots B_{j_n}(\varphi)$, where $j_1 < j_2 \cdots < j_n$ define a weight space decomposition of the representation of V^2 on X_φ.

(6) The representation of V^2 on X_φ is irreducible in both cases. Of course if the representation of V^2 on X is irreducible, then $X_\varphi = X$.

PROOF: Statements (1) and (2) are immediate consequences of Lemma 7. To prove (3) and (4) we prove first that $B_1(\varphi), B_2(\varphi) \cdots$ are linearly independent. Suppose that $c_1(B_1(\varphi) + \cdots + c_n B_n(\varphi) = 0$. Then for each k, $c_1 A_k B_1(\varphi) + \cdots + c_n A_k B_n(\varphi) = 0$. Since $A_k B_j(\varphi) = -\delta B_j A_k(\varphi) = 0$ for $k \neq j$ we have $c_k A_k B_k(\varphi) = 0$. But $A_k B_k(\varphi) = (1 - \delta B_k A_k(\varphi) = \varphi - 0 = \varphi$. Hence $c_k(\varphi) = 0$ or $c_k = 0$ for all k.

Thus the $B_j(\varphi)$ are linearly independent as asserted. Now suppose we have proved that for some $m = 1, 2 \cdots$ the vectors $B_{j_1} B_{j_2} \cdots B_{j_n}(\varphi)$ for all $r \leq m$ have been proved to be linearly independent for all $j_1, \cdots, j_n$. (We assume $j_1 \leq j_2 \cdots \leq j_n$ and $j_1 < j_2 \cdots < j_n$ when $\delta = 1$). We shall prove that the same must be true when we replace m by $m + 1$. It will then follow by mathematical induction that the $B_{j_1}B_{j_2}\cdots B_{j_n}(\varphi)$ are linearly independent is stated. Given a linear relationship

$$\Sigma c_{j_1 j_2 \cdots j_n} B_{j_1} B_{j_2} \cdots B_{j_n}(\varphi) = 0,$$

where $j_1, j_2 \cdots j_n$ varies over all sequences with $j_1 \leq j_2 \cdots \leq j_n$ and $n \leq m+1$ (and $j_1 < j_2 \cdots j_n$ when $\delta = 1$). Applying A_k to both sides of this relationship we obtain a similar relationship in which each term in which no $j_t = k$ is eliminated and each term in which some $j_t = k$ is replaced by a term in which the number of $\mathcal{B}$ factors is reduced by one and the coefficient of $c_{j_1} \cdots c_{j_n}$ is multiplied by ± 1. By the inductive hypothesis all coefficients $c_{j_1} \cdots c_{j_n}$ in which some $j_t = k$ must be zero. Since we can apply this argument for any k all $c_{j_1} \cdots c_{j_n}$ must be zero and the proof is complete.

To prove (6) we note that it follows from (2), (3) and (4) that the $B_{j_1}B_{j_2}\cdots B_{j_n}$ (φ) (with $j_1 \leq j_2 \cdots \leq j_n$ and $j_1 < j_2 \cdots < j_n$ when $\delta = 1$) form a basis for X_φ and from (1) that each basis vector is a weight vector.

To prove (φ) one uses Lemmas 6 and 7 to show that starting with any non zero vector in φ one can turn it into a non zero multiple of φ by acting on it with a suitable product $A_{j_1}A_{j_2}\cdots A_{j_n}$.

THEOREM 7. *Let $X, V, A_1, B_1 \cdots \varphi$ be as in Theorem 6. For each $r = 1, 2, \cdots$ let X_r denote the linear span of all $B_{j_1}B_{j_2}\cdots B_{j_n}(\varphi)$ with $j_1 \leq j_2 \cdots \leq j_n$ ($j_1 < j_2 \cdots < j_n$ if $\delta = 1$) and let X_0 be the set of all scalar multiples of φ. Then if $\delta = -1$ the vector space X_r is naturally isomorphic to the symmetric r-th tensor power of X_1 and if $\delta = 1$, the vector space X_r is naturally isomorphic to the anti symmetric r-th tensor power of X_1.*

PROOF: The proof is a routine verification which may be left to the reader.

This decomposition of X as a direct sum of symmetrized or anti symmetrized tensor powers of the subspace X_1 is known as the Fock space decomposition and has played an important role in quantum mechanics throughout most of its history – first in quantum field theory and since the late 1950's in many body theory. One of the main purposes of this paper is to point out its connection with weight and root space decompositions in the (purely mathematical) theory of Lie algebras. Once one has noted that V^2 is a Lie algebra and asks for the weight space decomposition of its natural representation on X one is led to it automatically. In this way we have found a purely mathematical motivation for an important construction in quantum physics.

As far as providing a purely mathematical motivation for the concept of δ algebra is concerned notice that the simplest Lie algebras are the commutative ones and perhaps the next simplest are those with a one dimensional center and a commutative quotient. A bosonic δ algebra with the constant operators adjoined is essentially the most general Lie algebra having this structure. Moreover a fermionic δ algebra with the constant operators adjoined is essentially the most general Lie superalgebra of operators whose even part is one dimensional.

8. Vacuum vectors and change of basis. The analysis of the natural representation of V^2 on X given in the preceding section are dependent not only on choosing a particular canonical basis $A_1, B_2, A_2, B_2, \cdots$ but also on choosing a particular vacuum vector φ in X. Such a vector need not exist and its existence or non existence may depend not only on V but on which canonical basis has been chosen. When it does exist the set of possible vacuum vectors may or may not change when the canonical basis changes and such changes play a crucial role in certain applications. We devote this section to an account of some of the more important facts.

The easiest case is that in which V is two dimensional and here the facts are simple and the proofs short and elementary. However they depend sharply on whether V is bosonic or fermionic.

THEOREM 8. *Let A, B be a canonical basis for the two dimensional δ algebra V. Then the most general canonical basis for V is A', B', where $A' = aA + bB$, $B' = cA + dD$ and a, b, c, d satisfy*

$$ad + \delta bc = 1, \quad ab(1+\delta) = 0 \quad \text{and} \quad cd(1+\delta) = 0.$$

(a) *If $\delta = 1$ then * reduces to $ad + bc = 1$, $ab = cd = 0$ so that either $a = 0, d = 0, bc = 1$ or $b = 0, c = 0, ad = 1$. Thus either $A' = bB, B' = \frac{1}{b}A$ or $A' = aA$, $B' = \frac{1}{a}B$. If φ is a vacuum vector for A, B then X_φ is the linear space φ and $B(\varphi)$ is invariant under V and V^2 and every vacuum vector in X_φ for A' and B' is either a multiple of φ or a multiple of $B(\varphi)$ depending upon whether $A' = aA$ or $A' = bB$.*

(b) *If $\delta = 1, 1+\delta = 0$ and * reduces to the single equation $ad - bc = 1$. Thus there is a canonical basis for every a, b, c, d satisfying this equation. If φ is a vacuum vector for A, B the linear space X_φ of $\varphi, B(\varphi), B^2(\varphi), \cdots$ is invariant under V and V^2 and the canonical basis A', B' defined by a, b, c, d has a vacuum vector if and only if $a \neq 0$. When $a \neq 0$ the possible vacuum vectors in X_φ are just the non zero multiples of φ.*

PROOF: The proof is short and results from obvious calculations. We leave details to the reader.

We begin to get an inkling of what can happen in the general finite dimensional case by looking at the four dimensional case. Let A_1, B_1, A_2, B_2 be a canonical basis for V having a vacuum vector φ. Then it is easy to write down three further canonical bases by simply permuting the given ones and inserting a δ at appropriate places. They are as follows:

(a)	$A_1' = B_2$	$A_2' = B_1$	$B_1' = \delta A_2$	$B_2' = \delta A_1$
(b)	$A_1' = \delta B_1$	$A_2' = A_2$	$B_1' = A$	$B_2' = B_2$
(c)	$A_1' = A_1$	$A_2' = \delta B_2$	$B_1' = B$	$B_2' = A_2$.

In the fermionic case, where $\delta = 1$, one knows from Theorem 6 that if φ is a vacuum vector for A_1, B_1, A_2, B_2 then $\varphi, B_1(\varphi), B_2(\varphi), B_1B_2(\varphi)$ is a basis for X_φ. Using this, a simple computation allows us to find the most general vacuum vector for A_1, B_1, A_2, B_2 and each of the three permuted bases (a), (b) and (c) listed above. For the original canonical basis it is a non zero multiple of φ. For basis (a) it is a non zero multiple of $B_1B_2(\varphi)$, for basis (b) it is a non zero multiple of $B_1(\varphi)$ and for basis (c) it is a non zero multiple of $B_2(\varphi)$. Notice that these are precisely the weight vectors of the natural representation of V^2 on X_φ.

While A_1, B_1, A_2, B_2 and the permutations (a), (b) and (c) are very special canonical bases they are interesting in that their vacuum vectors are the only possible vacuum vectors for *any* canonical basis in V.

THEOREM 9. *When $\delta = 1$ and V is four dimensional and admits a canonical basis A_1, B_1, A_2, B_2 with a vacuum vector φ then every canonical basis A'_1, B'_1, A'_2, B'_2 for V has a vacuum vector. Moreover in the invariant subspace X_φ this vacuum vector is unique up to to a multiplicative constant and is a multiple of one of the weight vectors $\varphi, B_1(\varphi), B_2(\varphi), B_1B_2(\varphi)$.*

PROOF: The only proof known to the author at the moment is too long and computational to present here. He hopes to find a simpler proof later.

Now let us investigate the four dimensional bosonic case. In this case X_φ is infinite dimensional and has a basis consisting of all $B_1^{n_1}B_2^{n_2}(\varphi)$ where $n_1 = 0, 1, 2\cdots, n_2 = 0, 1, 2\cdots$. The three canonical bases defined above by permuting A_1B_1, A_2B_2 and inserting δ can be easily seen to admit no vacuum vectors whatever and it is natural to conjecture that the proof of Theorem 9 can be adapted to prove that every canonical basis either has no vacuum vector in X_φ or has the multiples of φ as its only vacuum vectors. We have not yet attempted a proof of this conjecture.

The results proved and conjectured about the cases in which V is one or two dimensional suggest plausible conjectures about the general finite dimensional case.

CONJECTURE 1: Let n be a positive integer and let V be an n-dimensional δ algebra in the vector space X. If $\delta = -1$ and for some canonical bases $A_1, B_1, \cdots, A_n, B_n$ in X there exists a vacuum vector φ there also exist canonical bases $A'_1, B'_1, \cdots, A'_n, B'_n$ for which there is no vacuum vector and every canonical basis for V either has no vacuum vector in X_φ or every vacuum vector in X_φ is a multiple of φ.

If $\delta = 1$ and for some canonical basis $A_1, B_1, A_2, B_2, \cdots, A_n, B_n$ there exists a vacuum vector φ, then for every canonical basis $A'_1, B'_1, \cdots, A'_n, B'_n$ in V there exists a vacuum vector ψ in X_φ and this vacuum vector ψ is a multiple of one of the vectors $B_{i_1}, B_{i_2}, \cdots, B_{i_n}(\varphi)$ where $1 \leq i_1 < i_2 \cdots < i_n \leq n$; that is, it is one of the weight vectors for the natural represention of V^2 in X_φ.

The author already knows how to prove certain easy parts of this conjecture but the proof of the full conjecture awaits a suitable generalization of the difficult part of Theorem 9 as well as of its analogue in the bosonic case. He does not believe that finding these proofs is a really serious problem but it does seem to be one with no obvious easy solution.

9. Connections with the physical literature via "creation" and "annihilation" operators. The passage from a classical mechanical system of n particles to the corresponding quantum system consists in replacing the classical phase space by a separable infinite dimensional Hilbert space $\mathcal{H}$ and associating the classical coordinates and momenta with certain self adjoint operators in $\mathcal{H}$. This association is determined by postulating that these operators satisfy the celebrated Heisenberg commutation relations $Q_iQ_j - Q_jQ_i = P_iP_j - P_jP_i = 0$ and $P_jQ_j - Q_jP_j = \frac{h}{2\pi i}$ for all i and j where h is the celebrated Planck's constant. Here the Q_j, $j = 1, 2, 3 \cdots 3n$ are the self adjoint operators associated with the 3n rectangular position coordinates of the n particles and the P_j are the self adjoint operators associated with the linear momentum components of these particles. This picture is complicated by the fact that many of the particles such as electrons, protons, etc., to which one applies quantum mechanics have an internal structure with no classical counterpart. This is reflected in the fact that there then exist non constant self adjoint operators which commute with all the P_j and all the Q_j. The most important case is that in which the particles have "spin $\frac{1}{2}$" and in that case the set $\mathcal{A}$ of all bounded linear operators in $\mathcal{H}$ which commute with all the P_j and Q_j is a finite dimensional vector space whose dimension is 2^n. It is also of course a linear associative algebra in the sense that the product of any two elements in $\mathcal{A}$ is in $\mathcal{A}$ and this algebra is isomorphic to the algebra of all complex $2^n \times 2^n$ matrices or equivalently to the algebra of all linear transformations in complex vector space of dimension 2^n.

The P_j and Q_j for such particles have to be supplemented by certain self adjoint operators in $\mathcal{A}$ in order to have operator counterparts to enough observables to construct all others. It turns out that a linear basis for $\mathcal{A}$ may be constructed starting with $2n$ self adjoint operators σ_1^j, σ_2^j, where $j = 1, 2, \cdots, n$ and each operator has 1 and -1 as its only eigenvalues. These operators satisfy commutation relations slightly different from those satisfied by the P_j and Q_j. They are:

$$(****) \qquad \begin{aligned} &\sigma_a^j\sigma_b^k - \sigma_b^k\sigma_a^j = 0 \quad \text{whenever } j \neq k \\ &\sigma_a^j\sigma_b^k + \sigma_b^k\sigma_a^j = 0 \quad \text{whenever } j = k \text{ and } a \neq b \end{aligned}$$

For many purposes it turns out to be useful to replace these operators by $2n$ others defined by the formulae

$$\widehat{P}_j = \sigma_1^1\sigma_1^2 \cdots \sigma_1^{j-1}\sigma_2^j \qquad\qquad \widehat{Q}_j = \sigma_1^1\sigma_1^2 \cdots \sigma_1^{j-1} i\sigma_1^j\sigma_2^j.$$

One checks easily that $\frac{1}{i}\widehat{P}_j\widehat{Q}_j = \sigma_1^j$ and $\sigma_2^j = \sigma_1^1\sigma_1^2 \cdots \sigma_1^{j-1}\widehat{P}_j$ so that not only is each $\widehat{P}_j$ and $\widehat{Q}_j$ a product of σ_1^j and σ_2^j but conversely every σ_1^j and σ_2^j is a product of $\widehat{P}_j$ and $\widehat{Q}_j$. A simple calculation now shows that the σ_1^j and σ_2^k satisfy the relations (****) above if and only if the $\widehat{P}_j$ and $\widehat{Q}_j$ satisfy the *anti commutation relations*

$$(****)' \qquad \widehat{Q}_j\widehat{P}_k + \widehat{P}_k\widehat{Q}_j = \widehat{Q}_j\widehat{Q}_k + \widehat{Q}_k\widehat{Q}_j = \widehat{P}_j\widehat{P}_k + \widehat{P}_k\widehat{P}_j = 2\delta_i^j$$

or equivalently

(****)'' $\quad \widehat{P}_j^2 = \widehat{Q}_j^2 = 1$ and any two distinct members of the sequence $\widehat{Q}_1, \widehat{P}_1, \widehat{Q}_2, \widehat{P}_2, \widehat{Q}_3, \widehat{P}_3, \cdots$ anti commute with one another. The chief difference between (****) and (****)'' is that mixed commutation and anti commutation relations are replaced by pure anti commutation relations. On the other hand, the σ_1^j and σ_2^j have a direct interpretation as operators corresponding to the internal motion of the j-th particle.

It is obvious that the linear span of all the Q_j and P_j is a finite dimensional bosonic δ algebra and that the linear span of all the $\widehat{Q}_j$ and $\widehat{P}_j$ is a finite dimensional fermionic δ algebra. This fact underlies the interest for physics of the purely mathematical theory developed in the first eight sections of this paper. We remark next that we can at once write down canonical bases for these two δ algebras. Indeed we can do so in a uniform way since the operators $A_j = \sqrt{\frac{\pi}{h}}(Q_j + iP_j)$, $B_j = \sqrt{\frac{\pi}{h}}(Q_j - iP_j) = A_j^*$ constitute a canonical basis in the bosonic δ algebra spanned by the Q_j and P_j and $\widehat{A}_j = \widehat{Q}_j + i\widehat{P}_j$, $\widehat{B}_j = \widehat{Q}_j - i\widehat{P}_j = \widehat{A}_j$ do the same for the fermionic δ algebra spanned by the $\widehat{Q}_j$ and $\widehat{P}_j$. For reasons which will be explained below the operators B_j and $\widehat{B}_j$ are called *creation operators* and the operators A_j and $\widehat{A}_j$ are called *destruction* operators or *annihilation* operators. These creation operators and destruction operators are not self adjoint and hence do *not* correspond to quantum mechanical observables. However their real and imaginary parts do and up to constant multiples are the $Q_j, P_j, \widehat{Q}_j$ and $\widehat{P}_j$ introduced at the outset.

We remind the reader that the P_j and Q_j are *unbounded* self adjoint operators and hence not everywhere defined. One can overcome many of the difficulties that this produces by proving that there is a dense subspace of the Hilbert space on which they are all defined and which they carry into itself. However this paper is not the place in which to go into such technical points. The $\widehat{P}_j$ and $\widehat{Q}_j$ in the fermionic case are bounded operators for which the problem does not arise.

Since the bosonic δ algebra $\mathcal{B}$ generated by the Q_j and P_j and the fermionic δ algebra $\mathcal{F}$ generated by the $\widehat{Q}_j$ and $\widehat{P}_j$ commute with one another it follows that their linear sum $\mathcal{F} \dotplus \mathcal{B}$ is an example of a δ superalgebra as defined in section 4. Hence its square $(\mathcal{F} \dotplus \mathcal{B})^2 = \mathcal{F}\mathcal{B} \dotplus (\mathcal{F}^2 \dotplus \mathcal{B}^2)$ is an example of a Lie superalgebra whose "odd part" $\mathcal{F}\mathcal{B}$ is just the linear span of all products $Q_j\widehat{Q}_k, Q_j\widehat{P}_k, P_j\widehat{P}_k, P_j\widehat{Q}_k$. At the same time $\mathcal{F}\mathcal{B}$ is a concrete example of a notion introduced by the author some years ago (Czech Journal of Physics, vol. B37, 1987) and called a *Lie algebra square root*. That is, it is a vector space V of linear operators with the property that $AB^2 - B^2A$ is in V whenever A and B are in V.

The theorem that justifies the name asserts that $V^{2,s}$, the linear span of all squares of elements of V (or equivalently of all anti commutators of pairs of elements of V), is a Lie algebra of linear operators. In this case, $(\mathcal{F}\mathcal{B})^{2,s}$ is just the Lie algebra $\mathcal{F}^2 \dotplus \mathcal{B}^2$.

The results of section 6 on the construction of roots and weights for $\mathcal{F}^2$ and $\mathcal{B}^2$ and their representations on $\mathcal{F}$ and $\mathcal{B}$ from canonical bases in $\mathcal{F}$ and $\mathcal{B}$ suggest at once the following:

(a) That a canonical basis in a δ superalgebra be defined as the union of a canonical basis in the odd part with a canonical basis in the even part.

(b) How to define roots, root spaces, weights and weight spaces for Lie superalgebras and their representations in such a way that analogues of the results of section 6 hold for Lie superalgebras. We leave details to the reader. Of course these facts about Lie superalgebras could have been formulated in section 6.

Theorem 1 (Section 3), the discovery of which was the initial stimulus for the approach of this paper, was stated above to be unexpected; and does not seem to be known to mathematicians working in Lie theory. However, as the author learned from comments made after his lecture, its essence has been well known to physicists working in many body theory (as well as a few others) for several decades. In these circles it is known in the form of a principle or working rule which may be stated as follows. Let A, B, C and D be creation or annihilation operators which are all fermionic or all bosonic. Then the commutator $(AB)(CD)-(CD)(AB)$ is a linear combination of other such binary products.

Theorem 1 suggests a question which we have yet to consider in this paper. Given that V^2 is a Lie algebra for any δ algebra V and that much is known about the structure of Lie algebras, can one identify the Lie algebra V^2 – at least when V is finite dimensional. Giving the facts about the root structure of V^2 found in section 6 it is straightforward to verify the following. If V is finite dimensional and has a canonical basis then V^2 is simple (modulo its one dimensional center) and is one of the classical simple Lie algebras. When V is bosonic V^2 is the complex symplectic Lie algebra of the appropriate dimension and when V is fermionic V^2 is the complex Lie algebra of orthogonal matrices in an even dimensional space.

The author was rather pleased at having found such an elegant way of describing two of the principal classes of classical simple Lie algebras. However at the meeting he learned that a subset of the physicists already knew about this and had even gone a little further. In particular, let V be a finite dimensional δ algebra with a canonical basis $A_1, B_1, A_2, B_2, \cdots, A_n, B_n$ and consider the subspace of V^2 spanned by all binary products of the form B_jA_k (and omitting those of the form B_jB_k and A_jA_k). Then the subspace is a Lie subalgebra and whether V is bosonic or fermionic this subalgebra is isomorphic to the classical Lie algebra of all $n \times n$ complex matrices. That this is so was pointed out by P. Jordan in 1935. In those days physicists did not think in terms of Lie algebras as such and Jordan was not interested in the $n \times n$ complex matrices as an example of a Lie algebra. Thus there was no motivation for seeking similar representations of other classical Lie algebras until much later.

Jordan's paper was pointed out to the author by L. Biedenharn shortly after the talk and subsequently the author made a careful search of the literature. The earliest relevant reference he could find, after Jordan's paper, occurs in H. Lipkin's delightful booklet, "Lie groups for pedestrians", published over three decades later in 1967. There one finds it very clearly set forth that one can construct the classical complex Lie groups out of products of annihilation and creation operators in the manner indicated. Lipkin does not, however, mention Jordan's paper and declares that this part of his book is original. According to Biedenharn, Jordan's paper is not well known to modern physicists. In the meantime, Lie algebras had been popularized amongst particle physicists by the work of Gell-Mann on "current algebra" in the early 1960's and various aspects of many body theory had been invaded with great effect, by the adoption of the methods of quantum field theory in the latter half of the 1950's, especially the systematic use of creation and annihilation operators. Lipkin wrote his booklet in the spirit of a recent convert striving to bring the light to the still unconverted.

The systematic use of creation and annihilation operators to solve problems in the many body problem is often referred to as "The method of second quantization" because of the way it is involved in carrying the field notion over into quantum mechanics. Another missionary type booklet (published also in 1967) is Brian Judd's "Second quantization and atomic spectroscopy". Here the "converts to be" are atomic spectroscopists and the emphasis is not on Lie algebras as such but on the utility of expressing operators in terms of creation and annihilation operators. The connection with Lie algebras is mentioned briefly in the concluding remarks as an "important" omitted topic. In the introduction a (1933) paper of M.H. Johnson is cited as containing an early application of the method. However it is asserted that its full value was not appreciated until after the work of Racah in 1942, 1943, and 1949. It seems that the work of Racah made it feasible for atomic spectroscopists to attack much more complicated problems than previously and it is only in dealing with these that the method of second quantization makes a really significant difference. Indeed, in particular the method illuminates some of the rather arcane notions introduced by Racah. According to Judd this new appreciation of the method of second quantization began only in 1956 with a paper of Brink and Satchler. This paper, published in Nuovo Cimento, includes a discussion of Racah's "coefficients of fractional parentage" in terms of creation and annihilation operators.

Shortly after the end of the conference the author received a message from W. Klink giving him a reference to a paper of Fulton published in 1985 in the *Journal of Physics* and describing work of himself and others on the construction of classical Lie algebras out of creation and annihilation operators. References go back to around 1971 but there is no mention of the work of Jordan or Lipkin.

However the work goes beyond that of Lipkin in that exceptional Lie algebras are also treated.

10. Particle indistinguishability and the physical meaning of the main theorem of section 7.

The summary given in section 9 of the quantum mechanics of n particles is incomplete in that nothing has been said about dynamics and is only valid as far as it goes when the n particles are "mutually distinguishable". It turns out that electrons (as well as neutrons and protons) are "all alike" in a rather more profound sense than was realized before quantum mechanics was discovered and having to do with their "wave aspects". Speaking somewhat metaphysically the various electrons in a collection of n electrons do not have permanent individual identities. It makes no physical sense to speak of two of them as having changed places. This manifests itself in the mathematical model in a completely well defined way with no metaphysics involved. Let $\mathcal{H}$ be the Hilbert space of states for a single electron. If it were not for the indistinguishability of electrons the Hilbert space for a collection of n electrons would be the tensor product $\mathcal{H} \otimes \mathcal{H} \cdots \otimes \mathcal{H}$ of n replicas of this Hilbert space.

Actually it is a proper subspace of this Hilbert space called its ***anti symmetric subspace.*** It is easy to see that each member π of the group S_n of all permutations of n objects acts on $\mathcal{H} \otimes \mathcal{H} \cdots \otimes \mathcal{H}$ in a natural way and defines a unitary representation W of π in this Hilbert space. The subspace on which W reduces to a multiple of the identity is called the ***symmetric subspace*** or the ***symmetrized n-fold tensor power*** of $\mathcal{H}$. Now S_n has one other one-dimensional irreducible representation. It is the one taking each even permutation into the identity and each odd permutation into -1 times the identity. The subspace on which W reduces to ***this*** one-dimensional representation is called the ***anti symmetric subspace*** or the ***anti symmetrized n fold tensor power of*** $\mathcal{H}$. The indistinguishability of electrons is manifested by the fact that all states of a system of n electrons lie in the anti symmetric subspace of $\mathcal{H} \otimes \mathcal{H} \cdots \otimes \mathcal{H}$. One consequence of this is that many self adjoint operators in $\mathcal{H} \otimes \mathcal{H} \cdots \otimes \mathcal{H}$ which one would have expected to define observables of the system do not do so. Only self adjoint operators which carry the anti symmetric subspace into itself will do – and they must be replaced by their restrictions to this subspace. This means in particular that no one of the Q_j or the P_j can correspond to an observable. On the other hand, the sum of all the Q_j and the sum of all the P_j do leave the anti symmetric subspace invariant and correspond to some observable. This is not the place to go into detail but one finds in the end that any ***symmetric*** function of all the Q_j or all of the P_j corresponds to some observable. Now if one knows $x_1 + x_2$ and $x_1^2 + x_2^2$ for two real numbers one can compute $x_1 x_2$ and hence the coefficients of a quadratic equation whose roots are x_1 and x_2. Hence one can find the two numbers x_1 and x_2 but ***not*** which one is x, and which is x_2. Analogous results hold for n real numbers and this is how our model describes n

electrons but does not allow us to ask questions about which is which. Protons and neutrons behave like electrons in this respect and all particles which do so are called fermions after the physicist Fermi. However there are also particles, for example photons, whose indistinguishability is reflected in all states lying in a different subspace of $\mathcal{H} \otimes \mathcal{H} \otimes \cdots \mathcal{H}$; namely the symmetric subspace, defined above using the trivial representation of S_n. Such particles are called *bosons* after the physicist Bose.

With these facts in mind let us return to Theorem 7 in Section 7 where we found the weight space decomposition of the natural representation of V^2 in X_φ where φ is a vacuum vector for a canonical basis in the δ algebra V. We found that X_φ is a direct sum $X_0 \oplus X_1 \oplus X_2 + \cdots$, where X_0 is the one-dimensional vector space of all complex multiples of φ and X_r for $r > 1$ is the linear span of all $B_{j_1} B_{j_2} \cdots B_{j_r}(\varphi)$ with $j_1 \leq j_2 \cdots j_r$ or $j_1 < j_2 \cdots < j_r$ according as $\delta = -1$ or 1. It is not hard to show, as already stated in Section 7, that for $r = 2, 3, \cdots$, X_r is canonically isomorphic to a subspace of the r-fold tensor prime of X_1. This will be the anti symmetric subspace or the symmetric subspace according as $\delta = 1$ or -1. To the extent that X_1 can be interpreted as a dense subspace of the Hilbert space of a single fermion (boson if $\delta = -1$) X_r can be interpreted as a dense subspace of a system of r of these fermions (bosons). The direct sum of all X_r with $r = 0, 1, 2, \cdots$ can then be interpreted as a dense subspace of the Hilbert space of a system with a variable number of particles and then φ will correspond to a state in which there are no particles. This is the rationale for calling φ a vacuum vector. Similarly one speaks of the B_j as "creation operators" because B_j takes the r particle subspace into the $r+1$ particle subspace and if one thinks of the basis element $B_j(\varphi)$ of X_1 as representing a particle in a certain state, then B_j applied to a basis element $B_{j_1} B_{j_2} \cdots B_{j_r}$ of X_r takes it into a basis element of X_{r+1} which defines an $r+1$ particle state in which one additional particle is in the state defined by $B_j(\varphi)$. One speaks of the A_j as annihilation operators for analogous reasons involving the fact that A_j maps X_r with X_{r-1} and φ into 0.

It was Dirac, in 1927, who introduced such considerations by showing that the Maxwell electromagnetic field could be treated quantum mechanically by regarding it as a system of countably many harmonic oscillators (after confining the field to a box and introducing periodic boundary conditions). The resulting quantum mechanical system admits a bosonic δ algebra with a canonical basis and a vacuum vector such that each X_r is invariant in time and one is led to a new interpretation of the system as a system of an indefinite number of bosonic particles. These are the photons and this is an outline of how Dirac showed that electromagnetic radiation could be particles and waves at the same time.

It is more or less straightforward to extend the analysis of section 7 to the case in which V is a δ super algebra $\mathcal{F} \oplus \mathcal{B}$ and the Lie algebra V^2 is replaced by the Lie superalgebra $\mathcal{F}\mathcal{B} \dotplus (\mathcal{F}^2 \dotplus \mathcal{B}^2)$. In this case the analogues of the n particle space in the Fock space decomposition is a subspace corresponding to $n_1 + n_2$ particles of which n_1 are fermions and n_2 are bosons and it appears as a tensor product of two spaces one of which is an antisymmetrized tensor power of n_1 replicas of a one particle space and the other a symmetrized tensor power of n_2 replicas of another one particle space. We leave details to the reader. This sort of structure appears in quantum electrodynamics when one realizes electrons as quanta of a hypothetical field and considers the interaction between the photon, electron and positron fields.

The post war development of quantum electrodynamics in the hands of Tamagawa, Schwinger, Feynman and Dyson produced powerful techniques for making calculations using the properties of creation and annihilation operators. Eventually it was realized that one could use these same techniques in problems with a fixed number of particles by simply considering all possible numbers of particles at the same time.

11. Concluding remarks. In addition to the applications of the abstract theory of the first eight sections to many body theory and quantum field theory hinted at above, there are also significant applications to the theory of elementary particles (both through string theory and otherwise) and to lattice statistics. Moreover the interest of the author is not limited to calling attention to the connection of the paper's title. He has reasons for believing that this connection can be exploited to clarify and unify various known methods in the various fields of application and even to suggest new methods. He hopes to return to the subject at a later time.

Department of Mathematics, Harvard University,
Cambridge, Massachusetts 02138.

APPLICATIONS OF SP(3,R) IN NUCLEAR PHYSICS

David J. Rowe

ABSTRACT. A brief overview is given of the way the non-compact symplectic group Sp(3,R) is used as a dynamical group in the microscopic theory of nuclear collective motion. Two unfamiliar concepts arise in the theory: the concept of an *embedded representation* and the concept of a *quasi-spinor representation*. These concepts are explained and illustrated.

1. Introduction

In this brief review, I outline some of the remarkable properties of the non-compact symplectic group Sp(3,R) as a dynamical group in the microscopic theory of nuclear collective motion. First, I indicate how and why Sp(3,R) is a dynamical group for nuclear collective motion and explain its importance for giving the phenomenological collective model a microscopic interpretation. Next, I present the concept of an *embedded representation* and show that it makes it possible to use dynamical groups to describe collective phenomena even when there is strong mixing of irreps. Finally, I show that groups, such as SU(3), which have no spinor represention (i.e., representations with half-odd integer angular momentum) do have *quasi-spinor representations* and that the concept is important for describing the strong coupling of intrinsic spin to nuclear rotations.

2. The Sp(3,R) group as a dynamical group for nuclear collective motion

If G is a dynamical group for a Hamiltonian H, then the eigenstates of H can be arranged in subsets that span irreps of G. A subset of eigenstates which span an irrep will be referred to as a *G-band* of states. The states of a G-band do not, in general, have a common energy as they would if the group G were a full symmetry group of the Hamiltonian. Nevertheless, they carry the labels of the irrep to which they belong. A necessary condition for a model to have a dynamical group G is

E. A. Tanner and R. Wilson (eds.), Noncompact Lie Groups and Some of Their Applications, 285–300.

that its Hamiltonian should not mix states belonging to different irreps of G. A particularly useful situation is when the Hamiltonian and all the other observables of a model are polynomials of the Lie algebra of G (assuming it is a Lie group). Knowledge of the irreps of the Lie algebra can then be used to facilitate model calculations. Moreover, the dynamical group then provides a natural expression of the dynamical content of the model.

Symplectic groups are fundamental in physics because they are the most general groups of linear canonical transformations that leave the Heisenberg commutation relations,

$$[x_{mi}, p_{ni}] = i\hbar\delta_{mn}\delta_{ij}\,, \tag{1}$$

invariant. Thus, a symplectic transformation is basically a linear transformation

$$\begin{aligned} x_{mi} &\rightarrow x'_{mi} = \textstyle\sum_j (a_{ij}x_{mj} + b_{ij}p_{mj}) \\ p_{mi} &\rightarrow p'_{mi} = \textstyle\sum_j (c_{ij}x_{mj} + d_{ij}p_{mj}) \end{aligned} \tag{2}$$

for which $[x'_{mi}, p'_{ni}] = i\hbar\delta_{mn}\delta_{ij}$.

A second important property of the symplectic group Sp(3,R) is that it maps ellipsoidal density distributions into new ellipsoidal densities of different deformation and orientation. Consider, for example, the GL(3,R) subgroup of canonical point transformations

$$x_{mi} \rightarrow x'_{mi} = \sum_j a_{ij}x_{mj}\,, \quad p_{mi} \rightarrow p'_{mi} = \sum_j a^{-1}_{ji}p_{mj}\,. \tag{3}$$

Such a transformation maps the quadratic equation for an equidensity surface

$$\sum_n \alpha_{ij}x_{ni}x_{nj} = \text{const.} \tag{4}$$

into an equation of the form

$$\sum_n \alpha_{ij}x'_{ni}x'_{nj} = \sum_n \alpha'_{ij}x_{ni}x_{nj} = \text{const.} \tag{5}$$

In extending the point transformations of GL(3,R) to general linear canonical transformations, the Sp(3,R) group not only deforms and rotates an ellipsoidal density distribution, it also gives it rotational and vibrational current flows as illustrated in Fig. 1. Thus, it is an appropriate dynamical group for a model of nuclear quadrupole vibrations and rotations.

The historic path to the discovery of Sp(3,R), as a dynamical group for nuclear collective motions [1; 2; 3; 4; 5; 6; 7] (cf., ref. [8] for a more complete list of references), was by a consideration of observables. In seeking rotor model observables, it was recognized that one first needs a subset of observables which define the ellipsoidal shape of a nucleus. Suitable shape observables are the 6 Carte-

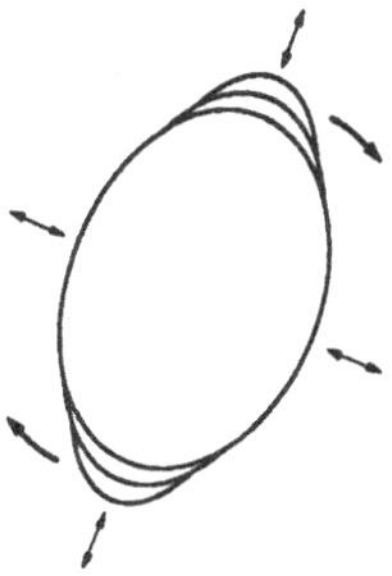

Fig. 1. Rotations and quadrupole vibrations of an ellipsoidal density distribution generated by Sp(3,R) transformations.

sian quadrupole moments $\{Q_{ij}\}$, which are given in terms of nucleon coordinates $\{x_{ni}; n = 1, \ldots, A, i = 1, 2, 3\}$ by

$$Q_{ij} = \sum_n x_{ni} x_{nj}\,, \quad i,j = 1,2,3\,, \tag{6}$$

or, if one thinks of the nucleus as a fluid, in terms of the nuclear matter density ρ

$$Q_{ij} = \int x_i x_j \rho(x)\, dv\,, \quad i,j = 1,2,3\,. \tag{7}$$

A nice feature of this choice is that the two expressions of the quadrupole moments become identical in form if one discretizes the continuous density of nuclear matter and regards it as an aggregate of small cells. The two scenarios are illustrated in Fig. 2.

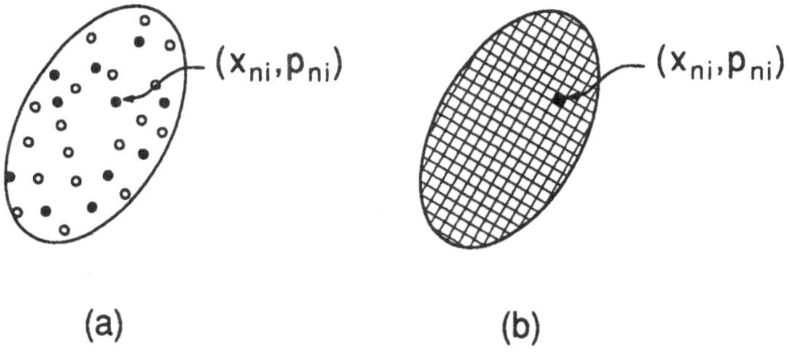

Fig. 2. Assignment of position and momentum coordinates: (a) to a nucleon and (b) to an element of the nuclear matter distribution.

For a model of nuclear rotations, we also need angular momentum observables. These are given by

$$L_k = \sum_n (x_{ni} p_{nj} - x_{nj} p_{ni})\,, \tag{8}$$

with the understanding that i, j and k are cyclic. One now observes that the set of quadrupole moments and angular momenta $\{Q_{ij}, L_k\}$ close under commutation. The angular momenta span the Lie algebra so(3) of the group of rotations, SO(3). The six quadrupole moments span an abelian Lie algebra isomorphic to R^6. The quadrupole moments also transform under rotations as the components of a one- plus a five-component tensor. The one-component tensor is spanned by the trace $\sum_i Q_{ii}$ and is rotationally invariant. The five-component tensor is traceless and transforms as a tensor of angular momentum 2. Thus, the traceless quadrupole moments and the angular momentum components span a semi-direct sum Lie algebra isomorphic to $[R^5]$so(3). Since this Lie algebra is a dynamical algebra for a rotational model in three-dimensional space [9; 10], we shall refer to it simply as rot(3).

Now, it is possible to construct a Hamiltonian for rotations in which the moments of inertia are functions of the quadrupole moments. Indeed, the phenomenological nuclear rotor model proceeds in this way. However, one of our objectives in constructing an algebraic model of nuclear collective motions is to learn, from first principles, how nuclei rotate; i.e., to find out what current flows are compatible with the microscopic structure of nuclear matter. One does not learn such things by burying the dynamics in a model with adjustable moments of inertia. We therefore include the full nuclear kinetic energy

$$T = \frac{1}{2m} \sum_{ni} p_{ni}^2 \tag{9}$$

in the dynamical algebra of the model we are constructing. We next observe that, although the rot(3) observables together with the kinetic energy, do not close under commutation, they generate a larger Lie algebra that does. It is the 21-dimensional Lie algebra sp(3,R) of the Sp(3,R) symplectic group. A phenomenological Hamiltonian for a symplectic model can be constructed by adding to the kinetic energy a suitable function of the quadrupole moments to give a restoring force against deformation [17]. A more fundamental Hamiltonian can be constructed by restricting a realistic microscopic Hamiltonian to the space of a single sp(3,R) irrep.

The sp(3,R) Lie algebra has many nice properties:

(i) It is semisimple and its representation theory is easy.

(ii) Its physical content is transparent.

(iii) The whole shell model Hilbert space is a sum of unitary sp(3,R) irreps.

The latter property is important because, by considering only irreps contained in the shell-model Hilbert space, we restrict to collective motions that are fully compatible with the microscopic many-nucleon structure of the nucleus and fully take into account the antisymmetry and other requirements of quantum mechanics. In return, we gain the possibility of discovering the dynamical content of the corresponding collective motions.

3. Induced representations of sp(3,R)

To induce a representation of Sp(3,R) it is convenient to regard Sp(3,R) as a subgroup of SU(3,3) by means of the isomorphism Sp(3,R) $\sim$ SU(3,3) $\cap$ Sp(3,C). A maximal compact subgroup for Sp(3,R) is then the U(3) $\subset$ SU(3,3) subgroup of block diagonal matrices

$$\mathrm{U}(3) \sim \left\{ \begin{pmatrix} g & 0 \\ 0 & g^* \end{pmatrix} ; g \in \mathrm{U}(3) \right\} \tag{10}$$

and we have the Cartan decomposition

$$\mathrm{sp}(3)^{\mathrm{C}} = \mathrm{u}(3)^{\mathrm{C}} + \mathbf{p}_+ + \mathbf{p}_- , \tag{11}$$

where $\mathbf{p}_\pm$ are nilpotent subalgebras of raising and lowering operators, respectively. Holomorphic discrete series representations of sp(3,R) are now induced from an irrep of u(3) [11; 12] in the standard way [13]. By restricting to discrete series irreps, induced from irreps of u(3) with appropriate highest weights, we can ensure that we construct only representations which belong to the shell-model Hilbert space of the nucleus under consideration.

Suppose we induce a representation of sp(3,R) from a u(3) irrep of highest weight $N(\lambda,\mu)$, where N is a u(1) weight and (λ,μ) is an su(3) weight. It is known that the $\mathbf{p}_+$ Lie algebra transforms as a u(3) irrep of highest weight 2(2,0). Thus, the module for the representation of sp(3,R) induced from the u(3) irrep $N(\lambda,\mu)$ is a direct sum of modules for the set of u(3) representations

$$N(\lambda,\mu), \quad (N+2)((\lambda,\mu)\times(2,0)), \quad (N+4)((\lambda,\mu)\times(2,0)\times(2,0)), \quad \text{etc.} \tag{12}$$

The corresponding u(3) subspaces are shown, schematically, in Fig. 3 in a manner that characterizes the way their states appear in the spectrum of a collective band of rotational-vibrational states.

An efficient algorithm for inducing the holomorphic discrete representations and calculating their explicit matrix elements has been given in terms of Vector Coherent State (VCS) theory [14; 15; 16] (cf., also ref. [8]).

4. The sp(3,R) energy spectrum

The details of the energy level spectrum depend on the Hamiltonian. The spectrum shown in Fig. 3 is the kind of spectrum one would obtain for a simple Hamiltonian of the form

$$H_0 = H_{\mathrm{HO}} - \chi(N)\tilde{Q}\cdot\tilde{Q} , \tag{13}$$

which is diagonal in a basis $\{|N(\lambda\mu), \alpha N'(\lambda'\mu')KLM\rangle\}$ which reduces the subgroup chain

$$\mathrm{Sp}(3,R) \supset \mathrm{U}(3) \supset \mathrm{SO}(3) \supset \mathrm{SO}(2) . \tag{14}$$

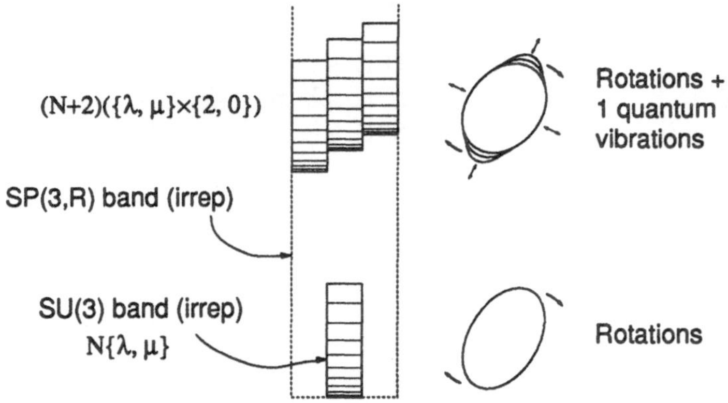

Fig. 3. A symplectic band of energy levels for the Hamiltonian H_0.

The Hamiltonian H_{HO} is the harmonic oscillator shell model Hamiltonian and $\tilde{Q}$ is an su(3) quadrupole tensor. In the given basis, H_0 has eigenvalues given by

$$H_0|N(\lambda\mu),\alpha N'(\lambda'\mu')KLM\rangle = E(N',\lambda',\mu',L)|N(\lambda\mu),\alpha N'(\lambda'\mu')KLM\rangle\,, \quad (15)$$

where

$$E(N,\lambda,\mu,L) = N\hbar\omega - 4\chi(N)(\lambda^2+\mu^2+\lambda\mu+3\lambda+3\mu)+\tfrac{1}{3}\chi L(L+1)\,. \quad (16)$$

One sees that the spectrum is of the rotational-vibrational type shown in Fig. 3.

A more realistic model Hamiltonian would be of the form

$$H = H_0 - \kappa \mathcal{Q}\cdot\tilde{Q}\,, \quad (17)$$

where $\mathcal{Q}$ is the quadrupole tensor with components

$$\mathcal{Q}_\nu = A_\nu + B_\nu \quad (18)$$

which are sums of sp(3,R) raising and lowering operators. Such a Hamiltonian admits coupling between the rotational and vibrational degrees of freedom. Its spectrum is qualitatively similarly to that of H_0 but the mixing of different U(3) irreps that results from the coupling interaction brings about a renormalization of the lowlying rotational states and enhances the electromagnetic transition rates between states to better agree with experimental data.

A more fundamental Hamiltonian is given by restricting a realistic microscopic Hamiltonian for the nucleus to a single sp(3,R) irrep [18; 19; 20; 21] . Better still, one would allow mixing between different sp(3,R) irreps in diagonalizing an unrestricted microscopic Hamiltonian . In practice, such microscopic calculations are not easy and, so far, have only been attempted for light nuclei [22].

5. The rotational-vibrational interpretation of the sympletic model

The energy spectrum given by eq. (15) is characteristic of that of a rotor-vibrator, with harmonic vibrational energies given by $N\hbar\omega$ and rotational energies proportional to $L(L+1)$. A physical interpretation of the dynamical content of the symplectic model as that of a coupled rotor-vibrator [23; 24] is given by considering the contraction limit of the sp(3,R) Lie algebra in the limit of large quantum numbers. It has been shown [25] that, as $N \to \infty$, the $u(1)^C + p_+ + p_-$ subalgebra of $sp(3)^C$ (cf., eq. (11)) contracts to a Heisenberg-Weyl algebra hw(6) for a vibrator with six vibrational degrees of freedom and (boson) commutation relations

$$[a_\mu, a_\nu^\dagger] = \delta_{\mu\nu} \,. \tag{19}$$

It has also been shown [26; 27] that, as $2\lambda+\mu \to \infty$, the su(3) subalgebra of sp(3,R) contracts to the rotor algebra rot(3). Thus, in the limit when both N and $2\lambda+\mu$ are large, the root diagram for sp(3,R) separates and becomes the sum of rotor and vibrator parts as shown in Fig. 4.

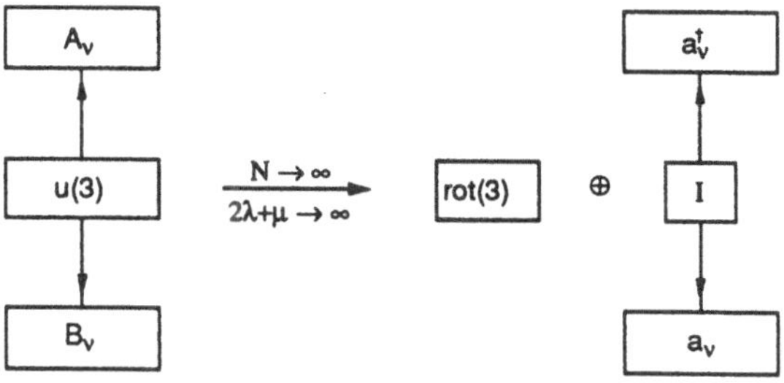

Fig. 4. The contraction, sp(3,R) $\to$ rot(3) + hw(6), of the sp(3,R) Lie algebra to a rotor-vibrator algebra

The contraction limits are, in fact, the leading terms in a Vector Coherent State expansion [14; 8] of the sp(3,R) Lie algebra. The VCS expansion can be carried to any order in practice and used to calculate the explicit matrix elements of the sp(3,R) irrep. The contraction limit of the sp(3,R) model has been called the CRV (Coupled-Rotor-Vibrator) model [24].

6. Some problems with the symplectic model

Two problems arise in the application of the symplectic model. The first is a strange one. It is that the model is more successful in fitting experimental data, than, at first sight, one thinks it ought to be. It is a problem because one knows there should be large sp(3,R) mixing interactions in a realistic microscopic Hamiltonian; e.g., large spin-orbit and pairing interactions. One might expect these interactions to mix irreps so strongly that the model would be rendered useless, especially in heavy nuclei where the mixing interactions are strongest. The second problem is that rotational bands are as prevalent in odd-mass nuclei as they are in even. However, odd nuclei have states of half-odd integer angular momentum as a consequence of the fact that the intrinsic spin of a nucleon is one half (in units of $\hbar$). This requires that one find a generalization of the symplectic model to incorporate the intrinsic spin degrees of freedom in a realistic manner.

The resolution of the these problems results in two new concepts in group representation theory which are of both practical and conceptual interest. They are the concepts of *embedded representations* [28] and *quasi-spinor representations* [29; 30].

7. Embedded representations

We first define an embedded representation, follwing ref. [28], and then discuss physical situations in which the concept is useful.

Let V be a module for a (possibly reducible) representation Γ of a group G and let $P : V \to U$ denote a projection of V to a subspace $U \subset V$. Then each $\Gamma(g)$ projects to a linear operator $M(g) = P\Gamma(g)P$ on U. The set of matrices $\{M(g); g \in G\}$ is not, in general, a representation of the group G. Exceptions occur when P projects V to a G-invariant subspace. The representation is then a subrepresentation. However, examples can be found in which M is a representation but U is not G-invariant. We then call M an *embedded representation*.

An embedded representation can also be defined in terms of matrices. Suppose that, in some basis, the matrices of the representation Γ are of the general form

$$\Gamma(g) = \begin{pmatrix} M(g) & * \\ * & * \end{pmatrix}, \tag{20}$$

where the off-diagonal blocks are all non-zero for at least some group element. If it should happen that the set of matrices $\{M(g); g \in G\}$ is a representation of G, then it is an embedded representation..

The simplest example of an embedded representation is for the group T of translations along the real line. Translation groups are abelian; therefore their irreps are one-dimensional. An irrep of T is given by

$$\Gamma_k(t) = e^{ikt}, \quad t \in T, \tag{21}$$

for some real value of k. Let ψ_k be a basis vector for the irrep Γ_k and let φ be a linear superposition of such vectors

$$\varphi = \int f(k)\psi_k \, dk \tag{22}$$

with

$$\int |f(k)|^2 dk = 1. \tag{23}$$

Let P be the projection operator

$$P\psi_k = f^*(k)\varphi, \tag{24}$$

so that $P\varphi = \varphi$. We then find that

$$P\Gamma(t)P\varphi = P\int f(k)e^{ikt}\psi_k\,dk = \int |f(k)|^2 e^{ikt}\,dk\,\varphi. \tag{25}$$

Thus, φ is a basis function for a one-dimensional embedded representation of T

$$P\Gamma(t)P = \int |f(k)|^2 e^{ikt}\,dk = \int |f(k)|^2 \Gamma_k(t)\,dk. \tag{26}$$

It can be seen that $P\Gamma P$ is the weighted average of the irreps of which it is composed. Thus, an embedded representation could also be called an *average* representation.

It has been shown [28] that the dynamical group for the rotor, ROT(3), likewise admits embedded representations which are similar in many respects to those of a translation group. An irrep of ROT(3) is characterized by sharply-defined (eigen)values of the intrinsic quadrupole moments. However, unless nuclei are truly rigid bodies, one should expect them to exhibit vibrational shape fluctuations, to some degree, and corresponding spreads in the values of their intrinsic quadrupole moments. In other words, a realistic intrinsic state should be a wave packet with a vibrational distribution of intrinsic quadrupole shapes; realistic rotor states do not belong to irreps but to linear combinations of irreps. Nevertheless, for sufficiently slow rotational motion (i.e., for large moments of inertia and small angular momentum), the Coriolis and centrifugal interactions are small and the intrinsic vibrational wave function may remain unperturbed as a nucleus rotates; the rotations and vibrations become decoupled. Rotations which are decoupled from the vibrational and other intrinsic degrees freedom by virtue of the fact that the rotational frequencies are slow, are said to be *adiabatically decoupled*. One sees that a band of adiabatic rotational states is characteristic of an embedded representation; i.e., the weighted average of the irreps of which it is composed. Thus, it appears that adiabaticity of the rotational motion is more important than the absence of representation-mixing interactions for the realization of decoupled rotational bands.

Fig. 5 gives a schematic illustration of the physical meaning of an embedded representation for a rotor. Fig. 5(a) indicates a state which is an arbitrary mixture of states from different rigid-rotor irreps. The figure is intended to convey the idea that there are no particular phase relations between the different components and that in different rotor states the linear combinations occur in unrelated ways. In contrast, Figs. 5(b) and 5(c) illustrate a situation in which a rotor state has contributions from an equal number of irreps but with amplitudes and phases that are related to one another and are the same in different states of a rotor band.

It may be noted that physical rotations would always be adiabatic and exhibit decoupled rotational bands, belonging to embedded representations of rot(3), if rotating frames were inertial frames. In reality, rotating frames are not inertial

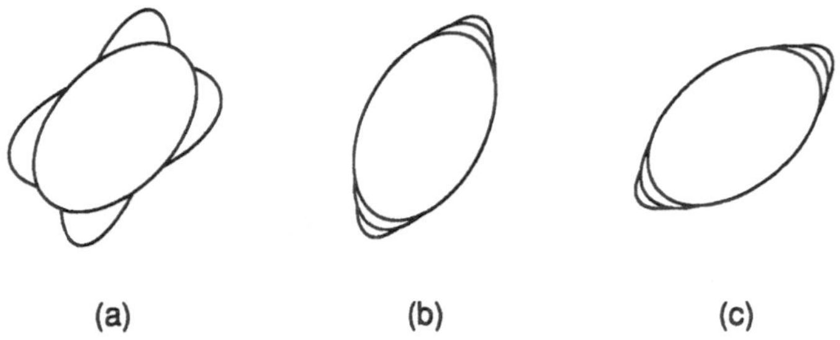

Fig. 5. Combinations of rigid-rotor irreps: (a) for an arbitrary mixture of irreps; (b) and (c) for different orientations of a state of an embedded representation.

but they become excellent approximations to inertial frames when the moments of inertia are large and the rotations are "slow" relative to other excitation modes. This is, of course, the only situation in which unperturbed rotational bands are, in any event, likely to be observed in any system. It should also be noted that, if one wishes, one may compute the mixing of different embedded representations to take into account the effects of Coriolis and centrifugal forces in higher order.

We now enquire if the symplectic model can similarly survive a huge mixing of its representations. The contraction limit, in which the sp(3,R) algebra reduces to a combination of a rotor and a vibrator algebra

$$\mathrm{sp}(3,\mathrm{R}) \rightarrow \mathrm{rot}(3) + \mathrm{hw}(6)\,, \tag{27}$$

shows immediately that it can. As pointed out by Von Neumann, all irreps of a Heisenberg-Weyl (vibrator) algebra are essentially equivalent. It is therefore an almost trivial operation to take superpositions of hw(6) irreps to form new irreps. Moreover, we have already found that the rot(3) algebra admits embedded representations. It follows that the sp(3,R) algebra likewise admits embedded representations in the contraction limit as N and $2\lambda + \mu$ become large. It turns out that the sp(3,R) representations relevant for heavy deformed nuclei have values of N and $2\lambda + \mu$ that are very large and the contraction limit of sp(3,R) is valid, to a high degree of accuracy, in application to such nuclei.

The reason for the existence of embedded representations of sp(3.R) for heavy nuclei is easy to understand. One simply observes, whenever one does a calculation, that the results become linear in the representation labels when the latter are large. Thus, if one takes the average of the results for different irreps, with similar highest weights, one will surely obtain the same result as would be obtained for an average value of the highest weight.

8. Quasi-spinor representations

Again we start by defining the concept of a quasi-spinor representation [30] and subsequently discuss the physical circumstances under which it is useful.

First recall that the group SO(3) has (double valued) spinor representations, i.e., representations in which the angular momentum takes half-odd integer values (in units of $\hbar$). Such representations are true representations of the double covering group SU(2). The group SU(3) has no covering group (other than itself); hence, it has no spinor representations. Thus, the states of an SU(3) irrep always have integer angular momentum values. Nevertheless, we shall show that SU(3) has quasi-spinor representations in a sense that will be defined.

Recall that, in the limit $2\lambda + \mu \to \infty$, an SU(3) irrep of highest weight (λ, μ) contracts to the dynamical group ROT(3) of the rotor model [26; 27]. The contraction limit means that, for large values of $2\lambda + \mu$, the states of angular momentum $L \ll 2\lambda + \mu$ have all the properties of corresponding states in a ROT(3) representation. In particular, they have ranges of angular momentum values and quadrupole matrix elements that are the same. However, we know that ROT(3) is isomorphic to the semi-direct product group $[\mathrm{R}^5]\mathrm{SO}(3)$ and that the latter has a double covering group $[\mathrm{R}^5]\mathrm{SU}(2)$. Thus, ROT(3) has spinor irreps with states of half-odd integer angular momentum. Moreover, such irreps feature in the rotor model description of the rotational bands of odd-mass nuclei [9; 10].

The construction of a spinor irrep of ROT(3) is straightforward. Let $\mathrm{SU}(2)_S$ denote a realization of the SU(2) group with angular momentum S, which takes half-odd integer values; think of S as an intrinsic spin. Thus, the infinitesimal generators of $\mathrm{SU}(2)_S$ are spin operators $\{S_k\}$. Let $[\mathrm{R}^5]\mathrm{SU}(2)_L$ denote a realization of $[\mathrm{R}^5]\mathrm{SU}(2)$ with integer-valued angular momentum states and infinitesimal generators $\{Q_\nu, L_k\}$. Then the direct product group $[\mathrm{R}^5]\mathrm{SU}(2)_L \times \mathrm{SU}(2)_S$ has a subgroup

$$[\mathrm{R}^5]\mathrm{SU}(2)_J \subset [\mathrm{R}^5]\mathrm{SU}(2)_L \times \mathrm{SU}(2)_S\,, \tag{28}$$

in which the angular momentum operators are the sums $J_k = L_k + S_k$, and whose irreps are rotor bands of half-odd integer angular momentum states. The reduction of a representation of the direct product $[\mathrm{R}^5]\mathrm{SU}(2)_L \times \mathrm{SU}(2)_S$ to its $[\mathrm{R}^5]\mathrm{SU}(2)_J$ subgroup is called, by nuclear physicists, *strong coupling*.

For example, if a so-called $K = 0$ axially symmetric irrep of $[\mathrm{R}^5]\mathrm{SU}(2)$ (with integer angular momentum values) is strongly-coupled to a spin 3/2 irrep of SU(2), then the resulting states span two spinor irreps of $[\mathrm{R}^5]\mathrm{SU}(2)$ with $K = 1/2$ and $K = 3/2$, respectively.

Now we can construct a direct product group $\mathrm{SU}(3)_L \times \mathrm{SU}(2)_S$ which has the group $[\mathrm{R}^5]\mathrm{SO}(3)_L \times \mathrm{SU}(2)_S$ as its contraction limit. However, although there is no SU(3) analog to $[\mathrm{R}^5]\mathrm{SU}(2)_J$, we can nevertheless strongly couple $\mathrm{SU}(3)_L$ and $\mathrm{SU}(2)_S$ states, as for the rotor. When we do this, we obtain states which look very much like the states of $\mathrm{SU}(3)_J$ spinor irreps [30].

For example, an SU(3) irrep (λ, μ), with λ even and $\mu = 0$, contracts in the limit as $\lambda \to \infty$ to a $K = 0$ axially symmetric irrep of $[\mathrm{R}^5]\mathrm{SO}(3)$. If we strongly couple the states of such an SU(3) irrep (for a finite value of λ) with spin 3/2 states, then the result is two sets of states which have almost zero matrix elements between them for values of $J \ll \lambda$. To illustrate how remarkably close the results are to irreps, we show in Fig. 6 some reduced quadrupole matrix elements between the states obtained by strongly-coupling the $L = 0, 2, 4, \ldots, 10$ states of a (10,0) SU(3) irrep to spin $S = 3/2$. Numerically computed matrix elements [30] are shown on the figure and beside them (in parentheses) the corresponding matrix elements obtained by analytically continuing the expressions for SU(3) irreps to spinor irreps with half-odd integer angular momentum states. It can be seen that, in spite of a relatively small value of λ (the values occuring in the description of rotations of heavy deformed nuclei are more like $\lambda \approx 100$) the two sets of numbers are remarkably similar. In particular, the matrix elements between states of the two quasi-spinor irreps are almost zero.

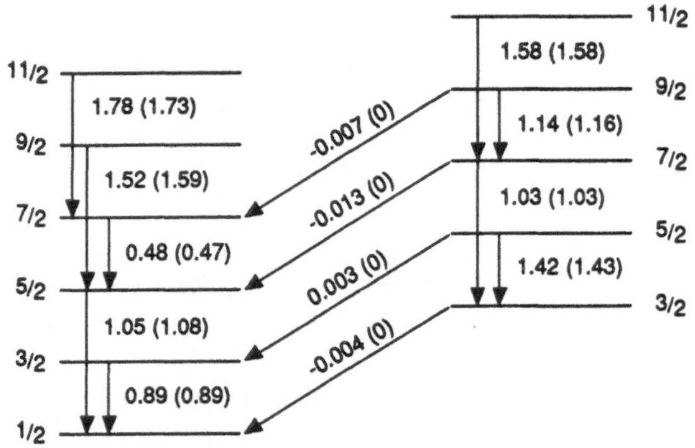

Fig. 6. The $j \leq 11/2$ states of two quasi-spinor irreps obtained by strongly coupling the states of a (10,0) SU(3) irrep to spin 3/2. The numbers attached to the arrows are the values of numerically computed quadrupole matrix elements. The numbers in parentheses are the corresponding values given by regarding the sets of states as comprising two quasi-spinor irreps.

It should be understood that one can strongly couple a spin of any value (integer or half-odd integer) to an SU(3) irrep to construct quasi-SU(3) irreps; the only requirement is that the value of the spin should be small in comparison to $2\lambda + \mu$. This makes it possible to bring intrinsic spin into the Sp(3,R) $\supset$ SU(3) description of collective states of both even and odd-mass nuclei.

9. Summary: The microscopic description of rotational nuclei with intrinsic spin

The symplectic algebra sp(3,R) contains the essential physical observables (angular momentum, quadrupole moments, etc.) by which one interprets a given sequence of experimentally observed states to be a rotational band. Thus, it is hard to avoid the inference that nuclear rotational motion is basically some kind of symplectic dynamics.

Symplectic dynamics include many possible rotational flows, ranging from irrotational to rigid-body flow. The flows appropriate for a nucleus, made up of interacting neutrons and protons, can be inferred from the fact that the sp(3,R) Lie algebra has a microscopic shell model realization. As a consequence, one can restrict its irreps to those carried by subspaces of the shell model Hilbert space of the nucleus under investigation. Shell model considerations then lead one to interpret rotations as arising, in the first instance, from the su(3) dynamics of the shell model renormalized by coupling to the high-energy (giant monopole and quadrupole) vibrational excitations contained within an sp(3,R) irrep (cf., ref. [8] for a review).

In this paper, we have reviewed the solution to two problems that arise in this interpretation of nuclear rotations: (i) that it does not take account of the large mixing of sp(3,R) irreps that must surely exist, and (ii) that it does not take into account the intrinsic spins of the nucleons.

In early applications of the symplectic model, the intrinsic spins were simply ignored. This would be reasonable, for an even nucleus, if all the nucleon spins were coupled to zero in lowlying states. It was then shown that, for adiabatic rotations, the dominant effect of allowing different ($S = 0$) sp(3,R) irreps to mix is rotational bands which have the average properties of the participating sp(3,R) irreps. This result is an expression of the fact that the shape vibrations of a rotating nucleus become decoupled from the rotational motions if the latter are slow enough. It can be understood by noting that rotational invariance of the Hamiltonian would always be sufficient to ensure a decoupling of rotations and intrinsic motions if it were not for the centrifugal and Coriolis interactions. Indeed, if a rotating frame were an inertial frame, rotations would be decoupled from all other degrees of freedom in the same way that centre-of-mass motions are decoupled.

This argument runs into problems when the intrinsic spins of the nucleons are allowed to participate in the dynamics. This is because, the conserved angular momentum is then the total angular momentum and not just its orbital angular momentum component. Thus, to exploit the rotational invariance of the nuclear Hamiltonian and the concept of adiabatic mixing of irreps it was necessary to replace the orbital angular momentum, L, in the symplectic model, with the total angular momentum, J. Specifically, we learned how to strongly couple a non-zero spin S to the states of an sp(3,R) irrep to form new quasi sp(3,R) irreps having all the essential properties of standard irreps.

We conclude by giving the results of a numerical calculation, which illustrates the utility of the above constructions. First, we remark that the kinds of sp(3,R) irreps relevant for the description of rotational bands in heavy (rare-earth) nuclei are ones with $N \sim 800$, $\lambda \sim 100$ and μ in the range 0-15. Consider, for example, an $SU(3)_L \times SU(2)_S$ irrep with $(\lambda, \mu) = (100, 0)$ and $S = 0$. Among the irreps which mix with this irrep, as a result of a spin-orbit interaction, will be one with $(\lambda, \mu) = (101, 1)$, ($S = 1$). The lowest L and J values for such a representation are shown in Fig, 7(a). Now, by diagonalizing a Hamiltonian of the type

$$H = H_{\mathrm{u}(3)} + V_{\mathrm{spin-orbit}}\,, \tag{29}$$

within the span of the $(\lambda, \mu) = (101, 1)$, $(S = 1)$ representation, one can obtain an energy spectrum of the type shown in Fig. 7(b). One finds that, provided the spin-orbit interaction is strong enough, the states fall into bands with extremely small quadrupole matrix elements between the states of different bands. Moreover, it is found (numerically) that the states obtained by diagonalizing the above Hamiltonian are the same as those obtained by strong coupling the spin, $(S = 1)$, to the $(\lambda, \mu) = (101, 1)$ $\mathrm{su}(3)_L$ states. The remarkable fact is that the states obtained are virtually indistinguishable (to within small fractions of a percent), from those of three $\mathrm{su}(3)_J$ irreps with (λ, μ) equal to (100,2), (102,1) and (101,0), respectively. In fact, if one observed these states experimentally, one could not distinguish them from $\mathrm{SU}(3)_L \times \mathrm{SU}(2)_S$ irreps with (λ, μ) equal to (100,2), (102,1) and (101,0) and $(S = 0)$.

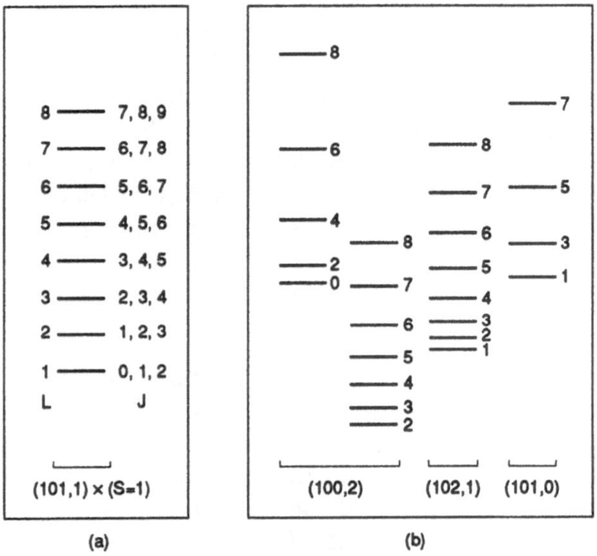

Fig. 7. The states of a $(\lambda, \mu) = (101, 1)$ su(3) irrep with intrinsic spin $S = 1$: (a) shows the range of values of the total angular momentum J for the lowest L values; (b) shows the energy spectrum and the way the states fall into strongly-coupled bands when a Hamiltonian with a strong spin-orbit interaction is diagonalized.

When we diagonalize the Hamiltonian (29) within the span of both the $(\lambda, \mu) = (100, 0)$, $(S = 0)$ and $(\lambda, \mu) = (101, 1)$, $(S = 1)$ representations, we obtain results as shown in Fig. 8. Because we used a strong spin-orbit interaction, the (100,0) and (100,2) states are strongly mixed. Nevertheless, the results are indistinguishable from those of four $\mathrm{su}(3)_J$ irreps with (λ, μ) equal to (100,0), (100,2), (102,1) and

(101,0), respectively. Detailed results for mixing the states of (12,0) $S = 0$ and (10,0) $S = 1$ SU(3) representations can be found in ref. [29].

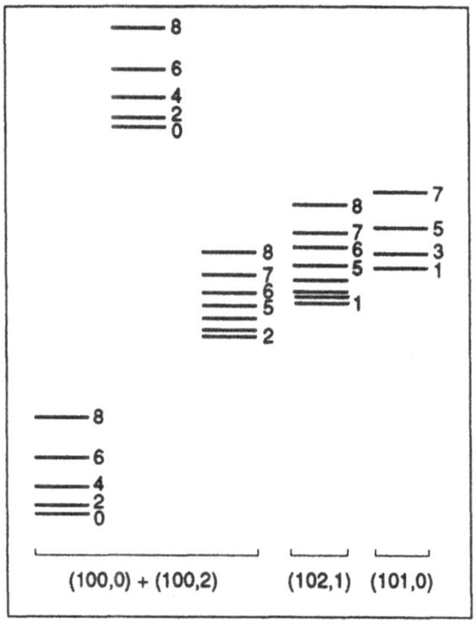

Fig. 8. The energy spectrum resulting from diagonalizing a Hamiltonian with a strong spin-orbit interaction in the space of $(\lambda,\mu) = (100,0)$, $(S = 0)$ and $(\lambda,\mu) = (101,1)$, $(S = 1)$ representations.

Although we do not have the space to pursue it here, it is worth mentioning that, when pairing and other representation-mixing interactions are also taken into account, the rotational structure of the bands continues to survive, provided the rotations are adiabatic. However, one can have significant spreads of λ and μ values contributing to the resulting rotational bands which can be interpreted in terms of so-called beta and gamma vibrations [27].

References

[1] S. Tomonaga, *Prog. Theor. Phys.* **13** (1955) 467, 482.
[2] R.Y. Cusson, *Nucl. Phys.* **A114** (1968) 289.
[3] G.N. Afanas'ev, E.N. Mikhailov and P.R. Raychev, *Yad. Fiz.* **14** (1971) 734 (*Sov. J. Nucl. Phys.* **14** (1971) 413);
G.N. Afanas'ev and P.R. Raychev, *Fiz. Elem. Chastits. At. Yadra*, **3** (1972) 436 (*Sov. Part. Phys.* **3** 229).
[4] R.M. Asherova *et al.*, *Yad. Fiz.* **21** (1975) 1126 (*Sov. J. Nucl. Phys.* **21** (975) 580).
[5] F. Arickx, P. Van Leuven and M. Bouten, *Nucl. Phys.* **A252** (1975) 416.
[6] G. Rosensteel and D.J. Rowe, *Phys. Rev. Lett.* **38** (1977) 10.
[7] G. Rosensteel and D.J. Rowe, *Ann. Phys., N.Y.*, **126** (1980) 343.
[8] D.J. Rowe, *Rep. Prog. in Phys.* **48** (1985) 1419.
[9] H. Ui, *Prog. Theor. Phys.* **44** (1970) 153.
[10] L. Weaver, L. C. Biedenharn and R. Y. Cusson, *Ann. Phys., NY* **77** (1973) 250.

[11] G. Rosensteel, *J. Math. Phys.* **21** (1980) 924.
[12] G. Rosensteel and D.J. Rowe, *J. Math. Phys.* **24** (1983) 2461.
[13] Harish-chandra, *Am. J. Math.* **78** 564 (1956) 1;
R. Godement, *Seminaire Cartan.* (École Normale Superieure, Paris, 1958).
[14] D.J. Rowe, *J. Math. Phys.* **25** (1984) 2662.
[15] D.J. Rowe, G. Rosensteel and R. Carr, *J. Phys. A: Math. Gen.* **17** (1984) L399.
[16] D.J. Rowe and J. Repka, *J. Math. Phys.* **32** (1991) 2614.
[17] P. Park *et al*, *Nucl. Phys.* **A414** (1984) 93.
[18] G. F. Filippov, L. L. Chopovsky and V. S. Vasilevsky, *Nucl. Phys.* **A388** (1982) 47.
[19] M. G. Vassanji and D. J. Rowe, *Phys. Lett.* **127B** (1983) 1; *Nucl. Phys.* **A426** (1984) 205.
[20] K. T. Hecht and J. P. Draayer, private communication;
E. J. Reske, *Ph.D. thesis*. (University of Michigan, 1984);
[21] A. L. Blokhin,*Collective Nuclear Potential within the Microscopic Sp(6,R) Model,* Kiev preprint ITP-91-74E (1992).
[22] J.P. Draayer, K.J. Weeks and G. Rosensteel, *Nucl. Phys.* **A413** (1984) 215.
[23] R. Le Blanc, J. Carvalho and D.J. Rowe, *Phys. Lett.* **B140** (1984) 155;
R. Le Blanc, J. Carvalho, M. G. Vassanji and D.J. Rowe, *Nucl. Phys.* **A452** (1986) 263;
D.J. Rowe, P. Rochford and R. Le Blanc, *Nucl. Phys.* **A464** (1987) 39.
[24] D.J. Rowe, M.G. Vassanji and J. Carvalho, *Nucl. Phys.* **A504** (1989) 76.
[25] G. Rosensteel and D. J. Rowe, *Phys. Rev. Lett.* **47** (1981) 223;
G. Rosensteel and D. J. Rowe, *Phys. Rev.* **C25** (1982) 3236.
[26] J. Carvalho, *Ph.D. thesis* (University of Toronto,1984).
[27] J. Carvalho, R. Le Blanc, M. Vassanji, D.J. Rowe and J.B. McGrory, Nucl. Phys. **A452** (1986) 240.
[28] D.J. Rowe, P. Rochford and J. Repka, *J. Math. Phys.* **29** (1988) 572.
[29] P. Rochford and D.J. Rowe, *Phys. Lett.* **B210** (1988) 5.
[30] D. J. Rowe and H. de Guise, *Nucl. Phys.* **A542** (1992) 173-194.

Department of Physics
University of Toronto
Toronto
Ont. M5S 1A7
Canada

NILPOTENT GROUPS AND ANHARMONIC OSCILLATORS

W. H. KLINK

ABSTRACT. Representations of nilpotent groups relate different quantum systems having polynomial interactions. In this paper the nilpotent group associated with the quartic anharmonic oscillator is analyzed in detail and the relationship between the quartic anharmonic oscillator Hamiltonian and irreducible representations of Lie algebra elements of the nilpotent group is given. Scaling operators are used to partially determine the functional form of the eigenvalues. The Hamiltonian for a particle in a nonconstant magnetic field, as well as a "heat" Hamiltonian, are shown to be related to reducible representations of the nilpotent group. Generalizations to other nilpotent groups for which there are scaling operators are also given.

1. Introduction

That there is a connection between the harmonic oscillator and the simplest nilpotent group, namely the Heisenberg group, is well known. It is less well known that every anharmonic oscillator is related to some nilpotent group [1]. The purpose of this paper is to exploit this relationship for the simplest anharmonic oscillator, namely the quartic anharmonic oscillator. We will show that the quartic anharmonic oscillator Hamiltonian is related to a certain nonconstant magnetic field Hamiltonian, as well as other quantum systems of interest, and that if the eigenfunctions and eigenvalues of one system are known, the eigenfunctions and eigenvalues for all the related systems can be given. In particular, we will use scaling operators to partially determine the functional form of the eigenvalues.

For the harmonic oscillator the connection to the Heisenberg group can be given in the following way. Let G be the nilpotent (Heisenberg) group with elements

$$\left\{\begin{pmatrix} 1 & a & c \\ & 1 & b \\ & & 1 \end{pmatrix}\right\},$$

$a, b, c \in \mathbf{R}$ and irreducible representations on $L^2(\mathbf{R})$ given by

$$(U^{\beta\gamma}_{(a,b,c)}\varphi)(x) = \exp i(\beta b + \gamma x b + \gamma c)\varphi(x+a)\ , \quad \varphi \in L^2(\mathbf{R})\ .$$

E. A. Tanner and R. Wilson (eds.), Noncompact Lie Groups and Some of Their Applications, 301–313.

Then the (antihermitian) Lie algebra elements corresponding to one-parameter subgroups are

$$\begin{aligned} (a,0,0) &\to A = \frac{\partial}{\partial x} \\ (0,b,0) &\to B = i(\beta + \gamma x) , \quad \beta, \gamma \in \mathbf{R} \\ (0,0,c) &\to C = i\gamma , \end{aligned}$$

with commutation relation

$$[A, B] = C ,$$

and all other commutators zero.

The harmonic oscillator Hamiltonian is a quadratic polynomial in Heisenberg Lie algebra elements:

$$\begin{aligned} -2H^{\beta\gamma} &= A^2 + B^2 \\ &= \frac{d^2}{dx^2} - (\beta + \gamma x)^2 . \end{aligned}$$

Similarly, for a particle in a constant magnetic field, consider the (reducible) representation of G induced by the subgroup $(0,0,c) \to e^{i\gamma}$, γ real. In this case the representation space is $L^2(\mathbf{R}^2)$ and the group action is

$$(U^{\gamma}_{(a,b,c)}\psi)(x,y) = e^{i\gamma(bx+c)}\psi(x+a, y+b) , \quad \psi \in L^2(\mathbf{R}^2)$$

with Lie algebra elements

$$A = \frac{\partial}{\partial x} , \quad B = \frac{\partial}{\partial y} + i\gamma x , \quad C = i\gamma$$

and Hamiltonian

$$\begin{aligned} -2H^{\gamma} &= A^2 + B^2 \\ &= \frac{\partial^2}{\partial x^2} + \left(\frac{\partial}{\partial y} + i\gamma x\right)^2 . \end{aligned}$$

If a z direction is added this gives

$$2H^{\gamma} = -\frac{\partial^2}{\partial x^2} + \left(\frac{1}{i}\frac{\partial}{\partial y} + \gamma x\right)^2 - \frac{\partial^2}{\partial z^2}$$

which is the Hamiltonian for a particle in a constant magnetic field with (dimensionless) strength γ. If H^{γ} is Fourier transformed in y, the harmonic oscillator Hamiltonian $H^{\beta\gamma}$ results, a property exploited by Landau [2] to get the eigenfunctions and eigenvalues of H^{γ} in terms of harmonic oscillator eigenfunctions and eigenvalues. Group theoretically the Fourier transform decomposes the reducible representation U^{γ}_g into a direct integral of irreducible representations $U^{\beta\gamma}_g$.

Finally, the regular representation of G can be used to define a heat kernel satisfying the heat equation on G. Define the right regular representation of G on $L^2(G)$ as

$$(R_g F)(x) := F(xg)$$
$$\big(R_{(a,b,c)}F\big)(x,y,z) = F(x+a, y+b, z+c+xb) , \quad F \in L^2(G) .$$

The Lie algebra elements corresponding to this representation are

$$A = \frac{\partial}{\partial x} , \quad B = \frac{\partial}{\partial y} + x\frac{\partial}{\partial z} , \quad C = \frac{\partial}{\partial z} ;$$

now the "Hamiltonian" is called the sublaplacian and is

$$\Delta = A^2 + B^2$$
$$= \left(\frac{\partial^2}{\partial x^2}\right) + \left(\frac{\partial}{\partial y} + x\frac{\partial}{\partial z}\right)^2 .$$

A great deal is known about sublaplacians of nilpotent groups [3], and in fact the (generalized) eigenfunctions of Δ are obtained by Fourier transforming in both y and z, which results in the harmonic oscillator Hamiltonian. Using this fact makes it possible to solve the heat equation

$$\Delta p_t = \frac{\partial p_t}{\partial t} , \quad p_{t=0} = \delta^3(\vec{x}) ,$$

as was first shown by Hulanicki [4]. It is also possible to solve the heat equation directly as shown in Ref. [5].

In this paper, we will generalize these ideas to a more complicated nilpotent group, dubbed the quartic group Q because its irreducible representations generate the Hamiltonian of the quartic anharmonic oscillator. Q will be analyzed in detail in section 2, where it is shown that certain reducible representations of Q generate Hamiltonians of particles in nonconstant magnetic and electric fields that are related to the quartic anharmonic oscillator in a similar way as constant magnetic field Hamiltonians are related to harmonic oscillators. In section 3 these ideas are generalized to arbitrary nilpotent groups for which there is a scaling operator.

2. The Quartic Anharmonic Oscillator and Its Associated Nilpotent Group

Let Q be the (quartic) nilpotent group defined by

$$Q := \left\{ \begin{pmatrix} 1 & b & b^2/2 & b_3 \\ & 1 & b & b_2 \\ & & 1 & b_1 \\ & & & 1 \end{pmatrix} \right\} = \{(b,\vec{b})\} , \quad b, b_i \in \mathbf{R} \tag{2.1}$$

with group operations given by

$$(b,\vec{b})(b',\vec{b}') = \left(b+b', b_1+b_1', b_2+b_2'+bb_1', b_3+b_3'+bb_2'+\frac{b^2}{2}b_1'\right)$$

$$(b,\vec{b})^{-1} = \left(-b, -b_1, -b_2+bb_1, -b_3+bb_2-\frac{b^2}{2}b_1\right). \tag{2.2}$$

The Heisenberg group is a subgroup of Q as can be seen by setting $b_1 = 0$.

The irreducible representations of Q are obtained by inducing from the invariant Abelian subgroup

$$(0,\vec{b}) \to \pi^{\vec{\beta}}(\vec{b}) = e^{i\vec{\beta}\cdot\vec{b}}\,; \tag{2.3}$$

then

$$(U^{\vec{\beta}}_{(b,\vec{b})}\varphi)(x) = \exp\left\{i\left[\beta_1 b_1 + \beta_2(b_2+xb_1) + \beta_3\left(b_3+xb_2+\left(\frac{x^2}{2}\right)b_1\right)\right]\right\}\varphi(x+b)$$

$$(b,\vec{b})\in Q\,,\quad \varphi\in L^2(\mathbf{R})\,. \tag{2.4}$$

From these irreducible representations on $L^2(\mathbf{R})$ it is possible to compute the (antihermitian) infinitesimal operators corresponding to one-parameter subgroups of Q:

$$(b,\vec{0}) \to X_0 = \frac{\partial}{\partial x}$$

$$(0,b_1 00) \to X_1 = i\left(\beta_1+\beta_2 x+\beta_3\frac{x^2}{2}\right)$$

$$(0,0b_2 0) \to X_2 = i(\beta_2+\beta_3 x)$$

$$(0,00b_3) \to X_3 = i(\beta_3)\,. \tag{2.5}$$

The commutation relations are

$$[X_0,X_1] = X_2\,,\quad [X_0,X_2] = X_3 \tag{2.6}$$

with all other commutators zero; these commutation relations agree with those coming from the Lie algebra of Q, as is easily checked by making use of the matrix realization of Q, Eq. (2.1).

The relationship between Q and quartic anharmonic oscillators is given by writing

$$\begin{aligned}-2H_c^{\vec{\beta}} &:= X_0^2 + X_1^2 + icX_2\\ &= \frac{\partial^2}{\partial x^2} - \left(\beta_1+\beta_2 x+\frac{\beta_3}{2}x^2\right)^2 - c(\beta_2+\beta_3 x)\end{aligned} \tag{2.7}$$

where c is a real constant. In terms of generators of Q, $H_c^{\vec{\beta}}$ is the most general quartic Hamiltonian having certain scaling properties which will be discussed later in this section.

The eigenvalue-eigenvector problem to be solved is

$$2H_c^{\vec\beta}\varphi_n^{\vec\beta}(x) = E_n(\vec\beta)\varphi_n^{\vec\beta}(x)\,, \quad \varphi_n^{\vec\beta} \in L^2(\mathbf{R}) \tag{2.8}$$

and some information of the dependence of the $n^{\underline{\text{th}}}$ energy eigenvalue E_n on $\vec\beta$ can be obtained by looking at orbits, which give equivalent representations:

$$\pi^{\vec\beta}(a\cdot\vec b\cdot a^{-1}) = \pi^{\vec\beta_a}(\vec b)\,, \qquad a := (a,\vec 0\,)$$

$$\vec\beta_a = \begin{pmatrix} \beta_1 + a\beta_2 + \frac{a^2}{2}\beta_3 \\ \beta_2 + a\beta_3 \\ \beta_3 \end{pmatrix}\,, \quad \vec b := (0,\vec b\,)\,, \tag{2.9}$$

where $\pi^{\vec\beta}(\vec b\,) = e^{i\vec\beta\cdot\vec b}$ is the one-dimensional representation of the invariant subgroup $\{(0,\vec b\,)\}$. From the commutation relations, Eq. (2.6), it is clear that the two Casimir operators X_3 and $C := X_1X_3 - {}^1\!/\!_2\,X_2^2$ commute with all Lie algebra elements and hence their eigenvalues can be used to label irreducible representations of Q, namely

$$\begin{aligned} X_3 &\to \beta_3 \\ C &\to \beta_1\beta_3 - \tfrac{1}{2}\beta_2^2\,. \end{aligned} \tag{2.10}$$

If β_2 is set equal to zero,

$$U_aU_{(b,\vec b\,)}^{\beta_1 0 \beta_3}U_a^{-1} = U_{(b,\vec b\,)}^{\beta_1+\frac{a^2}{2}\beta_3, a\beta_3, \beta_3} \tag{2.11}$$

which means that all representations $(\vec\beta\,)$ with the same value of $\beta_3, \beta_1\beta_3 - {}^1\!/\!_2\,\beta_2^2$ are equivalent.

Applied to the eigenfunctions $\varphi_n^{\vec\beta}(x)$, this gives

$$\begin{aligned} U_a 2H_c^{\vec\beta}\varphi_n^{\vec\beta} &= E_n(\vec\beta\,)U_a\varphi_n^{\vec\beta} \\ U_aH_c^{\vec\beta}U_a^{-1} &= H_c^{\vec\beta_a} \\ 2H_c^{\vec\beta_a}\varphi_n^{\vec\beta_a} &= E_n(\vec\beta_a)\varphi_n^{\vec\beta_a}\,, \end{aligned} \tag{2.12}$$

with $H_c^{\vec\beta}$ defined in Eq. (2.7); thus $U_a\varphi_n^{\vec\beta} = \varphi_n^{\vec\beta_a}$ and $E_n(\vec\beta_a) = E_n(\vec\beta\,)$, where $\vec\beta_a$ is given in Eq. (2.9). If a is made infinitesimally small, $E_n(\vec\beta_a) = E_n(\vec\beta\,)$ gives a functional equation whose solution is

$$E_n(\vec\beta) = E_n(\beta_3, \beta_1\beta_3 - \tfrac{1}{2}\beta_2^2)\,, \tag{2.13}$$

as expected, since β_3 and $\beta_1\beta_3 - {}^1\!/\!_2\,\beta_2^2$ are the eigenvalues of the Casimir invariants.

More properties of $E_n(\vec\beta\,)$ can be obtained from the scaling operator

$$(V_t\varphi)(x) := \sqrt{t}\,\varphi(tx)\,, \quad t > 0\,; \tag{2.14}$$

the factor $\sqrt{t}$ makes V_t a unitary operator on $L^2(\mathbf{R})$. Applying V_t to both sides of Eq. (2.8) gives

$$2H_c^{\vec{\beta}_t} V_t \varphi_n^{\vec{\beta}} = t^2 E_n(\vec{\beta}) V_t \varphi_n^{\vec{\beta}} \tag{2.15}$$

where $\vec{\beta}_t := (t\beta_1, t^2\beta_2, t^3\beta_3)$. As a consequence of Eq. (2.15), $\varphi_n^{\vec{\beta}_t} = V_t\varphi_n^{\vec{\beta}}$ and

$$E_n(\vec{\beta}_t) = t^2 E_n(\vec{\beta})\,. \tag{2.16}$$

Differentiating Eq. (2.16) with respect to t, evaluating at $t = 1$ and using Eq. (2.13) gives a functional equation whose solution is given by

$$E_n(\vec{\beta}) = (\beta_3)^{2/3} e_n \left[\frac{\left(\beta_1\beta_3 - \frac{1}{2}\beta_2^2\right)^3}{\beta_3^4} \right], \tag{2.17}$$

where $e_n(\alpha)$ is a real function to be determined. Thus by using scaling and translation invariance, the energy eigenvalues $E_n(\vec{\beta})$ are seen to depend on one function $e_n(\alpha)$, where $\alpha = (\beta_1\beta_3 - {}^1\!/_2\,\beta_2^2)^3/\beta_3^4$.

To conclude this section we want to show that Q is not only related to quartic anharmonic oscillators but also to heat equations and Hamiltonians for charged particles in curved electromagnetic fields. Consider first the right regular representation of Q,

$$\begin{aligned}\left(R_{(b,\vec{b})}F\right)(x,\vec{x}) &:= F\left((x,\vec{x})(b,\vec{b})\right)\\ &= F\left(x+b, x_1+b_1, x_2+b_2+xb_1, x_3+b_3+xb_2+\frac{x^2}{2}b_1\right)\end{aligned} \tag{2.18}$$

with $F \in L^2(Q)$. Representations of the Lie algebra of Q on $L^2(Q)$ are obtained with the same one-parameter subgroups as in Eq. (2.5):

$$\begin{aligned} X_0 &= \frac{\partial}{\partial x}\\ X_1 &= \frac{\partial}{\partial x_1} + x\frac{\partial}{\partial x_2} + \frac{x^2}{2}\frac{\partial}{\partial x_3}\\ X_2 &= \frac{\partial}{\partial x_2} + x\frac{\partial}{\partial x_3}\\ X_3 &= \frac{\partial}{\partial x_3} \end{aligned} \tag{2.19}$$

and from these operators a Hamiltonian on $L^2(Q)$ is given as before:

$$\begin{aligned}-2H_c &= X_0^2 + X_1^2 + icX_2\\ &= \frac{\partial^2}{\partial x^2} + \left(\frac{\partial}{\partial x_1} + x\frac{\partial}{\partial x_2} + \frac{x^2}{2}\frac{\partial}{\partial x_3}\right)^2 + ic\left(\frac{\partial}{\partial x_2} + x\frac{\partial}{\partial x_3}\right)\,. \end{aligned} \tag{2.20}$$

If $c = 0$, H_c is called a sublaplacian; sublaplacians on nilpotent groups have been extensively studied in the mathematics literature [3]. They are a generalization of ordinary Laplacians which are related to the translation group. Of particular interest is the heat equation

$$H_0 p_t(x,\vec{x}) = \frac{\partial p_t}{\partial t} , \quad p_{t=0}(x,\vec{x}) = \delta(x)\delta^3(\vec{x}) , \tag{2.21}$$

which can be solved if the (generalized) eigenfunctions

$$H_0 F_\lambda = \lambda F_\lambda \tag{2.22}$$

are known. The spectrum of H_0 is $-\mathbf{R}^+$, as can be seen by defining the scaling operator on $L^2(Q)$,

$$(V_t F)(x,\vec{x}) = t^{7/2} F(tx, tx_1, t^2 x_2, t^3 x_3) , \tag{2.23}$$

where the factor $t^{7/2}$ makes V_t unitary. Then

$$V_t H_0 V_t^{-1} = t^2 H_0 , \quad t > 0 , \tag{2.24}$$

which along with the antihermitian properties of X_0 and X_1 implies that H_0 has a negative real spectrum.

The eigenfunctions F_λ are intimately related to the eigenfunctions $\varphi_n^{\vec{\beta}}$ of the quartic anharmonic oscillator. This is most easily seen by Fourier transforming in $\vec{x}$, which in effect decomposes the right regular representation of Q into a direct integral of irreducible representations:

$$\begin{aligned} (\mathcal{F}F_\lambda)(x,\vec{\beta}) &:= \frac{1}{(2\pi)^{3/2}} \int d^3x\, e^{-i\vec{\beta}\cdot\vec{x}} F_\lambda(x,\vec{x}) \\ \mathcal{F}(-2H_c)\mathcal{F}^{-1} &= \frac{\partial^2}{\partial x^2} + \left(i\left(\beta_1 + \beta_2 x + \beta_3 \frac{x^2}{2}\right)\right)^2 + ici(\beta_2 + \beta_3 x) \\ &= -2H_c^{\vec{\beta}} \end{aligned} \tag{2.25}$$

which is the anharmonic oscillator Hamiltonian, Eq. (2.7). Thus, the eigenfunctions $F_\lambda(x,\vec{x})$ can be written as

$$F_{\lambda n}(x,\vec{x}) = \frac{1}{(2\pi)^{3/2}} \int d^3\beta\, e^{i\vec{\beta}\cdot\vec{x}} \delta(\lambda - E_n(\vec{\beta}))\varphi_n^{\vec{\beta}}(x) , \tag{2.26}$$

with $E_n(\vec{\beta})$ given by Eq. (2.17).

New Hamiltonians can be generated by looking at different representations of Q, induced by subgroups other than $\{(0,\vec{b})\}$ (anharmonic oscillator) or the identity (regular representation, heat equation). In particular, if the inducing subgroup is chosen to be

$$(0,0,b_2 b_3) \to e^{i(\beta_2 b_2 + \beta_3 b_3)} , \tag{2.27}$$

the induced representation is

$$(U^{\beta_2\beta_3}_{(b,\vec{b})}\psi)(x,y) = \exp\left\{i\left[\beta_2(b_2+xb_1)+\beta_3\left(b_3+xb_2+\frac{x^2}{2}b_1\right)\right]\right\}$$
$$\times\,\psi(x+b,y+b_1)\,,\quad \psi\in L^2(\mathbf{R}^2) \tag{2.28}$$

and produces operators

$$X_0 = \frac{\partial}{\partial x}$$
$$X_1 = i\left(\beta_2 x+\beta_3\frac{x^2}{2}\right)+\frac{\partial}{\partial y}$$
$$X_2 = i(\beta_2+x\beta_3)$$
$$X_3 = i\beta_3 \tag{2.29}$$

with the related Hamiltonian

$$-2H_c^{\beta_2\beta_3} = X_0^2+X_1^2+ic\,X_2$$
$$= \frac{\partial^2}{\partial x^2}+\left[i\left(\beta_2 x+\beta_3\frac{x^2}{2}\right)+\frac{\partial}{\partial y}\right]^2-c\,(\beta_2+x\beta_3)\,. \tag{2.30}$$

If a "z" direction is added, this becomes

$$2H_c^{\beta_2\beta_3} = -\frac{\partial^2}{\partial x^2}+\left(\frac{1}{i}\frac{\partial}{\partial y}+\left(\beta_2 x+\beta_3\frac{x^2}{2}\right)\right)^2-\frac{\partial^2}{\partial z^2}+c\,(\beta_2+x\beta_3)\,, \tag{2.31}$$

the Hamiltonian for a charged particle in a nonconstant magnetic field given by $A_x = A_z = 0$, $A_y = \beta_2 x+\beta_3(x^2/2)$,

$$B_z = \beta_2+\beta_3 x \tag{2.32}$$

and electrostatic potential $\Phi = c\,(\beta_2+x\beta_3)$, that is a constant electric field in the x direction.

To get the eigenfunctions $\psi_E^{\beta_2\beta_3}(x,y)$ of the Hamiltonian (2.30),

$$2H_c^{\beta_2\beta_3}\psi_E^{\beta_2\beta_3}(x,y) = E\psi_E^{\beta_2\beta_3}(x,y)\,, \tag{2.33}$$

we Fourier transform in the y variable,

$$(\mathcal{F}\psi)(x,\beta_1) = \frac{1}{\sqrt{2\pi}}\int dy\,e^{-i\beta_1 y}\psi(x,y) \tag{2.34}$$

so that

$$\mathcal{F}(-2H_c^{\beta_2\beta_3})\mathcal{F}^{-1} = \frac{\partial^2}{\partial x^2}-\left(\beta_1+\beta_2 x+\beta_3\frac{x^2}{2}\right)^2-c\,(\beta_2+\beta_3 x)$$
$$= -2H_c^{\beta_1\beta_2\beta_3}\,, \tag{2.35}$$

the Hamiltonian for the quartic anharmonic oscillator, Eq. (2.7). Then the eigenfunctions are

$$\psi_{E,n}^{\beta_2\beta_3}(x,y) = \frac{1}{\sqrt{2\pi}} \int d\beta_1\, e^{i\beta_1 y}\delta(E - E_n(\vec{\beta}))\varphi_n^{\vec{\beta}}(x)\ , \tag{2.36}$$

and are completely determined by the eigenfunctions $\varphi_n^{\vec{\beta}}(x)$ and eigenvalues of Eq. (2.17). We see that β_1 plays the role of momentum in the y direction, while β_2, β_3 give the strength of the magnetic field.

Further Hamiltonians are produced by inducing with other subgroups of Q. If the subgroup is chosen as the Heisenberg subgroup, whose (infinite dimensional) representations on $L^2(\mathbf{R})$ are given in the Introduction, a new magnetic field representation arises which is gauge equivalent to the Hamiltonian, Eq. (2.30), namely $A_x = y(\beta_2 + \beta_3 x)$, $A_y = A_z = 0$, which gives the same magnetic field as in Eq. (2.32). In any event it is clear that the partial differential operators resulting from different representations of Q all have eigenfunctions related by appropriate transforms that decompose a reducible representation to irreducible ones.

3. General Setup

In this section we will generalize the results of the Heisenberg group in the Introduction and the quartic group of section 2 to an arbitrary nilpotent Lie group G.

The right regular representation of G is given by

$$(R_g F)(x) := F(xg)\ , \quad F \in L^2(G) \\ g \in G\ , \tag{3.1}$$

with one-parameter subgroups of G generating operators X_i [e.g., Eq. (2.19) for Q]. There is a scaling operator V_t on $L^2(G)$ defined by

$$(V_t F)(x) = t^s F(\delta_t(x))\ , \tag{3.2}$$

where the exponent s is chosen to make V_t unitary, and $\delta_t(x)$ scales each of the variables x. For the Heisenberg group

$$(V_t F)(x,y,z) = t^2 F(tx, ty, t^2 z)\ ,$$

while for Q, V_t is given by Eq. (2.23).

Consider now those operators X_i scaling like t^{-1}; that is

$$V_t X_i V_t^{-1} = t^{-1} X_i\ . \tag{3.3}$$

The sublaplacian is defined to be the quadratic sum of such operators:

$$\Delta := \sum X_i^2\ , \tag{3.4}$$

where X_i scales like t^{-1}, from which it follows that

$$V_t \Delta V_t^{-1} = t^{-2}\Delta\ ; \tag{3.5}$$

using Eq. (3.5) and the fact that the X_i are antihermitian means that Δ has a continuous spectrum on the negative real line. It can also be shown that Δ is hypoelliptic [3] and is degenerate in the contravariant tensor $g^{ij}(x)$ (that is, $\det g^{ij}(x) = 0$). The definition of the sublaplacian can be generalized to include all operators from the Lie algebra that scale like t^{-2}, namely

$$\Delta \to \sum_{j,k} c_{jk} X_j X_k + i \sum d_\ell X_\ell \,, \tag{3.6}$$

where c_{jk} is a real symmetric matrix, X_j and X_k scale like t^{-1} as before and the X_ℓ scale like t^{-2}. The addition of terms like $i\sum d_\ell X_\ell$ is important in physics as it allows for the construction of Hamiltonians with external electric fields (see Eq. (2.7) for the quartic group) or Hamiltonians for particles with spin (see the Conclusion).

The sublaplacian has generalized eigenfunctions $F_{\lambda n}(x)$ which satisfy

$$\begin{aligned} (\Delta F_{\lambda n}(x) &= \lambda F_{\lambda n}(x) \,, \quad \lambda \in -\mathbf{R}^+ \\ (V_t F_{\lambda n})(x) &= t^s F_{t^2\lambda, n}(x) \,, \end{aligned} \tag{3.7}$$

where n is a discrete index labelling the multiplicity, and arises as the index of the eigenvalues of the Hamiltonian generated by the irreducible representations of G. For the quartic group n is the index of the energy eigenvalues of the quartic anharmonic oscillator, as seen in Eq. (2.26).

Of considerable interest is the heat equation

$$\Delta p_t(x) = \frac{\partial p_t(x)}{\partial t} \,, \quad p_{t=0}(x) = \delta(x) \,, \tag{3.8}$$

for as will be shown, from its solutions the time dependent Green's functions e^{-Ht} for all Hamiltonians related to Δ can be obtained. The heat kernel $p_t(x)$ can be given in terms of the generalized eigenfunctions $F_{\lambda n}(x)$. Such a solution for the heat kernel for the Heisenberg group is given in Ref. [5].

Consider next the irreducible representations of G. All irreducible representations of G are induced from a subgroup, say H, with irreducible representations $\pi^\beta(h)$. Then the unitary irreducible representations of G, U_g^β, with g restricted to one-parameter subgroups generate irreducible representation operators (of the Lie algebra of G) denoted by X_i^β. For Q these operators are given by Eq. (2.5). The Hamiltonian H^β is defined to be the same polynomial as the sublaplacian Δ, Eq. (3.4):

$$H^\beta := \sum \left(X_i^\beta\right)^2 \,, \tag{3.9}$$

and it is shown in Ref. [6] that this Hamiltonian always has a discrete spectrum. That is, for the eigenvalue equation

$$H^\beta \varphi_n^\beta = E_n(\beta)\varphi_n^\beta \,, \tag{3.10}$$

the eigenvalues $E_n(\beta)$ are labelled by a discrete index n.

Some of the properties of these eigenvalues can be determined from the equivalent irreducible representations of G. Let g be an element of $G - H$, where H is the inducing subgroup, and consider

$$\pi^{\beta}(ghg^{-1}) = \pi^{\beta_g}(h) , \quad h \in H \qquad g \in G - H , \tag{3.11}$$

where β_g is the transformed irreducible representation label generated from the element g [for the quartic group, see Eq. (2.9)]. Then the representations β_g are equivalent to β and the Hamiltonian is transformed to the new Hamiltonian

$$H^{\beta_g} = U_g H^{\beta} U_g^{-1} . \tag{3.12}$$

Applied to the eigenvalue equation, (3.10), this gives

$$\begin{aligned} U_g H^{\beta}\varphi_n^{\beta} &= E_n(\beta) U_g \varphi_n^{\beta} \\ H^{\beta_g} U_g \varphi_n^{\beta} &= E_n(\beta) U_g \varphi_n^{\beta} \\ H^{\beta_g}\varphi_n^{\beta_g} &= E_n(\beta_g)\varphi_n^{\beta_g} , \end{aligned} \tag{3.13}$$

from which it follows that $E_n(\beta_g) = E_n(\beta)$ and $\varphi_n^{\beta_g} = U_g\varphi_n^{\beta}$. Since β_g is a representation equivalent to β, $E_n(\beta)$ actually depends only on the Casimir invariants of G, as seen in Eq. (2.13) for the quartic group. If a scaling operator is also introduced, the form of the eigenvalues is even more determined. However, the conditions under which such scaling operators exist for general G requires further investigation.

Finally, if $\widetilde{G}$ is a subgroup "between" H, the subgroup that induces the irreducible representations of G and the identity subgroup (which induces the regular representation), then its representations $\pi^{\chi}(\tilde{g})$ can be used to induce reducible representations of G, U_g^{χ} on $L^2(G/\widetilde{G})$, whose one-parameter subgroups generate Lie algebra operators X_i^{χ} and Hamiltonians

$$H^{\chi} = \sum \left(X_i^{\chi}\right)^2 , \tag{3.14}$$

with the same polynomial dependence as the sublaplacian, Eq. (3.4), or its generalization, Eq. (3.6). Since all these reducible representations can be written as direct integrals of irreducible representations of G, the (generalized) eigenvalues and eigenfunctions of H^{χ} can be determined if the irreducible level eigenfunctions or the heat kernel are known. In fact, the heat kernel determines the time dependent Green's function of H^{χ}, as can be seen by defining the following operator on $L^2(G/\widetilde{G})$:

$$\begin{aligned} O_t &:= \int_G dg\, p_t(g) U_g^{\chi} \\ O_t U_{g_0}^{\chi} &= \int_G dg\, p_t(g) U_g^{\chi} U_{g_0}^{\chi} , \quad g_0 \in G \\ &= \int_G dg\, p_t(g g_0^{-1}) U_g^{\chi} \\ &= \int_G dg (R_{g_0^{-1}} p_t)(g) U_g^{\chi} , \end{aligned} \tag{3.15}$$

where the right regular representation R_g is defined in Eq. (3.1). Choosing one-parameter subgroups for g_0 and differentiating to get Lie algebra operators then gives

$$e^{-H^\times t} = \int_G dg\, p_t(g) U_g^\times . \tag{3.16}$$

Here use has been made of the heat equation, (3.8). If t is analytically continued to it, Eq. (3.16) gives a representation for the time dependent Green's function, $e^{-iH^\times t}$, a representation determined by the heat kernel $p_t(g)$. Some of the properties of the time dependent Green's function are discussed in Ref. [1].

4. Conclusion

All quantum mechanical anharmonic oscillators, in one or more dimensions, are related to some nilpotent group in the sense that the Hamiltonian for the anharmonic oscillator can be written as a quadratic or linear sum of representations of Lie algebra elements coming from the irreducible representation of the nilpotent group. Once the anharmonic oscillator Hamiltonian is written as a polynomial in Lie algebra elements, connections can be made with other quantum systems whose Hamiltonians are the same polynomial, but in Lie algebra elements coming from reducible representations. In particular, the regular representation induced from the identity element gives a Hamiltonian for heat equations on nilpotent groups. In this paper we have restricted our attention to Hamiltonians that have certain scaling properties which are used to restrict the form of the spectrum of the Hamiltonians. For the heat equations the spectrum is the negative real line, while for anharmonic oscillators the energy eigenvalues must obey certain functional equations.

The various Hamiltonians are all related via transforms that take reducible representations to direct sums or integrals of irreducible ones. If the eigenfunctions and eigenvalues of any Hamiltonian are known, the transforms can be used to obtain the eigenfunctions and eigenvalues for all the other Hamiltonians. Thus, large classes of differential equations are related by an underlying group structure which, in general, need not even be nilpotent.

In this paper a definite nilpotent group dubbed the quartic group Q [Eq. (2.1)] has been analyzed in detail. This group was chosen because it is the simplest generalization beyond the Heisenberg group. In section 2 various Hamiltonians arising as polynomials with certain scaling properties were exhibited, including the quartic anharmonic oscillator, coming from irreducible representations of Q, the Hamiltonian for a particle in a nonconstant magnetic field in the z direction, $B_z = \beta_2 + \beta_3 x$, and a constant electric field in the x direction, induced from a subgroup of Q, and the "heat Hamiltonian," coming from the regular representation of Q. These Hamiltonians were shown to be linked to one another by Fourier transforms, which are the transforms decomposing reducible representations of Q into direct integrals of irreducible ones.

In section 3 the procedures given for the quartic group were generalized to arbitrary nilpotent groups. It remains to generalize the results for anharmonic oscillators to systems in polynomial external magnetic fields. Consider the commutators of the generalized momentum,

$$\left[\frac{1}{i}\frac{\partial}{\partial x_j} - A_j(\vec{x})\, ,\, \frac{1}{i}\frac{\partial}{\partial x_k} - A_k(\vec{x})\right] = \epsilon_{jk\ell} B_\ell(\vec{x})$$

$$\left[\frac{1}{i}\frac{\partial}{\partial x_j} - A_j(\vec{x})\, ,\, B_\ell(\vec{x})\right] = \frac{1}{i}\frac{\partial B_\ell}{\partial x_j} .$$

If the vector potential $A_k(\vec{x})$ is polynomial in $\vec{x}$, then these commutation relations will close to give a nilpotent Lie algebra. The Hamiltonian for a spinless particle in an external magnetic field is

$$2H = \sum_{j=1}^{3} \left[\frac{1}{i} \frac{\partial}{\partial x_j} - A_j(x) \right]^2 ;$$

if this representation of the underlying nilpotent Lie algebra is irreducible, H has a purely discrete spectrum [6]. This raises the possibility of constructing "magnetic bottles" wherein the particle is trapped in its ground state by the magnetic field $\vec{B} = \vec{\nabla} \times \vec{A}$. The conditions under which polynomial vector potentials generate irreducible representations of some nilpotent group requires further investigation. In particular, since most particles have spin, a term coupling the magnetic moment to the external magnetic field must be added to the above Hamiltonian. However, in the context of the nilpotent Lie algebra, this is a natural term to add, since the magnetic field is the commutator of the generalized momenta.

ACKNOWLEDGEMENTS. This work was supported in part by the U.S. Department of Energy.

5. References

[1] Jorgensen, P.E.T. and Klink, W. H. (1985) 'Quantum mechanics and nilpotent groups: I. The curved magnetic field', *Publ. RIMS* **21**, 969–999.

[2] Landau, L. D. and Lifshitz, E. M. (1958) Quantum Mechanics, Pergamon Press, London, Chap. 16.

[3] Jorgensen, P.E.T. and Moore, R. T. (1984) Operator Commutation Relations, D. Reidel Publishing Co., Dordrecht-Boston; Goodman, R. (1976) Nilpotent Lie Groups: Structure and Applications to Analysis, Lecture Notes in Mathematics 562, Springer, Berlin.

[4] Hulanicki, A. (1976) 'The distribution of energy in the Brownian motion in Gaussian fields and analytic hypoellipticity of certain subelliptic operators on the Heisenberg group', *Studia Mathematica* **56**, 165–173.

[5] Jorgensen, P.E.T. and Klink, W. H. (1988) 'Spectral transform for the sub-Laplacian on the Heisenberg group', *Journal d'Analyse Mathématique* **50**, 101–121.

[6] Jorgensen, P.E.T. (1988) Operators and Representation Theory, North Holland Mathematics Studies 147, North-Holland, Amsterdam, New York.

Department of Physics and Astronomy
Department of Mathematics
University of Iowa, Iowa City, Iowa 52242

Extensions of the Mass 0 Helicity 0 Representation of the Poincare Group

Charles H. Conley

ABSTRACT.
Wigner's "little group" description of the irreducible representations of the Poincare group associated to the foward light cone is extended to smooth representations of finite length. As an application, we prove that there is a unique indecomposable representation of this group composed of n copies of the mass 0 helicity 0 representation.

Introduction

The purpose of this paper is to construct an equivalence between the category of smooth representations of finite length of a real semidirect product Lie group $G = H \times_s A$ associated to an orbit of H in the dual A^* of A, and a certain category of representations of the semidirect product of the H-stabilizer of the orbit with A. Here A is a vector space and the orbits of H in A^* are locally closed. A proof that the functor we construct is actually a category equivalence will appear elsewhere (4). As an application of this result, we prove that there is a unique indecomposable representation $\mathcal{V}^n$ of the Poincare group composed of $n+1$ copies of the smooth mass 0 helicity 0 representation $\mathcal{V}^0$, for any n. The existence of $\mathcal{V}^n$ was proven independently by A. Guichardet (9) and G. Rideau (11); here we show that $\mathcal{V}^n$ can be realized in the space of smooth compactly supported functions on the nth order infinitesimal neighborhood of the foward light cone. The uniqueness was conjectured to the author by Rideau, and we remark that it may be possible to obtain an alternate proof of it using cohomological methods, starting from the fact that $\text{Ext}^q(\mathcal{V}^0, \mathcal{V}^0)$ is one dimensional for q equal to 0 or 1.

We give a brief history of the problem. The irreducible unitary representations of G are described by Mackey's generalization of Wigner's result for the Poincare group: they are in bijection with pairs consisting of an orbit $\mathcal{O}$ of H in A^* and an irreducible unitary representation of the H-stabilizer of $\mathcal{O}$. Perhaps the most natural representations of finite length of G to study are bounded representations in a Hilbert space composed of unitary irreducible representations, but so far our methods are suited only to the study of smooth representations of G of finite length, composed of irreducible representations acting in the smooth compactly supported sections of H-vector bundles of finite rank over orbits of H in A^*, so we have begun with these. One finds in the introduction to the paper (7) of Guichardet that if the requirement of compact support is dropped, the same results are obtained, but if one takes the C^∞ vectors of irreducible unitary representations as composition series elements, somewhat pathological results can occur (the example given is due

E. A. Tanner and R. Wilson (eds.),
Noncompact Lie Groups and Some of Their Applications, 315–324.

to Blanc (1)).

In the case of the Poincare group, to my knowledge representations such as we study were first considered by Rideau, one of whose results is that there is a unique indecomposable representation composed of two mass 0 helicity 0 representations (10). Later Guichardet generalized Rideau's work to any G as above by constructing an exact sequence arising from representations of length 2 (7), and F. du Cloux made a study of the Ext^n groups in this setting (5). A corollary of Guichardet's work is that a smooth indecomposable representation of finite length is composed of irreducible representations which are all associated to the same orbit (this is false in the example of Blanc's), so here we shall consider only the catgory of representations of finite length composed of irreducible representations associated to a fixed orbit $\mathcal{O}$. In (8) Guichardet obtains a complete description of this category when the tangent bundle of $\mathcal{O}$ has an H-complement in the flat bundle $\mathcal{O} \times A^*$. His result is a generalization of the little group functor, which we generalize here for all orbits.

Rideau's work was built upon in another direction by Cassinelli, Truini, and Varadarajan, who observed that the indecomposable representation composed of two mass 0 helicity 0 representations of the Poincare group is realized in the smooth compactly supported functions on the first order infinitesimal neighborhood of the foward light cone (2). This enabled them to complete it to a bounded representation in a Hilbert space composed of two unitary mass 0 helicity 0 representations. It would be of interest to generalize their completion to all lengths, and also to generalize their observation to see which representations of finite length associated to a fixed orbit can be realized in vector bundles over infinitesimal neighborhoods of the orbit.

We have organized this paper so that in Section 1 we construct the functor, and in Section 2 we apply it to extensions of the mass 0 helicity 0 representation of the Poincare group.

1. Representations of Finite Length of Semidirect Product Lie Groups

We begin by defining the irreducible representations composing our representations of finite length. We use Schwartz' notations $\mathcal{E}$ and $\mathcal{D}$ in place of C^∞ and C_c^∞, and when the argument of $\mathcal{E}$ or $\mathcal{D}$ is a vector bundle it is understood that we are referring to its sections.

DEFINITIONS

Let K a Lie group, $\mathcal{O}$ a homogeneous space for K. We define the category $\mathrm{Geo}_K\mathcal{O}$ of *geometric* representations of K associated to $\mathcal{O}$ as follows. An object is the canonical representation $\mathcal{U}$ of K in $\mathcal{D}(B)$, where B is a complex K-homogeneous vector bundle of finite rank over $\mathcal{O}$: if k in K, p in $\mathcal{O}$, s in $\mathcal{D}(B)$, then $\mathcal{U}_k s(p) = ks(k^{-1}p)$. We say $\mathcal{U}$ is the representation *associated* to B. A morphism between two objects $\mathcal{U}_1$, $\mathcal{U}_2$ associated to bundles B_1, B_2 is a smooth section of $\mathrm{Hom}(B_1, B_2)$ intertwining their actions.

When S is the K-stabilizer of some point p_0 in $\mathcal{O}$ (the "little group" of $\mathcal{O}$), the restriction functor $\mathcal{R}_S^K$ from $\mathrm{Geo}_K\mathcal{O}$ to the category of complex finite dimensional representations of S is defined as follows. If $\mathcal{U}$ and B are as above, the action of S on B leaves the fiber B_{p_0} invariant, and $\mathcal{R}_S^K\mathcal{U}$ is the resulting representation of S

on it. It is elementary that $\mathcal{R}_S^K$ is an equivalence of categories whose inverse is the smooth induction functor Ind_S^K (13).

Let $G = H \times_s A$ be as in the introduction, and let $\mathcal{O}$ be an orbit of H in A^*. We have assumed $\mathcal{O}$ locally closed, so it is a submanifold of A^*. We remark that this is the setting in which the Mackey machine describes the irreducible unitary representations of G (6; 12). Let p_0 in $\mathcal{O}$, S the H-stabilizer of p_0. Notice that $\mathcal{O}$ is a G-space under the trivial action of A, so we have the categories $\mathrm{Geo}_H\mathcal{O}$ and $\mathrm{Geo}_G\mathcal{O}$. We will always view $\mathrm{Geo}_H\mathcal{O}$ as a full subcategory of $\mathrm{Geo}_G\mathcal{O}$ as follows. If B is an H-vector bundle over $\mathcal{O}$ and $\mathcal{U}$ is the associated representation in $\mathrm{Geo}_H\mathcal{O}$, then for a in A, s in $\mathcal{D}(B)$, and p in $\mathcal{O}$, $\mathcal{U}$ is extended to A by the character action $(\mathcal{U}_a s)(p) = e^{i\langle a,p\rangle} s(p)$. The G-stabilizer of p_0 is SA, and the above injection goes down to the little groups as the map from finite dimensional representations of S to finite dimensional representations of SA given by tensoring with e^{ip_0}.

Let $\mathrm{Ext}_G\mathcal{O}$ be the category of *extensions* of representations in $\mathrm{Geo}_H\mathcal{O}$: objects are C^∞ representations of G, given with a specified finite topologically split composition series of representations in $\mathrm{Geo}_H\mathcal{O}$, and morphisms are continuous linear intertwining maps (which need not respect the specified composition series).

Similarly, for any group S let Ext_S be the category whose objects are complex finite dimensional representations of S given with a specified composition series of representations, and whose morphisms are intertwining maps.

We remark that it is clumsy and unnecessary to to require that representations be given with a composition series; in (4) this requirement is removed. Note that composition series elements are not required to be irreducible; nevertheless, it is clear that objects in $\mathrm{Ext}_G\mathcal{O}$ and Ext_S are of finite length.

EXTENSIONS AS TRIANGULAR MATRICES

Fix n, and for $0 \le i \le n$ let $\mathcal{U}^i$ be representations of $\mathrm{Geo}_H\mathcal{O}$ associated to H-bundles F_i over $\mathcal{O}$, and let $\mathcal{U}$ in $\mathrm{Ext}_G\mathcal{O}$ have (topologically split) composition series $(\mathcal{U}^0, \ldots, \mathcal{U}^n)$. Then up to equivalence $\mathcal{U}$ acts in $\oplus_0^n \mathcal{D}(F_i)$, leaving the flag $\oplus_j^n \mathcal{D}(F_i)$ invariant and defining $\mathcal{U}^j$ in the subquotient $\mathcal{D}(F_j)$. Thus we may write $\mathcal{U}$ as a lower triangular matrix $\mathcal{U}^{ij} : \mathcal{D}(F_j) \to \mathcal{D}(F_i)$, where $\mathcal{U}^{ii} = \mathcal{U}^i$.

Similarly, if $\mathcal{U}'$ is another representation in $\mathrm{Ext}_G\mathcal{O}$ acting in $\oplus_0^{n'} \mathcal{D}(F_i')$ and $T : \mathcal{U} \to \mathcal{U}'$ is a morphism, then T is a n' by n matrix $T^{ij} : \mathcal{D}(F_j) \to \mathcal{D}(F_i')$. In general, $\mathcal{U}^{ij}$ $(i > j)$ and T^{ij} can be complicated continuous linear maps, but our first theorem shows that up to equivalence they are differential operators.

DIFFERENTIAL OPERATORS

Let B, C, and E be vector bundles over $\mathcal{O}$, f and f' diffeomorphisms of $\mathcal{O}$, s in $\mathcal{D}(B)$, p in $\mathcal{O}$. We define *differential operators of order $\le m$ above f from $\mathcal{D}(B)$ to $\mathcal{D}(C)$* (we sometimes say from B to C) to be continuous linear maps $F : \mathcal{D}(B) \to \mathcal{D}(C)$ such that $Fs(p)$ depends only on the m-jet of s at $f^{-1}(p)$. This definition behaves well with respect to composition: if F' is an order $\le m'$ differential operator above f' from C to E, $F' \circ F$ is an order $\le m' + m$ differential operator above $f' \circ f$ from B to E. Note that if f is the identity, F is an ordinary order $\le m$ differential operator from B to C, and if $m = 0$, F defines a unique fiberwise linear map $\tilde{F} : B \to C$ mapping $B_p \to C_{f(p)}$ such that $Fs(p) = \tilde{F}(s \circ f^{-1}(p))$. When $m = 0$ we refer to F as a *bundle map above f*.

Let $\Delta^r B$ be the vector bundle over $\mathcal{O}$ of order $\leq r$ differential operators on B: a smooth section of $\Delta^r B$ is an order $\leq r$ differential operator above the identity from $\mathcal{D}(B)$ to $\mathcal{D}(\mathcal{O})$. Let f and F be as above, and write $\lambda_f : \mathcal{D}(\mathcal{O}) \to \mathcal{D}(\mathcal{O})$ for the map $\beta \mapsto \beta \circ f^{-1}$, β in $\mathcal{D}(\mathcal{O})$; then λ_f is a bundle map above f. For $r \geq m$, D in $\mathcal{D}(\Delta^{r-m}C)$, the map $\lambda_{f^{-1}} \circ D \circ F : \mathcal{D}(B) \to \mathcal{D}(\mathcal{O})$ is an order $\leq r$ differential operator above the identity, and the map $D \mapsto \lambda_{f^{-1}} \circ D \circ F$ is a bundle map above f^{-1} from $\Delta^{r-m}C$ to $\Delta^r B$.

We remark that when $r = m$ the converse holds: if $\overline{F}$ is a bundle map above f^{-1} from $\Delta^0 C = C^*$ to $\Delta^r B$, there is a unique order $\leq r$ differential operator F above f from B to C such that for all ω in $\mathcal{D}(C^*)$, $\lambda_{f^{-1}} \circ \omega \circ F = \overline{F}\omega$. This fails for $r \geq m$, and one of the main difficulties of this paper arises from the problem of telling when a bundle map above f^{-1} from $\Delta^{r-m}C$ to $\Delta^r B$ arises from an order $\leq m$ differential operator above f as above.

RESULTS

Fix a real subbundle C of the flat bundle $\mathcal{O} \times A^*$, complementary to the tangent bundle $T\mathcal{O}$.

Theorem 1 (3). *Let $\mathcal{U}$ and $\mathcal{U}'$ be objects and $T : \mathcal{U} \to \mathcal{U}'$ a morphism in $\mathrm{Ext}_G\mathcal{O}$ as above, h in H, a in A, s in $\oplus_0^n \mathcal{D}(F_i)$. Then for each i, j, up to equivalence $\mathcal{U}_h^{ij}$ is an order $\leq i - j$ differential operator above $h : \mathcal{O} \to \mathcal{O}$, and $\mathcal{U}_a$ is a bundle map above the identity (note a acts as the identity on $\mathcal{O}$) of the following special form: for each a there is a continuous endomorphism l_a of $\oplus_0^n \mathcal{D}(F_i)$, linear in a, such that $l_a^{ij} = 0$ if $i \leq j$, l_a^{ij} is a bundle map above the identity if $i > j$, $l_a(p) = 0$ if $\langle a, C_p \rangle = 0$, and*

$$\mathcal{U}_a(p) = e^{i\langle a,p\rangle} \exp l_a(p).$$

Note that $\exp l_a$ is polynomial in l_a, as l_a^{ij} is strictly lower triangular.

If $\mathcal{U}$ and $\mathcal{U}'$ are both of this form, the entries T^{ij} of the morphism T are bundle maps above the identity.

Since all representations in $\mathrm{Ext}_G\mathcal{O}$ are equivalent to representations as in theorem 1, henceforth we shall fix C and restrict $\mathrm{Ext}_G\mathcal{O}$ to such representations and their morphisms. Roughly, our goal is to extend the domain of the little group functor $\mathcal{R}_S^H$ from $\mathrm{Geo}_H\mathcal{O}$ to $\mathrm{Ext}_G\mathcal{O}$. In general H may act by differential operators, so representations in $\mathrm{Ext}_G\mathcal{O}$ are not induced even from the inhomogeneous little group SA; the main point of this paper is to overcome this problem. However, when C can be chosen to be H-covariant, representations of $\mathrm{Ext}_G\mathcal{O}$ are induced from SA:

Theorem 2 (8). *Let $\mathcal{U}$ in $\mathrm{Ext}_G\mathcal{O}$, with composition series $\mathcal{U}^0, \ldots, \mathcal{U}^n$. If h in H leaves C invariant, then the $\mathcal{U}_h^{ij}$ are bundle maps above h. Consequently, if C is H-covariant and we define $\tilde{S} = S \times_s T\mathcal{O}_{p_0}^{\perp}$, then $\mathrm{Ext}_G\mathcal{O}$ is a full subcategory of $\mathrm{Geo}_G\mathcal{O}$, and is equivalent to the full subcategory of $\mathrm{Ext}_{\tilde{S}}$ of objects having a composition series of representations such that $T\mathcal{O}_{p_0}^{\perp}$ acts by e^{ip_0}. The equivalence is given by $\mathcal{R}_{SA}^G$ followed by ordinary restriction from SA to $\tilde{S}$, applied both to $\mathcal{U}$ and to its specified composition series $\mathcal{U}^0, \ldots, \mathcal{U}^n$.*

When C cannot be chosen to be H-covariant, we proceed as follows. Let $\mathcal{U}$ in $\mathrm{Ext}_G\mathcal{O}$ have composition series $\mathcal{U}^0, \ldots, \mathcal{U}^n$ associated to bundles F_i as above. Then

$\mathcal{U}^i$ gives rise to a representation $\Delta^r\mathcal{U}^i$ on $\mathcal{D}(\Delta^r F_i)$: if D in $\mathcal{D}(\Delta^r F_i)$ is an order $\leq r$ differential operator on F_i,

$$(\Delta^r\mathcal{U}^i)_g D = \lambda_g \circ D \circ \mathcal{U}^i_{g^{-1}}$$

(recall that for β in $\mathcal{D}(\mathcal{O})$, $\lambda_g\beta = \beta\circ g^{-1}$). Since $(\Delta^r\mathcal{U}^i)_g$, for which we write $\Delta^r_g\mathcal{U}^i$, is a bundle map above g, the representation $\Delta^r\mathcal{U}^i$ is in $\mathrm{Geo}_G\mathcal{O}$ (although not $\mathrm{Geo}_H\mathcal{O}$). The action of the complexified Lie algebra $\mathfrak{g}$ of G (all our Lie algebras are complexified) under $\mathcal{U}$ gives rise to the following well known description of $\mathcal{R}^G_{SA}\Delta^r\mathcal{U}^i$, the representation of SA on the fiber $\Delta^r F_i(p_0)$. Let $\mathfrak{U}(\mathfrak{g})$ be the universal enveloping algebra of $\mathfrak{g}$ with the usual filtration $\mathfrak{U}_r(\mathfrak{g})$, and let

$$\overline{\mathfrak{U}}_r = \mathfrak{U}_r(\mathfrak{g})\mathfrak{U}(\mathfrak{s}\oplus\mathfrak{a}).$$

For convenience, let $\mathfrak{s}' = \mathfrak{s}\oplus\mathfrak{a}$. Then for Z in $\overline{\mathfrak{U}}_r$, $\mathcal{U}^i_Z$ is a differential operator above the identity from F_i to F_i of order $\leq r$ at p_0, and for ω in $F_i^*(p_0)$, $\omega\otimes Z \mapsto \omega\circ\mathcal{U}^i_Z$ defines a projection

$$F_i^*(p_0)\otimes\overline{\mathfrak{U}}_r \to \Delta^r F_i(p_0)$$

that factors to an isomorphism from $F_i^*(p_0)\otimes_{\mathfrak{s}'}\overline{\mathfrak{U}}_r$ to $\mathcal{R}^G_{SA}(\Delta^r\mathcal{U}^i)$. This projection intertwines both the $(\mathcal{R}^G_{SA}\mathcal{U}^i)^*\otimes\mathrm{Ad}$ action of SA and the right action of $\mathfrak{s}'$ on $F_i^*\otimes\overline{\mathfrak{U}}_r$ with $\mathcal{R}^G_{SA}\Delta^r\mathcal{U}^i$. It is because we will be generalizing the right action of $\mathfrak{s}'$ below that we use $\overline{\mathfrak{U}}_r$ instead of the more usual $\mathfrak{U}_r(\mathfrak{g})$ here.

Recall that for all g in G, $\mathcal{U}^{ij}_g$ is an order $\leq i-j$ differential operator above $g:\mathcal{O}\to\mathcal{O}$. Consider the vector bundle $\oplus_0^n\Delta^{n-i}F_i$. We say that a smooth section D is a *graded differential operator*, as when restricted to $\oplus_j^n\mathcal{D}(F_i)$ it is of order $\leq n-j$. We define a representation $\Delta^{\mathrm{gr}}\mathcal{U}$ of G on $\oplus_0^n\mathcal{D}(\Delta^{n-i}F_i)\ni D$ such that the action of $(\Delta^{\mathrm{gr}}\mathcal{U})_g$, which we write as $\Delta^{\mathrm{gr}}_g\mathcal{U}$, is

$$\Delta^{\mathrm{gr}}_g\mathcal{U}(D) = \lambda_g\circ D\circ\mathcal{U}_{g^{-1}}.$$

It is clear that $\Delta^{\mathrm{gr}}\mathcal{U}$ is a representation; one checks that it acts by bundle maps above g and has composition series $\{\Delta^0\mathcal{U}^n,\ldots,\Delta^n\mathcal{U}^0\}$ (note the order reversal), so it lies in both $\mathrm{Geo}_G\mathcal{O}$ and $\mathrm{Ext}_G\mathcal{O}$ (for $i\neq 0$, the composition series element $\Delta^i\mathcal{U}^{n-i}$ is not in $\mathrm{Geo}_H\mathcal{O}$, but it has a canonical composition series of representations on symbol bundles that are). If $T:\mathcal{U}\to\mathcal{U}'$ is a morphism, T is a bundle map, and for D' in the space where $\Delta^{\mathrm{gr}}\mathcal{U}'$ acts, we define $\Delta^{\mathrm{gr}}T(D') = D'\circ T$; this makes Δ^{gr} a contravariant functor, and we have the following:

Theorem 3 (4). *The functor $\mathcal{R}^G_{SA}\circ\Delta^{\mathrm{gr}}$ from $\mathrm{Ext}_G\mathcal{O}$ to Ext_{SA}, mapping the object $\mathcal{U}$ with composition series $\mathcal{U}^0,\ldots,\mathcal{U}^n$ to the object $\mathcal{R}^G_{SA}\circ\Delta^{\mathrm{gr}}\mathcal{U}$ with composition series $\mathcal{R}^G_{SA}\Delta^0\mathcal{U}^n,\ldots,\mathcal{R}^G_{SA}\Delta^n\mathcal{U}^0$, is contravariant and faithful.*

Recall that a functor is faithful if it is injective on morphisms. We will spend the rest of Section 1 describing the image of $\mathcal{R}^G_{SA}\circ\Delta^{\mathrm{gr}}$, which we do by generalizing the description of $\mathcal{R}^G_{SA}\Delta^r\mathcal{U}^i$ above.

Lemma 4 (4). *Let $\mathcal{U}$ in $\mathrm{Ext}_G\mathcal{O}$, X in $\mathfrak{h}$. The lower triangular matrix $\mathcal{U}^{ij}_X$ is an order ≤ 1 differential operator above the identity.*

We remark that the proof depends on the special form of the $\mathcal{U}$-action of A given in theorem 1. Using only the graded form of the action of H leads only to $\mathcal{U}^{ij}_X$ order $\leq i-j$ for $i>j$, order ≤ 1 for $i=j$.

If X in $\mathfrak{s}$, it is clear that $\mathcal{U}_X^{ii} = \mathcal{U}_X^i$ is order 0 at p_0, but for $i > j$, $\mathcal{U}_X^{ij}$ may be order 1 at p_0. Indeed, $\mathcal{U}$ itself is induced from SA if and only if $\mathcal{U}_X^{ij}$ is order 0 at p_0 for all X in $\mathfrak{s}$, $i \geq j$. Observe that if α is in $\mathfrak{a} = A \otimes \mathbf{C}$, $\mathcal{U}_\alpha$ is order 0 by theorem 1: $\mathcal{U}_\alpha(p) = i\langle \alpha, p\rangle + l_\alpha(p)$.

In order to describe the image of $\mathcal{R}_{SA}^G \circ \Delta^{\mathrm{gr}}$, we must define several flags of vector spaces built from the fibers $F_i(p_0)$. For $0 \leq j \leq n$, let

$$E_j = \bigoplus_0^j \Delta^{j-i} F_i(p_0), \qquad C_j = \bigoplus_0^j F_i^*(p_0) \otimes \overline{\mathfrak{U}}_{j-i}.$$

The flag $C_0 \subset \ldots \subset C_n$ is a split $\mathfrak{s}'$ representation under the right action of $\mathfrak{s}'$ on $\overline{\mathfrak{U}}_{j-i}$. Let us write σ for the representation $\mathcal{R}_{SA}^G \circ \Delta^{\mathrm{gr}}\mathcal{U}$; then σ acts in E_n so as to leave E_j invariant, and $\mathcal{U}$ defines an $\mathfrak{s}'$-projection of flags $C_j \to E_j$:

$$\omega \otimes Z \mapsto \omega \circ \mathcal{U}_Z, \quad \omega \in F_i^*(p_0), Z \in \overline{\mathfrak{U}}_{j-i}, 0 \leq i \leq j \leq n.$$

Note that this projection respects the direct sum structure of the C_j and the E_j if and only if $\mathcal{U}$ is the split representation $\oplus_0^n \mathcal{U}^i$, in which case it is the sum of the projections describing $\mathcal{R}_{SA}^G \Delta^{n-i} \mathcal{U}^i$ above.

Let K_j be the kernel of the projection $C_j \to E_j$; then K_j is a $\mathfrak{s}'$-subspace of C_j, and the problem of describing σ amounts to that of describing the flag $\{K_j\}$. In order to do so it is convenient to fix a cross section for E_j in C_j.

Choose a subspace $\mathfrak{m}$ of $\mathfrak{h}$ complementary to $\mathfrak{s}$, let $(m_1, \ldots, m_p)$ be an ordered basis of $\mathfrak{m}$, and let M_r be the finite dimensional subspace of $\overline{\mathfrak{U}}_r$ spanned by the ordered monomials in the m_k of degree $\leq r$. Then it is well known that $F_i^*(p_0) \otimes M_r$ is a cross section for $\Delta^r F_i(p_0)$ in $F_i^*(p_0) \otimes \overline{\mathfrak{U}}_r$, and similarly one proves by induction that if we define

$$D_j = \bigoplus_0^j F_i^*(p_0) \otimes M_{j-i},$$

then $C_j = D_j \oplus K_j$, independent of $\mathcal{U}$.

At this point, notice that the endomorphism $\Delta_g^{\mathrm{gr}}\mathcal{U}$ of $\oplus_0^n \mathcal{D}(\Delta^{n-i} F_i)$ is completely determined by its restriction to the (non-invariant) subspace $\oplus_0^n \mathcal{D}(F_i^*)$, because it comes from the differential operator $\mathcal{U}_{g^{-1}}$. Less obviously, an analogous result holds at the fiber p_0: for $0 \leq j \leq n$ let

$$B_j = \bigoplus_0^j F_i^*(p_0),$$

and note that B_j is in both E_j and C_j, and that the projection $C_j \to E_j$ is the identity on it. The representation σ is determined by its restriction to the (non-invariant) subspace B_j, as follows.

For each X in $\mathfrak{s}'$, ω in B_n, we define a linear map

$$\eta_X : B_n \to D_n$$

by defining $\eta_X \omega$ to be the unique element of D_n that projects to $\sigma_X \omega = -\omega \circ \mathcal{U}_X$ in E_n. We define also a generalization of the right action of $\mathfrak{U}(\mathfrak{s}') = \overline{\mathfrak{U}}_0$ on C_j: an element Z' of $\overline{\mathfrak{U}}_{k-j}$ defines a map from $C_j \ni \omega \otimes Z$ to C_k by

$$(\omega \otimes Z)Z' = \omega \otimes (ZZ').$$

Lemma 5 (4). *For all X in $\mathfrak{s}'$, $\eta_X(B_j)$ is in D_j, so η is a linear map from $\mathfrak{s}'$ to* $\mathrm{Hom}(\{B_j\},\{D_j\})$. *Furthermore η, which is the lift to D_j of the restriction of σ to B_j, determines K_j and hence σ:*

$$K_j = \mathrm{Span}_{\mathbf{C}}\{\omega \otimes XZ + (\eta_X\omega)Z : \omega \in B_i,\ X \in \mathfrak{s}',\ Z \in \overline{\mathfrak{U}}_{j-i},\ 0 \leq i \leq j \leq n\}.$$

Now any linear map $\tilde{\eta} : \mathfrak{s}' \to \mathrm{Hom}(\{B_j\},\{D_j\})$ defines $\mathfrak{s}'$-subspaces $\tilde{K}_j$ of C_j as above, and thus a representation $\tilde{\sigma}$ of $\mathfrak{s}'$ on the flag $\{C_j/\tilde{K}_j\}$. It turns out that a representation such as $\tilde{\sigma}$ is in the image of $\mathcal{R}^G_{SA} \circ \Delta^{\mathrm{gr}}$ essentially if and only if $D_j \oplus \tilde{K}_j = C_j$ and $\tilde{\sigma}$ lifts from $\mathfrak{s}'$ to SA. We make this precise by defining the subcategory Gr_{SA} of Ext_{SA} that is the image of $\mathrm{Ext}_G\mathcal{O}$ under the functor $\mathcal{R}^G_{SA}\circ\Delta^{\mathrm{gr}}$.

An object of Gr_{SA} is a complex finite dimensional representation σ of SA, given with specified vector spaces $W_0, \ldots, W_n$ and a certain map η which defines σ as follows. For $0 \leq j \leq n$, define flags C_j, D_j, and B_j as above, but with the W_i replacing the $F_i^*(p_0)$. The map η is linear from $\mathfrak{s}'$ to $\mathrm{Hom}(\{B_j\},\{D_j\})$, and it defines $\mathfrak{s}'$-subspaces K_j of the C_j just as above. We require η to be such that $K_j \oplus D_j = C_j$ for all j, and the representation of $\mathfrak{s}'$ in C_n/K_n lifts to SA; this lift is σ. We require also that for any α in $\mathfrak{a}$, $\eta_\alpha(B_j)$ is in B_j, the action of η_α on the subquotients $B_j/B_{j-1} = W_j$ is multiplication by $-i\langle\alpha, p_0\rangle$, and if $\langle\alpha, C_{p_0}\rangle = 0$, then $\eta_\alpha = -i\langle\alpha, p_0\rangle$ (compare to the action of $\mathcal{U}|_A$ in theorem 1).

If $\{\sigma', W_0', \ldots, W'_{n'}, \eta'\}$ is another object of Gr_{SA}, a morphism between the two is a linear map $\tau : B_n \to B'_{n'}$ such that for all X in $\mathfrak{s}'$,

$$(\tau \otimes 1) \circ \eta_X = \eta'_X \circ \tau,$$

where $(\tau \otimes 1)(\omega \otimes Z) = \tau\omega \otimes Z$. Then $\tau \otimes 1 : C_n \to C'_{n'}$ maps K_n into $K'_{n'}$ and projects to an intertwining map $t : \sigma \to \sigma'$ (compare to the fact that morphisms between objects of $\mathrm{Ext}_G\mathcal{O}$ as in theorem 1 are order 0 differential operators).

The contravariant functor $\mathcal{R}^G_{SA} \circ \Delta^{\mathrm{gr}}$ from $\mathrm{Ext}_G\mathcal{O}$ to Gr_{SA} takes an object $\{\mathcal{U}, \mathcal{U}^0, \ldots, \mathcal{U}^n\}$ to the object $\{\sigma, F_0^*(p_0), \ldots, F_n^*(p_0), \eta\}$, where η and σ are defined from $\mathcal{U}$ as above, and a morphism $T : \mathcal{U} - \mathcal{U}'$ to the dual $T^*(p_0) : B'_{n'} \to B_n$ of $T(p_0)$.

Theorem 6 (4). *The functor $\mathcal{R}^G_{SA} \circ \Delta^{\mathrm{gr}}$ from $\mathrm{Ext}_G\mathcal{O}$ to Gr_{SA} is a contravariant equivalence of categories.*

In order to apply this theorem, we need two more lemmas. The first gives a useful condition on η guaranteeing that $K_j \oplus D_j = C_j$, and the second restricts the form of η.

Let $W_0, \ldots, W_n$ be arbitrary vector spaces, and define from them C_j, D_j, and B_j as above. Suppose $\tilde{\eta} : \mathfrak{s}' \to \mathrm{Hom}(\{B_j\},\{D_j\})$ is any linear map; we can still use it to define $\mathfrak{s}'$-subspaces $\tilde{K}_j$ of C_j as above. For $\omega \otimes Z$ in C_j, X, Y in $\mathfrak{s}'$, we define endomorphisms $\tilde{\eta}_X \otimes 1$ and $1 \otimes \mathrm{ad}_X$ of C_j by

$$(\tilde{\eta}_X \otimes 1)(\omega \otimes Z) = (\tilde{\eta}_X\omega)Z, \quad (1 \otimes \mathrm{ad}_X)(\omega \otimes Z) = \omega \otimes (XZ - ZX).$$

We define $\theta_{\tilde{\eta}}(X,Y) : B_j \to D_j$ by

$$\theta_{\tilde{\eta}}(X,Y) = (\tilde{\eta}_X \otimes 1 + 1 \otimes \mathrm{ad}_X) \circ \tilde{\eta}_Y - (\tilde{\eta}_Y \otimes 1 + 1 \otimes \mathrm{ad}_Y) \circ \tilde{\eta}_X - \tilde{\eta}_{[X,Y]}.$$

Lemma 7. *For any $\tilde{\eta}$ as above, $\tilde{K}_j + D_j = C_j$ for $0 \leq j \leq n$. We have $\tilde{K}_j \oplus D_j = C_j$ if and only if for all X, Y in $\mathfrak{s}'$, $0 \leq j \leq n$, $\theta_{\tilde{\eta}}(X,Y)(B_j)$ is in $\tilde{K}_{j-1}$.*

Lemma 8. *Let $\{\sigma, W_0, \ldots, W_n, \eta\}$ be an object of* Gr_{SA}. *It is a consequence of the form of $\eta|_{\mathfrak{a}}$ that for all X in $\mathfrak{s}'$ $0 \leq j \leq n$,*

$$\eta_X(B_j) \subset (B_{j-1} \otimes M_1) \oplus W_j$$

(compare to lemma 4), and if X leaves C_{p_0} invariant, $\eta_X(B_j)$ is in B_j (compare to theorem 2). Furthermore, for each j, η_X projects to a map

$$\eta_X^j : B_j/B_{j-1} \to (B_{j-1} \otimes M_1 \oplus W_j)/(B_{j-1} \otimes M_1).$$

Both of these spaces are canonically isomorphic to W_j, and η^j is a representation of $\mathfrak{s}'$ on W_j. If an object $\mathcal{U}$ with composition series $\mathcal{U}^0, \ldots, \mathcal{U}^n$ in $\mathrm{Ext}_G\mathcal{O}$ *goes to $\{\sigma, W_0, \ldots, W_n, \eta\}$ under $\mathcal{R}^G_{SA} \circ \Delta^{\mathrm{gr}}$, then η^j is the dual of the representation $\mathcal{R}^G_{SA}\mathcal{U}^j$.*

2. Extensions of the Mass 0 Helicity 0 Representation of the Poincare Group

Throughout this section, let $G = H \times_s A$ be the universal cover of the Poincare group: $H = \mathrm{SL}_2\mathbf{C}$ acts on $A = \mathbf{R}^{1,3}$ by the standard double cover $\mathrm{SL}_2\mathbf{C} \to \mathrm{SO}_{1,3}$. Here the standard inner product of signature $1,3$ gives an H-isomorphism of A^* with A, so we view A^* as equal to A. The theorem of Guichardet's (theorem 2 above) describes $\mathrm{Ext}_G\mathcal{O}$ for any H-orbit $\mathcal{O}$ in A except the two light cones. Henceforth let $\mathcal{O}$ be the foward light cone and take p_0 to be $(1,0,0,1)$; then the H-stabilizer S of p_0 is

$$S = \left\{ \begin{pmatrix} z & a \\ 0 & z^{-1} \end{pmatrix} : z \in S^1, a \in \mathbf{C} \right\}.$$

Let $\mathcal{V}$ be the representation in $\mathrm{Geo}_H\mathcal{O}$ induced from the trivial representation of S, so that $\mathcal{V}$ is the canonical representation of G in $\mathcal{D}(\mathcal{O})$.

Theorem 9 (4). *Up to equivalence there is a unique indecomposable representation in* $\mathrm{Ext}_G\mathcal{O}$ *composed of $n+1$ copies of $\mathcal{V}$, for all $n \geq 0$.*

Remark. As stated in the introduction, the existence was proven independently by Guichardet (9) and Rideau (11), and it may be possible to prove the uniqueness by cohomological methods. Our approach is to apply the "little group method" for describing representations of finite length associated to the foward light cone that was developed in Section 1.

Outline of Proof. Suppose $\mathcal{U}$ together with its composition series $\mathcal{U}^0, \ldots, \mathcal{U}^n$ is an object of $\mathrm{Ext}_G\mathcal{O}$, and $\mathcal{U}^i = \mathcal{V}$ for all i. The representation $\mathcal{V}$ is associated to the trivial bundle $\mathcal{O} \times \mathbf{C}$ over $\mathcal{O}$, and so the functor $\mathcal{R}^G_{SA} \circ \Delta^{\mathrm{gr}}$ maps $\mathcal{U}$ with its composition series to an object $\{\sigma, W_0, \ldots, W_n, \eta\}$ of Gr_{SA}, where W_i is $\mathbf{C}$ for all i. Recall that if we define C_j, D_j, and B_j as before, then η defines the $\mathfrak{s}' = \mathfrak{s} \oplus \mathfrak{a}$-subspaces K_j of C_j, and $\sigma = \mathcal{R}^G_{SA} \circ \Delta^{\mathrm{gr}}\mathcal{U}$ is the representation of $\mathfrak{s}'$ in C_n/K_n, lifted to SA.

Here $\mathcal{R}^G_{SA}\mathcal{U}^i$ restricts to the trivial representation of S for all i, and so by lemma 8 the representations $\eta^j|_{\mathfrak{s}}$ on $W_j = \mathbf{C}$ are all trivial, or in other words for $0 \leq j \leq n$, X in $\mathfrak{s}$,

$$\eta_X(B_j) \subset B_{j-1} \otimes M_1.$$

Since $\mathcal{R}^G_{SA} \circ \Delta^{\rm gr}$ is an equivalence of categories, we need only prove that there is a unique such indecomposable object of Gr_{SA}.

Let e_0, e_1, e_2, e_3 be the standard orthogonal basis of $\mathbf{R}^{1,3}$: $\langle e_0, e_0\rangle = 1$, $\langle e_i, e_i\rangle = -1$ for $i > 0$. For a basis of $\mathfrak{a} = A \otimes \mathbf{C}$ we take the vectors

$$x_0 = p_0 = e_0 + e_3, x_3 = e_0 - e_3, x_+ = e_1 - ie_2, x_- = e_1 + ie_2.$$

In Section 1 we had to choose a real vector bundle C complementary to $T\mathcal{O}$ in $\mathcal{O} \times A^*$, but here we need only choose C_{p_0}. The vectors x_0, e_1, e_2 are a basis of $T\mathcal{O}_{p_0}$, so we may take C_{p_0} to have basis x_3. Then the annihilator $C^\perp_{p_0}$ of C_{p_0} in $\mathfrak{a}$ has basis x_+, x_-, x_3, and by the definition of Gr_{SA}, $\eta_a = -i\langle a, x_0\rangle$ for a in $C^\perp_{p_0}$, and

$$\eta_{x_0} = -i\langle x_0, x_0\rangle + N = N,$$

where N is a (nilpotent) endomorphism of B_n such that $N(B_j)$ is in B_{j-1}.

We must now describe $\mathfrak{s} \subset \mathfrak{h}$, which we identify with $\mathfrak{so}_{1,3}$. Let $e_0^*, \ldots, e_3^*$ be the basis of A^* dual to $e_0, \ldots, e_3$ (of course, under the canonical isomorphism $A = A^*$, $e_0 = e_0^*$, etc.). The usual basis of $\mathfrak{h}$ is

$$M_{0j} = e_0 \otimes e_j^* + e_j \otimes e_0^*, \quad L_{ij} = e_i \otimes e_j^* - e_j \otimes e_i^*,$$

where $1 \le i, j \le 3$. Another basis of $\mathfrak{h}$ is given by the vectors

$$L = -2iL_{12}, L_\pm = L_{13} \mp iL_{23}, M_\pm = M_{01} \mp iM_{02}, M_3 = M_{03}.$$

The vectors

$$L, X = (M_+ - L_+)/2, Y = (M_- - L_-)/2$$

are a basis of $\mathfrak{s}$, and we will take M_+, M_-, and M_3 as our ordered basis of $\mathfrak{m}$ (in fact, in light of lemma 8 we really need only specify the space $\mathfrak{m}$).

We must calculate the possibilities for $\eta|_{\mathfrak{s}}$. Since C_{p_0} is L-invariant, lemma 8 gives that $\eta_L(B_n)$ is in B_n, and so $\eta_L = \sigma_L|_{B_n}$. Now the representations η^j of lemma 8 are trivial here, and σ lifts to SA, so σ_L is semisimple and hence $\eta_L = 0$.

We next consider η_X. Notice that $M_\pm$ and M_3 are eigenvectors of ad_L with eigenvalues ± 2, 0, respectively. From this one checks that if θ_η is as in lemma 7, then $\theta_\eta(L, X)$ maps B_n to D_n. But by lemma 7 its image is in K_n, so it is 0. Since $[L, X] = 2X$,

$$0 = \theta_\eta(L, X) = (1 \otimes \mathrm{ad}_L) \circ \eta_X - 2\eta_X,$$

i.e. η_X maps B_n to a $+2$-eigenspace of $1 \otimes \mathrm{ad}_L$ in D_n. Therefore

$$\eta_X(B_j) \subset B_{j-1} \otimes M_+,$$

and so there is a nilpotent endomorphism γ of B_n sending B_j to B_{j-1} such that for ω in B_n, $\eta_X \omega = \gamma\omega \otimes M_+$. We write $\eta_X = \gamma \otimes M_+$.

We can relate γ and $\eta_{x_0} = N$ by looking at $\theta_\eta(X, x_-)$. Since $[X, x_-] = Xx_- = x_0$, and $\eta_{x_-} = -i\langle x_-, x_0\rangle = 0$, we have for ω in B_j that

$$\theta_\eta(X, x_-)\omega = -(1 \otimes \mathrm{ad}_{x_-}) \circ (\gamma \otimes M_+)\omega - N\omega \in K_{j-1}.$$

Now $\mathrm{ad}_{x_-} M_+ = -M_+ x_- = -x_0 - x_3$, so we find that

$$\gamma\omega \otimes (x_0 + x_3) - N\omega \in K_{j-1}.$$

But by definition, $(\eta_{x_0} + \eta_{x_3})\gamma\omega + \gamma\omega \odot (x_0 + x_3)$ is in K_{j-1}, as ω is in B_{j-1}, so

$$(\eta_{x_0} + \eta_{x_3})\gamma\omega + N\omega = ((N - 2i)\gamma + N)\omega \in K_{j-1}.$$

This also lies in $B_{j-1} \subset D_{j-1}$, so it must be 0. Using N nilpotent gives $\gamma = N(2i - N)^{-1}$.

Similar calculations show that $\{\sigma, W_0, \ldots, W_n, \eta\}$ is an object of Gr_{SA}, such that $W_i = \mathbf{C}$ for all i and the representations $\eta^i|_S$ of lemma 8 are all trivial, if and only if $\eta_a = -i\langle a, x_0\rangle$ for a in the span of $\{x_\pm, x_3\}$, $\eta_{x_0} = N$ is an arbitrary nilpotent endomorphism of B_n mapping B_j to B_{j-1}, $\eta_L = 0$, $\eta_X = \gamma \otimes M_+$, and $\eta_Y = \gamma \otimes M_-$, where $\gamma = N(2i - N)^{-1}$.

It is easy to see from the definition of Gr_{SA} that such an object decomposes if and only if the action of N on B_n decomposes, and that two such objects η and η' are equivalent if and only if η_{x_0} and η'_{x_0} are similar. This proves theorem 9.

Corollary 10. *The unique indecomposable representation in* $\mathrm{Ext}_G\mathcal{O}$ *composed of* $n+1$ *copies of the mass 0 helicity 0 irreducible representation* $\mathcal{V}$ *is realized in the smooth compactly supported functions on the* n^{th} *order infinitesimal neighborhood of the foward light cone* $\mathcal{O}$.

Outline of Proof. In order to view the canonical representation of G in the space of such functions as an object in $\mathrm{Ext}_G\mathcal{O}$, it is necessary first to define such functions (this is a bit tricky because $\mathcal{O}$ is not closed), and second to use the bundle C of theorem 1 to fix specific transversal derivatives on $\mathcal{O}$; we shall do these things in a later paper. Once they are done, we need only check that $\mathcal{R}^G_{SA} \circ \Delta^{\mathrm{gr}}$ sends the resulting object of $\mathrm{Ext}_G\mathcal{O}$ to an object η of Gr_{SA} such that η_{x_0} is indecomposable, and this is easy.

References

[1] P. Blanc, Sur la cohomologie continue des groupes localement compacts, *Ann. Sci. Ec. Norm. Sup.* (4) **12** (1979), *no.* 2, 137–168.

[2] G. Cassinelli, P. Truini, V. S. Varadarajan, Hilbert space representations of the Poincaré group for the Landau gauge, *J. Math. Phys.* **32** (1991), 1076–1090.

[3] C. Conley, Representations of finite length of semidirect product Lie groups, to appear in *J. Func. Anal.*.

[4] C. Conley, Representations of finite length of semidirect product Lie groups II, in progress.

[5] F. du Cloux, Sur les n-extensions des représentations induites des produits semi-directs, *Astérisque* **124-125** (1985), 49–128.

[6] J. Glimm, Locally compact transformation groups, *Trans. Amer. Math. Soc.* **101** (1961), 124–138.

[7] A. Guichardet, Extensions de représentations induites des produits semidirects, *J. für reine angew. Math.*, **310** (1979), 7–32.

[8] A. Guichardet, Représentations de longeur finie des groupes de Lie inhomogènes, *Astérisque* **124-125** (1985), 212–252.

[9] A. Guichardet, Extensions and Deformations of Representations, *Symposia Mathematica* **31** (1988), 31–44.

[10] G. Rideau, Quantization of the Maxwell field and extensions of zero mass representations of Poincaré group, *Lett. Math. Phys.* **2** (1978), 529–536.

[11] G. Rideau, Cohomology of Extension for the Poincaré Group Representations: Applications to Quantum Mechanics with Indefinite Metric and to Non-Linear Representation Theory, *Symposia Mathematica* **31** (1988), 95–108.

[12] V. S. Varadarajan, "Geometry of Quantum Theory," 2nd ed., Springer-Verlag, New York, 1985.

[13] N. Wallach, "Harmonic Analysis on Homogeneous Spaces," Marcel Dekker, New York, 1973.

Massachussets Institute of Technology

Invariant causal propagators in conformal space

W. F. Heidenreich

Abstract

Invariant causal propagators in conformal space for scalar massless particles and for scalar fields with anomalous dimension are given. Some loop-integrals are calculated and the results compared with the usual calculations in Euclidean spacetime.

Introduction

The attitude of physicists concerning the conformal group SO(4.2) is divided: All the physically interesting renormalizable field theories are formally conformal invariant before masses are introduced. But it is generally accepted that renormalization introduces necessarily a scale which destroys conformal invariance [1]. This conclusion comes from perturbation theory in Euclidean space. The motivation for this work is the opinion, that this should be confirmed by calculations in conformal space.

A similar attitude has been expressed recently by Segal [2] who is probing in conformal space the Euclidean statement, that the conformal scalar model—massless ϕ^4 theory—exists as free theory only. An example for substantial differences between a formulation in Euclidean and conformal space is the free photon propagator: while the Euclidean version propagates pure gauge only, the formulation in conformal space does propagate physical photons [3]. As is generally known, causal propagators are represented by the internal lines of Feynman graphs. These are used in quantum field theory for most calculations which can be compared with experiments. The divergent loop integrals are in many approaches dealt with in a Euclidean setting.

I report here about some preliminary work which shows that perturbation theory in conformal space is possible and may lead to results different from Euclidean calculations.

Conformal spaces

In physics three different conformal compactifications of Minkowski space are used, $(S^3 \times S^1)/Z_2$, and its covering spaces $S^3 \times S^1$ and $S^3 \times R$. In the first one, spacelike and timelike are not invariant notions, only lightlike is [4]; for this reason it cannot be

E. A. Tanner and R. Wilson (eds.), Noncompact Lie Groups and Some of Their Applications, 325–331.

used here. To do calculations in the other two cases we use polar coordinates χ, θ, ϕ for the S^3 and $-\pi \leq \alpha < \pi$ resp. $-\infty < \alpha < \infty$. The space $S^3 \times S^1$ can also be easily expressed in conformal coordinates $x_1^2+x_2^2+x_3^2-x_4^2+x_5^2-x_6^2 = 0$, $x \equiv \lambda x$, with $\lambda > 0$. The coordinate systems are connected by $x_i = \rho(\chi, \theta, \phi)$, $x_4 = \rho \sin \alpha$, $x_6 = \rho \cos \alpha$. The coordinate ρ allows us to keep track of the degree of homogeneity of fields $\phi(x)$ in conformal coordinates, as $x \cdot \partial = \rho \partial_\rho$.

Pauli-Jordan commutation functions

The massless scalar SO(4.2) representation of positive energy D(1,0,0) acts on the space of scalar functions with the basis

$$\phi_{jlm}(x) = \frac{\sqrt{2}\pi}{\rho\sqrt{j+1}} e^{i(j+1)\alpha} Y_{jlm}(\chi, \theta, \phi), \tag{1}$$

which are normalized with respect to the scalar product

$$\int_{S^3} \phi^* \overleftrightarrow{\partial}_\alpha \phi d^3\Omega. \tag{2}$$

The positive frequency commutation function is obtained by summing over all these states; using $\tau = \alpha - \alpha'$, the addition theorem of spherical harmonics, and the generating function of Gegenbauer polynomials we find

$$\begin{aligned} D_+^1(x, x') &= \sum_{jlm} \phi_{jlm}(x)\phi^*_{jlm}(x') = (\rho\rho')^{-1} \sum_{j=0}^{\infty} C_j^1(\cos\chi) e^{i(j+1)\tau} \\ &= (\rho\rho')^{-1}[2(\cos\tau + i\epsilon - \cos\chi)]^{-1} \\ &= [-2(x \cdot x' - i\epsilon(x_4 x_6' - x_6 x_4'))]^{-1}. \end{aligned} \tag{3}$$

If we change the exponents (-1) in the next to last formulation to $-\lambda$ with $\lambda > 1$ then we get

$$\begin{aligned} D_+^\lambda(x, x') &= (\rho\rho')^{-\lambda}[2(\cos\tau + i\epsilon - \cos\chi)]^{-\lambda} \\ &= (\rho\rho')^{-\lambda} \sum_{n=0}^{\infty} C_n^\lambda(\cos\chi) e^{i(n+\lambda)\tau} \\ &= (\rho\rho')^{-\lambda} \sum_{j=0}^{\infty}\sum_{p=0}^{\infty} \frac{(\lambda)_{j+p}(\lambda-1)_p}{p!(j+p)!}(1+j)C_j^1(\cos\chi)e^{i(j+\lambda+2p)\tau}, \end{aligned} \tag{4}$$

where we expand C_n^λ in C_{n-2p}^1 and rearrange the double sum. The last expression is a sum over eigenstates of the "massive" scalar representations of positive energy D(λ, 0, 0) of the conformal group. They are the limit to the surface $S^3 \times R$ of normalized basis states on the interior $D^4 \times R$. These commutation functions appear in the quantization of objects with a continuous mass spectrum, which do not have a field equation on 4-dimensional spacetime (but on the five dimensional one) and which have "anomalous dimension" λ. For integer λ they are also defined on $S^3 \times S^1$.

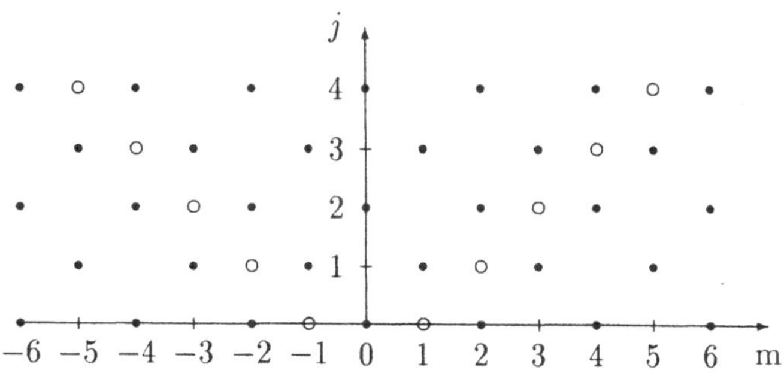

Figure 1: Part of the weight diagram of D^1_c.

Causal propagators

Causal propagators should propagate positive energy modes in the future and negative energy modes in the past. In the static Einstein spacetime there is no problem to define global causality: the points which can be reached by future-directed (past-directed) timelike geodesics belong to the timelike future (past) of a point; of the remaining points, those which can be reached by future- or past- directed lightlike geodesics belong to the future or past lightcone, and the remaining points are spacelike. For all values $\lambda \geq 1$ we use the causal propagator

$$D^\lambda_c = \theta(\tau)D^\lambda_+ + \theta(-\tau)(D^\lambda_+)^*. \tag{5}$$

This is a Green's function of the massless conformal Klein-Gordon equation in Einstein spacetime for $\lambda = 1$ only.

For $S^3 \times S^1$ the situation is less obvious. Spacelike ($x{\cdot}x' > 0$), lightlike ($x{\cdot}x' = 0$) and timelike ($x{\cdot}x' < 0$) are invariant notions, but the ordering in time is not global. Yet there exists a not transitive invariant ordering for lightlike events, given by [4]

$$\mathrm{sign}(x_4x'_6 - x_6x'_4) = \mathrm{sign}(\sin\tau). \tag{6}$$

This suggests as causal propagator for the massless particle

$$\begin{aligned} D^1_c(x,x') &= [-2(xx'+i\epsilon)]^{-1} \\ &= \theta(\sin\tau)D^1_+(x,x') + \theta(-\sin\tau)(D^1_+(x,x'))^* \end{aligned} \tag{7}$$

$$= (\rho\rho')^{-1}\sum_{j=0}^{\infty} C^1_j(\cos\chi)\left[\underbrace{\frac{1}{2}\left(e^{i(j+1)\tau}+e^{-i(j+1)\tau}\right)}_{\text{on mass shell}} + \underbrace{\frac{2i}{\pi}\sum_{m-j\ \text{even}}\frac{j+1}{(j+1)^2-m^2}e^{im\tau}}_{\text{off mass shell}}\right].$$

Part of the weight diagram ($j-m$-diagram) of this function is given in figure 1. It is a Green's function with two sources, at antipodal points on $S^3 \times S^1$,

$$D^2D^1_c(x,x') = \delta^4(x,x') + \delta^4(x,-x'), \tag{8}$$

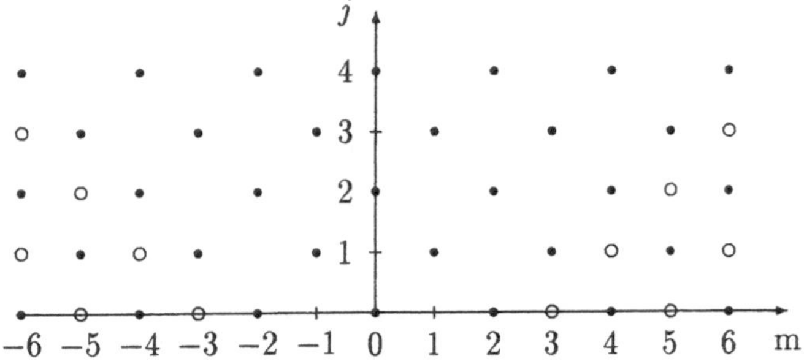

Figure 2: Part of the weight diagram of D_c^3.

where D^2 denotes the massless conformal Klein-Gordon operator on our spacetime.

We want to stress that this seems to be the smallest spacetime in which we can formulate a candidate for a conformally invariant causal propagator. The usual expression $1/((y-y')^2 - i\epsilon)$ in Minkowski space cannot be extended invariantly to the conformal compactification $(S^3 \times S^1)/Z_2$. Perturbation theory in Minkowski space is breaking conformal invariance already at this stage.

Also for the fields with anomalous dimension $\lambda = 2, 3, \ldots$ we can write a causal propagator as

$$\begin{aligned} D_c^\lambda(x,x') &= [-2(xx'+i\epsilon)]^{-\lambda} \\ &= \theta(\sin\tau)D_+^\lambda(x,x') + \theta(-\sin\tau)(D_+^\lambda(x,x'))^* \end{aligned} \tag{9}$$

$$= (\rho\rho')^{-\lambda}\sum_{n=0}^{\infty} C_n^\lambda(\cos\chi)\left[\underbrace{\frac{1}{2}\left(e^{i(n+\lambda)\tau}+e^{-i(n+\lambda)\tau}\right)}_{\text{on mass shell}} + \underbrace{\frac{2i}{\pi}\sum_{m-j-\lambda \text{ odd}}\frac{n+\lambda}{(n+\lambda)^2-m^2}e^{im\tau}}_{\text{off mass shell}}\right]$$

$$= (\rho\rho')^{-\lambda}\sum_{j=0}^{\infty}\sum_{p=0}^{\infty}\frac{(\lambda)_{j+p}(\lambda-1)_p}{p!(j+p)!}(1+j)C_j^1(\cos\chi)\,[n\to j+2p \text{ in last expression}].$$

Part of the weight diagram of this function is given in figure 2. The last line is formal: the p-sum is divergent for the off shell amplitudes, e.g. for $\lambda = 2$ and big p we sum over $\approx (1+j)$. So we find that D_c^λ is not Fourier-integrable for $\lambda = 2, 3, \ldots$ on $S^3 \times S^1$. But it would be Fourier-integrable on the 5-dimensional space $x^2 < 0$.

The reason why we consider these propagators of "massive" representations is, that they provide a regularization of the product of massless propagators,

$$D_c^n = \left(D_c^1\right)^n_{\text{regularized}}\,. \tag{10}$$

Some "loop"-integrals

In Euclidean calculations (see e.g. [5]) the "fish"-diagram diverges like $\int (D(x,x')^2 dx'$, and the main divergence of the "nut"-diagram is contained in $\int (D(x,x')^3 dx'$. These loop-diagrams can be easily calculated in our spaces. In $S^3 \times R$ we find a formal expression for all values of λ [6]:

$$\int D_c^\lambda \propto \sum_{n=0}^{\infty} \int_0^{\pi} C_n^\lambda(\cos\chi)\sin^2\chi d\chi \int_{-\infty}^{+\infty} \left[\theta(\alpha)e^{i(n+\lambda)\alpha} + \theta(-\alpha)e^{-i(n+\lambda)\alpha}\right] d\alpha$$

$$= \frac{i\pi}{\Gamma(2\lambda)} \sum_{n=0}^{\infty} \frac{\Gamma(n+2\lambda)}{(\lambda+n)n!} \, {}_3F_2(-n, n+2\lambda, 3/2; \lambda+1/2, 3; 1). \tag{11}$$

For $\lambda = 2$ there is an explicit formula for the generalized hypergeometric functions [7], which gives

$$\int D_c^2 \propto \sum_{n \text{ even}} \frac{1}{n+2}. \tag{12}$$

From this and numeric calculations we find, that the integral Eq. (11) is divergent for $\lambda \geq 2$ and finite for $\lambda < 2$.

On $S^3 \times S^1$ the α-integral has to be performed from $-\pi$ to $+\pi$. It gives zero if $\lambda + n$ is even and $2i/(\lambda + n)$ if $\lambda + n$ is odd. For $\lambda = 2$ the χ-integral vanishes for odd n and so we find

$$\int D_c^2 = 0 \text{ for } \lambda = 2. \tag{13}$$

For $\lambda = 1$ the integral is finite, for $\lambda = 3, 4, \ldots$ it remains divergent.

In order to compare with the corresponding expressions in Minkowski space, we employ the usual causal propagator $D_c(y, y') = 1/((y - y')^2 - i\epsilon)$ and get after performing the angular and the t-integrals

	Minkowski	$S^3 \times R$	$S^3 \times S^1$
$\int D_c dy' \propto \int_0^\infty r dr$	IR divergent	finite	finite
$\int (D_c)^2 dy' \propto \int_0^\infty 1/r dr$	IR and UV divergent	divergent	0
$\int (D_c)^3 dy' \propto \int_0^\infty 1/r^3 dr$	UV divergent	divergent	divergent

There are no infrared divergencies in the conformal spaces, as they are spatially compact.

The first correction to the propagator

Loop-corrections to the propagator can be calculated at least formally in conformal space using a variant of the Gegenbauer polynomial x-space technique [8]. As an example we consider the Feynman integral

$$\tilde{D}(x, y) = \int\int_{S^3 \times S^1} D_c^1(x, x') D_c^3(x', x'') D_c^1(x'', y) dx' dx''. \tag{14}$$

We use the orthogonality of $C_j^1(\cos\chi)e^{im\alpha}$ on $S^3 \times S^1$. Looking at the figures 1 and 2 it is clear that only the off shell amplitudes give a nonvanishing contribution.

After some rearrangement we find formally

$$\tilde{D} \propto \sum_{j=0}^{\infty} \sum_{p=0}^{\infty} \sum_{m-j \text{ even}} \frac{1}{[(j+1)^2 - m^2]^2} \left(\text{off shell expansion of } D_c^3 \right). \tag{15}$$

The amplitudes of this expression diverge, because the p-sum diverges. Performing the p-sum up to a finite value $p_{\mathbf{max}}$ could act as a SO(4)$\times$ SO(2)-invariant cutoff. So we find that $\tilde{D}(x,y)$ is not Fourier-integrable. But we do not yet know whether it exists as D_c^3 exists. If it does, then it is conformally invariant, as the integrals in Eq. (14) act on fields with degree of homogeneity -4 in x', x'' and thus are conformally invariant.

We conclude that it is possible to construct conformally invariant causal propagators. At least some loop-integrals in conformal space can be calculated. Infinities occur which need to be removed. As a next step we should find out, whether there is a conformally invariant way to separate them or whether renormalization breaks conformal invariance also in this formulation. Modifications of analytic renormalization, dimensional renormalization, differential regularization [9] or an extension to 5-dimensional space come to mind. Specifically one could try to calculate the "fish"-diagram and see whether a scale dependent coupling constant appears. Even in this case conformal invariance would not necessarily be lost, as the radius of the S^3 does provide a conformally invariant scale, a phenomenon not available in flat Minkowski space.

Acknowledgements

I have the pleasure to thank Prof. L. Castell for valuable discussions.

References

[1] S. Coleman and R. Jackiw. *Ann. Phys. (N.Y.)*, 67:552, 1971.

[2] J. Pedersen, I. E. Segal, and Z. Zhou. *Nucl. Phys. B*, 376:129, 1992.

[3] B. Binegar, C. Fronsdal, and W. Heidenreich. *J. Math. Phys.*, 24:2828, 1983.

[4] L. Castell. *Nucl. Phys. B*, 13:231, 1969.

[5] I. T. Drummond. *Nucl. Phys. B*, 94:115, 1975.

[6] I. S. Gradshteyn and I. M. Ryzhik. *Table of Integrals Series and Products.* Academic Press, New York, 1965.

[7] A. Erdelyi. *Higher Transcendental Functions.* McGraw-Hill, New York, 1953.

[8] K. G. Chetyrkin, A. L. Kataev, and F. V. Tkachov. *Nucl. Phys. B*, 174:345, 1980.

[9] D. Z. Freedman, K. Johnson, and J. I. Latorre. *Nucl. Phys. B*, 371:353, 1992.

Institut für Theoretische Physik A
TU Clausthal
D-3393 Clausthal-Zellerfeld
Germany

Gauge groups, anomalies and non-abelian cohomology

R.F.Streater

Abstract

We summarise the arguments leading to the existence of gauge groups, stressing the dependence of the group on the representation of the observables. We show that anomalies are two-cocycles in a non-abelian cohomology. At non- zero temperature the anomaly is a coboundary. We find general conditions that rule out the possibility that the total baryon-number is approximated by local gauge-invariant operators.
Key words: gauge groups, anomalies.

1 Symmetries, Wigner symmetries and gauge transformations

In the algebraic formulation of quantum mechanics,[12] the bounded observables are hermitian elements in a unital C^*-algebra $\mathcal{A}$ and the time- evolution is given by a group homomorphism $\tau : \mathbf{R} \to AUT\mathcal{A}$; here, $AUT\mathcal{A}$ is the group of $\mathbf{C}$-linear *- automorphisms of $\mathcal{A}$. Let $AUT_{\mathbf{R}}\mathcal{A}$ be the group of linear and anti-linear *-automorphisms of $\mathcal{A}$. A symmetry is then an element $g \in AUT_{\mathbf{R}}\mathcal{A}$ that commutes with $\{\tau_t\}$. The set G of symmetries is easily seen to be a subgroup of $AUT_{\mathbf{R}}\mathcal{A}$. In interesting cases the elements of G will not be inner: an inner automorphism is one of the form $g(A) = AdU\,A$ for some $U \in \mathcal{A}$, where for any unitary U, $AdU\,A = UAU^{-1}$.

A generalisation of the density matrix is the idea of a **state** on $\mathcal{A}$; it is any element ω of the complex linear dual space $\mathcal{A}^d$ obeying

$$\omega(1) = 1 \quad \text{and } \omega(A^*A) \geq 0 \text{ for all } A \in \mathcal{A}. \tag{1}$$

E. A. Tanner and R. Wilson (eds.),
Noncompact Lie Groups and Some of Their Applications, 333–340.

The set of states is a convex subset $\Sigma \subseteq \mathcal{A}^d$. In elementary quantum mechanics we use a Hilbert space $\mathcal{H}$ and $\mathcal{A}$ is taken to be $\mathbf{B}(\mathcal{H})$, the set of all bounded operators on $\mathcal{H}$. A normalised vector $\psi \in \mathcal{H}$ defines a state on $\mathbf{B}(\mathcal{H})$, denoted ω_ψ, by the formula

$$\omega_\psi(A) = \langle \psi, A\psi \rangle_{\mathcal{H}} \tag{2}$$

where $\langle \bullet, \bullet \rangle_{\mathcal{H}}$ is the scalar product in $\mathcal{H}$. Such a state is called a "vector state" relative to $\mathcal{H}$. A density matrix ρ defines a state ω_ρ by the formula

$$\omega_\rho(A) = Trace(\rho A) \tag{3}$$

The C^*-algebra version is more general; it allows us to consider more than one representation $\pi, \mathcal{H}_\pi$ of $\mathcal{A}$: π is a *-homomorphism from the *-algebra $\mathcal{A}$ into $\mathbf{B}(\mathcal{H}_\pi)$. In this way the operators $\{\pi(A) : A \in \mathcal{A}\}$ act on the representation space $\mathcal{H}_\pi$ and we recover examples of the concrete theory.

The Gelfand-Naimark-Segal construction allows us to find a cyclic representation π_ω acting on $\mathcal{H}_\omega$ corresponding to any given state $\omega \in \Sigma$. There is a cyclic vector $\Omega_\omega \in \mathcal{H}_\omega$ such that

$$\omega(A) = \langle \Omega_\omega, \pi_\omega(A)\Omega_\omega \rangle_{\mathcal{H}_\omega} \tag{4}$$

Thus ω is a vector state in $\mathcal{H}_\omega$. It is known that π_ω is irreducible if and only if ω is pure, that is, is an extreme point of Σ. In theories with superselection rules we use reducible representations. Also, the representation corresponding to the temperature state ρ_β at beta β is reducible, as ρ_β is mixed. That ρ_β is represented as a vector in a Hilbert space in "thermofield dynamics" is nothing more than this general result of Gelfand, Naimark and Segal.

We shall distinguish between Wigner symmetries and spontaneously broken symmetries. Both concepts depend on the representation we choose for $\mathcal{A}$. We say that a symmetry g is implemented in π if there exists a unitary or anti-unitary operator U_g acting on $\mathcal{H}_\pi$ such that

$$\pi\left(g(A)\right) = AdU(\pi(A)) \text{ for all } A \in \mathcal{A}; \tag{5}$$

then g is a symmetry in the sense of Wigner. If no such U_g exists we say that the symmetry is spontaneously broken (in the representation π). In that case the usual consequences of Wigner's theory of symmetries do not follow. We shall be concerned only with Wigner symmetries, and will denote by G_π the subgroup of G of symmetries implemented in π.

If the representation π of $\mathcal{A}$ is reducible, by Schur's lemma, then there exist non-trivial elements of the commutant $\pi(\mathcal{A})'$, which is the set of bounded operators on $\mathcal{H}_\pi$ commuting with all the representing operators $\pi(A)$. The unitary elements of $\pi(\mathcal{A})'$ form a group, which we shall call the π-gauge group $\mathcal{G}_\pi$, to emphasise its dependence on the representation. At zero temperature, we can define the representation π to be the direct sum of all the irreducible representations "close" to the vacuum state; then $\mathcal{G}_\pi$ is compact and is what physicists call the rigid gauge group; its existence can be derived under very general conditions [3]. If π is irreducible, as in Wigner's original theory, then $\mathcal{G}_\pi$ is just the circle group $U(1)$. For a temperature state, $\mathcal{G}_\pi$ is enormous, being the unitaries in a type III von Neumann algebra. Not all physicists agree that the gauge group can be different at non-zero temperature from what it is at zero temperature. With our definition, we do not regard the gauge transformations as symmetries. Nor can we envisage the gauge group's being spontaneously broken; it might happen that an element of $\mathcal{G}$ is present in one representation, but is simply not there in another. The unphysical nature of gauge transformations in our theory has its parallel in thermofield theory, in the unphysical nature of the thermofield.

We can now proceed as in [1] with the study of the mapping $g \mapsto U_g$. It is ambiguous up to a factor in $\mathcal{G}_\pi$, instead of up to a phase, as in Wigner's theory. It follows that any choice of U_g for all $g \in G_\pi$ leads to a projective representation with multiplier ω in $\mathcal{G}_\pi$ thus:

$$U_g U_h = \omega(g.h) U_{gh}. \tag{6}$$

It is shown in [1] that $\gamma_g = Ad\, U_g$ maps $\mathcal{G}_\pi$ to itself, and that ω obeys the cocycle property

$$\omega(g,h)\omega(gh,k) = \gamma_g\left(\omega(h,k)\right)\omega(g,hk). \tag{7}$$

Note that γ_g is not a true action of G on $\mathcal{G}_\pi$, but a projective action, namely

$$\gamma_g \gamma_h = Ad\omega(g,h)\,\gamma_{gh}. \tag{8}$$

This information is summarized by saying that the set $G_\pi \times \mathcal{G}_\pi = \mathcal{E}$ is made into a group if given the multiplication law

$$(g,u) \times_\omega (h,v) = (gh, u(\gamma_g v)\omega(g,h)). \tag{9}$$

This may be written neatly by saying that

$$1 \rightarrow \mathcal{G}_\pi \rightarrow \mathcal{E} \rightarrow G_\pi \rightarrow 1 \tag{10}$$

is an exact sequence of groups. Two implementations $g \mapsto U_g$ and $g \mapsto u(g)U_g$, where $u(g) \in \mathcal{G}_\pi$, give rise to the same action on $\pi(\mathcal{A})$; we say that the corresponding multipliers ω and ω' are equivalent. This is true if and only if the corresponding extensions $\mathcal{E}$ and $\mathcal{E}'$ are isomorphic in such a way that there exists an isomorphism $\phi : \mathcal{E} \rightarrow \mathcal{E}'$ such that

$$\begin{array}{ccccccc} & & & \mathcal{E} & & & \\ & & \nearrow & & \searrow & & \\ 1 \longrightarrow & \mathcal{G}_\pi & & \downarrow \phi & & G_\pi & \longrightarrow 1 \\ & & \searrow & & \nearrow & & \\ & & & \mathcal{E}' & & & \end{array}$$

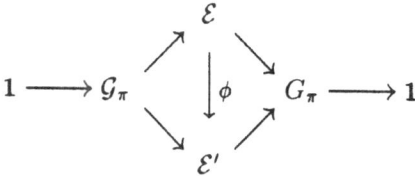

is commutative. This notion of equivalence, which arises naturally in our context, happens to coincide with that originally defined by Eilenberg and Mclean [11]. Nevertheless it is not common in the physics literature. Whereas in Wigner's theory the generator of a one-parameter symmetry group is ambiguous up to a scalar multiple of the identity, in our theory its ambiguity is the possible addition of any self-adjoint operator commuting with $\pi(\mathcal{A})$. We might try to resolve this ambiguity by requiring that the generators be observable, in that they are gauge invariant. At zero temperature, Borchers has shown that for the space-time translation group, this is always possible [6]. In [1] a similar-looking but less deep result is obtained for any symmetry, if the representation is a factor of type I. Then the ambiguity is reduced to the addition of a multiple of the identity. In Sect.(3) we show that in thermal states it is rarely possible to find gauge-invariant generators of one-parameter symmetry groups. But anomaly-free generators are easy to come by. Both these statements have generated controversy. We next describe anomalies in the context of our general theory.

2 Anomalies

Haag [12] has argued that while all representations π of $\mathcal{A}$ may be interesting, some are more interesting than others. For example, it is difficult to interpret the states of a representation in which time- evolution is spontaneously broken. We shall confine our discussion to representations in which for each t, τ_t is implemented by a unitary operator U_t, continuous in time. This rules out "quark" representations, which may be defined as those for which time-evolution, if implemented, is not continuous. But we will not require that the generator of time-evolution, the Hamiltonian, be bounded below. This will allow us to include infinite thermal systems.

According to our general analysis, it is possible for a Wigner symmetry U_g to fail to commute with U_t, but to acquire a multiplier

$$U_g U_t = \omega(g,t) U_{g\tau_t} \quad \omega \in \pi(\mathcal{A})'. \tag{11}$$

If ω is not a coboundary we say that the theory has an anomaly in its symmetry group. In particular, if g lies in a one-parameter group of Wigner symmetries, and has an anomaly, then the generator will not commute with the Hamiltonian, and so fails to be conserved. It is surprising that this phenomenon was not recognised until 1969 [4, 5]. In [1] we construct a lattice field theory in which the translation group of the lattice has an anomaly, and momentum is not conserved. This strange fact cannot be resolved by changing the definition of the Wigner translation to an equivalent one: the cocycle is not a coboundary. The anomaly in this model is caused by the presence of a constant magnetic field. Schrader [7] has studied a relativistic electron in a constant electromagnetic field and has shown that the kinematic group is a central extension of the Poincaré group. The field is translation-invariant, yet momentum is not conserved: the electron is accelerated along the electric field, and suffers the Lorentz force due to the magnetic field. The conventional explanation for the violation of translation symmetry is that it is not possible to choose the magnetic potential to be independent of position. We rather prefer to say that the dynamics has an anomaly. Also, consider the motion of a quantum particle in a space-time constant gravitational field. The gravitational potential is not translation-invariant, and momentum is not conserved. This can also be described by an anomaly in the space-time symmetry group. We can regard anomalies in symmetry groups as a useful generalisation of Wigner's theory allowing the presence of external fields.

Anomalies in gauge groups, on the other hand, are generally agreed to be a bad thing, indicating perhaps an inconsistency in the theory. Hence the physicists' desperate search for anomaly-free gauge theories. 'tHooft has remarked that there is an anomaly in the $U(1)$ rigid gauge group associated with baryon number in the standard model, but this does not lead to internal contradictions, since it is not "gauged", that is, the group is not coupled to a gauge boson. He suggested that this baryon non-conservation might lead in cosmological models to an explanation of the present excess of baryons in the universe. But his detailed calculation showed that the effect was too small by a huge factor. The idea was revived in [8], where it was suggested that the effect might be enhanced by the (presumed) high temperature of the early universe. In this connection, we note in [10] that in the thermal representation π_β the algebra is of type III, and that means that all symmetry groups can be implemented without multiplier. In other words, the cocycle $\omega(g, h)$ is a coboundary. This is applied to the baryon $U(1)$ group by working with the field algebra and adding this $U(1)$ to the symmetry group rather than taking it as part of the gauge group. We then add the assumption that the automorphisms of the field algebra generated by the group $U(1)$ are symmetries, i.e. they commute with the time-evolution automorphisms of the field algebra. This is formally true in the absence of an external baryon potential. We call the triviality of the multipliers in a thermal state the phenomenon of anomaly meltdown [10].

The unitary operators used to achieve a representation of G_π without multipliers in a thermal state are not gauge invariant. But their mathematical origin is impeccable: they are obtained by the Gelfand-Naimark-Segal construction from an invariant state. Like the KMS Hamiltonian, the generators of one-parameter symmetry groups are not weak limits of observables, and are not universally accepted by physicists as having a physical meaning. Some physicists maintain that our proof that the anomaly is a coboundary at non-zero temperature does not mean that the physical baryon number is conserved. By its construction, our baryon number N annihilates the cyclic vector Ω_β, the thermal state-vector. It therefore measures the baryon number relative to Ω_β rather than the total baryon number, which is infinite. Since Ω_β is a separating vector for the representation, it is not possible for an operator like N, obeying

$N\Omega_\beta = 0$, to be affiliated to the observable algebra; that is its spectral resolutions cannot lie in $\pi(\mathcal{A})''$. Physicists would like to replace our N, which is conserved in time, by a gauge invariant generator that is a weak limit of the physical fields (which includes the baryon field), and to be a sum of contributions from local regions, so that it is the integral of the baryon density. However we show in the next section that under quite general conditions, no such local gauge-invariant symmetry-generators can exist.

3 Non-existence of a quasi-local number operator

At a naive level, the total number operator for an infinite volume of material at non-zero density cannot exist, as its eigenvalues are infinite. We formulate this idea by showing that at non-zero temperature, symmetry groups in general cannot be inner automorphisms of the von Neumann algebra $\pi(\mathcal{A})''$, under the assumption that 0 is the only point eigenvalue of the KMS Hamiltonian. We first remark that any implemented automorphism of $\mathcal{A}$ can be extended to an implemented automorphism of $\pi(\mathcal{A})''$.

Theorem. *Let $\pi(\mathcal{A})''$ be a factor with cyclic and separating vector Ω. Let U_t be the KMS time-evolution (as given by the Tomita-Takesaki theory) and suppose that Ω is the only eigenvector of U_t in the Hilbert space. Suppose that $V \in \pi(\mathcal{A})''$ is an internal symmetry, that is, $Ad\,V$ commutes with $Ad\,U_t$ on $\pi(\mathcal{A})''$. Then V is a multiple of the identity!*

Proof. Since $Ad\,V$ is a symmetry we have

$$U_t V A V^{-1} U_t^{-1} = V U_t A U_t^{-1} V^{-1} \text{ for all } A \in \pi(\mathcal{A})''. \tag{12}$$

This says that A commutes with $U_t^{-1} V^{-1} U_t V$, which therefore lies in the commutant $\pi(\mathcal{A})'$. But $Ad\,U_t$ maps the observable algebra to itself, and so, as V is in the observable algebra, so is $U_t^{-1} V^{-1} U_t V$. Since it also lies in the commutant, and we have a factor, it must be a multiple of the identity. So we get

$$V^{-1} U_t V = U_t \lambda \text{ for some } \lambda \in U(1). \tag{13}$$

Apply this to Ω; we get $U_t(V\Omega) = \lambda(V\Omega)$. Hence $V\Omega$ is a stationary state. Since Ω is the only normalisable eigenvector, $V\Omega = \mu\Omega$ for some number μ. But Ω is separating, so $V = \mu$. This proves the theorem. □

This result shows that quasi-local number operator cannot be achieved in the usual theory, where the energy spectrum is continuous except for the ground state. The same applies to the total momentum operator, or the energy operator itself. This contrasts the case of zero temperature [6]. We cannot get round this by assuming a discrete spectrum of the Hamiltonian, such as would occur in a finite volume, if we are talking not about one symmetry operator, but a one-parameter group. For, in a separable Hilbert space the number of such energy-eigenstates is countable, but we can find an uncountable number of elements of the symmetry group which must be trivial. We note that any eigenvector of the energy is a cyclic and separating vector in a thermal state.

The actual universe is finite and contains a finite number of baryons. Our baryon number N measures the number relative to the thermal state. Suppose that the average density is ρ and the volume is Λ. Then the operator $N + \rho\Lambda$ is a conserved operator measuring this number, and Ω is an eigenstate of this, with eigenvalue $\rho\Lambda$. We have to rather far from the usual formalism to get a quasilocal number operator, which, if it existed, might well not be conserved. This is what shows up in perturbation theory when there is an anomaly at zero temperature.

Subsequent to [8], Shaposhnikov has considered [9] a gauge model with an anomaly *and* an infinitely degenerate ground state. This case is outside our formalism.

In the preprint [13], (5 May 1993, University of Vienna), Grosse, Maderner and Reitberger study the spin-chain anomaly of [1] at non-zero temperature, and construct a different implementer for the space translation than that suggested in [10]. Its multiplier is not the identity, but must be a coboundary (with respect to the equivalence relation used in [1]). This translation operator does not commute with the time evolution, and this fact expresses that it costs energy to move a soliton in the presence of a magnetic field. This is the correct choice for isolated dynamics, where the system, having been in contact with a heat bath to arrive at its non-zero temperature, is not coupled to the bath in its subsequent dynamics (Kubo dynamics). The choice of implementer for space-translation given in [10] commutes with time-evolution, and represents what happens in isothermal dynamics: local cooling causes heat to instantly flow in from the heat bath, replacing the lost energy. In this sense the anomaly is "at the boundary". A preprint by Hussin, del Olmo and Negro [14] describes a "smooth" version of the main theorem of [1]; the authors verify that a non-abelian multiplier can be reduced to a central multiplier in various classical gauge models, and moreover, that this reduction can be done using *smooth* gauge transformations. Smoothness is the relevant condition, because the arbitrariness of the gauge transformations is limited to those leaving the domain of the Hamiltonian fixed. Otherwise we can, for example, transform the Bohm-Aharonov effect away entirely by a singular gauge transformation. The effect is still there, in the "free" Hamiltonian subjected to singular boundary conditions.

Acknowledgements

This work was partially supported by the NATO workshop on Group Theory in Physics, St. Antonio, and was completed at Torun. The author thanks R. Wilson and A. Jamiolkowski for the hospitality of the Universities of Texas and

Torun. Thanks are due to R. Jackiw for helpful and informative correspondence, and to M. Michalski for TEXnical help.

References

[1] R.F.Streater, Symmetry groups and non-abelian cohomology. Commun. Math. Phys., **132**, 201-215 (1990).

[2] R.Haag and D. Kastler, An algebraic approach to quantum field theory. J. Mathematical Phys. **5**, 848-861 (1964).

[3] S.Doplicher and J.E.Roberts .Why there is a field algebra with a compact gauge group describing the superselection structure in particle physics. Commun. Math. Phys. **131**, 51-107 (1990).

[4] J.S.Bell and R.Jackiw, Nuovo Cimento **51**, 47 (1969).

[5] S.Adler, Phys. Rev. **177**, 2426 (1969).

[6] H.J.Borchers, Energy and Momentum as Observables in Quantum Field Theory,Commun. Math. Phys. **2**, 49-54, (1966).

[7] R.Schrader, Fortschritte der Phys.**20**, 701 (1972). See also: J.Negro ang M.A. del Olmo,Local realization of kinematical groups with a constant electromagnetic field I. The relativistic case. J. Math. Phys.**31**, 568-578 (1990)

[8] V. A. Kuzmin, V. Rubakov and M. E. Shaposhnikov, Phys. Lett. **B 155**, 36 (1985).

[9] M. E. Shaposhnikov, Structure of high temperature gauge ground state and electroweak production of the baryon asymmetry, Nucl. Phys. **B 299**, 797-817 (1988).

[10] C.Gajdzinski and R.F.Streater, Anomaly meltdown, J. Math. Phys. **32**, 1981-1983 (1991).

[11] S. Eilenberg and S. Mclean, Cohomology theory in abstract groups, Annals of Math. **48** 51-78 (1947); ibid, 326-341.

[12] R. Haag, **Local Quantum Physics**, Springer, (1992).

[13] H. Grosse, W. Maderner, and C. Reitberger, On Spin Chains, Charges and Anomalies, Preprint, Institute for Theoretical Physice, Vienna, 1993.

[14] V.Hussin, J. Negro and M. A. del Olmo, Invariant Connections in a Non-abelian Principal Bundle, Preprint CRM-1872, Universite de Montreal, April, 1993.

Dept. of Mathematics,
King's College,
The Strand,
London. WC2R 2LS.

THE E_8 FAMILY OF QUASICRYSTALS

R. V. Moody and J. Patera

Abstract. An explicit method for building of families of quasicrystals from two directions simultaneously: (i) by way of a projection from periodic structures of twice higher dimension and (ii) without any reference to higher dimensional structures. Both constructions are two interpretations of the same formalism. Our construction naturally contains the 2-, 3-, and 4-dimensional quasicrystals in a uniform formalism. Moreover it contains as special cases practically all the cornucopia of constructions found in literature involving 5-fold symmetry as well as many others.

The purpose of this talk is to overview a new approach to the structure of quasicrystals [1]. By starting from the root lattice of E_8, the largest exceptional simple Lie algebras, and treating it as the generic "mother of all quasicrystals"—certainly of all quasicrystals with 5-fold symmetry as well as many others—one gains an appealing unifying view of the diversity of special cases found in the literature. Indeed, the number of special cases and the avenues through which quasicrystals arise from E_8 is rather large and has not yet been investigated systematically.

There are several ingredients of our approach which we want to point out first.

i) Besides the real numbers $\mathbb{R}$, the rational numbers $\mathbb{Q}$ and integers $\mathbb{Z}$, one needs the extension $\mathbb{F}$ of $\mathbb{Q}$ by $\sqrt{5}$. The constants $\tau = \frac{1}{2}(1+\sqrt{5})$ and $\sigma = \frac{1}{2}(1-\sqrt{5})$ and the identities

$$\tau + \sigma = 1, \quad \tau^2 = 1 + \tau, \quad \sigma^2 = 1 + \sigma \tag{1}$$

are frequently used. A crucial role is played by the $\mathbb{F}$-conjugation denoted by prime:

$$x' = (a + b\sqrt{5})' = a - b\sqrt{5}, \qquad a, b \in \mathbb{Q}. \tag{2}$$

ii) It is advantageous to use the quaternions $\mathbb{H}_\mathbb{F}$ over $\mathbb{F}$:

$$\mathbb{H}_\mathbb{F} \ni x = x_1 + ix_2 = jx_3 + kx_4 =: (x_1, x_2, x_3, x_4), \qquad x_1, x_2, x_3, x_4 \in \mathbb{F}, \tag{3}$$

Work supported in part by grants from the Natural Sciences and Engineering Research Council of Canada, by the Fonds FCAR du Québec, and by the Killam Research Fellowship (J.P).

E. A. Tanner and R. Wilson (eds.), Noncompact Lie Groups and Some of Their Applications, 341–347.

and the standard operations

$$\begin{aligned} \mathbb{H}\text{-conjugation of } x &: \quad \bar{x} = (x_1, -x_2, -x_3, -x_4) && (4)\\ \text{multiplication in } \mathbb{H} &: \quad xy \in \mathbb{H},\ x, y \in \mathbb{H} && (5)\\ \text{scalar product in } \mathbb{H}_\mathbb{F} &: \quad x \cdot y = \frac{1}{2}(x\bar{y} + y\bar{x}) \in \mathbb{F} && (6)\\ \mathbb{H}_\mathbb{F}\text{-norm} &: \quad N(x) = x \cdot x = |x|^2 = x\bar{x} \in \mathbb{F}. && (7) \end{aligned}$$

A crucial role is played by the rational form $(x, y)_\tau$ relative to τ on $\mathbb{H}_\mathbb{F}$, defined as follows. If $x \cdot y = a + \tau b$, $a, b \in \mathbb{Q}$, then

$$(x, y)_\tau = a. \tag{8}$$

Analogously one defines a rational form relative to σ.

iii) Next is a special set of 120 quaternions of norm 1, called **icosians**. The set I of icosians consists of the following elements of $\mathbb{H}_\mathbb{F}$:

$$\begin{aligned} &\tfrac{1}{2}(\pm 1, \pm 1, \pm 1, \pm 1) && && (9)\\ &(\pm 1, 0, 0, 0) && \text{\& all permutations} && (10)\\ &\tfrac{1}{2}(0, \pm 1, \pm\sigma, \pm\tau) && \text{\& all even permutations} && (11) \end{aligned}$$

The icosians form the group H_3, $|H_3| = 120$, under the quaternionic multiplication. It is isomorphic to the binary icosahedral group.

iv) Finally, we use the isometric isomorphism $\pi_\parallel$ between the space V spanned by the roots of E_8 and $\mathbb{H}_\mathbb{F}$. The mapping $\pi_\parallel$ establishes the correspondence between the root system Δ of E_8 and the icosians and their τ-multiples

$$\begin{aligned} &\pi_\parallel : \Delta \longleftrightarrow I \cup \tau I \\ &(\alpha|\beta) = 2(\pi_\parallel(\alpha), \pi_\parallel(\beta))_\tau, \qquad \alpha, \beta \in \Delta. \end{aligned} \tag{12a}$$

Similarly one defines

$$\begin{aligned} &\pi_\perp : \Delta \longleftrightarrow I \cup \sigma I \\ &(\alpha|\beta) = 2(\pi_\perp(\alpha), \pi_\perp(\beta))_\sigma. \end{aligned} \tag{12b}$$

Note that one has $(\pi_\parallel(\alpha))' = \pi_\perp(\alpha)$.

Our departure point is an explicit realization of $\pi_\parallel$. On Fig. 1 the simple roots of E_8 are mapped into specific icosians a_1, a_2, a_3, a_4 and their τ-multiplies. The rational form $(\,,\,)_\tau$ in (8) assures that the geometric relations (relative lengths and angles) of E_8 roots are preserved under $\pi_\parallel$.

An instructive illustration of the action of $\pi_\parallel$ on the roots of A_4 is found on Fig. 2. The E_8 simple roots $\alpha_3, \alpha_4, \alpha_5, \alpha_8$ of Fig. 1 are renamed according to the A_4 convention as $\alpha_1, \alpha_2, \alpha_3, \alpha_4$ respectively.

DEFINITION OF QUASICRYSTALS

Consider the $\mathbb{Z}$-span of the icosians I, called the icosian ring $\mathbb{I}$. Clearly one has $\mathbb{I} \subset \mathbb{H}_\mathbb{F}$.

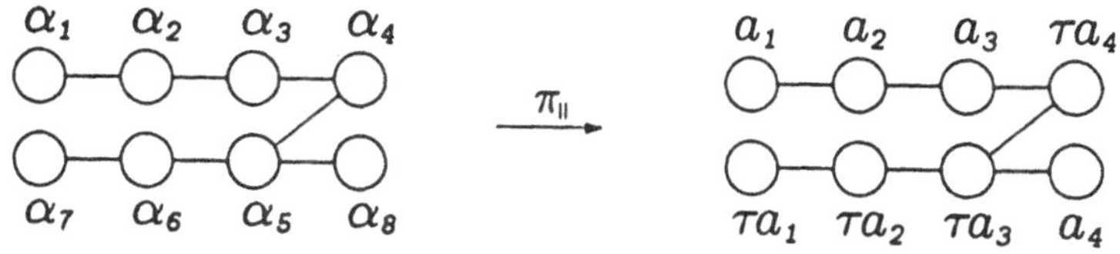

$$
\begin{aligned}
a_1 &= \tfrac{1}{2}(-\sigma, -\tau, 0, -1), & \tau a_1 &= \tfrac{1}{2}(1, -\tau^2, 0, -\tau),\\
a_2 &= \tfrac{1}{2}(0, -\sigma, -\tau, 1), & \tau a_2 &= \tfrac{1}{2}(0, 1, -\tau^2, \tau),\\
a_3 &= \tfrac{1}{2}(0, 1, -\sigma, -\tau), & \tau a_3 &= \tfrac{1}{2}(0, \tau, 1, -\tau^2),\\
a_4 &= \tfrac{1}{2}(0, -1, -\sigma, \tau), & \tau a_4 &= \tfrac{1}{2}(0, -\tau, 1, \tau^2).
\end{aligned}
$$

FIGURE 1. Mapping of the simple roots of E_8 into specific icosians (quaternions).

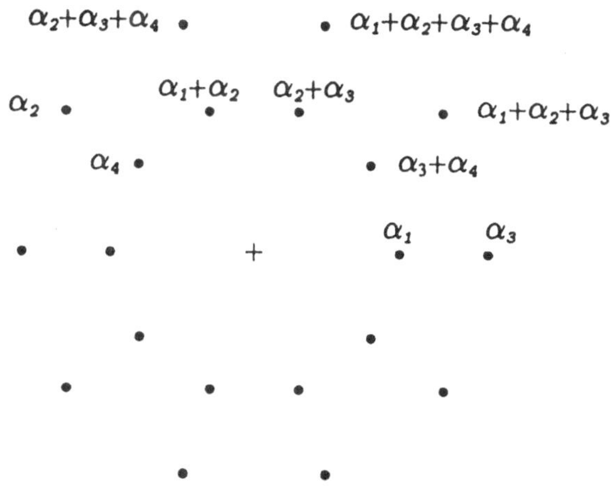

FIGURE 2. The root system of of A_4 after the projection $\pi_{\|}$. The E_8 simple roots $\alpha_3, \alpha_4, \alpha_5, \alpha_8$ of Fig. 1 are renamed according to the A_4 convention as $\alpha_1, \alpha_2, \alpha_3, \alpha_4$ respectively.

For any fixed $r > 0$ we define a generic quasicrystal as the set Σ^r of points x determined as follows:

$$\Sigma^r = \{x \mid x \in \mathbb{I} \text{ and } N(x') < r^2\}. \tag{13}$$

The interior of the sphere in $\mathbb{H}_{\mathbb{F}}$, containing the $\mathbb{F}$-conjugates of $x \in \mathbb{I}$, given by the inequality in (13) is called the **acceptance domain** of Σ^r.

The following properties are immediate consequences of the definition

$$\Sigma^r\Sigma^r \subset \Sigma^{rs}; \tag{14}$$

$$\Sigma^r + \Sigma^r \subset \Sigma^{r+s}; \tag{15}$$

$$\Sigma^r \subset \Sigma^s \qquad \text{if } r < s; \tag{16}$$

$$\bigcup_{r\geq 0} \Sigma^r = \mathbb{I}; \tag{17}$$

$$\text{If } x \in \Sigma^r \text{ then } \tau x \in \Sigma^{r/\tau} \subset \Sigma^r. \tag{18}$$

The last property is called the **inflation symmetry** of Σ^r.

The elements of $\mathbb{I}$ are quaternions. Hence they can be percieved as points of a four dimensional real space. Consequently the quasicrystals of (13) are four dimensional. Note that our definition makes no reference to any higher dimensional space. However, in view of the maps $\pi_\parallel$ and $\pi_\perp$ of (12) and Fig. 1, the points of $\mathbb{I}$ can also be taken to be the result of the mapping $\pi_\parallel$ of the points of the E_8 root lattice ('cut and projection method').

POINT GROUP SYMMETRIES OF Σ^r.

Suppose $x \in \Sigma^r$ and let $s, t \in I \times I$. There are 120^2 transformations of x

$$\begin{aligned} \phi_{(s,t)}x &= sxt^{-1} \\ \gamma x &= \bar{x} \end{aligned} \tag{19}$$

preserving the norm $N(x)$ and hence also $N(x')$. Consequently these are symmetries of Σ^r.

It can be shown [1], that the finite group of the transformations (19) of Σ^r is the largest non-crystallographic irreducible finite Coxeter group, called H_4. Its Coxeter diagram and the implied conventions provide a succinct exhaustive description of the group. [1]

PHASONS

Let $\Phi \in V$ be a fixed vector called **phason**, and let us generalize the quasicrystal definition (13) using the projections $\pi_\parallel(\Phi)$ and $\pi_\perp(\Phi)$ as follows.

$$\Sigma^r_\Phi = \{x + \pi_\parallel(\Phi) \mid x \in \mathbb{I} \text{ and } N(x' + \pi_\perp(\Phi)) < r^2\}. \tag{20}$$

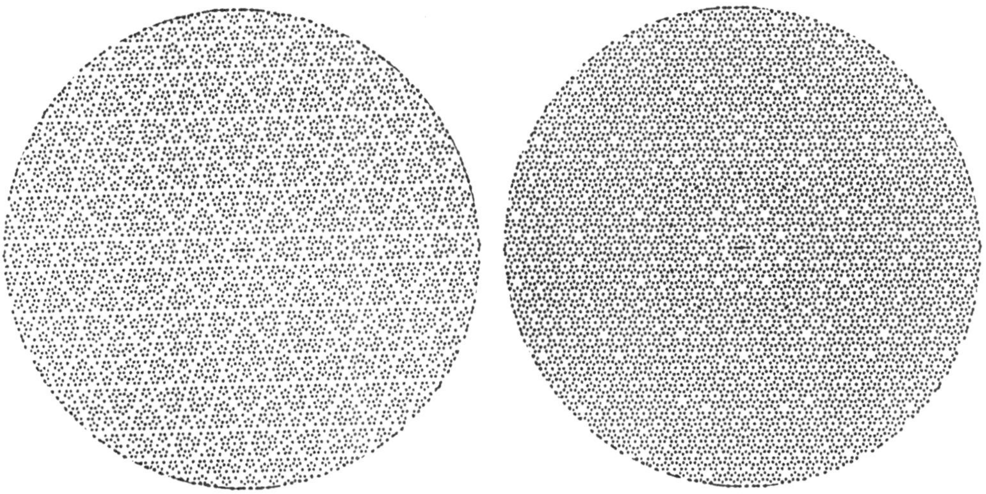

FIGURE 3. Circular window view of the A_4-quasicrystals Σ^8 and Σ^9.

Clearly Σ_0^r of (20) is Σ^r of (13). The acceptance domain of Σ_Φ^r for $\Phi \neq 0$ is not centered at the origin anymore. The fixed point of the H_4-symmetries is now $\pi_\parallel(\Phi)$ and, in general, it belongs neither to Σ_Φ^r nor to $\mathbb{I}$.

EXAMPLES

In order to be able to draw our examples in a plane, we consider here an A_4-subdiagram of the E_8-diagram of Fig. 1 consisting of the nodes $\alpha_3, \alpha_4, \alpha_5, \alpha_8$. Resulting quasicrystals are two dimensional. On Figs. 3-6 we draw a circular window view of some examples of such quasicrystals. This also serves as an illustration of how easily the definitions (13) and (20) apply to lower dimensional cases. Indeed, the quasicrystals arise by means of the same maps $\pi_\parallel$ and $\pi_\perp$ as before applied to a sublattice of the root E_8-lattice. All there is to do is to restrict $\mathbb{I}$ to the $\mathbb{Z}$-span of the corresponding subset of icosians. Note that the 3-dimensional quasicrystals with icosahedral symmetry arise in the same way by cutting off the E_8-diagram the simple roots α_1 and α_7. In our setup the later subset of icosians consists then of the pure icosians.

The first two examples are on Fig. 3. There two A_4-quasicrystals Σ^8 and Σ^9 are drawn (to the same scale) in order to illustrate the effect of changing the radius of the acceptance domain from $r = 8$ to $r = 9$. The origin, marked by a cross, is a quasicrystal point. Note the 10-fold rotation symmetry around the origin. Most striking feature is the presence/absence of the stright avenues across the quasicrystal.

The difference $\Sigma^9 \backslash \Sigma^8$ between the two quasicrystals of Fig. 3 is also quasiperiodic but in has no inflational symmetry. It is shown on Fig. 4.

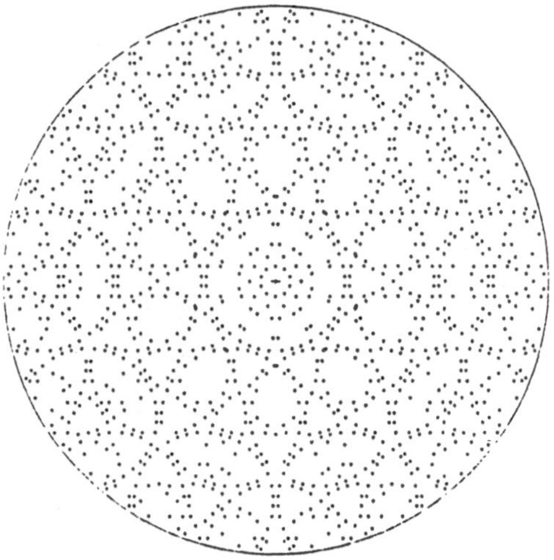

FIGURE 4. The set difference $\Sigma^9 \backslash \Sigma^8$ of the quasicrystals of Fig. 3 is quasiperiodic but has no inflation symmetry.

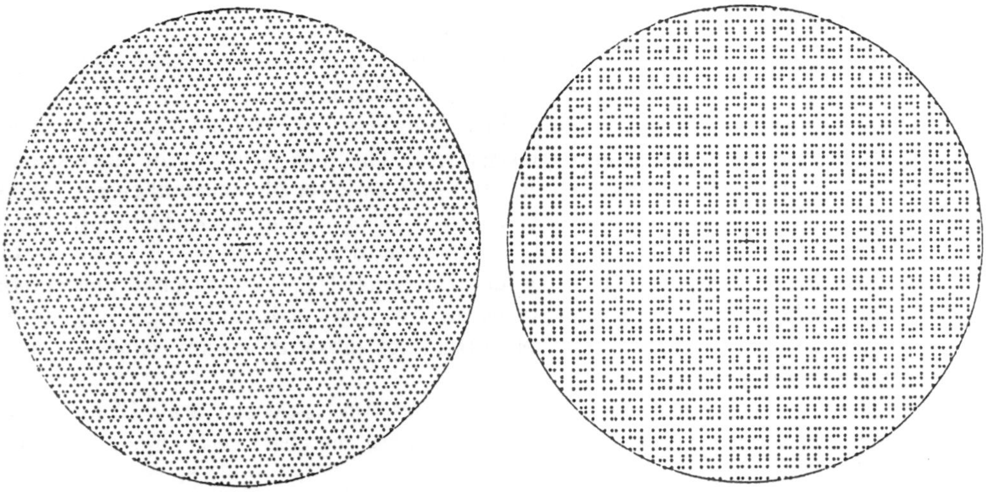

FIGURE 5. Examples of the Σ^5 quasicrystals of $2A_2$ and $4A_1$.

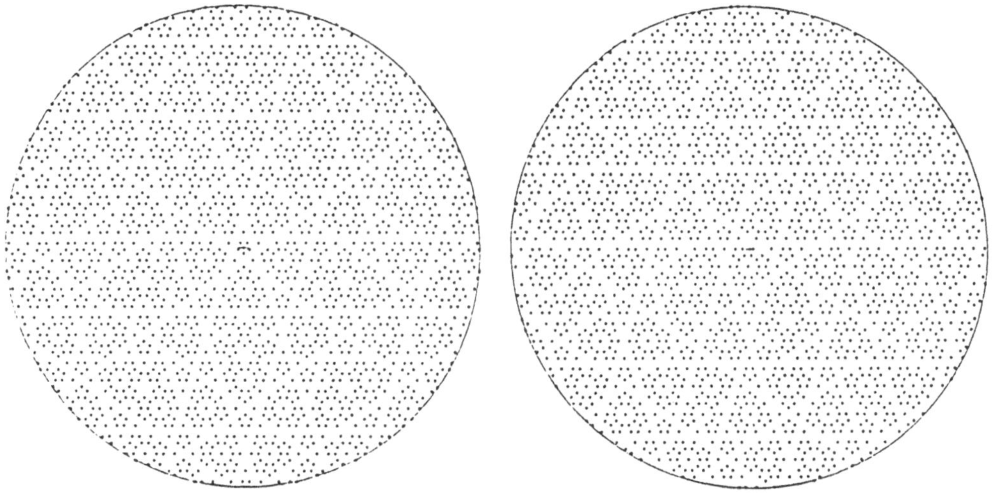

FIGURE 6. Two A_4-quasicrystals Σ^5_Φ. In the first one Φ is the fundamental weight ω_1, in the second case $\Phi = 0.7\omega_1 + 0.3\omega_2 + 0.4\omega_3 + 0.6\omega_4$.

Using another subdiagram of the E_8-diagram of pairs of vertically aligned nodes of the E_8-diagram and using the same mappings $\pi_\parallel$ and $\pi_\perp$ as before in the definition (13) or (20), yields different quasicrystals. Thus from the subdiagram $2A_2$ one gets quasicrystals with 6-fold symmetry as illustrated on Fig. 5. Similarly cutting the E_8-diagram so as to have two pairs of vertically aligned nodes, the $4A_1$-subdiagram, yields quasicrystals with rectangular symmetries. An example is shown in Fig. 5.

The examples of Fig. 6 illustrate the role of a phason in A_4-quasicrystals. In the first case $\Phi = \omega_1$, the first fundamental weight of A_4. That is the root lattice points are displaced to coincide with A_4-weight lattice points of congruence class 1. Note that a global 5-fold symmetry around the origin remains but the origin does not belong to Σ^r. The second case is a displacement of the root lattice by a phason with no special properties, namely $\Phi = 0.7\omega_1 + 0.3\omega_2 + 0.4\omega_3 + 0.6\omega_4$. No global point symmetry remains.

The examples of this talk were calculated and printed using the version 2 of the software [2].

REFERENCES

[1] R.V. Moody and J. Patera, *Quasicryatals and icosians*, J. Phys. A: Math. Gen. **26** (1993), 2829–2853.

[2] R.V. Moody, J. Patera, and D. Rand, *Macintosh software simpLie, v.2*, Publications CRM, Université de Montréal (1993).

R. V. MOODY, Department of Mathematics, University of Alberta, Edmonton, Alberta, Canada

J. PATERA, Centre de recherches mathématiques, Université de Montréal, CP 6128-A, Montréal, Québec, Canada

Wavelet Interpolation and Approximate Solutions of Elliptic Partial Differential Equations

Raymond O. Wells, Jr.
Xiaodong Zhou

Abstract

The paper formulates and proves a second order interpolation result for square-integrable functions by means of locally finite series of Daubechies' wavelets. Sample values of a sufficiently smooth function can be used as coefficients of a wavelet expansion at a fine scale, and the corresponding wavelet interpolation function converges in Sobolev norms of first order to the original function. This has applications to wavelet-Galerkin numerical solutions of elliptic partial differential equations.

1 Introduction

In 1988 Daubechies introduced in [5] a class of compactly supported wavelet functions which give a very nice decomposition of $L^2(\mathbf{R})$, and by tensor products, $L^2(\mathbf{R}^n)$. These have been studied by numerous authors and there have been a number of applications of these wavelet systems in engineering, technology as well as more abstract parts of mathematics and mathematical physics. For each positive integer N there is a family of compactly supported wavelets of dimension $N-1$ (see [13,14,19]). Out of this family one can choose wavelets which have specific properties for an application at hand. In her original paper Daubechies chose a particular wavelet system for

E. A. Tanner and R. Wilson (eds.), Noncompact Lie Groups and Some of Their Applications, 349–366.

each $N \geq 1$, which is distinguished by having the wavelet function having maximal vanishing moments, and by choosing a specfic choice of roots of certain polynomials which arise in the choice of the coefficients which define the wavelet system. These still seem to be most useful for many applications, especially in numerical approximation to solutions of differential equations, as they have important differentiability properties.

We will concentrate in this paper on this distinguished family of wavelets, which are defined by specific coefficient systems $\{a_0, a_1, \ldots, a_{2N-1}\}$, which define the scaling functions which we denote simply as $\phi(x)$ and which is denoted by ${}_N\phi(x)$ in [5]. The scaling function $\phi(x)$ satisfies the 2-scale difference equation

$$\phi(x) = \sum_{k=0}^{2N-1} a_k \phi(2x - k) \tag{1}$$

where the coefficients $\{a_0, a_1, \ldots, a_{2N-1}\}$ satisfy

$$\begin{gathered} \sum_{k=0}^{2N-1} a_k = 2 \\ \sum_{k=0}^{2N-1} a_k a_{k+2l} = 2\delta_{0,l}, \quad l \in \mathbf{Z} \end{gathered} \tag{2}$$

As mentioned above we are going to consider the specific choice of coefficients $\{a_n\}$ made by Daubechies for each N, and we will simply let ϕ denote this scaling function with the parameter N suppressed and understood. It is an important parameter as it determines the support of the scaling function (and associated wavelet functions), and this parameter plays a role in the calculations in this paper, as we shall see.

In Section 2 we review the well-known theory of the associated wavelet functions for the scaling functions $\phi(x)$ discussed above, their translations and rescalings and the corresponding orthonormal decompostion of $L^2(\mathbf{R})$ and $L^2(\mathbf{R}^2)$. In particular we recall the Mallat algorithm for determining wavelet series coefficients from one level to another, which is very useful in applications (see Mallat [11]), and plays an important role in our interpolation scheme and its applications in [8] and [20].

The Mallat algorithm leads to a natural interpolation by finite level wavelet series for a given function. In Section 3 we give a precise description of such a wavelet series interpolation, and with estimates which show that this interpolation approximates functions in $L^2(\mathbf{R}^2)$ and in the Sobolev space $H^1(\mathbf{R}^2)$ with convergence rates of the order of 2^{-2j} for the first case and 2^{-j} for the second case, where j is the level of the scaling functions involved in the interpolation series. The interpolation series can be expressed in different forms, but one of the simplest is given by:

$$f^j(x,y) = \sum_{p,q} f_{p,q} \phi_p^j(x) \phi_q^j(y), \tag{3}$$

where $\phi_p^j(x) := 2^{j/2}\phi(2^j x - p)$, for $p \in \mathbf{Z}$, is the standard rescaling of the given scaling function $\phi(x)$. The problem is: given a particular function f, what is the best choice of coefficients in (3) so that the series $f^j(x,y)$ best approximates the function f. There are many different versions of the word "best," and we present a specific choice which has a good convergence rate, and which is very useful in our applications. We demonstrate the convergence properties of this choice (which involves evaluating the function f at specific points when f is smooth), and apply it to proving estimates for wavelet-Galerkin solutions to elliptic differential equations in Section 4, which was the original motivation for these considerations. Our results are related to similar types of approximation results where the Fourier coefficient formulas of the form

$$c_k^j = \int f(x)\phi_k^j(x)dx$$

are approximated by evaluating f at specific points ([2,9,17,7]).

In related papers [20,8] we use this interpolation result in the context of specific Galerkin solutions of elliptic differential equation boundary value problems. In these papers we use a wavelet representation of the boundary data, the line integrals involved in the weak form of the differential equation, of the boundary itself, as well as the more traditional use (in the spirit of finite element methodology) for the solutions on the interior of the domain. The wavelet system provides a unifying common method of representing all of the elements of a boundary value problem (see [16]). The interpolation scheme presented in this paper is a key ingredient in going between "values of the wavelet expansion coefficients" and "values of the corresponding functions" which the wavelet series represent. The algorithms in [20,8] concentrate on the manipulation of the wavelet coefficients to find an approximation of the solution of the differential equation, just as in image processing the manipulation of the wavelet expansion coefficients plays the major role (see Mallat [11] and Zettler et al [21]).

2 Wavelet Series and the Mallat Algorithm

Let ϕ be the Daubechies scaling function of order N as discussed in Section 1, and let $\{a_0, \ldots, \alpha_{2N-1}\}$ be the corresponding coefficients for the 2-scale difference equation (1). The following summary contains results originally obtained in [5,11] (see also the books [12,3,15]).

The vector $(a_0, \ldots, a_{2N-1})$ is called the *scaling vector* for the system, and we define the associated *wavelet vector* $(b_0, \ldots, b_{2N-1})$ by defining

$$b_k := (-1)^{k+1} a_{2N-1-k} \tag{4}$$

This is the convention used in [10] and differs slightly from that used originally in [5]. We define the corresponding wavelet function $\psi(x)$ by

$$\psi(x) := \sum_{k=0}^{2N-1} b_k \phi(2x - k). \tag{5}$$

With this convention, both the scaling function $\phi(x)$ and the corresponding wavelet function $\psi(x)$ have the same compact support $[0, 2N-1]$. Moreover, we have the following properties of moments of these functions.

$$\begin{aligned} \textstyle\int \phi(x)\,dx &= 1, \\ \textstyle\int x\phi(x)\,dx &= \frac{1}{2}\sum_{k=0}^{2N-1} k a_k, \\ \textstyle\int x^m \psi(x)\,dx &= 0, \quad m = 0, \ldots, N-1. \end{aligned} \tag{6}$$

Let us denote by

$$c := \frac{1}{2}\sum_{k=0}^{2N-1} k a_k, \tag{7}$$

the first moment of the function $\phi(x)$, as this will play a role in our calculations later. Moreover, ϕ and ψ are both differentiable of a certain finite order $\alpha(N)$, where $\alpha = O(N)$, with $\alpha(2) \cong 0.55$, $\alpha(3) \cong 1.088$, etc. (see [5], [6], [18]), where the latter two references give asymptotic formulas for $\alpha(n)$.

For $j, k \in \mathbf{Z}$, let us define

$$\phi_k^j(x) = 2^{j/2}\phi(2^j x - k), \qquad \psi_k^j(x) = 2^{j/2}\psi(2^j x - k), \quad x \in \mathbf{R},$$

and as usual we define $\phi_k(x) := \phi_k^0(x)$. The following properties of these functions are valid, as is easy to verify:

$$\mathrm{supp}(\phi_k^j) = \mathrm{supp}(\psi_k^j) = \left[\frac{k}{2^j}, \frac{k+2N-1}{2^j}\right] \tag{8}$$

$$||\phi_k^j||_{L^2(\mathbf{R})} = 1 \tag{9}$$

$$\sum_{k\in\mathbf{Z}} \phi_k^j(x) = 2^{j/2} \quad x \in \mathbf{R}. \tag{10}$$

Property (10) is the partition of unity property (especially for $j = 0$). Moreover, the wavelet system

$$\{\phi_k(x), \psi_k^j(x), \quad k \in \mathbf{Z}, j \in \mathbf{Z}^+\} \tag{11}$$

is an orthonormal basis for $L^2(\mathbf{R})$, where we denote by $\mathbf{Z}^+$ the nonnegative integers (we denote the positive integers by $\mathbf{N}$).

Let, for $j \in \mathbf{Z}$,

$$V_j := \mathrm{closure}(\mathrm{span}\{\phi_k^j : k \in \mathbf{Z}\}). \tag{12}$$

Then $L^2(\mathbf{R}) = \bigcup_j V_j$, that is, for any function $f \in L^2(\mathbf{R})$, if we let P_j denote the orthogonal projection $L^2(\mathbf{R}) \to V^j$, then $P_j f$ converges to f in the L^2 norm. We recall that the coefficients of $P_j f$ are given by the classical formulas

$$P_j f(x) = \sum_{k\in\mathbf{Z}} f_k^{j,0} \phi_k^j(x), \tag{13}$$

where

$$f_k^{j,0} = \int_{\mathbf{R}} f(x)\phi_k^j(x)\, dx, \tag{14}$$

since the ϕ_k^j's are an orthonormal basis for V_j. We will denote coefficients of the scaled scaling functions generically by $f_*^{*,0}$ and of the scaled wavelet functions by $f_*^{*,1}$, letting the superscripts $\{0\}$ and $\{1\}$ distinguish between the two cases.

If we let W_j denote the orthogonal complement of V_j in V_{j+1}, then we see that, for a fixed $J \in \mathbf{Z}^+$,

$$V_0 \oplus W_0 \oplus W_1 \oplus \ldots \oplus W_{J-1} = V_J. \tag{15}$$

Moreover, if we let

$$f(x) = \sum_{k\in\mathbf{Z}} f_k^{0,0}\phi_k^0(x) + \sum_{j\in\mathbf{Z}+}\sum_{k\in\mathbf{Z}} f_k^{j,1}\psi_k^j(x),$$

for $f \in L^2(\mathbf{R})$, then we see that we have from (15),

$$\sum_{k\in\mathbf{Z}} f_k^{J,0}\phi_k^J(x) = \sum_{k\in\mathbf{Z}} f_k^{0,0}\phi_k^0(x) + \sum_{j=0}^{J-1}\sum_{k\in\mathbf{Z}} f_k^{j,1}\psi_k^j(x). \tag{16}$$

In this case all of the coefficients are given by integral formulas of the type:

$$f_k^{j,0} = \int_{\mathbf{R}} f(x)\phi_k^j(x)\,dx, \quad j = 0,\ldots,J-1, \tag{17}$$

$$f_k^{j,1} = \int_{\mathbf{R}} f(x)\psi_k^j(x)\,dx, \quad j = 0,\ldots,J-1. \tag{18}$$

Moreover, it follows from the 2-scale difference equation that one can determine the coefficients on the right hand side of (16) in terms of the coefficients on the left hand side, and conversely. This is the Mallat algorithm [11], and the formulas will be given below. First we want to remark that if we consider *any* expansion of the form

$$\tilde{f}^J(x) := \sum_{k\in\mathbf{Z}} f_k^{J,0}\phi_k^J(x), \tag{19}$$

where the coefficients $f_k^{J,0}$ are prescribed in some fashion (perhaps relating to some given L^2 function f or not), then because of (15), we see that there is a corresponding expansion of the form

$$\tilde{f}^J(x) = \sum_{k\in\mathbf{Z}} f_k^{0,0}\phi_k^0(x) + \sum_{j=0}^{J-1}\sum_{k\in\mathbf{Z}} f_k^{j,1}\psi_k^j(x), \tag{20}$$

and the coefficients in (20) are determined in terms of the coefficients in (19) and the converse is true. This is the Mallat algorithm in general. It doesn't have anything to do with the coefficients being generalized Fourier coefficients of some given function as in (17) and (18).

The Mallat algorithm is given as follows. Suppose we are given an expansion of the form (19), then we want to determine the coefficients of the corresponding lower order expansion of the type (20). By hypothesis we have that $\tilde{f}^J \in V_J$. By (15) we see that

$$\tilde{f}^j(x) = \sum_{k\in\mathbf{Z}} f_k^{0,0}\phi_k(x) + \sum_{j=0}^{j-1}\sum_{k\in\mathbf{Z}} f_k^{j,1}\psi_k^j(x) \tag{21}$$

for some coefficients $\{f_k^{0,0}, f_k^{j,1}\}$. By multiplying both sides of (21) by $\phi_k^j(x)$ and integrating , and using (1) and the orthonormality of the wavelet functions, we find that, for a fixed j and k,

$$\begin{aligned} f_k^{j,0} &= \sum_l f_l^{j+1}\int \phi_l^{j+1}(x)\phi_k^j(x)\,dx \\ &= \frac{1}{\sqrt{2}}\sum_{l,n} f_l^{j+1}a_n\int \phi_l^{j+1}(x)\phi_{2k+n}^{j+1}(x)\,dx \\ &= \frac{1}{\sqrt{2}}\sum_{l,n} f_l^{j+1}a_n\delta_{2k+n}^l \\ &= \frac{1}{\sqrt{2}}\sum_l f_l^{j+1}a_{l-2k}, \end{aligned} \tag{22}$$

and similarly by (5),

$$f_k^{j,1} = \frac{1}{\sqrt{2}}\sum_l f_l^{j+1}b_{l-2k}. \tag{23}$$

We note that we assume in these calculations that $a_i = 0$ for $i < 0$ or $i > 2N-1$.

Conversely, by knowing the expansion (20), we can determine the coefficients in (19) recursively by the intermediate expansions in terms of scaling functions at order less than J of the form

$$\tilde{f}^j(x) = \sum_k f_k^{j,0}\phi_k^j(x)$$

where, for $j = 0, \ldots, J$,

$$\begin{aligned} f_k^{j,0} &= \sum_l (f_l^{j-1,0}\int \phi_l^{j-1}(x)\phi_k^j(x)\,dx + f_l^{j-1,0}\int \psi_l^{j-1}(x)\phi_k^j(x)\,dx) \\ &= \frac{1}{\sqrt{2}}\sum_l (f_l^{j-1,0}a_{k-2l} + f_l^{j-1,0}b_{k-2l}). \end{aligned}$$

In the next section we will, for a given input function of a certain type, formulate a specfic expansion of the form (19) which will form a good approximation in the $L^2(\mathbf{R})$ norm to the given function. This expansion will be related to the lower order expansion of the form (20) by the Mallat algorithm. In general the coefficients in this expansion will not be given by Fourier coefficient formulas of the type (18) and (18), but will correspond instead to *evaluating the function* f at a well chosen lattice of points.

3 Wavelet Interpolation

In this section we prove the following interpolation theorem. The theorem is stated in $\mathbf{R^2}$ for simplicity, but is true in $\mathbf{R^n}$ as well, as is apparent from the proof. Recall that c is a constant depending on the wavelet system defined by (7) .

Theorem 3.1 *Assume the function* $f \in C^2(\bar{\Omega})$, *where* Ω *is a bounded open set in* $\mathbf{R^2}$. *Let, for* $j \in \mathbf{N}$,

$$f^j(x,y) := \frac{1}{2^j} \sum_{p,q \in \Lambda} f\left(\frac{p+c}{2^j}, \frac{q+c}{2^j}\right) \phi_p^j(x) \phi_q^j(y), \quad x, y \in \Omega, \tag{24}$$

where the index set $\Lambda = \{i \in Z : \operatorname{supp}(\phi_i^j) \cap \Omega \neq \emptyset\}$. *Then*

$$||f - f^j||_{L^2(\Omega)} \leq C(1/2^j)^2, \tag{25}$$

and

$$||f - f^j||_{H^1(\Omega)} \leq C/2^j,$$

where C *is a constant depending only on the diameter of* Ω, N, *and the maximum modulus of the first and second order derivatives of* f *on* $\bar{\Omega}$.

Proof: Without loss of generality, we can assume that Ω is a square with edge length L. Then the number of p's (and q's) in the summation in (24) is no more than $2^j L + (2N - 1)$.

We first show that

$$||f - f^j||_{L^2(\Omega)} \leq C/2^j. \tag{26}$$

For each $p, q \in \Lambda$, using the Taylor expansion for f at a point $(x, y) \in \Omega$, we have

$$f\left(\frac{p+c}{2^j}, \frac{q+c}{2^j}\right) = f(x,y) + \frac{\partial f}{\partial x}(\xi_p, \xi_q)\left(\frac{p+c}{2^j} - x\right) + \frac{\partial f}{\partial y}(\xi_p, \xi_q)\left(\frac{q+c}{2^j} - y\right), \tag{27}$$

for some (ξ_p, ξ_q) on the segment connecting $\left(\frac{p+c}{2^j}, \frac{q+c}{2^j}\right)$ and (x, y).

Substituting (27) into (24) and taking (10) into account, one derives

$$\begin{aligned} 2^j(f - f^j)(x,y) &= \sum_{p,q\in\Lambda} \frac{\partial f}{\partial x}(\xi_p, \xi_q)\left(\frac{p+c}{2^j} - x\right)\phi_p^j(x)\phi_q^j(y) + \\ &+ \sum_{p,q\in\Lambda} \frac{\partial f}{\partial y}(\xi_p, \xi_q)\left(\frac{q+c}{2^j} - y\right)\phi_p^j(x)\phi_q^j(y). \end{aligned} \tag{28}$$

Squaring both sides of (28),

$$\begin{aligned} 2^{2j}|f - f^j|^2(x,y) &\leq C2^j\left(\sum_{p\in\Lambda}\left|\phi_p^j\left(\frac{p+c}{2^j} - x\right)\right|\right)^2 + \left(\sum_{q\in\Lambda}\left|\phi_q^j\left(\frac{q+c}{2^j} - y\right)\right|\right)^2 \\ &= C2^j\left(\sum_{p,p'\in\Lambda}\left|\phi_p^j\phi_{p'}^j\right|\left|\left(\frac{p+c}{2^j} - x\right)\left(\frac{p'+c}{2^j} - x\right)\right| + \right. \\ &\left. + \sum_{q,q'\in\Lambda}\left|\phi_q^j\phi_{q'}^j\right|\left|\left(\frac{q+c}{2^j} - y\right)\left(\frac{q'+c}{2^j} - y\right)\right|\right., \end{aligned} \tag{29}$$

where the constant C depends only on the maximum modulus of the first derivatives of f on $\bar{\Omega}$. Therefore, by (9) and Hölder's inequality, (29) gives

$$\begin{aligned} 2^{2j}\|f - f^j\|_{L^2(\Omega)}^2 &\leq \frac{C2^j}{2^{2j}}\left\{\sum_{p\in\Lambda}\sum_{|p'-p|\leq 2N-2}\int|\phi_p^j\phi_{p'}^j|\,dx + \right. \\ &\left. + \sum_{q\in\Lambda}\sum_{|q'-p|\leq 2N-2}\int|\phi_q^j\phi_{q'}^j|\,dy\right\} \\ &\leq C, \end{aligned} \tag{30}$$

where the constant C depends on the constant in (29), N and the diameter of the domain Ω. Hence (26) follows from (30).

We now show that

$$||f - f^j||_{L^2(\Omega)} \le C(1/2^j)^2, \qquad (31)$$

and

$$||\nabla f - \nabla f^j||_{L^2(\Omega)} \le C/2^j. \qquad (32)$$

For each $p, q \in \Lambda$, using the Taylor expansion for f at point $(x, y) \in \Omega$, we have

$$\begin{aligned} f\left(\frac{p+c}{2^j}, \frac{q+c}{2^j}\right) &= f(x,y) + \frac{\partial f}{\partial x}(x,y)\left(\frac{p+c}{2^j} - x\right) + \frac{\partial f}{\partial y}(x,y)\left(\frac{q+c}{2^j} - y\right) + \\ &\quad + \tfrac{1}{2}R(x,y), \end{aligned} \qquad (33)$$

where

$$\begin{aligned} R(x,y) &= \frac{\partial^2 f}{\partial x^2}(\xi_p, \xi_q)\left(\frac{p+c}{2^j} - x\right)^2 + 2\frac{\partial^2 f}{\partial x \partial y}(\xi_p, \xi_q)\left(\frac{p+c}{2^j} - x\right)\left(\frac{q+c}{2^j} - y\right) \\ &\quad + \frac{\partial^2 f}{\partial y^2}(\xi_p, \xi_q)\left(\frac{q+c}{2^j} - y\right)^2 \end{aligned} \qquad (34)$$

for some (ξ_p, ξ_q) on the segment connecting $\left(\frac{p+c}{2^j}, \frac{q+c}{2^j}\right)$ and (x, y). Therefore,

$$\begin{aligned} 2^j(f^j - f)(x,y) &= \sum_{p,q\in\Lambda} \frac{\partial f}{\partial x}\left(\frac{p+c}{2^j} - x\right)\phi_p^j(x)\phi_q^j(y) + \\ &\quad + \sum_{p,q\in\Lambda} \frac{\partial f}{\partial y}\left(\frac{q+c}{2^j} - y\right)\phi_p^j(x)\phi_q^j(y) \\ &\quad + \sum_{p,q\in\Lambda} R(x,y)\phi_p^j(x)\phi_q^j(y). \end{aligned} \qquad (35)$$

To analyze the first two terms on the right hand side of (35), let us consider the wavelet expansion of the function x over the admissible x-slice of the domain Ω in terms of the scaling functions. Write

$$x = \sum_{k\in\Lambda} d_k \phi_k^j(x),$$

for some constants d_k's. Then a simple calculation implies that for each k,

$$d_k = 2^{-3j/2}[k + c],$$

where c is the first moment of ϕ (7). Hence

$$\sum_p p\phi_p^j(x) = 2^{3j/2}x - 2^{j/2}c. \tag{36}$$

This implies that the first two summations in (35) vanish, and we obtain,

$$2^j(f^j - f)(x,y) = \sum_{p,q\in\Lambda} R(x,y)\phi_p^j(x)\phi_q^j(y). \tag{37}$$

But

$$|R(x,y)| \le C\left\{\left|\frac{p+c}{2^j} - x\right|^2 + \left|\frac{p+c}{2^j} - x\right|\left|\frac{q+c}{2^j} - y\right| + \left|\frac{q+c}{2^j} - y\right|^2\right\}, \tag{38}$$

where the constant C depends only on the maximum of the second order derivatives of f. Therefore

$$\begin{aligned}
2^{2j}|f^j - f|^2 \;&\le\; C\left\{\sum_{p,q\in\Lambda}\left(\left|\frac{p+c}{2^j} - x\right|^2 + \left|\frac{p+c}{2^j} - x\right|\left|\frac{q+c}{2^j} - y\right|\right.\right. \\
&\quad \left.\left. + \left|\frac{q+c}{2^j} - y\right|^2\right)\phi_p^j(x)\phi_q^j(y)\right\}^2 \\
&= C\sum_{p,q\in\Lambda}\sum_{p',q'\in\Lambda}\left(\left|\frac{p+c}{2^j} - x\right|^2 + \left|\frac{p+c}{2^j} - x\right|\left|\frac{q+c}{2^j} - y\right| + \left|\frac{q+c}{2^j} - y\right|^2\right) \\
&\quad \cdot\left(\left|\frac{p'+c}{2^j} - x\right|^2 + \left|\frac{p'+c}{2^j} - x\right|\left|\frac{q'+c}{2^j} - y\right| + \left|\frac{q'+c}{2^j} - y\right|^2\right)\cdot \\
&\quad \cdot\phi_p^j(x)\phi_{p'}^j(x)\phi_q^j(y)\phi_{q'}^j(y).
\end{aligned} \tag{39}$$

Hence by (9),

$$2^{2j}||f - f^j||^2_{L^2(\Omega)} \le C\sum_{p.q\in\Lambda}(1/2^j)^4 \le C(1/2^{2j}),$$

and (31) is proved.

To prove estimate (32), we again use the Taylor expansion (33). From (24), we have

$$\frac{\partial f^j}{\partial x}(x,y) = \sum_{p,q\in\Lambda} f\left(\frac{p+c}{2^j}, \frac{q+c}{2^j}\right)\frac{\partial\phi_p^j(x)}{\partial x}\phi_q^j(y). \tag{40}$$

So (40) and (10) imply

$$
\begin{aligned}
2^j \frac{\partial f^j}{\partial x}(x,y) &= \sum_{p,q\in\Lambda} \frac{\partial f}{\partial x}\left(\frac{p+c}{2^j} - x\right) \frac{\partial \phi_p^j(x)}{\partial x} \phi_q^j(y) + \\
&+ \sum_{p,q\in\Lambda} \frac{\partial f}{\partial y}\left(\frac{q+c}{2^j} - y\right) \frac{\partial \phi_p^j(x)}{\partial x} \phi_q^j(y) + \sum_{p,q\in\Lambda} R(x,y) \frac{\partial \phi_p^j(x)}{\partial x} \phi_q^j(y).
\end{aligned}
\tag{41}
$$

Again by (10), one derives

$$
\begin{aligned}
2^j \frac{\partial f^j}{\partial x}(x,y) &= \sum_{p,q\in\Lambda} \frac{\partial f}{\partial x} \frac{p+c}{2^j} \frac{\partial \phi_p^j(x)}{\partial x} \phi_q^j(y) + \sum_{p,q\in\Lambda} R(x,y) \frac{\partial \phi_p^j(x)}{\partial x} \phi_q^j(y) \\
&= \frac{1}{2^{j/2}} \frac{\partial f}{\partial x} \sum_{p,q\in\Lambda} p \frac{\partial \phi_p^j(x)}{\partial x} + \sum_{p,q\in\Lambda} R(x,y) \frac{\partial \phi_p^j(x)}{\partial x} \phi_q^j(y).
\end{aligned}
\tag{42}
$$

Recalling (36), one has

$$
\begin{aligned}
\sum_{p\in\Lambda} p \frac{\partial \phi_p^j(x)}{\partial x} &= \frac{\partial}{\partial x}\left(\sum_{p\in\Lambda} p \phi_p^j(x)\right) \\
&= \frac{\partial}{\partial x}\left(2^{3j/2} x - c \sum_{p\in\Lambda} \phi_p^j(x)\right) = 2^{3j/2}.
\end{aligned}
\tag{43}
$$

Substituting (43) into (42), one has

$$
2^j \left(\frac{\partial f^j}{\partial x} - \frac{\partial f}{\partial x}\right)(x,y) = \sum_{p,q\in\Lambda} R(x,y) \frac{\partial \phi_p^j(x)}{\partial x} \phi_q^j(y). \tag{44}
$$

Therefore by (44) and (38),

$$
\begin{aligned}
2^{2j}\left|\frac{\partial f^j}{\partial x}-\frac{\partial f}{\partial x}\right|^2(x,y) &\le C\left\{\sum_{p,q\in\Lambda}\left(\left|\frac{p+c}{2^j}-x\right|^2+\left|\frac{p+c}{2^j}-x\right|\left|\frac{q+c}{2^j}-y\right|+\right.\right.\\
&\quad\left.\left.+\left|\frac{q+c}{2^j}-y\right|^2\right)\frac{\partial\phi_p^j(x)}{\partial x}\phi_q^j(y)\right\}^2\\
&= C\sum_{p,q\in\Lambda}\sum_{p',q'\in\Lambda}\left(\left|\frac{p+c}{2^j}-x\right|^2+\left|\frac{p+c}{2^j}-x\right|\left|\frac{q+c}{2^j}-y\right|+\right.\\
&\quad\left.+\left|\frac{q+c}{2^j}-y\right|^2\right)\left(\left|\frac{p'+c}{2^j}-x\right|^2+\left|\frac{p'+c}{2^j}-x\right|\left|\frac{q'+c}{2^j}-y\right|+\right.\\
&\quad\left.+\left|\frac{q'+c}{2^j}-y\right|^2\right)\frac{\partial\phi_p^j(x)}{\partial x}\frac{\partial\phi_{p'}^j(x)}{\partial x}\phi_q^j(y)\phi_{q'}^j(y).
\end{aligned}
\tag{45}
$$

Note that

$$\int\left|\frac{\partial\phi_p^j(x)}{\partial x}\right|^2 dx \le 2^{2j}\int|\phi\prime|^2\,dx. \tag{46}$$

Then combining (10) and (46), using the Hölder inequality, and again, as in (30), noticing the relation between p and p'; q and q', we obtain

$$2^{2j}\left\|\frac{\partial f}{\partial x}-\frac{\partial f^j}{\partial x}\right\|^2_{L^2(\Omega)} \le C\left(\sum_{p,q\in\Lambda}\left(\frac{1}{2^j}\right)^4(2^{2j}2^{2j})^{1/2}\right)\le C, \tag{47}$$

where the constant C depends on the constant in (40), N and the diameter of the domain Ω. That is,

$$\left\|\frac{\partial f}{\partial x}-\frac{\partial f^j}{\partial x}\right\|_{L^2(\Omega)}\le C/2^j.$$

Similarly, we can derive

$$\left\|\frac{\partial f}{\partial y}-\frac{\partial f^j}{\partial y}\right\|_{L^2(\Omega)}\le C/2^j.$$

Then (32) follows readily from above and the Theorem is proved. □

4 Estimates for Elliptic Partial Differential Equations

Using Theorem 3.1 as an interpolation one can easily derive some error estimates for the wavelet solutions of certain elliptic partial differential equations. In this section we derive one such estimate for the case of a Galerkin approximation to a solution of a specific boundary value problem as an example of this phenomenon. The result is easily generalized to other boundary value problems in higher dimensions provided that a general coercive system is considered. we recall that if W is a finite-dimensional subspace of some Sobolev space in which one looks for a weak solution of a differential equation, we look for a solution (the Galerkin solution) of the finite dimensional linear system of equations obtained from the weak form of the differential equation by restricting the test functions *and* the solution function to both be in the approximation space W (see [4] for a discussion of this).

More specifically let Ω be a bounded open set in $\mathbf{R}^2$ with a Lipschitz boundary. Let

$$W^j := \{f \in V^j : \text{supp} f \cap \bar{\Omega} \neq \emptyset\}.$$

Since Ω is bounded it's clear that W^j is a finite-dimensional subspace of V^j. This will be our Galerkin approximation space. If we consider the Sobolev space $H^1(\Omega)$, then we have clearly

$$W^j \subset H^1(\Omega),$$

and thus we can use elements of W^j to represent approximations to a solution of a differential equation which are in $H^1(\Omega)$.

Consider the elliptic equation

$$-\Delta u + u = f, \quad \text{in } \Omega, \tag{48}$$

with the Neumann boundary condition

$$\frac{\partial u}{\partial n} = g \quad \text{on } \partial\Omega \tag{49}$$

where n is the unit outward normal vector of $\partial\Omega$. If u is a solution of (48) and (49) and if h is a test function in $H^1(\Omega)$, then multiplying (48) by h and integrating by parts over Ω, one has from (49) that

$$\int_{\Omega} \nabla u \nabla h dx dy + \int_{\Omega} u h dx dy = \int_{\Omega} f h dx dy + \int_{\partial\Omega} g h ds. \tag{50}$$

Solving (48) and (49) is equivalent to finding $u \in H^1(\Omega)$ so that (50) is satisfied for all $h \in H^1(\Omega)$. Let u^j be a solution of (50) where $u^j \in W^j$ and (50) is satisfied for all $h \in W^j$ (the Galerkin approximation). Assume that the wavelet parameter $N \geq 2$.

Theorem 4.1 *If u is a solution to* (48) *and* (49), *then*

$$||u - u^j||_{H^1(\Omega)} \leq \frac{C}{2^j},$$

where C depends on the diameter of Ω, on N, and the maximum modulus of the first and second derivatives of u.

Proof: We show first that if $v \in W^j$, then

$$||u - u^j||_{H^1(\Omega)} \leq ||u - v||_{H^1(\Omega)}. \tag{51}$$

To see this, we let, for $v \in V^j$, $w = u^j - v$. Then since u and u^j both satisfy (50) for any $h \in W^j$ and $w \in W^j$, one has

$$\int_{\Omega} (\nabla u - \nabla u^j) \cdot \nabla w \, dx dy + \int_{\Omega} (u - u^j) w \, dx dy = 0. \tag{52}$$

Therefore, by (52) and the Schwarz inequality,

$$\begin{aligned} ||u - u^j||^2_{H^1(\Omega)} &= \int_{\Omega} (\nabla u - \nabla u^j)^2 \, dx dy + \int_{\Omega} (u - u^j)^2 \, dx dy \\ &= \int_{\Omega} (\nabla u - \nabla u^j) \cdot (\nabla u - \nabla v - \nabla w) \, dx dy + \\ &\quad + \int_{\Omega} (u - u^j)(u - v - w) \, dx dy \\ &= \int_{\Omega} (\nabla u - \nabla u^j) \cdot (\nabla u - \nabla v) \, dx dy \\ &\quad + \int_{\Omega} (u - u^j)(u - v) \, dx dy \\ &\leq ||u - u^j||_{H^1(\Omega)} ||u - v||_{H^1(\Omega)}. \end{aligned} \tag{53}$$

If $u = u^j$, (51) is obviously true, otherwise, (51) also holds because of (53). Let now $v = \tilde{u}^j$ be defined to be the interpolation of u as given by Theorem 3.1, then Theorem 4.1 follows from (51) and Theorem 3.1. □

We now generalize this last result by replacing the data f and g by perturbations of f and g before we make the Galerkin approximation. In practice, this is what one does, as it is then necessary to use numerical approximations for f and g in a given problem. Suppose that we are given f and g as in (48) and (49), where we suppose both f and g are C^2 functions in $\mathbf{R}^2$ which restrict to Ω and $\partial\Omega$ respectively, and let f^j and g^j be wavelet interpolations for f and g of order j as in Theorem 3.1. Let now u^j be the wavelet-Galerkin solution to

$$\int_\Omega \nabla u \nabla h dxdy + \int_\Omega uhdxdy = \int_\Omega f^j hdxdy + \int_{\partial\Omega} g^j hds, \tag{54}$$

and then we obtain

Corollary 4.2 *If u is a solution to* (48) *and* (49), *and u^j is the wavelet-Galerkin solution to* (54), *then*

$$||u - u^j|| \leq \frac{C}{2^j},$$

where C depends on the diameter of Ω, on N, and on the maximum modulus of the first derivatives of u.

Proof: From (50) and (52) we obtain

$$\int_\Omega (\nabla u - \nabla u^j)\cdot\nabla w \, dxdy + \int_\Omega (u - u^j) w \, dxdy = \int_\Omega (f - f^j) w \, dxdy + \int_{\partial\Omega} (g - g^j) w \, ds$$

Therefore, similar to the derivation of (53), one has

$$\begin{aligned} ||u - u^j||^2_{H^1(\Omega)} \leq \; & ||u - u^j||_{H^1(\Omega)} ||u - v||_{H^1(\Omega)} + \textstyle\int_\Omega |(f - f^j)(u^j - v)| \, dxdy + \\ & + \textstyle\int_{\partial\Omega} |(g - g^j)(u^j - v)| \, ds. \end{aligned} \tag{55}$$

Note that if $\partial\Omega$ is Lipschitz, then one has the following well-known inequality [1].

$$\int_{\partial\Omega} |f| \, ds \leq C ||f||_{H^1(\Omega)}, \tag{56}$$

where C depends only on the dimension and Ω. Combining the above two inequalities (55) and (56), estimates for f, g from Theorem 3.1, replacing v by $\tilde{u}^j$, and using the Schwarz inequality again, the Corollary follows. □

References

[1] R. A. Adams. *Sobolev spaces.* Academic Press, New York, 1975.

[2] G. Beylkin, R. Coifman, and V. Rokhlin. Fast wavelet transforms and numerical algorithms I. *Comm. Pure Appl. Math.*, 44:141–183, 1991.

[3] Charles Chui. *Wavelet Theory.* Academic Press, Cambridge, MA, 1991.

[4] Germund Dahlquist and Ake Bjorck. *Numerical Methods.* Prentice-Hall, Englewood Cliffs, NJ, 1974.

[5] I. Daubechies. Orthonormal bases of compactly supported wavelets. *Comm. Pure Appl. Math.*, 41:906–966, 1988.

[6] Timo Eirola. Sobolev characteristics of solutions of dilation equation. Technical Report A292, Helsinki University of Technology, Institute of Mathematics, Espoo, Finland, 1991.

[7] R. Glowinski, W. Lawton, M. Ravachol, and E. Tenenbaum. Wavelet solution of linear and nonlinear elliptic, parabolic and hyperbolic problems in one dimension. In R. Glowinski and A. Lichnewski, editors, *Proceedings of the Ninth International Conference on Computing Methods in Applied Sciences and Engineering*, Philadelphia, 1990. SIAM.

[8] Roland Glowinski, T. W. Pan, R. O. Wells, and Xiaodong Zhou. Wavelet and finite element solutions for the Neumann problem using fictitious domains. Technical Report 92-01, Rice University, 1992. Computational Mathematics Laboratory.

[9] R. Gopinath and C. S. Burrus. On the moments of the scaling function ϕ. Technical Report 91-05, Rice University, 1991. Computational Mathematics Laboratory; submitted to ISCAS '92.

[10] P. Heller, H. L. Resnikoff, and R. O. Wells. Wavelet matrices and the representation of discrete functions. In Charles Chui, editor, *Wavelets: A Tutorial.* Academic Press, Cambridge, MA, 1992.

[11] S. Mallat. Multiresolution approximation and wavelet orthonormal bases of $L^2(\mathbf{R})$. *Trans. Amer. Math. Soc.*, 315:69–87, 1989.

[12] Y. Meyer. *Ondelettes et Operateurs.* Hermann, 1990.

[13] D. Pollen. $SU_I(2, F[z, 1/z])$ for F a subfield of $\mathbf{C}$. *J. Amer. Math. Soc.*, 3(3):611–624, 1990.

[14] D. Pollen. Parametrization of compactly supported wavelets. Technical Report ATR 4, Aware, Inc., 1991.

[15] H. L. Resnikoff and R. O. Wells, Jr. *Wavelet Analysis.* Springer-Verlag, New York, 1992. In preparation.

[16] H. L. Resnikoff and R. O. Wells, Jr. Wavelet analysis and the geometry of Euclidean domains. *J. Geom. Phys.*, 8(1-4):273–282, 1992.

[17] G. Strang. Wavelets and dilation equations: A brief introduction. *SIAM Review*, 31:614–627, 1989.

[18] L. F. Villemoes. Energy moments in time and frequency for two-scale difference equation solutions and wavelets. Preprint, Technical University of Denmark.

[19] R. O. Wells. Parametrizing smooth compactly supported wavelets. *Trans. Amer. Math. Soc.*, 1992. to appear.

[20] R. O. Wells and Xiaodong Zhou. Wavelet solutions for the Dirichlet problem. Technical Report 92-02, Rice University, 1992. Computational Mathematics Laboratory.

[21] W. Zettler, J. Huffman, and D. Linden. Application of compactly supported wavelets to image compression. In *Proc. SPIE: Image Processing Algorithms and Techniques*, volume 1244, pages 150–160. Intl. Soc. for Optical Engineering, 1990.

Department of Mathematics, Rice University,
Houston, Texas 77251
E-mail: wells@rice.edu

FROM SUPER LIE ALGEBRAS TO SUPERGROUPS: MATRIX REALIZATIONS AND THE FACTORISATION PROBLEM

V. HUSSIN and L. M. NIETO[1]

ABSTRACT. Using a matrix realization, a generic element of the supergroup $OSP(m/2n)$ is obtained through the exponentiation of the corresponding super Lie algebra element. The emphasis is put on the contribution of the odd part, and the application to the factorisation problem is given.

1 Introduction

The appearance of the ideas of superalgebras and supergroups in theoretical physics (as well as other new concepts such as superspace, supermanifolds or Hilbert superspaces, which constitue what has been called by some authors [1] supermathematics) was brought about by the necessity of having a unified description of systems with a mixing of bosons and fermions. The techniques subsequently developped are now commonly used by a lot of theoretical physicists.

In particular, when the supersymmetry was introduced in quantum mechanics [2] it was possible to make a generalization of the concepts of kinematical and dynamical algebras to include superalgebras [3]. The generators of a superalgebra do not form an ordinary Lie algebra because they do not close under commutation relations. The new structure is determined by a set of both commutation and anticommutation relations between its generators. It will give detailed in the following section. Note that once the structure is clearly defined, the theory of Lie superalgebras is developped along similar lines to those of Lie algebras. A classification of simple superalgebras has been given [4] and the representation theory of superalgebras has been carefully studied [5].

Concerning the supergroup theory, the generalization was not so direct. In fact, the first attempts to construct such a theory [6] were not completely satisfactory from a physical point of view. Afterwards, a significant contribution to this question was given by Rogers [7]. This work deals with a global theory of supermanifolds based on an extension of real analysis to Grassmann algebras: a definition of supergroups both as abstract groups and superanalytic manifolds is given. From this point of view, it is important to notice that every supergroup is a noncompact group.

[1]On leave of absence from Dept. de Física Teórica, Universidad de Valladolid, 47011 Valladolid, Spain

E. A. Tanner and R. Wilson (eds.), Noncompact Lie Groups and Some of Their Applications, 367–372.

Taking into account such theories, the relation between superalgebras and supergroups can be investigated in detail. General results, without refering to a finite-dimensional representation, have been provided using their abstract structure. In particular, let us mention that there exists [8] a version of the Baker-Campbell-Haussdorff formula for supergroups, which has been the starting point of different applications [9]. In practice, such applications deal with a specific factorisation of a supergroup element, and involve the solution of non-linear differential equations [10].

The result we want to present here deals with the fact that the use of a particular matrix representation may help in solving and understanding this factorisation problem. We will show that the process of factorisation using a finite-dimensional representation is systematic, with no reference to solving non-linear differential superequations, and it also gives an explanation of the appearance of such superequations in the abstract theory. The method is developped for the supergroup $OSP(m/2n)$, but it is generalizable to other supergroups.

2 Some elements on superalgebras and supergroups

Let us first give a summary of the basic elements on superalgebras and supergroups (more details may be found in [11]). A Lie superalgebra $\mathcal{L} = \mathcal{L}_0 \oplus \mathcal{L}_1$ is defined as a real or complex graded vector space (the elements of $\mathcal{L}_0$ and $\mathcal{L}_1$ are assigned to have degree 0 and 1, and are called even and odd vectors, respectively), plus a generalized Lie product $[a,b]$ which satisfy the following properties:

- $[a,b] \in \mathcal{L}, \quad \forall a,b \in \mathcal{L}$;
- $[\alpha a + \beta b, c] = \alpha[a,c] + \beta[b,c], \quad \forall a,b \in \mathcal{L}, \quad \forall \alpha,\beta$ real or complex;
- $[a,b] \in \mathcal{L}_0$, if a,b are both even or odd;
  $[a,b] \in \mathcal{L}_1$, if one is even and the other is odd;
- $[b,a] = -(-1)^{(deg\,a)(deg\,b)}[a,b], \; a,b \in \mathcal{L}_0$ or $\mathcal{L}_1$; $deg\,a = 0\,(1)$ if a is even (odd);
- Generalized Jacobi identity

$$(-1)^{(deg\,a)(deg\,c)}[a,[b,c]] + (-1)^{(deg\,b)(deg\,a)}[b,[c,a]] + (-1)^{(deg\,c)(deg\,b)}[c,[a,b]] = 0.$$

As a simple example, let us consider the orthosymplectic superalgebra $osp(m/2n)$. It is formed by the matrices

$$M = M_0 + M_1 = \begin{pmatrix} A & 0 \\ 0 & D \end{pmatrix} + \begin{pmatrix} 0 & B \\ C & 0 \end{pmatrix}; \; A \in \mathbb{C}^{m\times m}, D \in \mathbb{C}^{2n\times 2n}, B, C^T \in \mathbb{C}^{m\times 2n}, \quad (1)$$

that satisfy

$$M^{ST}K + (-1)^{deg\,M}KM = 0, \quad K = \begin{pmatrix} I_m & 0 \\ 0 & J_{2n} \end{pmatrix}, \qquad J_{2n} = \begin{pmatrix} 0 & I_n \\ -I_n & 0 \end{pmatrix}. \quad (2)$$

The supertransposed matrix M^{ST} is defined to be

$$M^{ST} = \begin{pmatrix} A^T & (-1)^{\deg M} C^T \\ -(-1)^{\deg M} B^T & D^T \end{pmatrix}. \tag{3}$$

The matrices M_0 (M_1) are said to be even (odd). The generalized Lie product is given by

$$[M, N] = MN - (-1)^{\deg M \deg N} NM. \tag{4}$$

In order to go from superalgebras to supergroups, it is necessary to introduce the concept of supermatrices. They are essentially matrices with Grassmann numbers instead of ordinary complex numbers [11]. More precisely, we are working with even supermatrices

$$g = \begin{pmatrix} G_0 & 0 \\ 0 & G_1 \end{pmatrix} + \begin{pmatrix} 0 & \Gamma \\ \Lambda^T & 0 \end{pmatrix}, \tag{5}$$

where G_0 and G_1 are matrices whose elements are only even Grassmann numbers, and Γ, Λ^T are matrices whose elements are only odd Grassmann numbers (note that the square of an odd Grassmann number is zero, and its inverse does not exit; even Grassmann numbers are invertible if they have a non-zero body).

Now, we can say that a Lie supergroup is a set of such even supermatrices (5) which form a group under the product of matrices and verifies certain properties of regularity [11]. There is a Lie algebra of supermatrices (also called super Lie algebra) associated to such a group, but it is not always possible to associate a Lie superalgebra to the supergroup. Nevertheless, given a Lie superalgebra of matrices, it is always possible to construct a Lie algebra of supermatrices and, by exponentiation, we can get a supergroup. We restrict ourselves to those situations. In particular, this is the case for the supergroup $OSP(m/2n)$, whose elements are the supermatrices that verify

$$G^{ST} K G = K.$$

Hence, it is easy to see that the super Lie algebra corresponding to $OSP(m/2n)$ is nothing else that the set of even supermatrices $\mathcal{M}$ such that

$$\mathcal{M}^{ST} K + K\mathcal{M} = 0, \tag{6}$$

where

$$\mathcal{M} = \mathcal{M}_e + \mathcal{M}_o = \begin{pmatrix} R & 0 \\ 0 & S \end{pmatrix} + \begin{pmatrix} 0 & \beta \\ \gamma^T & 0 \end{pmatrix}, \quad \mathcal{M}^{ST} = \begin{pmatrix} R^T & \gamma \\ -\beta^T & S^T \end{pmatrix}. \tag{7}$$

Note the difference between this super Lie algebra and the $osp(m/2n)$ superalgebra.

Now, we will construct the supergroup matrix representation of $OSP(m/2n)$ by exponentiation of $\mathcal{M}$.

3 A general element of the supergroup $OSP(m/2n)$

From equations (6) and (7) it is easy to see that the Lie algebra has the dimension $d = (d_e, d_o)$, where the even (odd) dimension d_e (d_o) is the number of even (odd) Grassmann parameters in $\mathcal{M}$:

$$d_e = \frac{1}{2}m(m-1) + 2n^2 + n, \qquad d_o = 2mn.$$

The even part corresponds to the algebra $o(m) \oplus sp(2n)$ and the odd part corresponds to the matrices $\mathcal{M}_o$ of the form

$$\mathcal{M}_o = \sum_{l=1}^{2mn} \theta_l N_l \qquad \theta_l \text{ odd Grassmann numbers,} \tag{8}$$

with

$$N_l = \begin{pmatrix} 0 & B_l & 0 \\ 0 & 0 & 0 \\ -B_l^T & 0 & 0 \end{pmatrix}, \quad N_{mn+l} = \begin{pmatrix} 0 & 0 & B_l \\ B_l^T & 0 & 0 \\ 0 & 0 & 0 \end{pmatrix}, \ l = 1, \ldots, mn \tag{9}$$

and

$$(B_i)_{jk} = \delta_{j[i-(k-1)m]}, \quad 1 \le i \le mn, \quad 1 \le j \le m, \quad 1 \le k \le n. \tag{10}$$

To characterize the corresponding supergroup element $g = \exp(\mathcal{M})$ associated to the general matrix $\mathcal{M}$, we apply a usual technique of Lie group theory. Introducing the real parameter t, we solve $g(t)$ from

$$\frac{d}{dt}g(t) \equiv \dot{g}(t) = \mathcal{M}g(t), \quad g(0) = I. \tag{11}$$

Then g is obtained as $g = g(1)$.

The only difference with respect to group theory is the presence of the odd part, which possesses special properties. We will insist on how to deal with this complementary part. Our resolution is based on the fact that $g(t)$ can be decomposed on the form

$$g(t) = g_0(t) + \sum_{\substack{r=1 \\ 1 \le i_1 < \cdots < i_r \le 2mn}}^{2mn} \theta_{i_1} \cdots \theta_{i_r} \, g_{i_1 \cdots i_r}(t), \tag{12}$$

where $g_0(t)$ is block diagonal with even elements independent of the θ_j's, $g_0(t) = \exp(t\mathcal{M}_e)$, and $g_{i_1 \cdots i_r}(t)$ are supermatrices with even (odd) elements if r is even (odd). Note that formula (12) follows from the form of $\mathcal{M}_o$ in (8) and the fact that the θ_j's satisfy $\theta_j^2 = 0$, $\forall j$. Moreover, from (12) we deduce that the "initial condition" $g(0) = I$ becomes

$$g_0(0) = I, \quad g_{i_1 \cdots i_r}(0) = 0, \quad \forall\, i_k \in \{1, \ldots, 2mn\}. \tag{13}$$

Inserting (12) in (11) and identifying the coefficients of $\theta_{i_1} \cdots \theta_{i_r}$ ($r = 1, \ldots, 2mn$), we are able to solve $g \equiv g(1)$ (the proof is given in detail in Ref. [12]):

$$g_{i_1 \cdots i_r}(1) = (-1)^{r(r-1)/2} g_0 \int_0^1 d\tau_1 \int_0^{\tau_1} d\tau_2 \cdots \int_0^{\tau_{r-1}} d\tau_r \begin{Vmatrix} N_{i_1}^0(\tau_1) & \ldots & N_{i_r}^0(\tau_1) \\ \vdots & \vdots & \vdots \\ N_{i_1}^0(\tau_r) & \ldots & N_{i_r}^0(\tau_r) \end{Vmatrix}, \tag{14}$$

where

$$g_0 = g_0(1), \quad N_i^0(\tau) = g_0^{-1}(\tau)\; N_i\; g_0(\tau) \tag{15}$$

and the symbol $||\cdots||$ stands for the formal development of the determinant by the elements of the rows (beginning by the first one) or, if you prefer, is a "time-ordered product".

4 Factorisation of an $OSP(m/2n)$ supergroup element

We are going to deal with a factorisation based on the decomposition of an arbitrary element $g \in OSP(m/2n)$ as

$$g = pk, \tag{16}$$

where k is an element of a fixed sub-supergroup K of $OSP(m/2n)$. Therefore, p characterizes the coset space $OSP(m/2n)/K$. This last remark is important for two reasons. First, when we are concerned with the coherent states based on supergroups, one can show that these states are parametrized by the elements of a certain coset space, like the one given before. Second, the characterization of K as a sub-supergroup which is maximal in $OSP(m/2n)$ leads to the identification of the independent parameters in $p \in OSP(m/2n)/K$ as solutions of non-linear super-Riccati equations [10]. In fact, this last observation explains the appearance of such equations in the work of Fatyga *et al.* [8]. There, a factorisation of $OSP(1/2)$ is made, without referring to a particular representation of this group. It clearly appears to be on the form (16).

Here, we want to point out that the use of a matrix realization leads us to construct a group element g by solving only linear differential equations (as shown in the preceding section), and then to solve algebraic equations to proceed to the factorisation. The example of $OSP(1/2)$ is particularly simple from this point of view. We have been able to generalize it to $OSP(m/2n)$ for m and n arbitrary. In this way, a factorisation (16) is, for example, given by

$$pk = \begin{pmatrix} I & \Gamma_1^p & 0 \\ 0 & I & 0 \\ (\Gamma_1^p)^T & G_{21}^p & I \end{pmatrix} \begin{pmatrix} G^k & 0 & \Gamma_2 - \Gamma_1^p G_{12} \\ {\Lambda_1}^T & G_{11} & G_{12} \\ 0 & 0 & (G_{11}^{-1})^T \end{pmatrix}, \tag{17}$$

with

$$\Gamma_1^p = \Gamma_1 G_{11}^{-1}, \quad G_{21}^p = G_{21} G_{11}^{-1} \tag{18}$$

and

$$G^k = G - (G_{11}^{-1})^T {\Gamma_1}^T {\Lambda_1}^T. \tag{19}$$

Here, K is a maximal sub-supergroup leaving the following subspace $V \subset B_{L0}^m \times B_{L1}^{2n}$ (see Ref. [11]) invariant

$$V = \left\{ \begin{pmatrix} 0 \\ x_1 \\ 0 \end{pmatrix}, \; x_1 \in B_{L0}^n \right\}. \tag{20}$$

Since g is already known (cf (12) with (13)), p will be determined directly from (18), and then k from (17) and (19).

5 References

[1] Berezin F A, Sov. J. Nucl. Phys. **29**, 857 (1979)

[2] Witten E, Nucl. Phys. B **188**, 513 (1981)

[3] de Crombrugge M and Rittenberg V, Ann. Phys. **151**, 99 (1983); D'Hoker E, Vinet L and Kostelecký V A, *Dynamical Groups and Spectrum Generating Algebras* edited by A. Barut, A. Bohm and Y. Ne'eman (World Scientific Singapore 1987); Beckers J, Dehin D and Hussin V, J. Phys. A **21**, 651 (1988)

[4] Kac V G, Commun. Math. Phys. **53**, 31 (1977)

[5] Bars I, Lect. Appl. Math. **21**, 17 (1985)

[6] Berezin F A and Kac G I, Math. URSS Sb **11**, 311 (1970); Kostant B, Lecture Notes in Mathematics, Vol. 570 (Springer, N.Y.), p. 617 (1977)

[7] Rogers A, J. Math. Phys. **21**, 1352 (1980); **22**, 443 (1981); **22**, 939 (1981)

[8] Kostelecký V A, Nieto M M and Truax D R, J. Math. Phys. **27**, 1419 (1986); Fatyga B W, Kostelecký V A and Truax D R, J. Math. Phys. **30**, 291 (1989)

[9] Fatyga B W, Kostelecký V A, Nieto M M and Truax D R, Phys. Rev. D **43**, 1403 (1991)

[10] Beckers J, Gagnon L, Hussin V and Winternitz P, Lett. Math. Phys **13**, 113 (1987); J. Math. Phys. **31**, 2528 (1990)

[11] Cornwell J F, *Group Theory in Physics*, Vol III (Academic Press, 1989)

[12] Hussin V and Nieto L M, preprint CRM (1993)

Centre de recherches mathématiques
Université de Montréal
CP 6128-A, Montréal H3C 3J7
Québec, Canada

CURRENT ALGEBRAS AS HILBERT SPACE OPERATOR COCYCLES

JOUKO MICKELSSON

ABSTRACT Aspects of a generalized representation theory of current algebras in $3+1$ dimensions are discussed. Rules for a systematic computation of vacuum expectation values of products of currents are described. Their relation to gauge group actions in bundles of fermionic Fock spaces and to the sesquilinear form approach of Langmann and Ruijsenaars is explained. The regularization for a construction of an operator cocycle representation of the current algebra is explained. An alternative formula for the Schwinger terms defining gauge group extensions is written in terms of Wodzicki residue and Dixmier trace.

1. INTRODUCTION

In this talk I want to discuss generalized representations of certain group extensions associated to the mapping groups $Map(M, G)$, where M is a compact manifold of odd dimension and G is a compact Lie group. The group structure is given by point-wise multiplication of maps. A prototype is the *loop group* $LG = Map(S^1, G)$, which has nontrivial central extensions which, when G is simple, are labelled by an integer k (the level of the extension). These extensions are nontrivial both algebraically and topologically (when $k \neq 0$); the latter property means that the group extension is a nontrivial circle bundle over the base space LG. The Lie algebra $\widehat{LG}$, which is a central extension of the loop algebra $L\mathbf{g}$, is *an affine Lie algebra;* a particular case in the more general category of Kac-Moody algebras.

The Lie algebra can be interpreted in quantum field theory as the algebra of *charge densities* . If G is a group of symmetries of a relativistic field theory then to each basis element T_a $(a = 1, 2, \ldots N)$ of the Lie algebras $\mathbf{g}$ there corresponds a conserved current j^a_μ in the physical $(d+1)$-dimensional space-time. In the quantized theory the integrated charges (here f is a smooth test function on a space-like surface S)

$$j^a(f) = \int_S f(x) j^a(x) dS \tag{1.1}$$

E. A. Tanner and R. Wilson (eds.), Noncompact Lie Groups and Some of Their Applications, 373–390.

are supposed to satisfy the Lie algebra

(1.2) $$[j^a(f), j^b(g)] = \lambda_c^{ab} j^c(fg)$$

where the λ_c^{ab}'s are the structure constants of $\mathbf{g}$,

(1.3) $$[T_a, T_b] = \lambda_c^{ab} T_c.$$

However, in order to make sense of the quantum operators $j^a(f)$ one normally has to make certain regularizations. The effect of the regularizations is that the commutation relations (1.2) will be modified.

An important example of the quantum modifications is the case of free fermions in 1+1 dimensions. Suppose that the fermions in the one-particle representation transform according to a finite-dimensional unitary representation ρ of G. Assume $S = S^1$ and define the LG action on the fermion field $\psi(x)$ on the circle by $\psi'(x) = \rho(g(x))\psi(x)$, $g \in LG$. In the second quantized theory the charge densities are represented by

(1.4) $$T(f) = \sum f_{nm} : a_n^* a_m :,$$

where the indices refer to an orthonormal basis (consisting of eigenvectors of the one-particle hamiltonian), the $a^*{}_n$'s are fermions creation operators, the a_m's are annihilation operators, and the dots mean *normal ordering;* the operators lowering the energy are putted to the right. Let π_+ be the projection on the positive energy subspace in the one-particle Hilbert space H and $\epsilon = 2\pi_+ - 1$. The effect of the normal ordering is that the commutation relations for a pair X, Y of charge density operators become

(1.5) $$[X,Y]' = [X,Y] + \frac{1}{4}\mathrm{tr}\, \epsilon[\epsilon, X][\epsilon, Y].$$

The second term on the right in (1.5) is an antisymmetric bilinear form c on a Lie algebra $\mathbf{gl}_1$ which consists of bounded operators X in H such that $[\epsilon, X]$ is a Hilbert-Schmidt operator. The form c (Lundbergs's cocycle, [Lu]) satisfies the cocycle condition

(1.6) $$c(X,[Y,Z]) + c(Y,[Z,X]) + c(Z,[X,Y]) = 0$$

which guarantees that the Jacobi identity for the modified commutator $[\cdot,\cdot]'$ is satisfied.

In quantum field theory in space-time dimensions higher than two the normal ordering of the charge densities is not sufficient to make them finite; more radical regularizations are needed. The technical reason for this is that the currents are no more contained in $\mathbf{gl}_1$ but in a bigger Lie algebra $\mathbf{gl}_p$, where p must be greater or equal to $\frac{1}{2}(d+1)$, where d is the space dimension and $\mathbf{gl}_p$ is characterized by the property that $[\epsilon, X]$ belongs to the Schatten ideal L_{2p} consisting of operators T such that $(T^*T)^p$ has a finite trace. Note that the cocycle c is defined only on $\mathbf{gl}_1$ and not on any of the larger algebras $\mathbf{gl}_p$, $p > 1$.

There is a modified cocycle c_p which converges for $\mathbf{gl}_p$. The coefficient ring for these cocycles is not $\mathbb{C}$ but a certain space of complex valued functions on an infinite-dimensional manifold (which in this talk will be a Grassmannian manifold). For example, when $d = 3$ the cocycle takes the form, [MR],

$$(1.7) \qquad c_2(X, Y; F) = \frac{1}{8}\mathrm{tr}'(\epsilon - F)[[\epsilon, X], [\epsilon, Y]],$$

where F is a hermitean operator, with $F^2 = 1$ and $\epsilon F + F\epsilon - 2\epsilon \in L_2$, parametrizing points on a Grassmannian Gr_2. The prime over the trace means the following: Under the adjoint action of ϵ on bounded operators $L(H)$ in the one-particle space we have the decomposition $L(H) = L(H)_+ \oplus L(H)_0 \oplus L(H)_-$ to eigenspaces corresponding to the eigenvalues $0, \pm 2$. The trace tr$'$ picks up only the ϵ-diagonal part $L(H)_0$, that is, $\mathrm{tr}'T = \mathrm{tr}\frac{1}{2}(T + \epsilon T\epsilon)$. For $p > 2$ the cocycles c_p have been computed in [FT].

Whereas in the case of a loop group there is a extensive well developed representation theory, in the general case not much is known. In fact, the main point of this talk is that one should generalize the concept of a group representation when dealing with the extensions of $Map(M, G)$ in higher dimensions.

At the moment there are three different ways to generalize the representation theory and they all come from certain regularization processes in quantum field theory. In the examples considered so far all the three methods give equivalent descriptions of the same generalized representations.

In the first approach one considers an action of the current algebra not on a single quantum Hilbert space but in a bundle of Hilbert spaces parametrized by (nonquantized) external fields. A typical situation is the case of chiral fermions coupled to vector potentials. The fibers of the Hilbert bundle are then fermionic Fock spaces and the elements of $Map(M, G)$ act as gauge transformation on the base (=the space of smooth vector potentials $\mathcal{A}$ in M) and as certain linear isometries between the fibers. The base is flat, so the Hilbert bundle can be globally trivialized. This leads to the second point of view (which will be the main topic of this talk), namely the ***Hilbert space operator valued cocycles*** as representations of the current algebra. A representation of a group $\mathcal{G}$ is a homomorphism $\mathcal{G} \to U(V)$ to the group $U(V)$ of unitary operators in a vector space V. A cocycle of the gauge action of the gauge group $\mathcal{G} = Map(M, G)$ with values in $U(V)$ is a map $\omega : \mathcal{G} \times \mathcal{A} \to U(V)$ such that

$$(1.8) \qquad \omega(gg'; A) = \omega(g; g' \cdot A)\omega(g'; A).$$

If ω is constant as a function of A then we have a representation of $\mathcal{G}$ in V.

We shall do the construction in two steps. First we "regularize" the one- particle operators $T(g)$ in H; this leads to a U_1-valued cocycle (U_p is the unitary subgroup of the group GL_p corresponding to the Lie algebra $\mathbf{gl}_p$). Then we replace U_1 by a central extension $\hat{U}_1$ and we obtain a $\hat{U}_1$-valued cocycle for a certain abelian extension $\hat{\mathcal{G}}$. This is the group which can be "represented" in a Fock space $\mathcal{F}$ by a $U(\mathcal{F})$-valued cocycle which is fixed by a true representation of $\hat{U}_1$ in $\mathcal{F}$.

The third alternative description of the generalized representation of the current algebra was proposed by E. Langmann [L]. The starting point is the observation of S.N.M. Ruijsenaars that although the charges are not well-defined operators in the Fock space, they still make sense as sesquilinear forms, [R1]. The problem with sesquilinear forms is that one cannot multiply them like matrices; if one tries to do that one gets in general a divergent expression. Langmann showed that there is a regularized multiplication of the forms which leads to the cocycle (1.7). In fact, one can work backwards from the algebra to construct the regularized product, [M2].

2. FOCK BUNDLES AND REGULARIZED PRODUCTS OF SESQUILINEAR FORMS

Let H be an one-particle fermionic Hilbert space. For example, H could be the space of square-integrable spinor fields on a manifold M. Let h be the one-particle Hamiltonian acting in H. In the example, h would be the fixed time Dirac operator $\sum \gamma^k \nabla_k$. We shall denote by ϵ the sign of the energy, $\epsilon = \frac{h}{|h|}$. We use the convention that the zero energy correspond to eigenvalue $+1$ of ϵ. The canonical anticommutation relations based on H are defined by the only nonzero anticommutators

$$a^*(u)a(v) + a(v)a^*(u) = < u, v > . \tag{2.1}$$

The inner product $< \cdot, \cdot >$ is antilinear in the second argument.

Each closed subspace $W \subset H$ defines a representation of the CAR algebra in the following way. In the Fock representation in the Hilbert space $\mathcal{F}_W$ there is a unique (up to a phase) normalized vector $|W>$ such that

$$a^*(u)|W> = 0 = a(v)|W> \quad \text{for all } u \in W^\perp \text{ and } v \in W. \tag{2.2}$$

If $W = \{v \in H | \epsilon v = v\}$ then $|W>$ is the ***vacuum vector*** for the Hamiltonian h. The vacuum is the state of lowest energy (which is normalized to zero) and all other states have positive energy. The space $\mathcal{F}_W$ is generated by multiplying $|W>$ by polynomials of the creation and annihilation operators.

Let $\{e_n\}_{n\in \mathbf{Z}}$ be an orthonormal basis of H such that $\epsilon e_n = e_n$ for $n \geq 0$ and $\epsilon e_n = -e_n$ for $n < 0$. Denote $a_n = a(e_n)$ and $a_n^* = a^*(e_n)$. The normal ordering is defined by

$$: a_n^* a_m := \begin{cases} -a_m a_n^* \text{ if } n = m < 0 \\ a_n^* a_m \text{ otherwise} \end{cases} \tag{2.3}$$

All other products of the creation and annihilation operators are as usual. Let $X = (X_{nm}) \in \mathbf{gl}_1$. The second quantized operator $\hat{X}$ representing the one-particle operator X is

$$\hat{X} = \sum_{n,m\in \mathbf{Z}} X_{nm} : a_n^* a_m : \tag{2.4}$$

The commutation relations for the $\hat{X}'s$ are given by (1.5). In particular, when X is a gauge transformation and the physical space is the circle S^1, that is X

corresponds to a pointwise multiplication of the spinor field (with one space-time component) by a map $f : S^1 \to \mathbf{g}$, the commutation relations (1.5) become

$$[\hat{X}(f), \hat{X}(g)] = \hat{X}([f,g]) + \frac{i}{2\pi}\int_{S^1} \mathrm{tr} f(x)\frac{d}{dx}g(x)dx \tag{2.5}$$

where the trace is computed in the representation ρ of $\mathbf{g}$. This means that quantum algebra corresponding to the loop algebra $Map(S^1, \mathbf{g})$ is an affine Kac-Moody algebra, [BH], [F], [CH].

Let us denote by $\widehat{\mathbf{gl}}_1$ the *central extension* of $\mathbf{gl}_1$ defined by the cocycle term in (1.5). Let us write $H = H_+ \oplus H_-$, where H_+ is the subspace spanned by the positive energy eigenvectors e_n, $n \geq 0$, and H_- is its orthogonal complement. An operator X in H can then be split as

$$X = \begin{pmatrix} a & b \\ c & d \end{pmatrix} \tag{2.6}$$

where $a : H_+ \to H_+$, $b : H_- \to H_+$, etc. For $X \in \mathbf{gl}_1$ the off-diagonal blocks b, c are Hilbert-Schmidt. The Fock space representation of $\widehat{\mathbf{gl}}_1$ described above has the characteristic property that the vacuum vector $|H_+>$ is annihilated by all X's with $b(X) = 0$; we denote this subalgebra by $\mathbf{b}_+$.

Let p, q be a pair of polynomials in the creation and annihilation operators. The matrix element of a product of generators of $\mathbf{gl}_1$ between the states $p|H_+>$ and $q|H_+>$ can be computed in the following way. In the enveloping algebra of $\widehat{\mathbf{gl}}_1$ any element can be written (using the Poincare-Birkhoff-Witt theorem) as a linear combination of products

$$u = X_1 \ldots X_n Y_1 \ldots Y_m \tag{2.7}$$

where $Y_i \in \mathbf{b}_+$ and $X_i \in \mathbf{n}_-$. The algebra $\mathbf{n}_-$ consists of operators with $a = d = c = 0$. Using the commutation relations

$$[\hat{X}, a_n^*] = \sum X_{mn} a_m^*, \qquad [\hat{X}, a_n] = -\sum \overline{X}_{nm} a_m \tag{2.8}$$

and the properties $\hat{Y}_i|H_+>= 0$, $0 =< H_+|\hat{X}_i$ of the vacuum, the matrix element $< H_+|p^* uq|H_+ >$ is given as a linear combination of expressions of the type $< H_+|p'^* q'|H_+ >$ for some polynomials p', q'. These matrix elements are computed in the standard way by shifting energy decreasing operators to the right and energy increasing operators to the left.

The problem is then to generalize the above construction of the Fock space representation of $\widehat{\mathbf{gl}}_1$ to an appropriate extension $\widehat{\mathbf{gl}}_p$ of $\mathbf{gl}_p$ for $p > 1$. In fact, the representation does not exist in the usual sense. There is a theorem by D. Pickrell which says that $\widehat{\mathbf{gl}}_p$ does not have a faithful separable Hilbert space representation, [P]. There is however a simple procedure for defining any polynomial like (2.7) as a sesquilinear form in the Fock space. I shall explain this when $p = 2$. This gives a "representation" of the current algebra in three space dimensions.

The Lie algebra $\widehat{\mathbf{gl}}_2$ is an extension of $\mathbf{gl}_2$ by the abelian ideal I of functions on the Grassmannian Gr_2 of the form $\gamma(F) = \alpha + \mathrm{tr}(F - \epsilon)\xi$, where α is a complex constant and $\xi : H \to H$ is an operator such that $[\epsilon, \xi] \in L_{4/3}$ and $\epsilon\xi + \xi\epsilon \in L_2$. The action of $\mathbf{gl}_2$ on I is given by the Lie derivative, $[X, \gamma](F) = (\mathcal{L}_X\gamma)(F) = \mathrm{tr}[F, X]\xi$. The complete commutator is

$$[(X, \gamma), (X', \gamma')] = ([X, X'], \mathcal{L}_X\gamma' - \mathcal{L}_{X'}\gamma + c_2(X, X'; \cdot)), \tag{2.9}$$

where c_2 is defined in (1.7). The Jacobi identity is satisfied by the 2-cocycle property

$$c_2([X, Y], Z; F) - \mathcal{L}_X c_2(Y, Z; F) + \text{ cyclic permutations } = 0. \tag{2.10}$$

The universal enveloping algebra of $\widehat{\mathbf{gl}}_2$ is spanned by products of the form

$$u = X_1 \dots X_n \beta Y_1 \dots Y_m, \tag{2.11}$$

where β is some polynomial in the variable F (a product of the simple affine functions γ). The prescription for computing the matrix element $< H_+|p^* uq|H_+ >$ is now the same as in the case of $\widehat{\mathbf{gl}}_1$ except for the following additional rule: After shifting the elements of $\mathbf{b}_+$ to the right (annihilating the vacuum) and the elements of $\mathbf{n}_-$ to the left, we are left with an expression of the form $< H_+|p'^* \beta q'|H_+ >$. This is evaluated by computing β at the point $F = \epsilon$ and then proceeding as before.

There is a geometrical derivation for this rule. There is a bundle of Fock spaces parametrized by points $F \in Gr_2$. There is a special section ψ_0 of the Fock bundle, called the vacuum section, such that $\psi_0(F)$ for any F belongs to the one-dimensional vacuum subspace of $\mathcal{F}_F$. The vacuum subbundle Vac is twisted; it is equivalent to the determinant line bundle over Gr_2. The vacuum ψ_0 is (up to a multiplicative constant) the unique holomorphic section of Vac which is annihilated by $\mathbf{b}_+$. We normalize the vacuum section in such a way that $\psi_0(H_+)$ is a vector of unit norm in $\mathcal{F}_{H_+}$. One can prove following theorem, [M2],

Theorem. *The vacuum expectation values of the elements u of the enveloping algebra of $\widehat{\mathbf{gl}}_2$ can be computed as*

$$< H_+|u|H_+ > = < \psi_0, u\psi_0 >_{H_+},$$

where $< \cdot, \cdot >_{H_+}$ is the Fock space inner product in the fiber $\mathcal{F}_{H_+}$.

Thinking of $\mathbf{gl}_2$ as a subspace of $\widehat{\mathbf{gl}}_2$ we can derive, among others, the following formulas for the vacuum expectation values $< u > = < H_+|u|H_+ >$.

$$\begin{aligned} &< X > = 0 = < XY > \\ &< XYZV > = \mathrm{tr}\, V_{12}X_{21}Z_{12}Y_{21} + \mathrm{tr}\, X_{21}V_{12}Y_{21}Z_{12}, \text{ for all } X, Y, Z, V \in \mathbf{gl}_1 \end{aligned} \tag{2.12}$$

where $X_{12} : H_- \to H_+$ and $X_{21} : H_+ \to H_-$ are the upper left and lower right blocks of the operator $X : H \to H$. These formulas are closely related to cyclic cocycles in the Fredholm modules of Connes, [C]. The cyclic cocycles for even k are given by

$$\tau(X_1,\ldots,X_k) = \frac{1}{2^k}\mathrm{tr}\,\epsilon[\epsilon,X_1][\epsilon,X_2]\ldots[\epsilon,X_k]. \tag{2.13}$$

One has

$$\tau(X,Y,Z,V) + \tau(X,V,Z,Y) = \langle VYZX \rangle - \langle XZVY \rangle . \tag{2.14}$$

3. A BOSONIZATION IN 3 + 1 DIMENSIONS

In two space-time dimensions a system of quantized free fermions can be described in terms of bosonic field operators. There are several ways to construct the 'Bose-Fermi correspondence' but for the purposes of the present talk the following Segals 'blip' formulation is particularly well adapted, [S2],[PS]. He starts from a central extension $\widehat{LT}$ of the group LT of smooth maps $S^1 \to U(1) = T$. The basic observation is that if $f,g \in \widehat{LT}$ are elements of winding number $n(f), n(g)$, respectively, with disjoint support ($f(x) \neq 1$ implies $g(x) = 1$) then $fg = (-1)^{n(f)n(g)}gf$. This means that loops of odd winding number satisfy fermionic anticommutation relations. The second important observation is that in a positive energy representation D it makes sense to speak about the limit $\lim D(f_n)$ when f_n converges to 'blip', a loop which is constant except that the winding number is concentrated at a single point $x \in S^1$. The limit $\psi(x)$ should be interpreted as an operator valued distribution on S^1 and it satisfies the anticommutation relations

$$\begin{aligned} &[\psi(x),\psi(y)]_+ = 0 = [\psi^*(x),\psi^*(y)]_+ \\ &[\psi^*(x),\psi(y)]_+ = 2\pi\delta(x-y). \end{aligned} \tag{3.1}$$

The representation space of D carries an irreducible representation of the fermionic anticommutation relations (3.1). In the other direction, the bosonic 'currents', generators of the Lie algebra of $\widehat{LT}$, can be recovered as normal ordered fermionic bilinears

$$\begin{aligned} &\alpha(x) =: \psi(x)^*\psi(x) : \\ &[\alpha(x),\alpha(y)] = 2\pi\partial_x\delta(x-y) \end{aligned} \tag{3.2}$$

This is of course the content of (2.4) in the special case $T = G$. Again the α's should be understood as operator valued distributions. This method can be used to construct highest weight representations of affine simply laced Kac-Moody algebras, [F], [CH]. The circle group T is replaced by the maximal torus $T \subset G$ in a simple Lie group. A basis of $\widehat{LG}$ is obtained by taking blips in various directions on T, determined by the roots of G.

When trying to generalize the above to higher space-time dimensions one encounters great difficulties. The main problem is that the group of maps $Map(M,G)$ is not represented in the fermionic Fock space $\mathcal{F}$ when the dimension of the physical space M is greater than one. However, there has been already for some time indications that some modification of the Bose-Fermi correspondence should work also in $3+1$ space-time dimensions. The first such indication was the 'kink' construction of fermions by Finkelstein and Rubinstein, [FR]. To take a simple example consider the case $M = S^3$ and $G = SU(n)$. A kink is then a noncontractible map $f : S^3 \to G$. The homotopy classes of kinks are classified by an element of $\pi_3 G$, which is equal to $\mathbb{Z}$ when $n \geq 2$. When $n = 2$ the fourth homotopy group $\pi_4 G$ is $\mathbb{Z}_2$. In [FR] it was shown that in a configuration containing two nonoverlapping kinks a continuous rotation of one of the kinks is homotopically equivalent to a continuous interchange of the kink positions. This is a kind of spin-statistic theorem; the result has been later generalized by Sorkin, [So]. In particular, the case $G = SU(2)$ allows two types of field configurations. A contractible loop in $\mathcal{G} = Map(S^3, G)$ may be interpreted as a boson whereas a noncontractible loop (generator of $\pi_4 SU(2)$) is a fermion.

The possibility of realization of fermions in terms of seemingly bosonic fields really occurs in the WZW model, as shown by Witten in [W]. The quantum mechanical WZW action functional obtains a factor -1 each time as the field $f : S^3 \to SU(n)$ is adiabatically rotated by the angle 2π. On the other hand, it was proved in [M3] that if $\mathcal{G}$ acts projectively in the quantum vector space (the nontrivial projective phases are related to chiral anomaly) then $D(f)D(g) = (-1)^{n(f)n(g)} D(g)D(f)$. I want to stress that this relation does not depend on the existence of a unitary (projective) representation D (that may not exist) in the ordinary sense. The group has nonunitary linear representations which have some similarities with the unitary highest weight representations in the one-dimensional case, [MR]. The formula is purely a result of the structural relations in an extension $\hat{\mathcal{G}}$.

A more recent and important result on the Bose-Fermi relation in $3+1$ dimensions is due to Ruijsenaars, [R2]. As we have seen, although the operators $D(g)$ representing gauge transformations do not exist as linear operators in the fermionic Fock space, they still exist as sesquilinear forms in a suitable dense domain. Furthermore, one can take limits of kinks $f_n : S^3 \to G$ such that $\lim D(f_n)$ exists as a bilinear in the Fock space, when $\lim f_n$ is a 'pointlike kink', the winding being concentrated to a single point $x \in S^3$. The limiting bilinear is the fermion field $\psi(x)$.

4. HILBERT SPACE COCYCLES

A representation of a group G can be viewed as a special kind of 1-cocycle, with values in a unitary group $U(H)$ of a Hilbert space H. If we have in addition an action of G on a manifold $\mathcal{A}$, a 1-cocycle for this action with values in $U(H)$ is a map $\omega : G \times \mathcal{A} \to U(H)$ such that

$$\omega(g; g' \cdot A)\omega(g'; A) = \omega(gg'; A). \tag{4.1}$$

If ω does not depend on A we have an ordinary representation of G in H. We shall see that it is physically quite natural to study the Hilbert space cocycles ω when G is the current group in 3 dimensions.

The point of view developed here is closely related to the Fock bundle approach in [M4]. In fact, the Hilbert cocycle results from a special kind of trivialization of the Fock bundle over the space of vector potentials.

Let M be a compact oriented three-manifold with a fixed spin structure, G a compact Lie group, and $\mathcal{A}$ the space of smooth 1-forms on M with values in the Lie algebra $\mathbf{g}$ of G. The topology and differentiable structure of $\mathcal{A}$ can be defined using a family of seminorms give by L^2 norms of partial derivatives of the vector potentials, but the choice of the smooth structure does not play any essential role.

Fix a trivial complex vector bundle $M \times V$ over M with a unitary representation ρ of G in the fiber V. Let $E = S \otimes V$ be the tensor product of a (Weyl) spinor bundle S and V. Let $H = L^2(E)$ be the Hilbert space consisting of square integrable spinor fields. If λ is any real number not in the spectrum of D_A we denote by $H_+(A, \lambda)$ the closed subspace of H spanned by eigenvectors of D_A belonging to eigenvalues greater than λ and by $H_-(A, \lambda)$ the orthogonal complement of $H_+(A, \lambda)$. As a reference plane we fix $H_+ = H_+(0, \lambda_0)$ for some λ_0.

Because of the flow of eigenvalues of D_A as a function of A it is not possible to fix the vacuum level λ once and for all without introducing discontinuities in the function $A \mapsto H_+(A, \lambda)$.

Remember that for each closed subspace $W \subset H$ there is a unique Fock representation of CAR characterized by a vacuum vector $|W>$ subject to the conditions (2.2). One would be tempted to set $W = W(A) = H_+(A, \lambda)$ as a function of the vector potential. We would then have for each potential A a Fock space $\mathcal{F}_W$ with a vacuum (vector of lowest energy, the energy of the vacuum normalized to zero by a normal ordering principle). However, this does not quite work because of the discontinuity of $W(A)$.

In spite of the fact that we cannot fix the vacuum in a continuous way the equivalence class of the CAR representation is well-defined for each $A \in \mathcal{A}$. For a given potential A define the Grassmannian $Gr_1(A)$ consisting of all closed subspaces $W \in H$ such that the orthogonal projection $W \to H_-(A, \lambda)$ is Hilbert-Schmidt. This condition does not depend on the choice of λ since there are only a finite number of linearly independent eigenvectors of D_A in any finite interval of the spectrum: thus the difference of the projections for two different λ's is of finite rank and consequently Hilbert-Schmidt. It is known that a pair of vacuum representations of CAR, corresponding to vacua $|W>, |W'>$ such that the projection of W to the complement of W' is Hilbert-Schmidt, are equivalent,[A]. It follows that we may use any of the planes $W \in Gr_1(A)$ to define the CAR representation so long as we are only interested in the equivalence class of the representation.

The representations of CAR corresponding to vacua $|W>, |W'>$ with $W = H_+(A, \lambda), W' = H_+(A, \lambda')$ are equivalent but the equivalence is defined naturally only up to a phase. It is precisely this property of chiral fermions which is responsible to the anomalies of the gauge group action, [S], [NA]. In the case of four component fermions the phase ambiguities from the two different chiral sectors cancel and there is no anomaly.

The action of a gauge transformation $g \in \mathcal{G}$ defined by pointwise multiplication on square-integ spinor fields defines a homomorphism $T : \mathcal{G} \to GL_p$, where $p =$

$\frac{1}{2}(1 + dim M)$, see e.g. [MR]. Since we are here concerned with the case $dim M = 3$ we have $p = 2$.

The L_{2p} Grassmannians Gr_p are defined by $Gr_p = GL_p/B_p$, where B_p consists of operators with the lower left block $c = 0$. We can also write $Gr_p = U_p/K_p$, where K_p consists of the unitary operators with the off-diagonal blocks b, c equal to zero. It is easy to see that GL_p is the maximal subgroup of $GL(H)$ which acts on Gr_p.

The Grassmannians $Gr_1(A)$ form a smooth fiber bundle over $\mathcal{A}$. Since the base is flat this bundle is trivial. For any $A \in \mathcal{A}$ let P_A be the set of unitary operators g in H such that $g \cdot H_+ \in Gr_1(A)$. If $g \in P_A$ then $gh \in P_A$ if and only if $h \in U_1$. The fibers P_A fit together to form a principal U_1 bundle P over $\mathcal{A}$. The bundle of Grassmannians can be viewed as an associated fiber bundle, defined by the usual U_1 action on Gr_1.

Choose a section $A \mapsto h_A$ of the bundle P. We define the U_1 valued cocycle

$$\omega(g; A) = h_{g\cdot A}^{-1} T(g) h_A \tag{4.2}$$

It is easy to check that the cocycle condition (4.1) is indeed satisfied. Let $\mathcal{G}_o$ be the group if *based* gauged transformations, that is, elements $g \in \mathcal{G}$ such that in a given base point $x \in M$ we have $g(x) = 1$. The quotient $\mathcal{A}/\mathcal{G}_o$ is a smooth manifold and we may form the bundle $Q = P/\mathcal{G}_o$ over $\mathcal{A}/\mathcal{G}_o$, defined by the right action in the fibers and the gauge action in the base. There is an alternative way to define this bundle using the cocycle ω. Define a right $\mathcal{G}_o$ action in $\mathcal{A} \times U_1$ by

$$(A, g) \cdot h = \left(h^{-1}Ah + hd(h^{-1}), g\omega(h; A)\right) \tag{4.3}$$

Then $Q = (\mathcal{A} \times U_1)/\mathcal{G}_o$.

The formula (4.2) should be considered as a regularization of the nonquantizable operator $T(g)$. Taking a derivative at the unity along a one-parameter subgroup of $\mathcal{G}$ one obtains the Lie algebra cocycle

$$(d\omega)(X; A) = h_A^{-1} dT(X) h_A - h_A^{-1} \mathcal{L}_X h_A,$$

where $\mathcal{L}_X$ denotes the Lie derivative along the vector field generated by an element X in the Lie algebra of $\mathcal{G}$ and dT is the Lie algebra representation corresponding to the group representation T. Thus the regularization on the Lie algebra level (i.e., for charge densities) consists of two parts: 1) A conjugation by h_A, 2) a subtraction by $h_A^{-1}\mathcal{L}_X h_A$. These two operations do not make sense separately in the fermionic Fock space, but combined they produce an element in the Lie algebra $\mathbf{u}_1$, which is quantizable.

In general, we can choose any U_1 valued cocycle ω to define a U_1 bundle Q_ω over $\mathcal{A}/\mathcal{G}_o$. Two cocycles ω, ω' are said to be cohomologous if there exists a map $f : \mathcal{A} \to U_1$ such that

$$\omega'(g; A) = f(g \cdot A)^{-1} \omega(g; A) f(A). \tag{4.4}$$

It is not difficult to show that the bundles Q_ω and $Q_{\omega'}$ are equivalent if and only if the cocycles ω and ω' are cohomologous.

If h, h' is a pair of trivializations of P then $h'_A = h_A f(A)$ for some map $f : \mathcal{A} \to U_1$ and so the cocycles ω, ω' defined by (4.2) are cohomologous. Thus, as expected, the bundles $Q_\omega, Q_{\omega'}$ constructed using two different trivializations of P are equivalent.

Whereas P is always trivial, Q will typically be nontrivial. Namely, if Q is trivial then the cocycle ω has to be contractible as a map from $\mathcal{G}_o \times \mathcal{A} \to U_1$ because in the trivial case $\omega(g; A) = f(g \cdot A)^{-1} f(A)$ for some f and $\mathcal{A}$ is a contractible space. Now the function h in the definition of ω is contractible, again by contractibility of $\mathcal{A}$. But the embedding $T : \mathcal{G}_o \to U_2$ is not contractible, [PS], in the case of a nonabelian compact gauge group G.

The idea for constructing the bundle $\mathcal{F}$ of Fock spaces over $\mathcal{A}$, together with an (anomalous) gauge group action is to consider the Fock bundle as an associated bundle to P via a *projective* representation of U_1 in a standard Fock space $\mathcal{F}_0$. This would lead to a bundle of projective Fock space (as explained in [S]). In order to create a true vector bundle with the right physical properties we shall first lift the cocycle ω to a cocycle of an appropriate extension $\hat{\mathcal{G}}$ of $\mathcal{G}$, with values in the *central extension* $\hat{U}_1$ of U_1.

Recall from [PS] the construction of the central extension $\widehat{GL}_1$. This group consists of pairs $(g, q) \in GL_1 \times GL(H_+)$ such that $a(g)q^{-1} - 1 \in L_1$, modulo the equivalence relation $(g, q) \equiv (g, qt)$ for any $t \in GL(H_+)$ with $t - 1 \in L_1$ and $\det t = 1$. The multiplication is defined simply by $(g, q)(g', q') = (gg', qq')$ and the unitary subgroup $\hat{U}_1$ is obtained by taking the intersection of $\widehat{GL}_1$ with $U(H) \times U(H_+)$. A section near the unit element is given by the map $g \mapsto q = a(g)$. The local 2-cocycle on GL_1 corresponding to this choice is

$$\Omega_1(g, g') = \det(a(g)a(g')a(gg')^{-1}). \tag{4.5}$$

Locally the group $\widehat{GL}_1$ is the product $GL_1 \times \mathbb{C}^\times$ with the multiplication $(g, \alpha)(g', \alpha') = (gg', \alpha\alpha'\Omega_1(g, g'))$. Globally the group is a twisted line bundle (the zero section removed) and cannot be defined using a single cocycle.

The Lie algebra of $\widehat{GL}_1$ is $\mathbf{gl}_1 \oplus \mathbf{C}$, where the Lie algebra $\mathbf{gl}_1$ of GL_1 consists of all bounded operators with Hilbert-Schmidt off-diagonal blocks. The commutator is the Lie algebra sum commutator plus the Lundberg's cocycle, [Lu],

$$c_1(X, Y) = \frac{1}{4} \mathrm{tr}\,\epsilon[\epsilon, X][\epsilon, Y], \quad X, Y \in \mathbf{gl}_1 \tag{4.6}$$

For a given $g \in \mathcal{G}$ we may choose (since $\mathcal{A}$ is flat) a function

$$A \mapsto \hat{\omega}(g; A) \in \hat{U}_1, \qquad \pi(\hat{\omega}(g; A) = \omega(g; A). \tag{4.7}$$

where $\pi : \hat{U}_1 \to U_1$ is the canonical projection. Any two choices differ by a S^1 valued function on $\mathcal{A}$. But the circle bundle $\hat{U}_1 \to U_1$ is nontrivial, [PS], and therefore there is no continuous lift $\hat{\omega}$ to the whole space $\mathcal{A} \times \mathcal{G}$. If we make a choice of the lift in some open contractible neighborhood of unity in $\mathcal{G}$ then

$$\hat{\omega}(gg'; A) = \Omega(g, g'; A)\hat{\omega}(g; g' \cdot A)\hat{\omega}(g'; A) \tag{4.8}$$

for some local function Ω. This leads to an extension $\hat{\mathcal{G}}$ of $\mathcal{G}$ by the abelian group of maps $Map(\mathcal{A}, S^1)$. Elements of the extension are pairs (g, θ), where $\theta : \mathcal{A} \to \hat{U}_1$ is any lift of $A \mapsto \omega(g; A)$. The group multiplication is defined by

$$(g, \theta)(g', \theta') = (gg', \theta''). \text{ with } \theta''(A) = \theta(g' \cdot A)\theta'(A). \tag{4.9}$$

The extension $\hat{\mathcal{G}}$ is completely determined by the bundle Q, or in the other words, by the cocycle ω.

Lemma 4.10. *The extensions $\hat{\mathcal{G}}, \hat{\mathcal{G}}'$ defined by the cohomologous cocycles ω, ω' are isomorphic.*

Proof. Suppose $\omega'(g; A) = f(g \cdot A)^{-1}\omega(g; A)f(A)$ for some function f. Choose a lift $\hat{f} : \mathcal{A} \to \hat{U}_1$. We can now define a map

$$\phi : \hat{\mathcal{G}} \to \hat{\mathcal{G}}', \quad \phi(g, \theta) = (g, \theta') \text{ with } \theta'(A) = \hat{f}(g \cdot A)^{-1}\theta(A)\hat{f}(A) \tag{4.11}$$

By the choice of $\hat{f}$ we know that θ' is a lift of $A \mapsto \omega'(g; A)$. Likewise, the homomorphism property is an immediate consequence of the definitions. The inverse map of ϕ is obtained by replacing $\hat{f}(A)$ by $\hat{f}(A)^{-1}$. □

The bundle $\mathcal{F}$ of Fock spaces over $\mathcal{A}$ is now defined as follows. The free Fock space $\mathcal{F}_0$ carries a faithful representation of $\hat{U}_1$, [PS]. Set $\mathcal{F} = \mathcal{A} \times \mathcal{F}_0$, with the following action of $\hat{\mathcal{G}}$. Let $(g, \theta) \in \hat{\mathcal{G}}$. Then

$$(g, \theta)(A, \psi) = (g \cdot A, \theta(A) \cdot \psi) \tag{4.12}$$

To complete the picture we have to give the action of the Dirac Hamiltonian D_A in each fiber. By the trivialization h_A the operator D_A in the one-particle space over A is conjugated to $D'_A = h_A^{-1} D_A h_A$ in the one-particle space over $0 \in \mathcal{A}$. By the choice of h, D'_A is an unbounded selfadjoint operator with the additional property that the off-diagonal blocks of its sign operator with respect to the polarization $H = H_+ \oplus H_-$ are Hilbert-Schmidt.

After fixing a vacuum level $\lambda \notin Spec(D_A)$ there is a unique vacuum ray $\mathbb{C}\psi(A, \lambda)$ in the Fock space $\mathcal{F}_0$ for the conjugated operator D'_A. As usual, the quantum Hamiltonian is then

$$\hat{D}'_A = \sum_n \lambda_n : a^*(u_n)a(u_n) : \tag{4.13}$$

where $\{u_n\}$ is a complete set of eigenvectors for D'_A with eigenvalues λ_n. The normal ordering is defined such that $: a^*(u_n)a(u_n) := -a(u_n)a^*(u_n)$ when $\lambda_n < \lambda$ and no change of order otherwise; this means that the energy of the vacuum is normalized to zero.

Note that the vacuum $\psi(A, \lambda)$ is a continuous function of the potential A only locally, because of the crossing of the eigenvalues of the vacuum level. On the other hand, if we choose $\lambda = 0$ we obtain a (weakly) continuous family of quantum Hamiltonians $\hat{D}'_A$.

Near the unity in $\mathcal{G}$ the extension $\hat{\mathcal{G}}$ behaves like the product $\mathcal{G} \times Map(\mathcal{A}, S^1)$. It is convenient to consider the larger group with the complexified fiber $Map(\mathcal{A}, \mathbb{C}^\times)$. The product is given then by the formula

$$(4.14) \qquad (g, f)(g', f') = (gg', f''), \text{ with } f''(A) = f(g' \cdot A) f'(A) \Omega(g, g'; A),$$

where the cocycle Ω is

$$(4.15) \qquad \Omega(g, g'; A) = \Omega_1(\omega(g; g' \cdot A), \omega(g'; A)).$$

As a vector space, the Lie algebra of $\hat{\mathcal{G}}$ is the direct sum of $Map(M, \mathbf{g})$ and $Map(\mathcal{A}, i\mathbb{R})$. The latter part is an abelian ideal, the infinitesimal gauge transformations acting on it through Lie derivatives. Taking derivatives of (4.15) along 1-parameter subgroups of $\mathcal{G}$ one gets:

Theorem 4.16. *The commutator of two infinitesimal gauge transformations X, Y is the pointwise commutator plus the cocycle term $c_2(X, Y; A)$, computed through*

$$\begin{aligned} c_2(X, Y; A) &= c_1(d\omega(X; A), d\omega(Y; A)) \\ &= c_1\left(h_A^{-1} T(X) h_A - h_A^{-1} \mathcal{L}_X h_A, h_A^{-1} T(Y) h_A - h_A^{-1} \mathcal{L}_Y h_A\right). \end{aligned}$$

Remark. In the $1+1$ dimensional case one may take $h_A \equiv 1$ and then $c_2(X, Y; A) = c_1(T(X), T(Y))$ does not depend on the potential. When $M = S^1$ this expression reduces to the usual formula for the central term of a Kac-Moody algebra, [PS].

Generalizations. The method described above can be used in principle also for 1) the group of local spin transformations , 2) for diffeomorphisms. In the former case the base space of the Fock bundle is the space of spin connections and in the latter the space of metrics of the physical 3-space. In both cases the base space is flat and so the global trivializations corresponding to the function h_A above exist. The cocycles are then constructed as in the case of gauge transformations. However, I have not been able to compute explicite formulas for the cocycles.

Remark There is an ambiguity in the cocycle Ω which is due to the freedom to choose the local section for the bundle $\hat{\mathcal{G}} \to \mathcal{G}$. A change in the local section is effected by a function $g \mapsto \beta(g; A)$ from $\mathcal{G}$ to the group $Map(\mathcal{A}, S^1)$. The (local) 2-cocycle is then transformed by a coboundary to the new cocycle

$$(4.17) \qquad \Omega'(g, g'; A) = \beta(gg'; A)^{-1} \beta(g; g' \cdot A) \beta(g'; A) \Omega(g, g'; A).$$

Thus only the cohomology class of Ω has invariant meaning. The cocycle constructed above is nonlocal. There is another cocycle which is local; its infinitesimal form was given in [M4],[FS]:

$$(4.18) \qquad (d\Omega)(X, Y; A) = \frac{1}{24\pi^2} \int_M \operatorname{tr} A(dX dY - dY dX),$$

where the trace is evaluated in the representation ρ of G. This form of the commutator anomaly can be derived from the Atiyah-Singer index theory approach to determinants of Dirac operators, [AS], as explained in [M3]. It appears also in perturbation series computations in a Yang-Mills-Dirac system, [JJ]. It would be interesting to have a "simple" nonperturbative regularization h_A which would directly lead to the nice local form (4.18). A modern scattering theoretic treatment (in time dependent external fields) can be found in [IO].

5. EXTENSIONS IN TERMS OF OPERATOR RESIDUE

The problem with the simple expression (1.7) for the Lie algebra cocycle is that it does not lead to the nice local formula (4.18) when restricted to the subgroup $\mathcal{G} = Map(M, G)$ of GL_2. However, when using the operator residue of Wodzicki [Wo] instead of ordinary traces we shall see below that there is a version of the trace formula which leads to the local expresssion for the Lie algebra cocycle.

We shall recall some basic facts about the operator residue. For more information I recommend [Wo] and the review article [VG]. A classical pseudodifferential operator (PSDO) P on a compact oriented Riemannian manifold M of dimension n, acting on vector valued functions or sections of a complex vector bundle E, can be written in terms of its ***symbol***. Locally, the symbol is a function $p(x, \xi)$ in a coordinate patch T^*U in the phase space T^*M such that it has an asymptotic expansion

$$p \sim \sum_{j\geq 0} p_{k-j}(x, \xi) \tag{5.1}$$

where the p_{k-j}'s are smooth (local) matrix valued functions on T^*M, homogeneous of degree $k-j$ in the momentum variables $(\xi_1, \dots, \xi_n)$,

$$p_j(x, a\xi) = a^j p_j(x, \xi), \text{ for } a > 0 . \tag{5.2}$$

The largest index $k-j$ for which $p_{k-j} \neq 0$ is ***the order*** of P and the corresponding term in the asymptotic expansion is the ***principal symbol*** of P. The matrix indices of p_{k-j}'s refer to a (local) basis of sections in the vector bundle where P is acting.

When M is not an open submanifold of $\mathbf{R}^n$ one has to use several coordinate patches. There is a 'change of variables' formula relating the local symbols in the overlap of two coordinate patches which is a straight-forward generalization of the usual change of variables formula for differential operators, [Wo]. As in the case of differential operators, only the leading term transforms tensorially whereas the lower order terms transform inhomogeneously in coordinate transformations.

The ***residue*** of a PSDO given by asymptotic expansion p is defined as follows. Let $S^*M \subset T^*M$ be the cosphere bundle and μ the measure in S^*M defined by the Riemannian metric. Then

$$Res(P) = \frac{1}{n(2\pi)^n} \int_{S^*M} \operatorname{tr} p_{-n}(x, \xi) d\mu(x, \xi). \tag{5.3}$$

If M is not an open submanifold of $\mathbf{R}^n$ one has to use a partition of unity and take a sum of integrals over the various coordinate patches.

The residue has the fundamental property that

$$Res(PQ) = Res(QP), \tag{5.4}$$

and so it behaves like an ordinary operator trace. However, the operator corresponding to the symbol p is trace class only if $deg(p) < -dim M$, in which case the residue vanishes!

The symbol of a product PQ of two pseudodifferential operators is given by

$$p * q = \sum \frac{i^{-|\alpha|}}{\alpha!} \frac{\partial^\alpha p}{\partial \xi^\alpha} \frac{\partial^\alpha q}{\partial x^\alpha}, \tag{5.5}$$

where we use the multi-index notation $x^\alpha = x_1^{\alpha_1} x_2^{\alpha_2} \dots$. In particular, the principal symbol of the product operator is just the matrix product of the principal symbols of the factors.

Often it is sufficient to study PSDO's only modulo *infinitely smoothing* operators: these are operators corresponding to symbols which vanish at $|\xi| \mapsto \infty$ faster than any power $|\xi|^{-k}$.

As before, we denote by E the tensor product of a spin bundle and a trivial complex vector bundle V (with a fiber inner product) over the compact manifold M. A compact gauge group G acts through a unitary representation ρ in the fibers of V. The sign of a background Weyl-Dirac operator D_0 is denoted by ϵ; the background potential is chosen such that D_0 does not have zero modes. We shall consider PSDO's acting on sections of the bundle E. The PSDO's form a Lie algebra $\mathfrak{ps}$ with respect to the usual operator commutator, corresponding to the commutator

$$[p, q]_* = p * q - q * p \tag{5.6}$$

for the symbols.

Denote by K the group of all invertible order zero PSDO's on E, with the additional property

$$[D_0, g] \text{ is bounded for } g \in K. \tag{A}$$

One can define a Frechet space topology and a smooth structure in K using seminorms defined by partials derivatives of the symbols $p_j(x, \xi)$ in the asymptotic expansion of the symbol of an element in K, but actually the choice of the topology in K is not essential for the following discussion.

We shall now concentrate to the case when the dimension $n = 3$. Let $\mathcal{A}_s$ be the space of order zero $PSDO$'s on the vector bundle E. We think of elements $A \in \mathcal{A}_s$ as generalized vector potentials $\sigma_i A_i$, but now A is not necessarily a multiplication operator; the lower order terms of A are arbitrary. We define a 2-cocycle c_s, with coefficients in the abelian algebra of complex functions on $\mathcal{A}_s$, for the Lie algebra $\mathfrak{k}$ of K.

$$c_s(X, Y; F) = \frac{1}{8} Res \frac{1}{|D_0|^3} A[[D_0, X], [D_0, Y]] \tag{5.7}$$

The group K acts on functions of $A \in \mathcal{A}_s$ by $(\gamma \cdot h)(A) = h(g^{-1}Ag + g^{-1}[D_0, g])$. This action defines an infinitesimal action of the Lie algebra $\mathfrak{k}$ through the Lie derivative $\mathcal{L}_X h$, $X \in \mathfrak{k}$.

The cocycle property (2.10) for c_s is proven by a straight-forward computation using the property (5.4), the fact that $[|D_0|^{-3}, X]$ is at most of order -4, and $\mathcal{L}_X A = [A, X] + [D, X]$.

Remark Since the operator on the right in (5.7) is of order -3, the residue is actually equal to the Dixmier trace.

In the special case when the Lie algebra $\mathfrak{k}$ is restricted to multiplication operators by $\mathbf{g}$ valued functions on M and A is an ordinary vector potential we get a familiar formula for the Lie algebra cocycle c_s.

Theorem 5.8. *When the range of A in (5.7) is restricted to the smooth vector potentials in V, and X, Y are infinitesimal gauge transformations (=multiplication operators by elements of $Map(M, \mathbf{g})$), one has*

$$c_s(X, Y; F) = \frac{i}{24\pi^2} \int_M tr\, A[dX, dY].$$

Proof. The principal symbol of the operator on the right in (5.7) is of order -3 and is now equal to

$$\frac{1}{8} \frac{\sigma_k A_k}{|\xi|^3} [\sigma_i \partial_i X, \sigma_j \partial_j Y].$$

The integral over the unit sphere in the momentum space gives just the area 4π of the sphere. The trace over the spinor indices gives the factor $2i$, because of the property $\sigma_1\sigma_2\sigma_3 = i$ of Pauli matrices. From the normalization convention of the residue we get an additional factor $\frac{1}{3(2\pi)^3}$. Putting all together we get the claimed formula. □

The advantage with the generalization (5.7) as compared to the classical formula is that in the larger group K containing the gauge group $Map(M, G)$ one can use a Borel type decomposition with respect to the sign operator $\epsilon = \frac{D_0}{|D_0|}$. That is,

$$\mathfrak{k} = \mathfrak{k}_- \oplus \mathfrak{k}_0 \oplus \mathfrak{k}_+, \tag{5.9}$$

where $\mathfrak{k}_0$ commutes with ϵ and $[\epsilon, \mathfrak{k}_\pm] = \pm 2\mathfrak{k}_\pm$. The corresponding projection operators are given by $X \mapsto \frac{1}{2}(X + \epsilon X \epsilon)$ and $X \mapsto \frac{1}{4}(X - \epsilon X \epsilon) \pm \frac{1}{4}[\epsilon, X]$, respectively. Since ϵ commutes with D_0 and is of order zero, a multiplication by ϵ does not destroy the property (A). One can then define (generalized) highest weight representations for the group K using the splitting (5.9).

Acknowledgements It is pleasure to thank Hartmann Römer for hospitality at University of Freiburg, where this work was done, and Alexander von Humboldt Stiftung (Bonn) for financial support.

References

[A] Araki, H.: Bogoliubov automorphisms and Fock representations of canonical anticommutation relations. In: Contemporary Mathematics. American Mathematical Society, vol. 62 (1987).

[AS] Atiyah, M. and I. Singer: Dirac operators coupled to vector potentials. Proc. Natl. Acad. Sci. USA **81**,2597 (1984).

[BH] Bardakci, K. and M.B. Halpern, Phys. Rev. **D3**, p. 2493 (1971); M.B. Halpern, Phys. Rev. **D4** , p. 2398 (1971).

[CH] Carey, A. and C.A. Hurst: A note on boson-fermion correspondence and infinite-dimensional groups. Commun. Math. Phys. **98,** 435 (1985).

[C] Connes, A.: Noncommutative differential geometry. Publ. Math. IHES **62,** 81 (1986).

[F] Frenkel, I.: Two constructions of affine Lie algebra representations and boson-fermion correspondence in quantum field theory. J. Funct. Anal. **44,** 259 (1981).

[FR] Finkelstein, D. and J. Rubinstein: Connection between spin, statistics, and kinks. J. Math. Phys. **9,** 1762 (1968).

[FS] Faddeev, L., and S. Shatasvili: Algebraic and Hamiltonian methods in the theory of non-Abelian anomalies. Theor. Math. Phys. **60,** 770 (1984).

[FT] Fujii, K., and M. Tanaka: Universal Schwinger cocycles of current algebras in $(D+1)$ dimensions: Geometry and physics. Commun. Math. Phys. **129,** 267 (1990).

[IO] Itoh, T. and K. Odaka: A particle-picture approach to anomalies in chiral gauge theory. Fortschr. Phys. **39**,557 (1991).

[JJ] Jackiw, R. and K. Johnson: Anomalies of the axial vector current. Phys. Rev. **182**,1459 (1969).

[L] Langmann, E.: On Schwinger terms in $(3+1)$ dimensions. Proceedings of the XXVII Karpacz Winter School of Theoretical Physics, Feb. 1991. Also: Proc. of the colloquim "Topological and Geometrical Methods in Field Theory", Turku, Finland, May 1991 (eds. by J. Mickelsson and O. Pekonen, World Scientific, Singapore, 1992).

[Lu] Lundberg, L.-E.: Quasi-free second "quantization". Commun. Math. Phys. **50,** 103 (1976).

[M1] Mickelsson, J.: Commutator anomaly and the Fock bundle. Commun. Math. Phys. **127,** 285 (1990).

[M2] ——: Vacuum expectation values of products of chiral currents in $3+1$ dimensions. M.I.T. preprint CTP#2107, June 1992. To be publ. in Commun. Math. Phys.

[M3] ——: *Current Algebras and Groups.* Plenum Press, New York and London (1989).

[M4] ——: Chiral anomalies in even and odd dimensions. Commun. Math. Phys. **97**,361 (1985).

[MR] Mickelsson, J. and S. Rajeev: Current algebras in $(d+1)$ dimensions and determinant bundles over infinite-dimensional Grassmannians. Commun. Math. Phys. **116,** 365 (1988).

[P] Pickrell, D: On the Mickelsson-Faddeev extension and unitary representations. Commun. Math. Phys. **123,** 617 (1989).

[PS] Pressley, A. and G. Segal: *Loop Groups.* Clarendon Press. Oxford (1986).

[R1] Ruijsenaars, S.N.M.: On Bogoliubov transformations for systems of relativistic charged particles. J. Math. Phys. **18,** 517 (1977).

[R2] ——: Index formulas for generalized Wiener-Hopf operators and boson-fermion correspondence in $2N$ dimensions. Commun. Math. Phys. **124,** 553 (1989).

[S1] Segal, G.: Faddeev's anomaly in the Gauss' law. Preprint (unpublished) Oxford (1985).

[S2] —: Unitary representations of some infinite-dimensional groups. Commun. Math. Phys. **80,** 301 (1981).

[So] Sorkin, R.: A general relation between kink-exchange and kink-rotation. Commun. Math. Phys. **115,** 421 (1988).

[VG] Várilly, J., and J.M. Gracia-Bondia: Connes' noncommutative differential geometry and the standard model. Preprint, Madrid (1992).

[W] Witten, E.: Current algebra, baryons, and quark confinement. Nucl. Phys. **B223, 433** (1983).

[Wo] Wodzicki, M.: Noncommutative residue. In: *K-theory, Arithmetic and Geometry,*Springer Lecture Notes in Math. 1289 (ed. by Yu.I. Manin) p. 320- 399 (1987).

Department of Mathematics, University of
Jyväskylä, SF-40100. Jyväskylä, Finland

NON-LINEAR REALIZATION TECHNIQUE - THE MOST CONVENIENT WAY OF DERIVING N=1 SUPERGRAVITY

J. NIEDERLE

ABSTRACT: This invited talk is based on the results derived by E.A. Ivanov and the author [1,2]. In the Introduction motivations to formulate the Einstein gravity and supergravity as non-linear realizations are briefly summarized. Then, in Section 2, non-linear realizations of connected groups are discussed and the general method for construction a theory possessing a spontaneously broken symmetry via non-linear realizations is illustrated on the Einstein theory of gravity. Section 3 is devoted to N=1 supergravity (resp. the minimal Einstein version of it). It is shown that this theory can also be consistently formulated in terms of non-linear realizations of its supergauge group G or more precisely as a theory of simultaneous non-linear realizations of two complex finite-dimensional subgroups of G generating via their closure the whole infinite-dimensional supergroup G. It turns out that the only independent Goldstone superfield accompanying a spontaneous breaking of the infinite-dimensional supersymetry G down to the rigid N=1 supersymetry is an axial vector superfield $H^{\mu\dot{\mu}}(x,\theta,\bar{\theta})$ identified with the N=1 supergravity prepotential. All the other Goldstone superfields are expressed in terms of $H^{\mu\dot{\mu}}$ and its derivatives by imposing appropriate covariant constraints on the corresponding Cartan superforms (the inverse Higgs effect). Possible implications and new geometrical insights of the proposed formulation of N=1 supergravity are discussed in the Concluding remarks. In particular, the intriguing analogy between N=1 supergravity and the (super)p-branes theories is pointed out.

1. INTRODUCTION

Albert Einstein completed his theory of gravity, space and time - "The general theory of relativity" in 1916 [3]. First this theory met with successes largely in theoretical domain. However, the situation changed due to an enormous increase of experimental technique in the last 25 years so that at present Einstein's theory not only exhibits a beautiful theoretical structure but also predicts, and we should say very successfully, all known experimental results [5]. In fact several present-day experimental tests of Einstein's theory are measured with accuracy which is highest reached in physics so far.

E. A. Tanner and R. Wilson (eds.), Noncompact Lie Groups and Some of Their Applications, 391–404.

Moreover, modern telecommunication systems as well as some advanced technologies must take into account the Einstein theory so that it starts to be a science for engineers.

In spite of all these achievements there are many attempts to reformulate Einstein's theory (see e.g. [5, 6, 2]). They are stimulated by problems of gravity theory itself (by its quantization, singularities, non-Riemannian structure etc.) and, in particular, by a recent development of particle physics according to which all fundamental interactions in nature, including gravity, should be formulated in the framework of gauge theories and unified.

However, in contradistinction to QED, the Weinberg-Salam-Glashow theory of electroweak interactions and QCD, the theory of gravitation was not formulated by Einstein as a gauge theory but rather as a geometrical theory in which gravity was identified with a deviation of four-dimensional space-time geometry from that of a flat space-time.

In this article we shall, therefore, review a gauge formulation of the (super)gravitation theory, namely the formulation of (super)gravity theory as a generalized σ–model with appropriate (super)group coset spaces playing the role of target manifolds. Thus the (super)gravity theory as any theory possessing spontaneously broken symmetry will be constructed in terms of the corresponding non-linear realizations of its (super)gauge group. This gives new geometrical insights into this theory and indicates its profound relations with most of the theories of current interest ((super)particle, (super)string and (super)membrane theories [7]) and, in particular, with more customary σ–models associated with spontaneously broken internal symmetry.

2. NON-LINEAR REALIZATION OF CONNECTED GROUPS

2.1 Non-linear realization technique

In field theory two concepts of symmetries, i.e. transformations with respect to which the considered theory is invariant, are introduced: algebraic and dynamical [8].

The algebraic symmetries are realized by linear and homogeneus transformations of all fields in the theory. They lead to various conservation laws, to classifications of fields and state vectors according to representations of the corresponding group of transformations or to restrictions on S-matrix elements between states with fixed number of particles etc.. They are formed by transformations with respect to which not only the considered field theory is invariant but also solutions of its equations of motion or its vacuum.

In contradistinction to algebraic symmetries the dynamical symmetries are realized by inhomogenous and even non-linear trasformations of fields. They do not lead to classifications of particle states etc. but they influence strongly the possible form of the system dynamics instead (e.g. they lead to low energy theorems, equivalence principle in general relativity etc.). The group of such symmetries does not yield any new conservation law beside those generated by its possible algebraic subgroup. These are the symmetries we shall deal with. More precisely we shall consider theories of which dynamical symmmetry groups have subgroups of algebraic symmetries. Thus the dynamical symmetries leave the considered theory invariant but not solutions of its equations of motion or its vacuum. These theories as any theory possessing spontaneously broken symmetry can be constructed in terms of non-linear realizations of their dynamical groups.

Non-linear realization technique was first developed for finite-dimensional Lie groups by Coleman, Wess and Zumino and by Volkov [9] and then for some infinte-dimensional ones. In particular non-linear realization of conformal symmetry were considered by Isham, Salam, Strathdee [10] and Volkov [11]. Volkov [11] described also a general theory of non-linear realization using Cartan differential forms [12] and in a remarkable paper [13] the briefly described non-linear realization of supergroup. Non-linear realizations for supergroups were worked out in detail by Ivanov and Niederle [2]·

2.2 Gravity as non-linear realization

The main ideas of gauge theories formulated in terms of non-linear realizations can be illustrated on the gravitation theory and summarized as follows:

i) gravitational interaction like all other interactions has a dynamical symmetry group G.

ii) Group G is obtained from an algebraic symmetry group H yielding the physical conservation laws of the theory and forming a subgroup of G.

iii) Such enlargement of H to G is associated with an appearance of Goldstone particles. In other words, the non-linear realizations of G consist in identification of Goldstone fields with those parameters of G which are not connected with generators yielding the physical conservation laws, i.e. with the parameters of factor space G/H.

iv) From geometrical structure of G/H the gravitation Lagrangians (as well as the conserved currents) are explicitely constructed by methods of differential geometry in terms of group invarinats.

In the Borisov-Ogievetsky theory of gravitation the dynamical group G is the group Diff $R^{3,1}$ and the symmetry group H generating the physical conservation laws of the theory - the Poincaré group P. Then the gravity theory itself appears to be a non-linear realization of Diff $R^{3,1}$ and the gravitational field nothing else than the Goldstone field identified with parameters of Diff $R^{3,1}/P$.

In the case of non-linear realizations of infinite-dimensional group Diff $R^{3,1}$ the situation is greatly simplified due to the Ogievetsky theorem [15] according to which the algebra of Diff $R^{3,1}$ can be regarded as a closure of two finite-dimensional algebras: the 20-dimensional affine algebra $\mathrm{A}(4)$ and the 15-dimensional conformal algebra C. Thus non-linear realizations of Diff $R^{3,1}$ namely, Diff $R^{3,1}/P$, can be constructed by taking simultaneous non-linear realizations of $\mathrm{A}(4)$ and C, namely $\mathrm{A}(4)/P$ and C/P

respectively. The non-linear realizations on $A(4)/P$ gives rise to a 10-components Goldstone field $h_{\mu\upsilon}(x)$ and that on C/P yields the scalar Goldstone field $\varphi(x)$ as well as its gradient $\varphi_{\mu}(x)$. The requirement that both non-linear realizations should be realized simultaneously leads to the following identification of the Goldstone fields: $\varphi(x) \equiv \frac{1}{4} h_{\mu\mu}(x)$. Thus the resulting theory has only the 10-components gravitation field $h_{\mu\upsilon}(x)$. Since its Lagrangian gives rise to the same equations of motion as appear in the Einstein theory of gravitation, the resulting theory is equivalent to that of Einstein (for details see [14]).

3. N=1 MINIMAL EINSTEIN SUPERGRAVITY AS NON-LINEAR REALIZATIONS

3.1 Gauge supergroup of N=1 minimal Einstein supergravity and its structure

The Ogievetsky-Sokatchev formulation of N=1 minimal Einstein supergravity [10] is based on (4+2)-dimensional complex superspace

$$C^{4/2} = C^{4/4} \Big/ C^{0/2} = \left\{ \left(x_L^{\rho\dot{\rho}}, \theta_L^{\mu} \right) \right\} = \left\{ \left(x_R^{\rho\dot{\rho}}, \bar{\theta}_R^{\dot{\mu}} \right) = \left(x_L^{\rho\dot{\rho}}, \theta_L^{\mu} \right)^{+} \right\} \tag{3.1,1}$$

with $\left(x_L^{\rho\dot{\rho}}, \theta_L^{\mu}\right)$, and $\left(x_R^{\rho\dot{\rho}}, \bar{\theta}_R^{\dot{\mu}}\right)$ being its left- and right-handed parametrizations respectively.

In superspace $C^{4/2}$ an infinite-dimensional complex gauge supergroup G acts. Its infinitesimal action defined by

$$\delta x_L^{\rho\dot{\rho}} = \lambda^{\rho\dot{\rho}}(x_L, \theta_L), \tag{3.1,2}$$

$$\delta\theta_L^{\mu} = \lambda^{\mu}(x_L, \theta_L) \tag{3.1,3}$$

is specified by superfunction-parameters $\lambda^{\rho\dot{\rho}}, \lambda^{\mu}$ satisfying the relation

$$\frac{\partial\lambda^{\rho\dot{\rho}}}{\partial x_L^{\rho\dot{\rho}}} - \frac{\partial\lambda^{\mu}}{\partial\theta_L^{\mu}} = 0 \tag{3.1,4}$$

This formula expresses infinitesimally a correlation of transformation properties of supervolumes in $C^{4/2}$ with those in $C^{4/4}$ for the Gates-Siegel parameter $n = \frac{1}{3}$ (for detail see [17]).

By using [1] it can be shown that the following theorem is true.

Theorem: *Infinite-dimensional superalgebra A of supergroup G can be obtained by taking a closure of two finite-dimensional superalgebras namely* A_I *generated by*

$$\left\{Q_\mu,\ P_{L\rho\dot\rho},\ Q^\mu_{\rho\dot\rho},\ K_{\beta\dot\beta}\right\} \tag{3.1,5}$$

and A_{II} *generated by*

$$\left\{Q_\mu, P_{L\rho\dot\rho}, Q^\mu_{\rho\dot\rho}, R^{\alpha\dot\alpha}_{\beta\dot\beta}, T_{(\mu\nu)}, I^{\rho\dot\rho}_\mu, D\right\} \tag{3.1,6}$$

where generators

$$Q_\mu = -i\frac{\partial}{\partial\theta^\mu_L} \equiv -i\partial_\mu, \tag{3.1,7}$$

$$P_{L\rho\dot\rho} = -\iota\frac{\partial}{\partial x^{\rho\dot\rho}_L} \equiv -i\partial_{L\rho\dot\rho}, \tag{3.1,8}$$

$$Q^\mu_{\rho\dot\rho} = \theta^\mu\partial_{L\rho\dot\rho}, \tag{3.1,9}$$

$$K_{\beta\dot\beta} = -i\left(\theta^\mu_L\theta_{L\mu}\right)\partial_{L\beta\dot\beta}, \tag{3.1,10}$$

$$R^{\alpha\dot\alpha}{}_{\beta\dot\beta} = -i\left(x^{\alpha\dot\alpha}\partial_{\beta\dot\beta} - \tfrac{1}{4}\delta^\alpha_\beta\,\delta^{\dot\alpha}_{\dot\beta}\,(x\partial)\right), \tag{3.1,11}$$

$$T_{(\mu\nu)} = \tfrac{1}{2}\theta_{(\mu}\partial_{\upsilon)}, \tag{3.1,12}$$

$$I^{\rho\dot\rho}_\mu = x^{\rho\dot\rho}\partial_\mu, \tag{3.1,13}$$

$$D = -i\left(x^{\rho\dot\rho}\partial_{\rho\dot\rho} + 2\theta^\mu\partial_\mu\right) \tag{3.1,14}$$

fulfil relations specified in [2].

It turns out that

i) A_I is a superalgebra of all Grassmann vector fields (i.e. each element of A_I can be obtained by gauging ordinary 4-translations in purely Grassmann directions)

ii) A_{II} is a special linear superalgebra in $C^{4/2}$. It contains the Lorentz generators $M_{\alpha\beta}$ and $M_{\dot\alpha\dot\beta}$. For instance $M_{\alpha\beta}$ is given by

$$M_{\alpha\beta} = R_{(\alpha\beta)} + T_{(\alpha\beta)},\ R_{(\alpha\beta)} \equiv -ix_{(\alpha}{}^{\dot\beta}\partial_{\beta)\dot\beta} \tag{3.1,15}$$

iii) Graviton and gravitino is associated with $R^{\alpha\dot\alpha}_{\beta\dot\beta}$ and $I^{\rho\rho}_\mu$ respectively.

3.2 Non-linear realization of G_I

G_I denotes a complex supergroup the superalgebra of which is A_{II} defined in (3.1,5). Each element g_I of supergroup G_I can be parametrized in the following way:

$$g_I = g_1 \cdot g_2 \cdot g_3 \,, \tag{3.2,1}$$

where

$$\begin{aligned} g_1 &= \exp\left\{i\left(\theta^{\mu} Q_{\mu} + x_L^{\rho\dot\rho} P_{L\rho\dot\rho}\right)\right\}, \\ g_2 &= \exp\left\{i\psi_{\mu}^{\rho\dot\rho} Q^{\mu}_{\rho\dot\rho}\right\}, \\ g_3 &= \exp\left\{i a^{\beta\dot\beta} K_{\beta\dot\beta}\right\}. \end{aligned} \tag{3.2,2}$$

Transformation properties of the group parameters $\theta^{\mu}, x_L^{\rho\dot\rho}, \psi_{\mu}^{\rho\dot\rho}$ and $a^{\beta\dot\beta}$ follow from the group multiplication law

$$g_I^0 \cdot g_I = g_I' \tag{3.2,3}$$

that is from multiplication of g_I from left by fixed element $g_I^0 \in G_I$ yielding a new element g_I'. Assuming g_I^0 of the form

$$g_I^0 \approx I + i\left(\varepsilon^{\mu} Q_{\mu} + c^{\rho\dot\rho} P_{L\rho\dot\rho} + \beta_{\mu}^{\rho\dot\rho} Q^{\mu}_{\rho\dot\rho} + \gamma^{\alpha\dot\alpha} K_{\alpha\dot\alpha}\right) \tag{3.2,4}$$

we obtain

$$\begin{aligned} \delta\theta^{\mu} &= \varepsilon^{\mu}, \\ \delta x_L^{\rho\dot\rho} &= c^{\rho\dot\rho} + i\theta^{\mu}\beta_{\mu}^{\rho\dot\rho} + \left(\theta^{\mu}\theta_{\mu}\right)\gamma^{\rho\dot\rho}, \\ \delta\psi_{\mu}^{\rho\dot\rho} &= \beta_{\mu}^{\rho\dot\rho} - 2i\theta_{\mu}\gamma^{\rho\dot\rho}, \\ \delta a^{\rho\dot\rho} &= \gamma^{\rho\dot\rho}. \end{aligned} \tag{3.2,5}$$

Now, following the general theory [9-13], we introduce left-invariant Cartan's forms $\omega_I{}^{\alpha}, \omega_{IL}{}^{\beta\dot\beta}, \omega_{I\mu}{}^{\rho\dot\rho}, k^{\beta\dot\beta}$ via

$$g_I^{-1} dg_I = i\left\{\omega_I^{\alpha} Q_{\alpha} + \omega_{IL}^{\beta\dot\beta} P_{L\beta\dot\beta} + \omega_{I\mu}^{\rho\dot\rho} Q^{\mu}_{\rho\dot\rho} + k^{\beta\dot\beta} K_{\beta\dot\beta}\right\}. \tag{3.2,6}$$

By comparing both sides of this identity we obtain

$$\omega_I^{\alpha} = d\theta^{\alpha}, \tag{3.2,7a}$$

$$\theta_{IL}^{\beta\dot{\beta}} = dx_L^{\beta\dot{\beta}} + i\psi_\mu^{\beta\dot{\beta}} d\theta^\mu, \tag{3.2,7b}$$

$$\omega_{I\mu}^{\rho\dot{\rho}} = d\psi_\mu^{\rho\dot{\rho}} + 2ia^{\rho\dot{\rho}} d\theta_\mu, \tag{3.2,7c}$$

$$k^{\beta\dot{\beta}} = da^{\beta\dot{\beta}} . \tag{3.2,7d}$$

These left-invariant Cartan 1-forms can be used to eliminate some Goldstonians which are associated with the group parameters. More precisely according to [2,13] we shall identify the group parameters associated with $P^{\mu\dot{\mu}} \equiv P_L^{\mu\dot{\mu}} + P_R^{\mu\dot{\mu}} = P_L^{\mu\dot{\mu}} + \left(\overline{P_L^{\mu\dot{\mu}}}\right)$ with coordinates $x^{\mu\dot{\mu}}$ of the real superspace, while that with the $P_A^{\mu\dot{\mu}} \equiv i\left(P_L^{\mu\dot{\mu}} - P_R^{\mu\dot{\mu}}\right)$ with the Goldstone superfield $H^{\mu\dot{\mu}}(x,\theta,\bar{\theta})$. The group parameters $\theta^\mu, \bar{\theta}^{\dot{\mu}}$ are interpreted as Grassmanian coordinates of the real superspace. The remaining group parameters $\psi_\mu^{\rho\dot{\rho}}, a^{\rho\dot{\rho}}$ are associated with Goldstone superfields. They can be expressed in terms of $H^{\mu\dot{\mu}}\left(x,\theta,\bar{\theta}\right)$ via the inverse Higgs effect.

Namely expanding the appropriate Cartan 1-forms (3.2,7) in the covariant differentials $\nabla x^{\alpha\dot{\alpha}}, d\theta^\mu$ and $d\bar{\theta}^{\dot{\mu}}$ we can define covariant derivatives and spinor covariant derivatives of $H^{\alpha\alpha}$ and $\psi_\mu^{\alpha\dot{\alpha}}$. Equating these covariant derivatives to zero, i.e. ,

$$\mathcal{D}_\mu H^{\alpha\dot{\alpha}} = 0, \ \mathcal{D}_{\dot{\mu}} H^{\alpha\dot{\alpha}} = 0 \ \ and \ \ \mathcal{D}^\mu \psi_\mu^{\alpha\alpha'} = 0, \tag{3.2,8}$$

we obtain

$$\psi_\mu^{\alpha\dot{\alpha}} = 2\nabla_\mu H^{\alpha\dot{\alpha}}, \ \psi_{\dot{\mu}}^{\alpha\dot{\alpha}} = 2\nabla_{\dot{\mu}} H^{\alpha\dot{\alpha}} \ \text{and} \ a^{\alpha\dot{\alpha}} = \tfrac{-i}{2}\left(\nabla^\mu \nabla_\mu\right) H^{\alpha\dot{\alpha}} . \tag{3.2,9}$$

Consequently $\psi_\mu^{\alpha\dot{\alpha}}$ and $a^{\alpha\dot{\alpha}}$ are covariantly eliminated from the theory (for detail see [2]).

Thus the non-linear realization of G_I can be entirely given in terms of superfield $H^{\mu\dot{\mu}}(x,\theta,\bar{\theta})$.

3.3 Non-linear realization of G_{II}

Analogously to the previous case, G_{II} denotes a complex supergroup the superalgebra of which is A_{II} defined in (3.1,6). Each element g_{II} of supergroup G_{II} can be parametrised

$$g_{II} = \tilde{g}_{II}.l \ , \tag{3.3,1}$$

where $\tilde{g}_{II}$ denotes the element of the coset space G_{II}/L with L being the Lorentz group

$$L = \exp\{il^{\alpha\beta}M_{\alpha\beta}\}.\exp\{i\tilde{l}^{\dot{\alpha}\dot{\beta}}M_{\dot{\alpha}\dot{\beta}}\} \ . \tag{3.3,2}$$

here $M_{\alpha\beta}$ is defined in (3.1,5) and l is the element of the Lorentz group L.

The element $\tilde{g}_{II}$ of the coset space G_{II}/L can be parametrized in the following way

$$\tilde{g}_{II} = g_1.g_2.g_3.g_4.g_5 \ , \tag{3.3,3}$$

where g_1 and g_2 are defined in (3.2,2) and

$$\begin{aligned} g_3 &= \exp\{i\lambda^{\mu}_{\rho\dot{\rho}} I^{\rho\dot{\rho}}_{\mu}\}, \\ g_4 &= \exp\{i\pi^{\alpha\dot{\alpha}\beta\dot{\beta}} R_{\alpha\dot{\alpha}\beta\dot{\beta}}\}, \\ g_5 &= \exp\{i\varphi^{D}\}. \end{aligned} \tag{3.3,4}$$

The Cartan 1-forms are defined according to the general theory [9-13] by

$$\tilde{g}^{-1}_{II} d\tilde{g}_{II} = i\{\omega^{\alpha}_{II}Q_{\alpha} + \omega^{\alpha\dot{\alpha}}_{IIL}P_{L\alpha\dot{\alpha}} + \omega^{\alpha\dot{\alpha}}_{II\mu}Q^{\mu}_{\alpha\dot{\alpha}} + \Omega^{\mu}_{\rho\dot{\rho}} I^{\rho\dot{\rho}}_{\mu} + \omega^{\alpha\dot{\alpha}\beta\dot{\beta}}_{R} R_{\alpha\dot{\alpha}\beta\dot{\beta}} + \omega^{\alpha\beta}_{T} T_{(\alpha\beta)} + \omega_{D} D\}. \tag{3.3,5}$$

By comparing both sides of (3.3,5) we obtain

$$\begin{aligned} \omega^{\alpha}_{II} &= \left(d\theta^{\alpha} + i\lambda^{\alpha}_{\rho\dot{\rho}}\omega^{\rho\dot{\rho}}_{IL}\right)e^{2\varphi}, \\ \omega^{\alpha\dot{\alpha}}_{IIL} &= \left(dx^{\rho\dot{\rho}}_{L} + i\psi^{\rho\dot{\rho}}_{\mu} d\theta^{\mu}\right)B^{\alpha\dot{\alpha}}_{\rho\dot{\rho}} e^{\varphi} = \omega^{\rho\dot{\rho}}_{IL} B^{\alpha\dot{\alpha}}_{\rho\dot{\rho}} e^{\varphi}, \\ \omega_{II\mu}^{\ \ \rho\dot{\rho}} &= d\psi^{\lambda\dot{\lambda}}_{\mu} B_{\lambda\dot{\lambda}}^{\ \ \rho\dot{\rho}} e^{-\varphi}, \\ \Omega^{\mu}_{\rho\dot{\rho}} &= \left(d\lambda^{\mu}_{\beta\dot{\beta}} + \lambda^{\upsilon}_{\beta\dot{\beta}}\lambda^{\mu}_{\gamma\dot{\gamma}} d\psi^{\gamma\dot{\gamma}}_{\upsilon}\right)\left(B^{-1}\right)^{\beta\dot{\beta}}{}_{\rho\dot{\rho}} e^{\varphi}, \\ \omega^{\alpha\dot{\alpha}\beta\dot{\beta}}_{R} &= -\lambda^{\gamma}_{\rho\dot{\rho}} d\psi^{\upsilon\dot{\upsilon}}_{\gamma}\left(B^{-1}\right)^{\rho\dot{\rho}\alpha\dot{\alpha}} B^{\beta\dot{\beta}}{}_{\upsilon\dot{\upsilon}} + \left(B^{-1}\right)^{\alpha\dot{\alpha}\tau\dot{\tau}} dB^{\beta\dot{\beta}}{}_{\tau\dot{\tau}}, \\ \omega^{\alpha\beta}_{T} &= \tfrac{1}{4}\lambda^{(\alpha}_{\gamma\dot{\rho}} d\psi^{\beta)\gamma\dot{\rho}}, \end{aligned} \tag{3.3,6}$$

$$\omega_D = d\varphi - \tfrac{i}{4}\lambda^{\mu}_{\rho\dot\rho} d\psi_{\mu}{}^{\rho\dot\rho} ,$$

where B is defined by

$$g_5^{-1} R_{\alpha\dot\alpha\beta\dot\beta} g_5 = \left(B^{-1}\right)_{\alpha\dot\alpha}{}^{\tau\dot\tau} B_{\beta\dot\beta}{}^{\upsilon\dot\upsilon} R_{\tau\dot\tau\upsilon\dot\upsilon} \quad . \tag{3.3,7}$$

All these 1-forms, except those associated with $M_{\alpha\beta}$, $M_{\dot\alpha\dot\beta}$ which are hidden in ω_R and ω_T, undergo the induced Lorentz transformation with respect to their spinor indices when g_{II} acts on $\tilde g_{II}$ via the left multiplication

$$\overset{0}{g}_{II} \cdot \tilde g_{II} = \tilde g'_{II} . L^{ind} , \tag{3.3,8}$$

where

$$L^{ind} \approx I + i\delta h^{\alpha\beta}\left(x,\theta,\bar\theta\right) M_{\alpha\beta} + i\delta \bar h^{\dot\alpha\dot\beta}\left(x,\theta,\bar\theta\right) M_{\dot\alpha\dot\beta} \tag{3.3,9}$$

and

$$\overset{0}{g}_{II} \approx I + i\left(\varepsilon^{\alpha} Q_{\alpha} + c^{\alpha\dot\alpha} P_{L\alpha\dot\alpha} + \beta^{\rho\dot\rho}_{\mu} Q^{\mu}_{\rho\dot\rho} + \sigma^{\alpha\dot\alpha\beta\dot\beta} R_{\alpha\dot\alpha\beta\dot\beta} + \upsilon^{\alpha\beta} T_{\alpha\beta} + \rho^{\mu}_{\rho\dot\rho} I^{\rho\dot\rho}_{\mu} + cD\right). \tag{3.3,10}$$

Applying the general formula (3.3,8) we obtain the transformation properties of the coset parameters $\theta^{\mu}, x_L^{\rho\dot\rho}, \psi_{\mu}^{\rho\dot\rho}, \lambda^{\mu}_{\rho\dot\rho}, \bar h^{\rho\dot\rho}$ and $B_{\tau\dot\tau}{}^{\rho\dot\rho}$ under g_{II} (for detail see [2]).

Now we shall eliminate extra Goldstonians and single out covariants of G_{II} which are simultaneously also covariants w.r.t. G_I.

Looking at equations (3.3,6), (3.3,7) and using the fact that $\omega^{\alpha\dot\alpha}_{IIL}$ is covariant also under G_{II} since R and D can be added to G_I as extra automorphism generators we see that G_I does not transform B and φ. Now we should decompose $\omega^{\alpha\dot\alpha}_{IIL}$ into the covariant differentials of $x^{\alpha\dot\alpha}$ and $H^{\alpha\dot\alpha}$ $\left(\Delta x^{\alpha\dot\alpha} \text{ and } \Delta H^{\alpha\dot\alpha}\right)$ once again and then extract covariant derivatives of $H^{\alpha\dot\alpha}$ from $\Delta H^{\alpha\dot\alpha}$.

It turns out that conditions (3.2,9) for elimination of $\psi^{\rho\dot\rho}_{\mu}$ in G_I are simultaneously covariant under G_{II} so that $\psi^{\rho\dot\rho}_{\mu}$ given by (3.2,9) possesses correct transformation properties with respect to both, G_I and G_{II} (for the proof see [2]).

Eliminating $\psi_{\mu}^{\alpha\dot\alpha},\psi_{\dot\mu}^{\alpha\dot\alpha}$ by (3.2,9), the covariant differentials of $x^{\alpha\dot\alpha}$ and $H^{\alpha\dot\alpha}$, Δx and ΔH acquire the forms

$$\Delta x^{\rho\dot\rho} = \nabla x^{\alpha\dot\alpha}\left(b_{\alpha\dot\alpha}{}^{\rho\dot\rho} + \partial_{\alpha\dot\alpha}H^{\gamma\dot\gamma}c_{\gamma\dot\gamma}{}^{\rho\dot\rho}\right),$$
$$\Delta H^{\rho\dot\rho} = -\nabla x^{\gamma\dot\gamma}\left(c_{\gamma\dot\gamma}{}^{\rho\dot\rho} - \partial_{\gamma\dot\gamma}H^{\lambda\dot\lambda}b_{\lambda\dot\lambda}{}^{\rho\dot\rho}\right) \tag{3.3,11}$$

with

$$b_{\rho\dot\rho}^{\alpha\dot\alpha} \equiv \tfrac{1}{2}\left(B_{\rho\dot\rho}^{\alpha\dot\alpha}e^{\varphi} + \overline{B}_{\rho\dot\rho}^{\alpha\dot\alpha}e^{\overline{\varphi}}\right), \tag{3.3,12}$$

$$c_{\rho\dot\rho}^{\alpha\dot\alpha} = \tfrac{1}{2}\left(B_{\rho\dot\rho}^{\alpha\dot\alpha}e^{\varphi} - \overline{B}_{\rho\dot\rho}^{\alpha\dot\alpha}e^{\overline{\varphi}}\right) .$$

Their structure is completely specified by expressing the remaing Goldstonians in terms of $H^{\mu\dot\mu}$.

We begin with $B_{\tau\dot\tau}{}^{\rho\dot\rho}$. By inspecting the structure of the Cartan forms (3.3,6) we conclude that $B_{\tau\dot\tau}{}^{\rho\dot\rho}$ can be eliminated by imposing an appropriate constraint on one of the spinor covariant derivatives of the Goldstone field $\psi_{\mu}^{\alpha\dot\alpha}$. These are defined as the coefficients in front of ω_{II}^{α}, $\overline{\omega}_{II}^{\dot\alpha}$ in the G_{II}-covariant Cartan form $\omega_{II\mu}^{\rho\dot\rho}$. It turns out that only $\overline{\Delta}_{\dot\alpha}\psi_{\mu}^{\rho\dot\rho}$ can be used for implementing the thought covariant constraint since it is covariant w.r.t. G_I and G_{II} and hence w.r.t. the whole infinite-dimensional supergroup G. It is meaningless to equate $\overline{\Delta}_{\dot\alpha}\psi_{\mu}^{\rho\dot\rho}$ to zero because it would contradict to the flat-superspace limit. Thus one should equate $\overline{\Delta}_{\dot\alpha}\psi_{\mu}^{\rho\dot\rho}$ to a proper Lorentz-covariant constant metrics consistent with the flat limit namely

$$\overline{\Delta}_{\dot\alpha}\psi_{\mu}^{\rho\dot\rho} = -\delta_{\dot\alpha}^{\dot\rho}\delta_{\mu}^{\rho} . \tag{3.3,13}$$

From here by taking into account that $\det B = 1$ we can express superfields $B_{\tau\dot\tau}{}^{\rho\dot\rho}$ and $\varphi,\overline{\varphi}$ in terms of $H^{\mu\dot\mu}$.

Let us explain now how to eliminate the Goldstone superfield $\lambda^{\mu}_{\rho\dot\rho}$. The corresponding constraints arise from the requirement that only inhomogeniously tranformed components associated with the Lorentz generators $M_{\alpha\beta}$, $M_{\dot\alpha\dot\beta}$ in the $d\overline{\theta}$-projections of the Cartan forms standing in front of the generators R, D and T survive. These constraints are again manifestly G_{II}- and G_I-covariant and give $\lambda^{\mu}_{\rho\dot\rho}$ in terms of

$H^{\mu\dot{\mu}}$ (for detail see [2]). Hence we are eventualy left with a single Goldstone superfield $H^{\mu\dot{\mu}}$ which alone supplies non-linear realizations of G_I and G_{II}. This conferms its role as the fundamental geometric object of the minimal N=1 supergravity discussed in [16] from another point of view.

It remains to show how the minimal N=1 supergravity action reappears within the present framework.

3.4 Invariant action

After employing the inverse Higgs effect constraints, the remaining simultaneous G_I- and G_{II}-covariants are reduced to the covarinst differentials of the N=1 superspace coordinates, the covariant differential of $H^{\mu\dot{\mu}}(x,\theta,\bar{\theta})$ and to the $\bar{\theta}$-covariant derivative of the Goldstone field $\lambda^{\mu}_{\rho\rho}$ (the projection of the Cartan form $\Omega^{\mu}_{\rho\dot{\rho}}$ onto the covariant differential $\Delta\bar{\theta}^{\dot{\mu}}$). Sincethe last covariant gives rise to a higher derivative invariant and the covariant differentials $\Delta x^{\mu\dot{\mu}}$, $\Delta\theta^{\mu}$, $\Delta\bar{\theta}^{\dot{\mu}}$. An obvious simplest invariant is the supervolume of N=1 superspace ($x^{\mu\dot{\mu}}$, θ^{μ}, $\bar{\theta}^{\dot{\mu}}$) constructed as an integral of the Berezinean of the corresponding vielbeins E_M^N over $d^4x\, d^2\theta\, d^2\bar{\theta}$ namely

$$S = \frac{1}{\kappa^2}\int d^4x\, d^2\theta\, d^2\bar{\theta}\; Ber E_M^N \qquad (3.4,1)$$

with κ^2 being the dimensional gravitational coupling constant. As shown in [2] action S coincides up to a renormalization factor wtih the minimal Einstein N=1 supergravity action given in [16].

4. CONCLUDING REMARKS

i) The presented non-linear realization approach allows an algorithmic construction of N=1 supergravity based on the universal method of Cartan's forms augmented with the inverse Higgs phenomenon. The N=1 supergravity prepotential appears from the beginning as a Goldstone superfield describing the simultaneous spontaneous breaking of G_I and G_{II} supersymmetries. Many objects and relations introduced "by hand" or postulated in the Ogievetsky-Sokatchev approach [16] acquire a clear group-theoretical meaning. For instance objects F and $\bar{\text{F}}$ playing the crucial role in [16] turn out to be related to the Goldstone superfield associated with the spontaneously broken generator D_{II} of the supergroup G_{II}. The relations (4.25) in [16] prove to be a particular case of the inverse Higgs effect.

ii) It is worth mentioning that the inverse-Higgs-effect constraints are purely algebraic, in contradistinction to the standard N=1 supergravity constraints which are reduced to certain differential equations (vanishing some components of the torsion), the prepotential being a solution of the latter. In the present formulation these latter constraints are secondary, they can be shown to be a consequence of the Maurer-Cartan structure equations for G_I and G_{II}.

iii) It is interesting to see how the complex geometry of N=1 supergravity [16] (the preservation of chirality) reappears in the framework of the non-linear realization description. Primarily it manifests itself in that one deals with the complex supergroups G_I and G_{II} in a holomorphic parametrization. The $C^{4/2}$-coordinates $x_L^{\mu\dot\mu}$, θ^μ naturally arise as the parameters of the relevant complex coset spaces. The constraints of the inverse Higgs effect in the present case can also be interpreted as a kind of the covariant chirality conditions stating the absence of the $d\bar\theta$-projections in the corresponding Cartan forms.

iv) Let us stress the defining role of the non-linear realization of linear supergroup G_{II}. The structure of the basic building blocks of N=1 supergravity, the covariant differentials, $\Delta x^{\mu\dot\mu}$, $\Delta\theta^\mu$, $\Delta\bar\theta^{\dot\mu}$, is completely specified by this non-linear realization (together with the inverse Higgs effect). The role of G_I is in a sense subsidiary - it provides very simple criterions in what cases the G_{II}-covariant quantities and relations are covariant under the whole N=1 supergravity group G.

v) The construction of N=1 supergravity as a non-linear realization of the complex $(x^{\mu\dot\mu}, \theta^\mu, \bar\theta^{\dot\mu})$ as a real subspace and the N=1 supergravity action as a G_{II}-invariant supervolume of this subspace suggests an interesting analogy of N=1 supergravity with the (super)p-branes (strings, membranes, ...) in the treatment of references [18]. Actually, the minimal N=1 supergravity is recognized as a kind of "spinning" (super)p-brane of dimension (4/4) moving in the complex coset space G_{II}/L as the target space. The Goldstone superfields eliminated by the inverse Higgs effect are direct analogs of the Goldstone fields which parametrize the cosets of the relevant Lorentz groups in ordinary p-branes and are expressed there in terms of the translation Goldstone fields by the same procedure [18]. This similarity raises some questions, in particular, whether N=1 supergravity can be reproduced as a effective "low-energy" limit of some higher-dimensional superfield supersymmetric theories, by analogy with condensation of (super)p-branes in a field theory [19].

vi) Closely related to the latter remark is the problem of existence of theories with a "linearly realized" N=1 supergravity group. Such theories could be related to the non-linear relization formulation of N=1 supergravity much like linear sigma models with associated internal symmetrics are related to the corresponding nonlinear sigma models, via appearance of non-zero vacuum expectation values of some fields. Our construction gives a hint that these linear realizations should operate with linear representations of supergroup G_{II}. An analogous problem for the Einstein gravitation theory has been settled in [14]. As was suggested by Witten [20], the linear sigma model of this kind describes the phases with unbroken local symmetries in gauge theories and can be presumably understood as topological field theories.

vii) Finally we note that the non-linear realization treatment of the non-minimal N=1 supergravity theories can seemingly be constructed in an analogous way. However, it is a much more ambitious problem to find a general principle allowing us to construct higher N supergravities by the non-linear realization techniques. One might hope to obtain in this way the geometric prepotential formulations of supergravities with $N \geq 3$ which are unknown at present.

REFERENCE

[1] Ivanov, E.A. and Niederle, J. (1985), Class. Quantum Grav. **2**, 631

[2] Ivanov, E.A. and Niederle, J. (1992), Phys. Rev. **D45**, 4545

[3] Einstein, A. (1917), Sitz. Preuss Akad. Wiss. 142

[4] Shapiro, J.L. (1980), in A. Held (ed.), General Relativity and Gravitation, Plenum Press, New York, Vol. II p. 469
Reasenberg, P.D. (1980), in P.G. Bergmann and V. de Sabbata (eds.), Cosmology and Gravitation, Plenum Press, New York, p. 317
Fischbach, E. ib, p.359

[5] Gupta, S. (1957), Rev. Mod. Phys. **29**, 334
Thirring, W. (1963), Ann. Phys. **16**, 697
Feynman, R. (1963), Acta Phys. Polonica **24**, 697
Faddeev, L.D. and Popov, V.N. (1967), Phys. Lett. **B25**, 30
De-Witt, B.S. (1967), Phys. Rev. **160**, 1113
De-Witt, B.S. (1967), Phys. Rev. **162**, 1195
De-Witt, B.S. (1967), Phys. Rev. **162**, 1239
Weinberg, S. (1972) Gravity and Cosmology, John Wiley and Sons, Inc., New York
Trautman, A. (1980), in A. Held (ed.), Genral Relativity and Gravitation, Plenum Press, New York and London, Vol. 1 p. 287
Mehl, W., Nitsch, J. and von der Heyde, P. (1980), in A. Held (ed.), General Relativity and Gravitation, Plenum Press, New York and London, Vol. 1 p. 329 and references therein
Ivanov, E.A. and Niederle, J. (1982), Phys. Rev. **D25**, 976
Ivanov, E.A. and Niederle, J. (1982), Phys. Rev. **D25**, 982 and references therein

[6] Borisov, A.B. and Ogievetsky, V.I. (1974), Teor. i mat. fiziki **21**, 329 (in Russian)

[7] Henneaux, M. and Mezincescu (1985), Phys. Lett. **152B**, 340
Hughes, J. and Polchinski, J. (1986), Nulc. Phys. **B278**, 147
Hughes, J., Liu, J. and Polchinski, J. (1986), *ibid.* **180**, 370
Achuccaro, A., Gauntlett, J., Itoh, K. and Townsend, P.K. (1989), Nucl. Phys. B314, 129
Green, M.B. (1989), Phys. Lett. B223, 157

Bergshoeff, E., Pope, C.N., Romans, L.J., Sezgin, E., Shen, X. and Stelle, K. (1990), Phys. Lett. B243, 350

[8] Weinberg, S. (1970), in S. Deser, M. Grisaru and H. Pendleton (eds.), Proceeding of 1970 Brandeis University Summer Institute, MIT Press, Cambridge

[9] Coleman, S., Wess, J. and Zumino, B. (1969), Phys. Rev. **177**, 2239
Callan, C.G., Coleman, S., Wess, J. and Zumino, B. (1969), Phys. Rev. **177**, 2247
Isham, C. (1969), Nuovo cimento **LIXA**, 356
Volkov, D.B. (1969), Preprint IFT-69-75

[10] Salam, A. and Strathdee, J. (1969), Phys. Rev. **184**, 1750
Isham, C.J., Salam, A. and Strathdee, J. (1971), Ann. Phys. N.Y. **62**, 98

[11] Volkov, D.B. (1973), Fiz. Elem. Chastits At. Yadra **4**, 3
Ogievetsky, V.I. (1974) Proceedings of the X Winter School of Theoretical Physics, Karpacz, Poland, 1973, Acta Universitatis Wratislaviensis No. 1966, Wroclaw

[12] Cartan,E. (1949) Geometry of Lie groups and symmetrical spaces (in Russian), MIR, Moscow

[13] Volkov, D.B. (1972), Pis'ma JETF **16**, 621

[14] Borisov, A.B. and Ogievetsky, V.I. (1974), Theor. Mat. Phys. **21**, 329

[15] Ogievetsky,V.I. (1973), Lett. Nuovo Cimento **8**, 988

[16] Ogievetsky, V.I. and Sokatchev, E. (1980), Yadern. Fiz. **31**, 821
Ogievetsky, V.I. and Sokatchev, E. (1980), Yadern. Fiz. **32**, 862
Ogievetsky, V.I. and Sokatchev, E. (1980), Yadern. Fiz. **32**, 870
Ogievetsky, V.I. and Sokatchev, E. (1980), Yadern. Fiz. **32**, 1142
see also
Ogievetsky, V.I. and Sokatchev, E. (1978), Phys. Lett. **B79**, 222

[17] Siegel, W. and Gates Jr., S.J. (1979), Nucl. Phys. **B147**, 77

[18] Hughes, J. and Polchinski, J. (1986), Nulc. Phys. **B278**, 147
Hughes, J., Liu, J. and Polchinski, J. (1986), *ibid.* **180**, 370
Achuccaro, A., Gauntlett, J., Itoh, K. and Townsend, P.K. (1989), Nucl. Phys. B314, 129

[19] Townsend, P. (1988), Phys. Lett. **B202**, 53
Ivanov, E.A. and Kapustnikov, A.A. (1990), Phys. Lett. **B252**, 212

[20] Witten, E. (1988), Commun. Math. Phys. **117**, 353

Institute of Physics AS CR
Na Slovance 2
180 40 PRAGUE
the Czech Republic

TODA SYSTEMS AS CONSTRAINED LINEAR SYSTEMS

L.O'RAIFEARTAIGH

ABSTRACT. It is shown that Toda field theories, which are non-linear but integrable, can be viewed as constrained linear systems, in fact as Wess-Zumino-Witten (WZW) theories in which the Kac-moody currents are subject to linear constraints. The main advantages of taking this point of view is that the general solutions and symmetry algebras of the Toda theories are easily derived, the latter being W-algebras of the kind proposed some years ago by Zamolodchikov. In addition, the procedure can be generalized to find new integrable systems and can be used to study the general theory of second-class constraints, the quantization of non-linear systems and the gauge-fixing cohomology theory of Becchi-Rouet-Stora-Tyutin (BRST) in a simple but non-trivial context.

1. Wess-Zumino-Witten Systems.

As the WZW systems form the starting point for our considerations we begin by recalling that these systems are the generalizations of free 2-dimensional field theory to theories in which the fields $g(x)$ are elements of semi-simple Lie groups G. The Lagrangians take the form

$$\mathcal{L}_{WZW} = \frac{\kappa}{2}\int d^2x\, \mathrm{tr}(J_\mu(x)J^\mu(x)) + \frac{\kappa}{3}\int d^3x\, \epsilon^{\alpha\beta\gamma}\mathrm{tr}(J_\alpha(x)J_\beta(x)J_\gamma(x)) \qquad (1)$$

where κ is a constant and

$$J_\mu(x) = g^{-1}(x)\partial_\mu g(x), \qquad g(x) \in G. \qquad (2)$$

The field equations corresponding to the Lagrangian (1) are

$$\partial_- J_+(x) = 0 \qquad \partial_+ \tilde{J}_-(x) = 0 \quad \text{where} \quad \tilde{J}_-(x) = g(x)J_-(x)g^{-1}(x), \qquad (3)$$

and $\partial_\pm = \partial_0 \pm \partial_1$. It is easy to see that the general solution of these field equations is

$$g(x) = g_l(x_-)g_r(x_+), \qquad (4)$$

E. A. Tanner and R. Wilson (eds.), Noncompact Lie Groups and Some of Their Applications, 405–411.

where the group elements $g_l(x_-)$ and $g_r(x_+)$ are arbitrary second-differentiable functions of $x_\pm = x_0 \pm x_1$ respectively. The solution (4) is clearly the generalization of the solution of the free 2-dimensional wave-equation.
There are two remarkable features of the Lagrangian $\mathcal{L}_{WZW}$. First there is the appearance of a 3-dimensional integral in a two-dimensional theory. This integral is purely topological in the sense that its *variation* is a divergence and can therefore be converted into an integral over the boundary of the 3-dimensional space, which is supposed to be the 2-dimensional Minkowski-space of the first integral. Its role is to produce the chirality of the field-equations evident in (3). Indeed, a change of sign of the 3-dimensional integral in (1) produces the theory with the opposite chirality i.e. the theory in which J and $\tilde{J}$ depend on x_- and x_+ respectively. The second feature is the fact that $\mathcal{L}_{WZW}$ admits the rigid symmetry $G \times G$ defined by $g(x) \rightarrow hg(x)k$ where h and k are constant (x-independent) elements of $G \times 1$ and $1 \times G$ respectively , and that algebras of the Noether currents associated with these two symmetries are the two commuting Kac-Moody (KM) Poisson-bracket algebras

$$\{\mathrm{tr}(aK(x)), \mathrm{tr}(bK(y))\} = \mathrm{tr}([a, b]K(x))\delta(x-y) \pm \kappa \mathrm{tr}(ab)\partial_x \delta(x-y), \qquad (5)$$

where a and b are arbitrary elements of the Lie algebra of G, and $\{K, x, y\}$ denote either $\{J, x_+, y_+\}$ or $\{\tilde{J}, x_-, y_-\}$ according to the sign in front of κ. Indeed the WZW Lagrangians provide the most natural way of embedding KM algebras in physical systems.

2. Toda Systems

In contrast to the WZW system, for which the field equations are linear, there are a number of non-linear 2-D systems for which the field equations are non-linear but the systems are nonetheless integrable. Here we will be interested in a particular set of these systems called the Toda systems. The simplest Toda system is the Liouville system with Lagrangian density

$$\mathcal{L}_L(x) = \frac{1}{2}(\partial\phi(x))^2 + e^{\phi}(x), \qquad (6)$$

where $\phi(x)$ denotes a single scalar field. The Liouville system is the $sl(2, R)$ member of the class of *abelian* Toda systems with Lagrangian

$$\mathcal{L}_{AT}(x) = \frac{1}{2}C_{rs}\partial\phi^r(x)\partial\phi^s(x) + \sum_{r=1}^{r=l} e^{K_{rs}\phi_s(x)}, \qquad (7)$$

where l is the rank, and C and K the Coxeter and Cartan matrices, of the real semi-simple Lie group G which is obtained by exponentiating the Cartan basis of a semi-simple Lie algebra using only real parameters. The abelian Toda systems are, in turn, the abelian case of the general Toda systems with Lagrangian density

$$\mathcal{L}_T = \mathcal{L}_{WZW}(h(x)) + \mathrm{tr}(h(x)Mh^{-1}(x)M^t), \qquad h(x) \in H, \qquad (8)$$

where $H = H_1 \times H_2 \times ...H_n \subset G$ is the stabilizer of a diagonalizable element of a simple Lie algebra $\hat{G}$, and M and its transpose M^t are constant matrices connecting the neighbouring subgroups in the subgroup H. The abelian Toda systems are the special systems for which H is abelian ie. H is the Cartan subgroup.

3. Historical Context

Around 1980 Leznov and Saveliev in [1] found the general solution of the abelian Toda field equations, and generalized both the equations and the solutions to the non-abelian case. Around 1988 various authors (see ref. [2]) found the symmetry-algebras of the Noether currents of the abelian Toda systems. These algebras turned out to be W-algebras, defined in [3] as extensions of Virasoro algebras by primary fields. Here, a Virasoro algebra is defined as the (infinite-dimensional) Lie algebra of the 2-dimensional conformal group $\mathcal{C}$ (or to be more precise a central extension of one of the two identical, commuting, infinite-dimensional Lie algebras that form the Lie algebra of $\mathcal{C}$) and primary fields are defined as fields which transform as single-component tensors with respect to $\mathcal{C}$. The product in the W-algebras is defined as the Poisson-bracket and although the algebras are differential polynomial ($\mathcal{DP}$), they are highly non-linear. What limits their structure is that their Poisson-brackets preserve conformal dimension, where the conformal dimensions of ∂_+ and ∂_- are $(1,0)$ and $(0,1)$ respectively, and the conformal dimensions of the fields are defined by their dimensions with respect to dilations with respect to x_+ and x_-.
All of the above results were obtained at the classical level, the multiplication in the algebra being defined as Poisson-bracket multiplication, but at present considerable progress is being made in obtaining the analogous results for quantized fields [4].

What I wish to report on here is some recent work in [5] [6] which shows that the (classical) Toda theories, which as we have seen, are non-linear in both their field equations and their symmetry algebras, can be viewed as *constrained* linear systems. More precisely they can be viewed as constrained WZW systems, where the constraints are linear in the currents. The advantages of viewing the Toda systems as constrained WZW systems are

(1) The general solutions are easily derived by applying the constraints to the (trivial) general WZW solutions.
(2) The W-algebras are derived as constrained KM-algebras.
(3) Further generalizatiobns are obtained corresponding to new integrable systems (including supersymmetric ones).
(4) A new approach to quantization is obtained.
(5) The procedures provide a non-trivial but solvable situation in which various aspects of constraint theory, gauge-theory, BRST etc. can be worked out explicitly.
(6) The kind of reduction used relates to that used for KdV hierarchies and other integrable systems.

4. The Magic Constraints

The constraints that do all these wonderful things are actually quite simple. Let G be the real Lie algebra of the WZW theory and thus of the KM algebra, and let $\{M_-, M_0, M_+\}$ be the canonical generators of any $sl(2,R)$ which is tensorially embedded in G i.e. embedded so that the spectrum of M_0 is integral. (The fact that we use $sl(2,R)$ rather than $SU(2)$ implies, of course, that G must be non-compact, and in general it is highly non-compact). For arbitrary $sl(2,R)$ embedding the spectrum of M_0 will be half-integral, but for simplicity let me consider here only the case of tensor embeddings, defined as those for which the spectrum of M_0 is integral. All the results generalize to the half-integral case also—it is just that the presentation is more complicated in the half-integral case. Let $G = G_- + G_0 + G_+$ be the decomposition of the Lie algebra G of the WZW system with respect to M_0. Then the constraints are simply

$$J(x) = M_- + J_o(x_+) + J_+(x_+) \quad \text{and} \quad \tilde{J}(x_-) = M_+ + \tilde{J}_0(x_-) + \tilde{J}_-(x_-) \tag{9}$$

where the currents are the KM currents defined in (2) and (3). An alternative way to write these constraints is to say that

$$J(x_+) = M_- + j(x_+) \quad \text{where} \quad \mathrm{tr}(G_+ j(x_+)) = 0 \tag{10}$$

and similiarly for $\tilde{J}(x_-)$. Thus the constrained are defined by the subalgebras G_+ and G_-, which we call the *constraint algebras*, and it is easy to verify that they are first-class constraints in the sense of Dirac, namely

$$\{\mathrm{tr}(\gamma_1 j(x_+)), \mathrm{tr}(\gamma_2 j(x_+))\}_{km} \simeq 0 \qquad \gamma_1, \gamma_2 \in G_+ \tag{11}$$

and similiarly for the tilded variables, where the symbol $\simeq 0$ means weakly zero i.e, zero on the constrained surface.

5. Conformal Invariance

From the definition of the KM currents in (2) one sees that they are not conformal scalars but vectors of weight $(1,0)$ and $(0,1)$ respectively. So how can the constraints be made conformally invariant? The answer is to change the conformal group. For this we recall that the the conformal group for the KM system is generated by the Virasoro operators $L_{KM} = \mathrm{tr}(J^2(x_+))$ and $\tilde{L}_{KM} = \mathrm{tr}(\tilde{J}^2(x_-))$. The required change is to consider the conformal group generated by the modified Virasoro's

$$\Lambda(x_+) = L_{KM}(x_+) + \mathrm{tr}(M_0 \partial_+ J(x_+)), \tag{12}$$

and similiarly for the tilded variables with x_- instead of x_+. This changes the conformal weights of the current components to $(1+m, 0)$ and $(0, 1-m)$ respectively, where m are the weights with respect to M_0. In particular it changes the conformal weights of $M_\pm$ to $(0,0)$ and thus with respect to the conformal group generated by the Λ's the components of the currents that have been set equal to non-zero constants are scalars. So the conformal invariance of the constrained system is guaranteed. It might be worth remarking that the unmodified Virasoro is the energy-momentum tensor for the WZW system and that the modified Virasoro turns out to be the energy-momentum tensor of the constrained system that is obtained by making the traditional "improvement" of the canonical energy-momentum tensor.

6. Gauge-Fixing

It is well-known that any first-class constrained system is a gauge theory, with gauge-transformations generated by the constraints—which in our case are

$\mathrm{tr}(G_-j(x_+))$ and $\mathrm{tr}(G_+\tilde{j}(x_-))$. To obtain the physical content of the theory one has to choose a gauge and there are two natural choices.

I. Physical (Toda) Gauge: This gauge is defined by supplementing the constraints with the linear conditions $\mathrm{tr}(G_+j(x_+)) = 0$ and $\mathrm{tr}(G_-\tilde{j}(x_-)) = 0$. In this case the surviving current components are evidently those in the kernel of M_0.

II. W-Gauge: This gauge, which was first introduced for the abelian case in [7], is obtained by supplementing the constraints with the linear conditions $\omega_+(G_+, j(x_+)) = 0$ and $\omega_-(G_-, \tilde{j}(x_-)) = 0$, where the $\omega_\pm$ are the Kostant-Kirillov forms defined as $\omega_\pm(a,b) \equiv \mathrm{tr}(M_\pm[a,b])$. Thus the gauge-fixing constraints are $\mathrm{tr}([M_+, G_-]j(x_+)) = 0$ and $\mathrm{tr}([M_-, G_+]\tilde{j}(x_-)) = 0$. In this case the surviving components of the current are those in the kernel of M_+ i.e. are the highest weights with respect to $sl(2,R)$.

The gauge choices I and II are convenient for obtaining the reduced Lagrangian and field equations and for obtaining the reduced symmetry algebra respectively. In both cases the gauge-fixing is complete and thus the combined system of constraints is a second-class system in the sense of Dirac. That is to say, if we let $\mathrm{tr}(C_a j(x)) = 0$ where the C's denote *both* the first-class and gauge-fixing constraints, then the Dirac constraint matrices

$$\Delta_{ab}(x,y) \equiv \det\{\mathrm{tr}(C_a j(x)), \mathrm{tr}(C_b j(y))\}, \tag{13}$$

and similiarly for the tilded variables, are non-degenerate.

7. Reduced Lagrangian

There is a standard procedure for obtaining the reduced Lagrangian from the original one when the first-class constraints are linear in the momentum variables. The procedure is to (a) gauge the original Lagrangian in the standard manner, but leaving out the kinetic term for the gauge-fields (so that the gauge-fields play the role of Lagrange multipliers) (b) go to the physical gauge (c) eliminate the gauge-fields. This procedure can be applied in the present instance since the KM currents are the momenta, and in this instance the three steps proceed as follows:
(a) After gauging the WZW Lagrangian (1) becomes

$$\mathcal{L}_{WZW}(G) + \int d^2x\,\mathrm{tr}\Big(A_+(\tilde{J} - M_-) + A_-(J - M_+) + A_+ g A_- g^{-1}\Big), \tag{14}$$

where $A_\pm \in G_\pm$. Thus the $A_\pm$ are Lagrange multipliers but appear quadratically.
(b) On going to physical gauge, which means reducing $g(x) \in G$ to $g_0(x) \in G_0$, where G_0 is the stabilizer of M_0 in G, (14) becomes

$$\mathcal{L}_{WZW}(G_0) + \int d^2x\,\mathrm{tr}\Big(A_+ g_0 A_- g_0^{-1} - A_+ M_- - A_- M_+\Big). \tag{15}$$

(c) On eliminating the A-fields (15) becomes

$$\mathcal{L}_{WZW}(G_0) + \int d^2x\,\mathrm{tr}\Big(g_0 M_- g_0^{-1} M_+\Big). \tag{16}$$

This is just the Toda Lagrangian. In the case of half-integral embeddings the expression is much more complicated, in fact the Lagrangian is rational rather than polynomial, but the principal for deriving it is the same.

8 Reduced Symmetry Algebras : W-algebras

The reduced symmetry algebra can be defined in two equivalent ways, namely as

(i) the algebra of gauge-invariant functions of the first-class constrained current.

(ii) the Dirac star algebra of the second-class constrained current.

Using either definition one can show that the reduced algebra is (a) a differential polynomial ($\mathcal{DP}$) algebra and (b) has a basis that consists of the Virasoro operator and primary fields. Each definition has its advantages. Using (i) it is easy to prove even more, namely that the set of gauge-invariant functions has a differential polynomial basis i.e. a basis that consists of $\mathcal{DP}$'s of the first-class constrained current. Using (ii) the proofs of (a) and (b) are much shorter. Indeed the proof of (a) reduces to showing that the Dirac second-class constrained matrices, $\Delta_{ab}(x,y)$ in (13), and its tilded counterpart, are $\mathcal{DP}$-invertible, which, in turn, reduces to showing that $\det\Delta_{ab}(x,y) = \text{const.}\delta_{ab}\delta(x-y)$ and the proof of (b) reduces to showing that the Virasoro operator $\Lambda(x_+) + \frac{1}{2}\text{tr}(M_+\partial_+^2 j(x_+))$, which is gauge-equivalent to the Virasoro (12), and its tilded counterpart, commute with the W-gauge-fixing constraints.

9. Summary

The non-linear Toda systems (and a number of generalizations thereof) may be viewed as linearly constrained WZW systems. This explains the role of the Lie group G in the Toda systems and allows one to obtain the general solutions and the (non-linear) symmetry algebras of the Toda systems in a rather simple way. It also provides a new approach to the quantization of Toda systems, and provides a solvable but non-trivial testing ground for studying gauge-theories and second-class constrained theories in an explicit manner. Reviews of WZW-Reduction and W-Algebras can be found in references [8] and [9].

References

[1] A.N. Leznov and M.V. Savaliev, (1980) Comm. Math. Phys. **74**, 111.
[2] A. Bilal and J-L. Gervais, Phys. Lett. (1988) **206B**,412; Nucl. Phys. (1989) **B314** (1989), 646; **B318**, 579;O. Babelon, Phys. Lett. (1988) **215B**, 523.
[3] A.B. Zamolodchikov, Theor. Math. Phys. (1986) **65**, 1205; V.A. Fateev and A.B. Zamolodchikov, Nucl. Phys. (1987) **B280**, 644.
[4] V.A. Fateev and S.L. Lukyanov, Int. J. Mod. Phys. (1988) **A3**, 507; S.L. Lukyanov, Funct. Anal. Appl. (1989) **22** 255; P. Mansfield and B. Spence, Nucl. Phys. (1991) **B362**, 294; P. Bowcock and G. Watts, Nucl. Phys. (1992) **B379**.
[5] L. Feher et al. Ann. Phys. (1990) **203**, 76 and (1992) **213**, 1.
[6] F. A. Bais et al. Nucl. Phys. (1991) **B357**, 632; P. Bouwknegt and K. Schoutens, Phys. Reports (1993) **C** (in press).

[7] V. Drinfeld and V. Sokolov, J. Sov. Math. (1984) **30** , 1975.
[8] L. Feher et al. *On Hamiltonian Reductions of the Wess-Zumino-Novikov-Witten Theories* Phys. Reports C (1992) **222**, 1-64.
[9] P. Bouwknegt and K. Schoutens, *W-symmetry in Conformal Field Theory* Phys.Reports C (1993) **223**, 183-276.

Dublin Institute for Advanced Studies
10, Burlington Road, Dublin 4, Ireland

On the Definitions of the Quantum Group $\mathcal{U}_h(sl(2,k))$ and the Restricted Dual of $\mathcal{U}_h(sl(n,k))$

Alain Guichardet

Abstract

We prove that the quantum group $\mathcal{U}_h(sl(2,k))$ as usually defined in the specialized literature, is actually a formal deformation of $\mathcal{U}(sl(2,k))$ in the sense of Gerstenhaber; then, using the triviality of this deformation of algebra, we define its restricted dual, which is a formal deformation of that of $\mathcal{U}(sl(2,k))$, and we prove that it is given by the usual generators and relations; actually the latter proof also works for the case of $sl(N+1,k)$.

1 Introduction

Several authors give the following definition: $\mathcal{U}_h(sl(2,k))$ *is the* $k[[h]]$*-algebra generated in the h-adic sense by the generators* H, X, Y *with the relations*

$$[H,X] = 2X, \, , \, [H,Y] = -2Y, \, , \, [X,Y] = \frac{\sinh(\frac{hH}{2})}{\sinh(\frac{h}{2})}.$$

The first aim of these notes is to make the above definition more explicit. One possible way is as follows: First, recall the precise definition of $\mathcal{U}(sl(2,k))$. Let E be the set $\{H, X, Y\}$, $< E >$ the free monoid generated by E, $W = k < E >$ the vector space spanned by $< E >$ or free associative algebra generated by E; let $K^{(0)}$ be the two-sided ideal of W generated by the elements

$$[H,X] - 2X, \, , \, [H,Y] + 2Y, \; [X,Y] - H.$$

Then $\mathcal{U}(sl(2,k))$ is the quotient algebra W/K^0.

Now, to define $\mathcal{U}_h(sl(2,k))$, we can introduce the algebra $W[[h]]$, then its closed two-sided ideal K generated by the elements

$$[H,X] - 2X, \; [H,Y] + 2Y, \; [X,Y] - \frac{\sinh(\frac{hH}{2})}{\sinh(\frac{h}{2})},$$

E. A. Tanner and R. Wilson (eds.), Noncompact Lie Groups and Some of Their Applications, 413–422.

and finally set

$$\mathcal{U}_h(sl(2,k)) = W[[h]]/K.$$

[The multiplication in $W[[h]]$ is the natural one, namely

$$(\sum_p w_p h^p, \sum_q w'_q h^q) \to \sum_n (\sum_{p+q=n} w_p w'_q) h^n ;$$

the topology in $W[[h]]$ is the h-adic one].

But this definition is not entirely satisfactory; indeed, in order to study formal deformations of an algebra A, the right frame seems to be that of formal power series, i.e.the space $A[[h]]$; in particular this point of view gives very natural definitions and good properties for morphisms, tensor products, representations, etc (see also Gerstenhaber [4]); moreover it also allows to prove that, if A is the enveloping algebra of a semisimple Lie algebra, then every formal deformation of A is trivial and that its restricted dual, conveniently defined, is a formal deformation of the restricted dual of A. So it seems important to make sure that we have $W[[h]]/K = \mathcal{U}(sl(2,k))[[h]]$ as $k[[h]]$-modules. This is the first goal of these notes (see Theorem 4.1 below); the proof will heavily rely on the *diamond lemma*, proved by Bergman in [2], which we shall first recall. As for the case of $sl(N+1,k)$, we only indicate a possible way to proceed (see remark at the end of §4). Our second goal will be to give a similar presentation for the restricted dual of $\mathcal{U}_h(sl(N+1,k))$ (see Theorem 5.1); actually the passage from 2 to n seems more difficult for $\mathcal{U}_h$ than for its restricted dual because of the appearence of the Serre's relations.

Notational conventions: In order to simplify notations, we shall write $\widetilde{V}$ instead of $V[[h]]$ when V is a vector space over a field k; the elements of $\widetilde{V}$ will be denoted by

$$\widetilde{v} = (v_n) = \sum_{n\geq 0} v_n h^n.$$

In particular we have the ring of formal power series $\widetilde{k}$. Also $\mathrm{Hom}(V,W)$, $V \otimes W$, etc will stand for $\mathrm{Hom}_k(V,W)$, $V \otimes_k W$, etc.

2 The Diamond Lemma

In this paragraph, k is a commutative ring. Let E be a set. A *reduction system for E* is a family $S = (\xi_i, \phi_i)_{i\in I}$, I arbitrary set of indices, with $(\xi_i, \phi_i) \in < E > \times k < E >$. With each $i \in I, a, b \in < E >$ we associate the *reduction* $r_{a,i,b}$: k-linear operator in $k < E >$ transforming $a\xi_i b$ in $a\phi_i b$ and leaving fixed all other elements of the basis $< E >$. A monomial (=element of $< E >$) is said to be *irreducible* if it does not contain any ξ_i. A *covering ambiguity* is a quintuple (i, j, a, b, c) such that

$i, j \in I$, $a, b, c \in \langle E \rangle - \{1\}$, $\xi_i = ab$, $\xi_j = bc$; it is called *solvable* if there exist products of reductions r', r'' satisfying $r'(\phi_i c) = r''(a\phi_j)$. An *inclusion ambiguity* is a quintuple (i, j, a, b, c) such that $i, j \in I, i \neq j$, $a, b, c \in \langle E \rangle$, $\xi_i = b$, $\xi_j = abc$; it is called *solvable* if there exist products of reductions r', r'' satisfying $r'(a\phi_i c) = r''(\phi_j)$. The reduction system S is called *confluent* if all ambiguities are solvable.

Theorem 2.1: *We suppose that the following conditions are fulfilled:*

(a) *S is confluent.*

(b) *There exists on the set $\langle E \rangle$ a (partial) order relation, compatible with products, satisfying the descending chain condition, and such that for every $i \in I$, ϕ_i is a linear combination of monomials strictly less than ξ_i.*

Then

(i) *The irreducible monomials form a basis of $k \langle E \rangle$ modulo the two-sided ideal generated by the elements $\xi_i - \phi_i$, $i \in I$;*

(ii) *Let $\zeta \in k \langle E \rangle$; every sequence (ζ_n) obtained by successive reductions from ζ is stationary; its last element is a linear combination of irreducible monomials, which does not depend on the chosen sequence of reductions.*

For the proof, we refer to [2]; but we indicate a useful way to insure condition (b). Write each ϕ_i as a linear combination

$$\phi_i = \sum_{\alpha} \lambda_{i,\alpha} \eta_{i,\alpha} , \quad \lambda_{i,\alpha} \in k , \quad \eta_{i,\alpha} \in \langle E \rangle .$$

Assume we have defined a mapping $F = (F^1, ..., F^n)$ of $\langle E \rangle$ into some space $\mathbf{N}^n$ (which we endow with the lexicographical order), and that F satisfies the following condition:

(b′) *If $\theta, \zeta \in \langle E \rangle$ and if ζ is obtained from θ by replacing some ξ_i by some $\eta_{i,\alpha}$, then $F(\zeta)$ is strictly less than $F(\theta)$.*

Then condition (b) is fulfilled with the following order relation:

$\theta > \zeta$ iff there exists a sequence $\theta = \theta_0, \theta_1, ..., \theta_p = \zeta$ such that each θ_k is obtained from θ_{k-1} by replacing some ξ_i by some $\eta_{i,\alpha}$.

3 Some General Properties of Formal Deformations

Let $\mathfrak{g}$ be a semisimple Lie algebra over a field k, algebraically closed and of characteristic 0 ; set $\mathcal{U} = \mathcal{U}(\mathfrak{g})$ and consider a formal deformation $\widetilde{\mathcal{U}}$ of $\mathcal{U}$ considered as a bialgebra (cf. [5]); the multiplication and comultiplication in $\widetilde{\mathcal{U}}$ are given by sequences $\widetilde{\mu} = (\mu_n)$, $\widetilde{\Delta} = (\Delta_n)$ where

$$\mu_n \in \mathrm{Hom}(\mathcal{U} \otimes \mathcal{U}, \mathcal{U}) \quad , \quad \mu_0 = \mu \ (= \textit{multiplication in } \mathcal{U}),$$
$$\Delta_n \in \mathrm{Hom}(\mathcal{U}, \mathcal{U} \otimes \mathcal{U}) \quad , \quad \Delta_0 = \Delta \ (= \textit{comultiplication in } \mathcal{U}).$$

As noted in [5], $\widetilde{\mathcal{U}}$ is a trivial deformation of $\mathcal{U}$ as an algebra - but not as a coalgebra!; a detailed proof of this fact is given in [6]. Hence, in the sequel of this paragraph we shall assume that $\mu_n = 0 \ \forall n > 0$.

We call *finite rank representation* of $\widetilde{\mathcal{U}}$ any pair $(\widetilde{V}, \widetilde{\pi})$ where V is a finite dimensional k-vector space and $\widetilde{\pi}$ is a multiplicative $\widetilde{k}$-linear mapping from $\widetilde{\mathcal{U}}$ into the space of $\widetilde{k}$-linear mappings $\widetilde{V} \to \widetilde{V}$; $\widetilde{\pi}$ is given by a sequence of linear mappings $\pi_n \in \mathrm{Hom}(\mathcal{U}, \mathrm{End}\, V)$ satisfying

$$\pi_n(u'u'') = \sum_{p+q=n} \pi_p(u') \cdot \pi_q(u'');$$

this means that π_0 is a representation of $\mathcal{U}$ in V and that $\widetilde{\pi}$ is a formal deformation of π_0.

Lemma 3.1: *Such a formal deformation is always trivial, i.e. equivalent to one such that* $\pi_n = 0 \ \forall n > 0$.

This follows from the fact that $H^1(\mathcal{U}, \mathrm{End}\, V) = 0$ (cf.[6]). So we shall assume that $\pi_n = 0 \ \forall n > 0$.

At this point we recall that the *restricted dual* of an algebra A is the set of all coefficients of finite dimensional representations of A; this is a vector subspace of the dual A^*, denoted by A^0. We also recall that, if A is a bialgebra, A^0 is an algebra with multiplication $\mu^0 : A^0 \otimes A^0 \to A^0$ defined as the transpose of the comultiplication in A.

Returning to our formal deformation $\widetilde{\mathcal{U}}$, it is natural to call *coefficient* of a representation $(\widetilde{V}, \widetilde{\pi})$ any $\widetilde{k}$-linear functional $\widetilde{\phi}$ on $\widetilde{\mathcal{U}}$ of the form

$$\widetilde{\phi}(\widetilde{u}) = < \widetilde{\xi}, \widetilde{\pi}(\widetilde{u}) \cdot \widetilde{v} >, \quad \forall \widetilde{u} \in \widetilde{\mathcal{U}}$$

where

$$\widetilde{v} \ = \ (v_n) \in \widetilde{V}$$
$$\widetilde{\xi} \ = \ (\xi_n) \in \mathrm{Hom}_{\widetilde{k}}(\widetilde{V}, \widetilde{k}) = \widetilde{V^*};$$

the components $\phi_n \in \mathcal{U}^*$ of $\widetilde{\phi}$ are given by

$$\phi_n(u) = \sum_{p+q=n} < \xi_p, \pi_0(u) \cdot v_q >

which shows that $\phi_n \in \mathcal{U}^0$ and that the set $(\widetilde{\mathcal{U}})^0$ of all such $\widetilde{\phi}$ is included in $(\mathcal{U}^0)\widetilde{}$; actually it is easy to see that $(\mathcal{U}^0)\widetilde{}$ is the closure $((\widetilde{\mathcal{U}})^0)\bar{}$ of $(\widetilde{\mathcal{U}})^0$ in $\mathrm{Hom}_{\widetilde{k}}(\widetilde{\mathcal{U}},\widetilde{k}) = (\mathcal{U}^*)\widetilde{}$. It is this closure we shall call the *restricted dual* of $\widetilde{\mathcal{U}}$; it is clearly a formal deformation of the algebra $\mathcal{U}^0$.

Another consequence of Lemma 3.1 is the following:

Lemma 3.2: *Let us denote by* $\mathrm{Rep}(\mathcal{U})$ *(respectively* $\mathrm{Rep}(\widetilde{\mathcal{U}})$*) the set of all finite dimensional representations of* $\mathcal{U}$ *(respectively finite rank representations of* $\widetilde{\mathcal{U}}$*). Then there exists a natural bijective correspondence between* $\mathrm{Rep}(\mathcal{U})$ *and* $\mathrm{Rep}(\widetilde{\mathcal{U}})$*, preserving direct sums, indecomposability and tensor products. In particular every finite rank representation of* $\widetilde{\mathcal{U}}$ *is a direct sum of indecomposable ones.*

Warning: the conditions *indecomposable* and *irreducible* are equivalent in the case of $\mathcal{U}$, but not in that of $\widetilde{\mathcal{U}}$; indeed a representation $(\widetilde{V},\widetilde{\pi})$ of $\widetilde{\mathcal{U}}$ always contains the submodules $h^n \cdot \widetilde{V}$, and there exist no other submodules if and only if $(\widetilde{V},\widetilde{\pi})$ is indecomposable.

4 Presentation of the Algebra $\mathcal{U}_h(sl(2,k))$

We use the notations of §1 and we write $\mathcal{U}$ for $\mathcal{U}(sl(2,k))$.

Theorem 4.1: *The* $\widetilde{k}$*-modules* $\widetilde{W}/K$ *and* $\widetilde{\mathcal{U}}$ *are isomorphic.*

We first introduce some notations:

- For every k-vector space V we set

$$\widetilde{V}_n = \widetilde{V}/h^{n+1} \cdot \widetilde{V} = V \otimes \widetilde{k}_n \ \ \forall n \in \mathbf{N}\,;$$

  clearly $\widetilde{V}$ is the inverse limit of the $\widetilde{V}_n$'s.

- We denote by π_n the canonical mapping $\widetilde{W} \to \widetilde{W}_n$.

- We write

$$\frac{\sinh(\frac{hH}{2})}{\sinh(\frac{h}{2})} = \sum_q P_q(H) \cdot h^q$$

  where P_q is some polynomial with rational coefficients.

- We set $I = \{1,2,3\}$,

$$\text{(i) } \xi_1 = XH, \ \ \phi_1 = HX - 2X, \ \ \phi_1^q = \begin{cases} -2X & if \ \ q=0 \\ 0 & if \ \ q>0 \end{cases}.$$

(ii) $\xi_2 = YH, \quad \phi_2 = HY + 2Y, \quad \phi_2^q = \begin{cases} 2Y & if \quad q = 0 \\ 0 & if \quad q > 0 \end{cases}.$

(iii) $\xi_3 = YX, \quad \phi_3 = XY - \frac{\sinh(\frac{hH}{2})}{\sinh(\frac{h}{2})}, \quad \phi_3^q = \begin{cases} XY - P_0(H) & if \quad q = 0 \\ -P_q(H) & if \quad q > 0 \end{cases}.$

Proof of Theorem 4.1:

(a) We want to prove that $\widetilde{W}/K = \widetilde{\mathcal{U}}$; we have

$$\begin{aligned} \widetilde{\mathcal{U}} &= \lim_{\leftarrow} \widetilde{\mathcal{U}}_n = \lim_{\leftarrow} \mathcal{U} \otimes \widetilde{k}_n \\ \widetilde{W} &= \lim_{\leftarrow} \widetilde{W}_n = \lim_{\leftarrow} W \otimes \widetilde{k}_n; \end{aligned}$$

by a general property of inverse limits, we also have

$$\widetilde{W}/K = \lim_{\leftarrow} \widetilde{W}_n/\pi_n(K).$$

thus we are led to prove that

$$(W \otimes \widetilde{k}_n)/\pi_n(k) = \mathcal{U} \otimes \widetilde{k}_n.$$

(b) It is clear that $\pi_n(K)$ is the two-sided ideal of $W \otimes \widetilde{k}_n$ generated by the elements $\xi_i - \pi_n(\phi_i)$. We shall apply the diamond lemma with the base ring $\widetilde{k}_n$. Assume for a while that we can do so; then we get a $\widetilde{k}_n$-basis for each quotient space $(W \otimes \widetilde{k}_n)/\pi_n(K)$, and also for $\mathcal{U} \otimes \widetilde{k}_n$ (since $\mathcal{U} = W/\pi_0(K)$), by taking the irreducible monomials; and this clearly implies our assertion.

(c) We are now left to prove that the hypotheses of the diamond lemma are satisfied for a reduction system of the following form: $E = \{H, X, Y\}$, $I = \{1, 2, 3\}$,

$$\begin{aligned} \xi_1 = XH \quad &, \quad \phi_1 = HX - 2X \\ \xi_2 = YH \quad &, \quad \phi_2 = HY + 2Y \\ \xi_3 = YX \quad &, \quad \phi_3 = XY + P(H) \end{aligned}$$

where P is some polynomial with coefficients in $\widetilde{k}_n$.
There is no inclusion ambiguity, and only one covering ambiguity, namely $(YX)H = Y(XH)$; but $(YX)H$ can be successively reduced to $XYH + P(H)H$, $XHY + 2XY + P(H)H$, $HXY + P(H)H$, and $Y(XH)$ to $YHX - 2YX$, $HYX + 2YX - 2YX$, $HXY + HP(H)$.
Let us now check condition (b'). The various $\eta_{i,\alpha}$ are as follows:

- for ξ_1 : HX and X
- for ξ_2 : HY and Y
- for ξ_3 : XY and some powers of H.

We define $F :< E > \rightarrow \mathbf{N}^3$ by

$$\begin{aligned} F_1 &= \text{total degree with respect to } X \text{ and } Y \\ F_2 &= \text{total degree with respect to } H, X \text{ and } Y \\ F_3 &= \text{inversion number} \end{aligned}$$

[setting $H < X < Y$, the *inversion number* of a monomial $a_1 ... a_m$ with $a_i \in E$ is the number of pairs $i, j = 1, ..., m$ such that $i < j$ and $a_j < a_i$].
Then condition (b') is trivially satisfied, and the proof is complete.

Remark: The proof of the analogous result for $sl(N+1,k)$ seems to be considerably more difficult; indeed it is not clear whether or not one can find a confluent reduction system as in part (**c**) above. It seems possible to proceed as follows (but we have not checked the details): first to establish a triangular decomposition for the analogue of $\widetilde{W}_n/\pi_n(K)$, say

$$\widetilde{W}_n/\pi_n(K) = (\widetilde{W}_n/\pi_n(K))_- \otimes (\widetilde{W}_n/\pi_n(K))_0 \otimes (\widetilde{W}_n/\pi_n(K))_+$$

with $(\widetilde{W}_n/\pi_n(K))_0 = \tilde{k}_n[H_1, ..., H_N]$; and then, following [9], to construct for $(\widetilde{W}_n/\pi_n(K))_\pm$ a basis independant of n; a detailed proof of this result as well as of Theorem 5.1 will appear elsewhere.

5 Presentation of the Restricted Dual of $\mathcal{U}_h(sl(N+1,k))$

Here the field k is assumed to be algebraically closed and of characteristic 0. We write $\mathcal{U} = \mathcal{U}(sl(N+1,k))$. Moreover we admit that $\mathcal{U}_h(sl(N+1,k))$, defined in the usual way, is actually a formal deformation of $\mathcal{U}$. We have seen in §3 that the algebra $((\widetilde{\mathcal{U}})^0)\tilde{}$ is a formal deformation of $\mathcal{U}^0$; we shall write $(\mathcal{U}^0)\tilde{}$ instead of $((\widetilde{\mathcal{U}})^0)\tilde{}$.

Let ρ be the natural representation of $\mathcal{U}$ in k^{N+1} and $\rho_{ij} \in \mathcal{U}^0$ its coefficients; let $\tilde{\rho}$ be the correseponding representation of $\widetilde{\mathcal{U}}$; every finite rank representation of $\widetilde{\mathcal{U}}$ is included in some tensor power of $\tilde{\rho}$, since the same is true for $\mathcal{U}$; hence the coefficients $\tilde{\rho}_{ij}$ of $\tilde{\rho}$ generate (topologically) the algebra $(\mathcal{U}^0)\tilde{}$. Denoting by W the free associative k-algebra generated by abstract elements ρ'_{ij} (this in order to avoid confusions), we thus have a surjective morphism of algebras $F : \widetilde{W} \rightarrow (\mathcal{U}^0)\tilde{}$ defined by $F(\rho'_{ij}) = \tilde{\rho}_{ij}$. We want to describe its kernel ker$F$.

We shall denote by $\tilde{\varepsilon}$ the unit of $(\mathcal{U}^0)\tilde{}$ and by $\tilde{D}$ the following element of $(\mathcal{U}^0)\tilde{}$ (*quantum determinant*):

$$\tilde{D} = \sum_{s \in S_{N+1}} (-q^{-1})^{I(s)} \tilde{\rho}_{1,s(1)} \cdots \tilde{\rho}_{N+1,s(N+1)}$$

where $q = e^{\frac{h}{2}}$ and $I(s)$ is the inversion number of the permutation s.

Finally we denote by ε' and D' the corresponding elements of $\widetilde{W}$. We shall admit that ker$F$ contains the elements of the 5 following types (cf. e.g. [5]):

1. $\rho'_{ik}\rho'_{ij} - q\rho'_{ij}\rho'_{ik}$ when $j < k$
2. $\rho'_{jk}\rho'_{ik} - q\rho'_{ik}\rho'_{jk}$ when $i < j$
3. $\rho'_{ij}\rho'_{kl} - \rho'_{kl}\rho'_{ij}$ when $k < i$, $j < l$
4. $\rho'_{ij}\rho'_{kl} - (\rho'_{kl}\rho'_{ij} + (q - q^{-1})\rho'_{kj}\rho'_{il})$ when $k < i$, $l < j$
5. $D' - \varepsilon'$

Theorem 5.1: *The kernel of F is the closed two-sided ideal K of $\widetilde{W}$ generated by the above elements.*

NB: This result has been announced in [7] and proved in [5] in a slightly different context.

Proof: We shall proceed in two steps: firstly, using the diamond lemma, we describe the quotient $\widetilde{W}/K'$ where K' is the closed two-sided ideal generated by elements of the types (1),...,(4); secondly we divide $\widetilde{W}/K'$ by the last relation.

(a) Notations: K'_0 = set of components of degree 0 of elements of K' = two-sided ideal of W generated by all elements of the form $\rho'_{ij}\rho'_{kl} - \rho'_{kl}\rho'_{ij}$.
$V = W/K'_0 = k[(\rho'_{ij})]$
π = canonical mapping $\widetilde{W} \to \widetilde{W}/K'$.

(b) Claim: $\widetilde{W}/K'$ is isomorphic, as $\tilde{k}$-module, with $\widetilde{V}$; in other words the algebra $\widetilde{W}/K'$ is a formal deformation of V.

Sketch of proof: One proceeds as for Theorem 4.1, taking the reduction system containing the elements $(\rho'_{ik}\rho'_{ij} , q\rho'_{ij}\rho'_{ik})$, etc, and the mapping F_1 from $< (\rho_{ij}) >$ to $\mathbf{N}$ defined by the inversion number of a monomial for the lexicographical order on the set $\{\rho_{ij}\}$; it is tedious but not difficult to check that the hypotheses of the diamond lemma are satisfied.
In what follows we shall write $\widetilde{V}$ instead of $\widetilde{W}/K'$.

(c) Notations: L = closed two-sided ideal of $\widetilde{V}$ generated by $\pi(D' - \varepsilon')$
L_0 = set of components of degree 0 of L = two-sided ideal of V generated by $D - \varepsilon$ where D is the usual determinant and ε the unit of V.

We have to prove that the mapping $H : \widetilde{V}/L \to (\mathcal{U}^0)\tilde{}$ induced by F is injective.

(d) Claim: $V/L_0 = \mathcal{U}^0$. In fact this is the usual presentation of $\mathcal{U}^0$, which is well known at least when $k = \mathbf{C}$, since then $\mathcal{U}^0$ is isomorphic with the algebra of "regular functions" on the group $SL(N + 1, \mathbf{C})$.

(e) Claim: L is also the left ideal in V generated (algebraically) by $\pi(D'-\varepsilon')$.

Proof: We set $\tilde{\alpha}=\pi(D'-\varepsilon')$ and we remark that its component $\alpha_0 = D-\varepsilon$ is a regular (i.e. not divisor of 0) element of V. One knows (cf.[6],Theorem 3) that $\tilde{\alpha}$ is a central element of $\tilde{V}$; hence we only have to prove that the left ideal $\tilde{V}\cdot\tilde{\alpha}$ is closed.

Let $\tilde{v}^{(i)}$ a sequence of elements of $\tilde{V}\cdot\tilde{\alpha}$ converging to an element $\tilde{v}$ of $\tilde{V}$; we must construct $\tilde{y}$ in $\tilde{V}$ such that $\tilde{v}=\tilde{y}\cdot\tilde{\alpha}$. For each i, $\tilde{v}^{(i)}$ is of the form $\tilde{x}^{(i)}\cdot\tilde{\alpha}$, i.e.

$$v_n^{(i)} = \sum_{p+q+r=n} \mu_p(x_q^{(i)},\alpha_r)$$

where $\tilde{\mu}$ is the multiplication in $\tilde{V}$. There exists i_0 such that, for every $i\geq i_0$, we have $v_0=v_0^{(i)}=x_0^{(i)}\alpha_0$; since α_0 is regular, this implies $x_0^{(i)}=x_0^{(i_0)}$; we set $y_0=x_0^{(i_0)}$. Then there exists $i_1\geq i_0$ such that, for $i\geq i_1$ we have

$$v_1 = v_1^{(i)} = \mu_1(x_0^{(i_0)},\alpha_0)+x_1^{(i)}\alpha_0+x_0^{(i_0)}\alpha_1\,;$$

this implies $x_1^{(i)}=x_1^{(i_1)}$ and we set $y_1=x_1^{(i_1)}$; etc.

(f) Claim: The conditions $\tilde{v}\in\tilde{V}$, $h\tilde{v}\in L$ imply $\tilde{v}\in L$.

Proof: We can write $h\tilde{v}=\tilde{x}\cdot\tilde{\alpha}$, hence

$$\sum_{p+q+r=n} \mu_p(x_q,\alpha_r) = v_{n-1} \; if \; n>0 \; and \; 0 \; if \; n=0;$$

since α_0 is regular we have $x_0=0$ and we can write

$$h\tilde{v}=h\cdot(h^{-1}\tilde{x})\cdot\tilde{\alpha},\quad \tilde{v}=(h^{-1}\tilde{x})\cdot\tilde{\alpha}.$$

(g) The $\tilde{k}$-module $\tilde{V}/L$ is Hausdorff, complete (for the h-adic topology) and the action of h in it is injective by part (**f**); set $X=(\tilde{V}/L)/h\cdot(\tilde{V}/L)$; choosing a k-vector subspace in $\tilde{V}/L$ supplementary to $h\cdot(\tilde{V}/L)$, we can write

$$\tilde{V}/L = X\oplus h(\tilde{V}/L) = X\oplus h\cdot X\oplus h^2(\tilde{V}/L) = \ldots.$$

and we see that $\tilde{V}/L$ is equal to $\tilde{X}$ as a $\tilde{k}$-module.

The mapping $H:\tilde{X}\to(\mathcal{U}_0)\tilde{}$ can be written as

$$H(\tilde{x}) = \sum_n \sum_{p+q=n} H_p(x_q)\cdot h^n$$

with $H_p\in\mathrm{Hom}(X,\mathcal{U}^0)$; now identifying X with V/L_0, we see by part (**d**) that H_0 is injective, and this easily implies that H is injective.

BIBLIOGRAPHY

1. H. H. Anderson, P. Polo and W. Kexin, *Representations of quantum algebras*, Invent. Math. **104** (1991), 1-59.

2. G. M. Bergman, *The Diamond lemma for ring theory*, Adv. in Math. **29** (1978), 178-218.

3. V. G. Drinfeld, *Quantum groups*, Proc. ICM, pp. 798-820, Berkeley, (1986).

4. M. Gerstenhaber, *On the deformations of rings and algebras*, Ann. Math. **79** (1964), 59-103.

5. M. Gerstenhaber and S. D. Schack, *Bialgebra cohomology, deformations and quantum groups*, Contemporary Math. **134** (1992), 51-92.

6. A. Guichardet, *Introduction to quantum groups (formal point of view)*, Lectures at the ICTP Workshop, Trieste, March 15 - April 2, 1993.

7. N. Yu. Reshetikhin, L. A. Takhtadzhyan and L. D. Faddeev, *Quantization of Lie groups and Lie algebras*, Leningrad Math. J. **1** (1990), 193-225.

8. M. Rosso, *Finite dimensional representations of the quantum analog of the enveloping algebra of a complex simple Lie algebra*, Commun. Math. Phys. **117** (1988), 581-593.

9. M. Rosso, *An analogue of P.B.W. theorem and the universal* R-*matrix for* $\mathcal{U}_h(sl(N+1))$, Commun. Math. Phys. **124** (1989), 307-318.

Centre de Mathematiques (CNRS, URA 169)
Ecole Polytechnique, 91128 Palaiseau, France.

Universal T-matrix for Twisted Quantum $g\ell(N)$

C. Frønsdal

Abstract

The Universal T-matrix is the capstone of the structure that consists of a quantum group and its dual, and the central object from which spring the T-matrices (monodromies) of all the associated integrable models. A closed expression is obtained for the case of multiparameter (twisted) quantum $g\ell(N)$. The factorized nature of standard quantum groups, that allows the explicit expression for UT to be obtained with relative ease, extends to some nonstandard quantum groups, such as those based on $A_n^{(2)}$, and perhaps to all. The paper is mostly concerned with parameters in general position, but the extension to roots of unity is also explored, in the case of $g\ell(N)$. The structure of the dual is now radically different, and an interesting generalization of the q-exponential appears in the formulas for the Universal T- and R-matrices. The projection to quantum $sl(N)$ is simple and direct; this allows, in particular, to apply recent results concerning deformations of twisted $gl(N)$ to the semisimple quotient.

E. A. Tanner and R. Wilson (eds.), Noncompact Lie Groups and Some of Their Applications, 423–452.

1. Introduction

1.1. PICTURES

Quantum groups may be studied in two different "pictures", much as quantum mechanics has a Schroedinger picture and a Heisenberg picture. The Drinfeld picture is a quantum group from the point of view of integrable field theories, where the first examples were discovered. The development is due to Drinfeld [1], Jimbo [2] and many others. The Woronowicz-Manin picture appeared already in early work of Baxter [3] and Faddeev *et al.* [4], but the systematic study is due to Woronowicz [5] and Manin [6]. The term psuedo-group, proposed by Woronowicz in [5], deserves to be retained, in order that the terminology distinguish between this object and the quantum group in the sense of Drinfeld.

1.2. DUALITY

This paper is a contribution to the study of the duality between the two pictures. Much work has been done already [7], and a firm foundation that incorporatess reflexivity and deformation theory has been discovered very recently [8], but the object that we regard as the capstone of the structure seems not to have been calculated except in some simple cases. It appears as the universal bi-character in Woronowicz and as a "canonical element" or "dual form" in other places. Here it will be called the Universal T-matrix, in recognition of the fact that the transition matrices of integrable models appear upon specialization, in passing from structure to representations. This is the main reason why it is useful for physics.

Let A be a vector space with a (countable) basis $\{|\alpha>\}$. Let $\{<\beta|\}$ be the functions on A defined by

$$<\beta|\alpha>=\delta_{\alpha\beta}. \tag{1.1}$$

To avoid or at least postpone the thorny question of an appropriate and precise definition of the dual space A^*, let us just say that these functions, and (at least) all finite linear combinations of them, are in A^*. The Universal T-matrix is the unit operator in *End A*, namely

$$UT = \sum_\alpha |\alpha >< \alpha|, \tag{1.2}$$

the "resolution of the identity." It has been called the "canonical element of $A \otimes A^*$" and this expression can perhaps be justified once a convenient and rigorous definition of A^* has been agreed upon.

The notation was chosen in order to exhibit the connection between the object of main interest and familiar operations in quantum mechanics in Eq. (1.2). We now switch to a more convenient notation.

Let A be an associative algebra with basis $\{L_\alpha\}$, and $\{F_\alpha\}$ the functions defined by

$$F_\alpha(L_\beta) = \delta_{\alpha\beta}. \tag{1.3}$$

We shall suppose that the product $L_\alpha L_\beta$ is a finite linear combination of the basis elements; then a structure of coassociative coproduct is naturally induced on the linear span A^* of the functions F_α; it is defined by

$$F_\alpha \mapsto \Delta F_\alpha, \quad \Delta F_\alpha(L_\beta, L_\gamma) = F_\alpha(L_\beta L_\gamma). \tag{1.4}$$

The expression for UT is

$$UT = T_{L,F} = \sum_\alpha L_\alpha F_\alpha, \tag{1.5}$$

and the relation (1.4) can be expressed as

$$T_{L,F} T_{L,F'} = \sum_{\alpha\beta} L_\alpha L_\beta F_\alpha F'_\beta = \sum_\gamma L_\gamma \Delta F_\gamma. \tag{1.6}$$

Here F, F' stands for two copies of A, $F_\alpha F'_\beta$ has the same meaning as $F_\alpha \otimes F_\beta$, and the result is that

$$T_{L,F} T_{L,F'} = T_{L,\Delta F}. \tag{1.7}$$

This is the first of the two structural relations that characterize the universal T-matrix.

Keep the notations as above, and suppose that A has, in addition, a coproduct that turns it into a bialgebra. Then A^* also becomes a bialgebra, with algebraic structure uniquely defined by

$$F_\alpha F_\beta(L_\gamma) = F_\alpha \otimes F_\beta(\Delta L_\gamma). \tag{1.8}$$

This too can be expressed in terms of $\mathcal{UT}$:

$$T_{L,F} T_{L',F} = T_{\Delta L,F}. \tag{1.9}$$

This is the second structural relation satisfied by the Universal T-matrix; note that (1.7) and (1.9) are equivalent to (1.4) and (1.8).

Finally, suppose that A, as an algebra, is finitely generated by elements $\{\ell_i\}$ $i = 1, \cdots, n$, and that each basis element L_α is an ordered monomial. We would like to answer the following.

Question. Under what condition is A^* finitely generated (as an algebra), and under what additional conditions is each element of the dual basis $\{F_\alpha\}$ a polynomial in the generators?

Though we do not know the answer in general, we can always use the relations (1.7) and (1.9) to determine the structure of A^* and thus answer the question on a case by case basis. Furthermore, when A^* turns out to be finitely generated we expect to obtain a useful expression for $\mathcal{UT}$ in terms of the two sets of generators.

1.3. EXAMPLES

Let A be the unital algebra freely generated by a single element x, with

$$\Delta x = x \otimes 1 + 1 \otimes x,$$

and take the basis $1, x, x^2, \cdots$. Then

$$UT = \sum F_n x^n,$$

and the problem is to determine the dual structure. Eq. (1.9) gives

$$\sum F_m F_n x^m \otimes x^n = \sum F_k (x \otimes 1 + 1 \otimes x)^k$$

or

$$F_k F_1 = (k+1) F_{k+1} \tag{1.10}$$

with the unique solution ($p := F_1$)

$$F_k = p^k / k!$$

Hence A^* is the unital algebra freely generated by p, and

$$UT = e^{xp}. \tag{1.11}$$

Finally, (1.7) provides an easy evaluation of the coproduct:

$$e^{xp} e^{xp'} = e^{x\Delta p}$$
$$\Rightarrow \Delta p = p + p' = p \otimes 1 + 1 \otimes p.$$

Next, let G be a Lie group and $\mathcal{G}$ its Lie algebra with basis $\{\ell_i\}$ $i = 1, \cdots, n$. Let A be the universal enveloping algebra of $\mathcal{G}$, with a basis $\{L_\alpha\}$ of ordered monomials, and Δ the unique compatible coproduct generated by

$$\Delta \ell_i = \ell_i \otimes 1 + 1 \otimes \ell_i. \tag{1.12}$$

[Compatible, that is, with the structure; $\ell_i \mapsto \Delta \ell_i$ generates a homomorphism.] Let $\{F_\alpha\}$ be the terms of functions on G at the identity defined by

$$L_\alpha F_\beta \big|_{Id} = \delta_{\alpha\beta}.$$

Then a calculation that follows precisely the pattern of the first example leads to a unique structure of bialgebra on the space A^* spanned by $\{F_\alpha\}$. This structure depends on the choice of the basis $\{L_\alpha\}$.

(i) If L_α is the symmetric product of $\{\ell_i\}$ $i \in \alpha$, then

$$F_\alpha = (1/|\alpha|!) \prod_{i \in \alpha} p_i.$$

The set α includes repetitions, $i \in \{1, \cdots, n\}$, and $|\alpha|)$ is the cardinality of α. The structure of A^* is Abelian,

$$\mathcal{U}\mathcal{T} = e^{p \cdot \ell}, \quad p \cdot \ell := \sum_{i=1}^{n} p_i \ell_i \tag{1.13}$$

and the coproduct $p_i \mapsto \Delta p_i$ is given by the Campbell-Haussdorff formula.

(ii) If L_α is an ordered polynomial, defined in terms of an ordering of the ℓ_i, for example $\ell_1 < \ell_2 < \cdots < \ell_n$, then

$$F_\alpha = \prod_{i=1}^{n} (p_i)^{\alpha_i} / \alpha_i!,$$

where α_i is the incidence of i in α. The structure is Abelian,

$$\mathcal{U}\mathcal{T} = \prod_{i=1}^{n} e^{p_i \ell_i} \tag{1.14}$$

with the same ordering of factors, and the coproduct is expressed by another version of the Campbell-Hausdorff formula.

1.4. QUANTUM GROUPS

Our program is to obtain a formula for the Universal T-matrix of quantum groups analogous to (1.11) and (1.14). The problem of a quantum version of (1.13) seems to be more difficult [9].

For the quantum group $U_{q,q'}(g\ell_2)$ the formula

$$\mathcal{U}\mathcal{T} = e_a^{p_- x_-} e^{p_1 \rho_1} e^{p_2 \rho_2} e_{1/a}^{p_+ x_+}, \tag{1.15}$$

$$e_a^z := \sum z^n/[n!]_a, \; a = 1/qq',$$

was given in [10]. The p_i generate the two-parameter (twisted) version of quantum $g\ell(2)$ in the sense of Drinfeld, with

$$[p_-, p_+] = (q - 1/q')(q^{p_1} q'^{p_2} - q'^{-p_1} q^{-p_2})$$

and the rest; the generators $x_\pm, \rho_i$ of the dual generates a solvable Lie algebra with

$$[\rho_i, x_\pm] \propto x_\pm,$$

$$[\rho_1, \rho_2] = 0 = [x_+, x_-].$$

This is a special case of twisted, quantum $g\ell(N)$, treated in detail in this paper, with analogous results for all N.

The key to a simple generalization of (1.15) is to choose a preferred ordering of the generators. In this paper we take the Woronowicz-Manin pseudogroup (A) as our starting point, to end up (via the construction of $\mathcal{UT}$) with the Drinfeld quantum group (A^*). The strategy is exactly the same as that followed in the above examples. The important question of ordering is related to a unique factorization of A, carried out in two steps. First

$$A = A_- \otimes_{A_0} A_+, \tag{1.16}$$

where $A_\pm$ and A_0 are sub-bialgebras. Then

$$A_- \ni \tilde{X} = \prod_{i=1}^{k} \tilde{X}(i), \quad A_+ \ni \tilde{Y} = \prod_{i=1}^{k} \tilde{Y}(i) \tag{1.17}$$

where $\tilde{X}(i)$ and $\tilde{Y}(i)$ are nilpotent matrices arranged in a particular order, the choice of which is quite crucial.

The factorization, for the case of twisted, quantum $g\ell(N)$ is carried out with all details in Section 2, the Universal T-matrix is found in Section 3 and the structure of the dual in Section 4. In this case A_0 is Abelian and the structure expressed by (1.16) is essentially the famous quantum double.

In Section 5 we show how the Universal R-matrix is obtained by a simple projection from UT, and in Section 6 we discuss the generalizations of (1.15) and (1.16) to

(i) The standard quantization of any simple Lie algebra.

(ii) Nonstandard quantum groups $(A_n^{(2)}, D_n^{(2)})$.

(iii) Roots of unity.

Finally, in Section 7, we study the relation between twisted, quantum $g\ell(N)$ and $s\ell(N)$, and the question of rigidity of these quantum groups to further, essential deformations.

This paper does not include physical applications, but it may be worth repeating that the transfer matrices (monodromies) of solvable models (without spectral parameters) are obtained from UT by specialization to a representation of A. The choice of generators of A that is introduced via the factorization (1.16) and (1.17), gives a presentation of A as a deformed enveloping algebra, which greatly facilitates the construction of representations. This brings out the interesting problem of giving a sense to the Universal T-matrix for an affine quantum group (with spectral parameter). Applications without spectral parameter include ice-type models, knot theory and conformal field theory in two dimensions, but the inclusion of a spectral parameter seems to be required for more interesting physical applications. All this work on finite dimensional quantum groups must be regarded as preparation for an assault on the real problem.

1.5. TWISTED QUANTUM $g\ell(n)$

This quantum group was found independently by Reshetikhin [11], Schirrmacher [12] and Sudbery [13]. In [11] it is offered as an example of "gauge transformations" on quantum groups based on simple Lie algebras. Since $g\ell(N)$ is not simple it is of some interest to study the relationship between twisted $g\ell(N)$ and twisted $s\ell(N)$.

The R-matrix of twisted, quantum $g\ell(N)$ is an element $R \in End(V \otimes V)$, $V = N$-dimensional vector space, that in a particular basis is given by

$$R = \sum_i M_i^i \otimes M_i^i + \sum_{i<j} (q^{ji} M_j^j \otimes M_i^i + a q^{ij} M_i^i \otimes M_j^j + (1-a) M_j^i \otimes M_i^j). \quad (1.18)$$

The parameters are q^{ij} with $q^{ij}q^{ji} = 1$ and a. The q's are subject to change under "gauge transformation"; the parameter a is much more fundamental. The associated deformed permutation matrix P, often denoted $\check{R}$, is defined by

$$P_{ij}^{k\ell} = R_{ji}^{k\ell}; \quad (1.19)$$

it has two eigenvalues and satisfies the Hecke condition

$$(P-1)(P+a) = 0. \quad (1.20)$$

Reference to "roots of unity" means that $a^K = 1$ for some (minimal) positive integer K. The observation that the Hecke parameter a, and not the q's, is the relevant parameter, raises the question of identifying the important parameters in the case of quantum groups that do not satisfy the Hecke condition.

The matrix defined by (1.18) satisfies the Yang-Baxter relation

$$R_{12}R_{13}R_{23} = R_{23}R_{13}R_{12} \quad (1.21)$$

and the matrix P satisfies, in consequence, the braid relation

$$(braid)_{123} \equiv P_{12}P_{23}P_{12} - P_{23}P_{12}P_{23} = 0. \quad (1.22)$$

The next paragraph depends on (1.20) and (1.22), but not on the specific form of P that is implied by (1.18) and (1.19).

A quantum plane, in the sense of Manin, is an algebra generated by $\{x^i\}$ $i = 1, \cdots, N$, with relations

$$xx(P-1)=0 \quad (x^i x^j P_{ij}^{k\ell} = x^k x^\ell). \tag{1.23}$$

A differential calculus D on this quantum plane, as defined most succinctly in [14], is the algebra generated by $\{x^i\}$ and $\{\theta^i\}$ $i=1,\cdots,N$, with additional relations

$$\theta\theta(P+a)=0, \tag{1.24}$$

$$a\theta x = x\theta P. \tag{1.25}$$

Definition. The pseudogroup $A(P)$ is the unital algebra generated by the matrix elements Z_i^j of an N-dimensional square matrix Z, with relations

$$[P, Z\otimes Z]=0. \tag{1.26}$$

The following remarks also depend only on the validity of the Hecke relation and the braid relation.

Remark 1. One can look upon $A(P)$ as the algebra of quantum automorphisms of the differential calculus D, the action on D being generated by $x\mapsto xZ$, $\theta\mapsto\theta Z$ ($x^i\mapsto x^j Z_j^i$ etc.). The idea of "quantum automorphsims of the quantum plane" does not lead to $A(P)$.

Remark 2. The precise implication of the braid relation for the differential calculus D is contained in a theorem [15] that may be paraphrased as follows: The braid relation (1.22) is equivalent to the statement "x commutes with (1.24) and θ commutes with (1.23)."

Remark 3. An implication of (1.26) is the existence of a unique compatible coproduct on $A(P)$, such that

$$\Delta(z_i^j) = z_i^k \otimes z_k^j. \tag{1.27}$$

Henceforth we regard $A(P)$ as a bialgebra with this coproduct.

Remark 4. The statement (1.26) can be expressed as

$$ZZ(\mathcal{P}-1)=0, \tag{1.28}$$

where $\mathcal{P}$ is the endomorphism defined by the matrix P acting on $End\ V \otimes V$. This matrix does not satisfy the Hecke condition (1.20), but instead

$$(\mathcal{P}-1)(\mathcal{P}+a)(\mathcal{P}+a^{-1})=0. \tag{1.29}$$

thus $A(P)$ is a quantum plane with $\mathcal{P}$ now taking the place of P and (1.29) replacing (1.20). Of course $\mathcal{P}$ satisfies the braid relation.

It is natural to ask what is the algebra of quantum automorphisms of a quantum group, but the preceding remarks suggest something else. Regarding $A(P)$ as a quantum plane, one introduces a differential calculus $\mathcal{D}$ on $A(P)$. The interesting object is the algebra of quantum automorphisms of this differential calculus.

Here we shall obtain (Section 3), in the case that A^* is the twisted, or multi-parameter, quantum group $U_{<q>,a}(gl_N)$, for generic a, the formula

$$\mathcal{UT}=\prod_{\substack{i>j\\ m<n\\ k}} e_a^{X_i^j P_i^j}\, e^{\tau_k H_k}\, e_{1/a}^{Y_m^n Q_m^n} \tag{1.30}$$

The quantum group $A^*=U_{<q>,a}(gl_N)$ is generated by P_i^j (simple positive roots), Q_m^n (simple negative roots), and H_k (Cartan generators); the dual algebra $A(<q>,a)$ by X_i^j, τ_k and Y_m^n; e_a^x is a deformed exponential.

The structure of bialgebra on $A(A^*)$ determines the structure of bialgebra on the dual, so that $\mathcal{UT}$ can be calculated along with the structure of $A^*(A)$. The strategy followed in this paper was to begin from the structure of the pseudo-group $A=A(<q>,a)$ (Woronowicz-Manin picture), and use the structural properties of the dual form to determine, first $\mathcal{UT}$ and then the structure of the quantum group $A^* = U_{<q>,a}(gl_N)$ (Drinfeld picture). The structural relations are, once more

$$T(x,l)\;T(x,l') = T(x,\Delta(l)), \tag{1.31}$$

$$T(x,l)\;T(x',l) = T(\Delta(x),l). \tag{1.32}$$

Here x and x' refers to two copies of A and l,l' to two copies of A^*; $\Delta(x)$ is the coproduct of A and $\Delta(l)$ is the coproduct of A^*. The first relation determines the algebraic structure of A^* and the second one gives the coproduct (Section 4).

There are interesting homomorphisms into a quantum group from its dual, that allow us to obtain the Universal R-matrix from the expression for the Universal T-matrix. If Φ is such a homomorphism, then

$$\mathcal{UR} = (id \otimes \Phi)\mathcal{UT}; \tag{1.33}$$

Φ operates on the generators of $A(<q>,a)$. (Section 6)

The similarity of (1.3) to (1.2) seems to suggest that the formula may be useful in connection with Fourier transforms on quantum groups.

2. Factorization

Let R be the Yang-Baxter matrix (1.18), in the fundamental representation of multiparameter [12, 13] (twisted [11]) quantum $gl(N)$, V an N-dimensional vector space, $P \in End(V \otimes V)$ defined by

$$P^{kl}_{ji} = R^{kl}_{ij}. \tag{2.1}$$

Then P satisfies the braid relation,

$$(\text{braid})_{123} \equiv P_{12}P_{23}P_{12} - P_{23}P_{12}P_{23} = 0, \tag{2.2}$$

and the Hecke condition (with a in the field)

$$(P-1)(P+a)=0. \tag{2.3}$$

Let F be the algebra of formal power series finitely generated by (z_i^j), $i,j=1,\ldots,N$, $(z_i^i)^{-1}$, $i=1,\ldots,N$, and the unit, with relations

$$z_i^i(z_i^i)^{-1}=(z_i^i)^{-1}z_i^i=1,\ i=1,\ldots,N.$$

The quantum algebra (pseudogroup [5]) $A=A(<q>,a)$ is the quotient of F by the ideal generated by the relations

$$[P,Z\otimes Z]=0,\ Z\equiv \text{matrix }(z_i^j). \tag{2.4}$$

There is a unique algebra homomorphism $\Delta: A\to A\otimes A$, such that $\Delta(Z)=Z\otimes T$; that is,

$$\Delta(z_i^j)=\sum_k z_i^k\otimes z_k^j, \tag{2.5}$$

which gives A a structure of bialgebra.

There is more than one sense in which $A(<q>,a)$ is dual to the quantum group $U_{<q>,a}(gl_N)$, a deformation of the enveloping algebra $U(gl_N)$ of $gl(N)$. Recall that (the differential of) the classical r-matrix defines a Lie structure on the dual space $gl(N)^*$, turning $\{gl(N),gl(N)^*\}$ into a Lie bialgebra dual pair. Here we shall relate $A(<q>,a)$ to a deformation $U_{<q>,a}(gl_N{}^*)$ of the enveloping algebra of $gl(N)^*$. Then we show that there is a natural duality between the two deformed enveloping (bi-)algebras.

The first step is to justify the following factorization:

$$z_i^j=\sum_k X_i^k z_k Y_k^j, \tag{2.6}$$

$$X_i^j=\begin{cases}1, & i=j,\\ 0, & i<j,\end{cases}\quad Y_i^j=\begin{cases}1, & i=j,\\ 0, & i>j,\end{cases}. \tag{2.7}$$

Let F' be the algebra generated by X_i^j, z_k, $(z_k)^{-1}$ and Y_i^j (with unit), with relations $z_k(z_k)^{-1} = (z_k)^{-1}z_k = 1$ and (2.7). The formula (2.6) is invertible and allows the identification of F' with F. The relations (2.4) make sense in F' and the associated quotient is an alternative presentation of A. We shall obtain the alternative expressions for the relations and show, in particular, that the X's commute with the Y's.

Proposition. Let A_-, A_+ and A_0 be quotients of A by the ideals generated by

$$I_- = \{z_i^j,\ i < j\}, \quad I_+ = \{z_i^j,\ i > j\}, \quad I_0 = \{z_i^j,\ i \neq j\}, \tag{2.8}$$

then

$$A = A_- \underset{A_0}{\otimes} A_+. \tag{2.9}$$

Proof. The crucial ingredient is the fact that the sets $I_{\pm,0}$ actually generate ideals of A. This is an easy consequence of the relations, but there is a deeper reason for it. It is known that A has representations π and π' given by

$$(\pi(z_i^k))_j^l = R_{ij}^{kl}, \quad (\pi'(z_i^k))_j^l = (R^{-1})_{ji}^{lk}, \tag{2.10}$$

reflecting the existence of two homomorphisms from $gl(N)^*$ to $gl(N)$. The kernels of these two isomorphisms include the sets I_- and I_+, respectively. Now let $(\tilde{X}_i^j)$ and $(\tilde{Y}_i^j)$ be the images of (z_i^j) under the projections $A \to A_-$ and $A \to A_+$, then $\tilde{X}_i^j = 0$ for $i < j$ and $\tilde{Y}_i^j = 0$ for $i > j$, and define

$$\tilde{z}_i^j \equiv \sum_k \tilde{X}_i^k \otimes \tilde{Y}_k^j. \tag{2.11}$$

The relations among the $\tilde{X}$'s and among the $\tilde{Y}$'s are exactly the same as the relations among the z's and, because Δ is a homomorphism, the same relations are also obeyed by the $\tilde{z}$'s. The composite mapping given by the two projections of A on

A_+ and on A_- followed by (2.11) is one-one, so we can identify $\tilde{z}_i^j$ with z_i^j. Now set

$$\tilde{X}_i^j = X_i^j x_j, \quad i \geq j, \quad \tilde{Y}_i^j = y_i Y_i^j, \quad i \leq j, \quad X_i^i = Y_i^i = 1, \tag{2.12}$$

and identify X_i^j with $X_i^j \otimes 1$, Y_i^j with $1 \otimes Y_i^j$, so that

$$z_i^j = X_i^k z_k Y_k^j, \quad z_k \equiv x_k \otimes y_k. \tag{2.13}$$

This is the desired formula (2.6). It is now evident that the X's commute with the Y's, (2.13) is precisely what we mean (2.9), and the proposition is proved.

As we said, the relations among the $\tilde{X}$'s and among the $\tilde{Y}$'s are exactly the same as among the z's, and one easily derives the following relations among the generators of the decomposition (2.13). First,

$$z_i z_j = z_j z_i,$$

$$z_k X_i^j = C_{kij} X_i^j x_k, \quad z_k Y_i^j = C'_{kij} Y_i^j z_k. \tag{2.14}$$

The coefficients C_{kij} and C'_{kij} are given in the Appendix, Eq. (A.3-3'). Next, define "simple generators"

$$X_i \equiv X_{i+1}^i, \quad Y_i = Y_i^{i+1}, \quad i = 1, \ldots, N,$$

then there are quommutation relations and Serre relations

$$[X_i, X_j]_{k_{ij}} = 0, \quad [Y_j, Y_i]_{k_{ij}} = 0, \quad |i-j| > 1, \tag{2.15}$$

$$[X_{i+1}, X_i]_{k_i} = (1 - 1/a)\, X_{i+1}^{i-1}, \tag{2.16}$$

$$[Y_i, Y_{i+1}]_{k_i} = -(1 - 1/a)\, Y_{i+1}^{i-1}, \tag{2.17}$$

$$[X_i,[X_{i+1},X_i]_{k_i}]_{r_i}=0=[X_{i+1},[X_{i+1},X_i]_{k_i}]_{s_i}, \tag{2.18}$$

$$[[Y_i,Y_{i+1}]_{k_i},Y_i]_{r_i}=0=[[Y_i,Y_{i+1}]_{k_i},Y_{i+1}]_{s_i}. \tag{2.19}$$

Here $[A,B]_k \equiv AB-kBA$. The coefficients k_{ij} and k_i are in (A.4).

This alternative presentation of A, in terms of X_i^j, Y_i^j, z_k and $(z_k)^{-1}$, is very convenient for our purpose. In particular, the construction of the universal T-matrix for A reduces to the same problem for $A_\pm$. As an ultimate refinement, we introduce elements ρ_k of A_-, σ_k of A_+ and τ_k of A by setting

$$x_k=e^{\rho_k},\quad y_k=e^{\sigma_k},\quad z_k=e^{\tau_k}, \tag{2.20}$$

and adopt ρ_k,σ_k,τ_k as generators instead of x_k,y_k,z_k. By abuse of notation we still use the same names, $A_\pm$ and A. As is very well known, and obvious in the sequel, these algebras must be completed with some infinite series, including the series (2.20). Care must be taken to extend far enough to get closure under the algebraic operations, without making the algebras too large to be manageable. For such questions we refer to the papers [15] and [16].

In this new presentation, in which (τ_k) replace (z_k^k) and $(z_k^k)^{-1}$ as generators, A becomes a deformation $U_{<q>,a}(gl_N{}^*)$ of the enveloping algebra of the Lie algebra $gl(N)^*$ (with the Lie structure determined by the classical r-matrix). The Serre presentation of $U_{<q>,a}(gl_N{}^*)$ is easily obtained from (2.14-19).

3. The Universal T-Matrix

The construction has already been explained elsewhere [10]. For A_- we take the basis

$$X^{[a][\alpha]}\equiv\prod_{\substack{i>j\\k}}(X_i^j)^{a_{ij}}(\rho_k)^{\alpha_k},\quad e^{\rho_k}\equiv x_k. \tag{3.1}$$

To facilitate the manipulations that follow, it is crucial to adopt a good ordering of the factors in the definition of the basis elements; the good rule is that X_i^j precedes X_l^k if $j < k$ or $j = k, i < l$.

A rigorous definition of the dual is based on the natural basis that is provided by the functions defined by

$$P_{[a][\alpha]}(X^{[b][\beta]}) = \begin{cases} 1, & \text{if } [a][\alpha] = [b][\beta], \\ 0, & \text{otherwise.} \end{cases}$$

For details concerning completion of the dual algebra in terms of entire functions on A we refer to the papers [15] by and [17].

If $\{P_{[a][\alpha]}\}$ is the dual basis, then the universal T-matrix for A_- is

$$T^- = \sum_{[a][\alpha]} X^{[a][\alpha]} P_{[a][\alpha]}. \tag{3.2}$$

The important structural properties of this operator were discussed in the Introduction; Eq. (1.32) for T^- reads

$$\begin{aligned} \sum \Big(\prod_{\substack{i>j \\ k}} (X_i^j)^{a_{ij}} (\rho_k)^{\alpha_k}\Big)\Big(\prod_{\substack{i>j \\ k}} (X_i^{\prime j})^{b_{ij}} (\rho_k^\prime)^{\beta_k}\Big) P_{[a][\alpha]} P_{[b][\beta]} \\ = \prod_{\substack{i>j \\ k}} (\Delta X_i^j)^{c_{ij}} (\Delta \rho_k)^{\gamma_k} P_{[c][\gamma]}. \end{aligned} \tag{3.3}$$

We have abandoned the cumbersome notation with $\otimes$, writing X for $X \otimes 1$ and X' for $1 \otimes X$ from now on. Comparing coefficients of both sides one gets the relations sastisfied by the dual basis. As an example, consider the coefficients of $(X_i^j)^{n-1} X_i^{\prime j}$ for a fixed pair (i, j). On the left one has (dots standing for zeros) $P_{[\ldots,n-1,\ldots][0]} P_{[\ldots,1,\ldots][0]}$; and on the other side the only contributing term is

$$(\Delta X_i^j)^n P_{[\ldots,n,\ldots][0]}. \quad (a_i^j = n, \quad \text{all others zero}).$$

[This at first sight innocent statement is valid because of the particular order chosen.] We need to know that the coefficient of $(X_i^j)^{n-1} X_i^{\prime j}$ in $(\Delta X_i^j)^n$ is

$$[n]_a \equiv (a^n - 1)/(a-1), \tag{3.4}$$

to get a simple recursion relation for $P_{[\ldots,n,\ldots][0]}$, with the solution

$$P_{[\ldots,n,\ldots][0]} = (P_i^j)^n/[n!]_a, \tag{3.5}$$

provided a is not a root of 1.

The result is that the dual of A_- is generated by

$$(P_i^j), \quad i > j, \quad \text{and} \quad H_k = P_{[0][\ldots,1,\ldots]}, \quad \text{1 in k'th place,}$$

and that

$$T^- = \prod_{\substack{i>j \\ k}} e_a^{X_i^j P_i^j} e^{\rho_k H_k}.$$

A similar calculation gives T^+, and the product $T^- T^+$, with the identification $\rho_k + \sigma_k = \tau_k$ is the Universal T-matrix for $A(<q>, a)$:

$$UT = \prod_{\substack{i>j \\ m<n \\ k}} e_a^{X_i^j P_i^j} e^{\tau_k H_k} e_{1/a}^{Y_m^n Q_m^n}. \tag{3.6}$$

The structure of UT reflects that of (2.6) and (2.9).

4. Relations and Coproduct of Twisted, Quantum $gl(N)$

Such relations are of course known [18], but we want to show that they drop out of our formula for the universal T-matrix, and that the generators P_i^j, H_k and Q_i^j are precisely the generators of a conventional presentation. Actually, we have found earlier derivations difficult to understand, and we hope that the one given here may be an improvement.

First, in

$$T^-_{x,p}T^-_{x',p} = T^-_{\Delta x,p} \tag{4.1}$$

we compare the coefficients of (properly ordered) elements of the basis of A. We first notice that terms linear in X^j_i and ρ'_k, for a fixed triple (i,j,k), occur with the same coefficients on both sides, as a direct result of our construction. But terms involving $\rho_k X'^j_i$ are in the wrong order; the coefficient on the left is $H_k P^j_i$, and on the right side it is found by examination of the term

$$P^j_i \Delta(X^j_i) e^{H_k \Delta(\rho_k)} = P^j_i \Delta(X^j_i)(1 + H_k\Delta(\rho_k) + \ldots). \tag{4.2}$$

Applying (A.6-10) one sees that the relevant part of $\Delta(X^j_i)$ is contained in

$$(x_i/x_j)X'^j_i = (1 + \rho_i - \rho_j + \ldots)X'^j_i;$$

the coefficient we are looking for is thus $P^j_i(H_k + \delta_{ik} - \delta_{jk})$, and

$$[H_k, P^j_i] = (\delta_{ki} - \delta_{kj})P^j_i. \tag{4.3}$$

By these means one easily recovers the complete Serre presentation of the positive Borel subalgebra of the quantum group $U_{<q>,a}(gl_N)$. The "simple" generators are $P_i = P^i_{i+1}$, $i = 1, \ldots, N$, and the remaining relations are

$$\begin{aligned} &[P_j, P_i]_{k_{ij}} = 0, \quad \text{if} \quad |i-j| > 1, \\ &[[P_i, P_{i+1}]_{k_i}, P_i]_{r_i} = [[P_i, P_{i+1}]_{k_i}, P_{i+1}]_{s_i} = 0. \end{aligned} \tag{4.4}$$

Next, we get the commutator between the simple generators of $U_{<q>,a}(gl_N)$,

$$P_i \equiv P^i_{i+1} \quad \text{and} \quad Q_i \equiv Q^{i+1}_i, \tag{4.5}$$

by comparing coefficients of $X'^i_{i+1} Y^{i+1}_i$ on both sides of $T_{x,p}T_{x',p} = T_{\Delta x,p}$. For this we need to know parts of the expression for $\Delta(\rho_k)$. This is found in the Appendix; the result is that, for $i = 1, 2, ..., N-1$,

$$[P_i, Q_j] = \delta^j_i \frac{a}{1-a}(q^{i,i+1})^{1-H_{i+1}-H_i}(a^{-H_{i+1}} - a^{-H_i})C_i. \tag{4.6}$$

The relations

$$
\begin{aligned}
&[H_k, Q_i^j] = (\delta_{ki} - \delta_{kj})Q_i^j \\
&[Q_i, Q_j]_{k_{ij}} = 0, \quad \text{if} \quad |i-j| = 1, \\
&[Q_i, [Q_{i+1,}, Q_i]_{k_i}]_{s_i} = [Q_{i+1,}[Q_{i+1,}Q_i]_{k_i}]_{s_i} = 0,
\end{aligned} \tag{4.7}
$$

complete the structure. The dependence on $q^{i,i+1}$ can be removed by a slight redefinition of the simple generators to make the structure reduce to that of standard quantum $gl(N)$, as found in [18], but that would mess up the expression (3.6).

The coproduct is obtained from the other structural formula, Eq. (1.4),

$$
\prod_{\substack{i>j \\ m<n \\ k}} e_a^{X_i^j P_i^j} e^{\rho_k H_k} e^{Y_m^n Q_m^n} \prod_{\substack{i>j \\ m<n \\ k}} e_a^{X_i^j P_i'^j} e^{\rho_k H_k'} e^{Y_m^n Q_m'^n} = \prod_{\substack{i>j \\ m<n \\ k}} e_a^{X_i^j \Delta P_i^j} e^{\rho_k \Delta H_k} e^{Y_m^n \Delta Q_m^n}, \tag{4.8}
$$

it is the homomorphism generated by

$$
\Delta(H_k) = H_k \otimes 1 + 1 \otimes H_k,
$$

$$
\Delta(P_i) = P_i \otimes 1 + A_i \otimes P_i, \quad \Delta(Q_i) = Q_i \otimes B_i + 1 \otimes Q_i.
$$

$$
A_i = C_i(q^{i+1,i})^{H_i + H_{i+1}} a^{-H_i}, \quad B_i = C_i(q^{i+1,i})^{H_i' + H_{i+1}'} a^{-H_{i+1}'},
$$

$$
C_i = \prod_{k \neq i, i+1} (q^{i+1,k} q^{ki})^{H_k}. \tag{4.9}
$$

5. Applications of the Universal T-Matrix

We noted the existence of two representations of the quantum algebra $A(<q>, a)$ in $gl(N)$. If, in the formula (3.6), one takes the generators P, H and Q in the fundamental representation, then one recovers the original N-dimensional matrix $Z = (z_i^j)$, and if one then takes z_i^j in the representation (2.10), then one gets the original R-matrix in the fundamental representation. Our results concerning the structure of the dual algebra shows that the representations (2.10) [see Appendix, Eq.(A.9)] lift to the structure; there are a homomorphism $\Phi, \Phi' : A \to U_{<q>,a}(gl_N)$ given by

$$\Phi(X^i_j) = (1/a - 1)q^{ji}Q^j_i, \quad i < j, \quad \Phi(Y^j_i) = 0,$$

$$\Phi(z_k) = \Big(\prod_i (q^{ki})^{H_i}\Big)a^{H_{k+1}+\ldots+H_N}, \tag{5.1}$$

$$\Phi'(Y^j_i) = (a - 1)q^{ji}P^i_j, \quad i < j, \quad \Phi'(X^j_i) = 0,$$

$$\Phi'(z_k) = \Big(\prod_i (q^{ki})^{H_i}\Big)a^{-H_1-\ldots-H_{k-1}}. \tag{5.2}$$

One can view the universal T-matrix as an element of $U_{<q>,a}(gl_N{}^*) \otimes U_{<q>,a}(gl_N)$. Then we have

Theorem. The universal R-matrix for $U_{<q>,a}(gl_N)$ is found by applying the mapping $id \otimes \Phi$ to the universal T-matrix (3.6),

$$\mathcal{UR} = (id \otimes \Phi)\mathcal{UT} = \prod_{\substack{i>j \\ k}} e_a^{\Phi(X^j_i)P^j_i}\,[\Phi(z_k)]^{H_k}. \tag{5.3}$$

6. Further Generalizations

6.1. STANDARD QUANTUM GROUPS

These are the 1-parameter deformations given by Drinfeld in [1]. Much of their structure is summed up in the quantum double construction; it is basically the same as for twisted $gl(N)$. There are representations defined by (2.10), with kernels $I_\pm$ and associated quotients $A_\pm$, and the structure formula (2.9),

$$A = A_- \otimes_{A_0} A_+$$

holds generally, with A_0 the Abelian quotient by $I_+ \cup I_-$. As for twisted $gl(N)$, these representations lift to isomorphisms Φ and Φ' from $A_\pm$ to subalgebras $U_\pm$ of U. The Universal R-matrix is known [19] (hence, so are Φ and Φ'), actually in two forms, $R_+ \in U_- \otimes U_+$ and $R_- \in U_+ \otimes U_-$. The Universal T-matrices for $A_\pm$ are

$$T^- = (\Phi^{-1} \otimes 1)R_-, \quad T^+ = (\Phi'^{-1} \otimes 1)R_-$$

and for A, it is

$$\mathcal{UT} = T^- T^+.$$

6.2. NONSTANDARD QUANTUM GROUPS

These include the constructions on $A_n^{(2)}$ and $D_n^{(2)}$ of Jimbo [2], as well as "Esoteric Quantum $gl(N)$" and the other deformations of twisted quantum $gl(N)$. It seems that the general formulas that apply to the standard quantum groups have direct generalizations to these cases as well. However, the structure of the quantum double is very different and A_0 is no longer abelian. See [10], second paper.

6.3. ROOTS OF UNITY

The most interesting aspects of quantum groups are those that appear at special values of the parameters; up to this point, in order to postpone the discussion of these phenomena, we have assumed $(<q>, a)$ in generic position.

Recall that the exponential form $\mathcal{UT} = e^{xp}$ appeared in our first example, in subsection (1.3), as the series $\sum F_k x^k$ with F_k subject to the recursion relation (1.10),

$$F_k F_1 = (k+1) F_{k+1} \quad \Rightarrow \quad F_k = p^k / k!. \tag{6.1}$$

In the evaluation of $\mathcal{UT}$ for the quantum groups one encounters modified exponentials,

$$e_a^{xp} = \sum (xp)^n / [n!]_a, \tag{6.2}$$

arising from the solution of the recursion relation

$$F_kF_1 = [k+1]_a F_{k+1} \quad \Rightarrow \quad F_k = p^k/[k!]_a. \tag{6.3}$$

See Eq.(3.5). This is the only place that the restriction to generic parameters is relevant. There are no restrictions on the parameters q^{ij}. The limitation to parameters in generic position applies only to the Hecke parameter a, and the (till now) excluded values are those for which there is an integer K such that

$$a^K = 1, \quad a^n \neq 1, \quad n = 1, 2, ..., K-1.$$

We now suppose that this relation holds; thus a is a primitive root of unity.

Then $[K]_a = 0$, and the general solution of the recursion relation (6.3) is

$$F_{mK+n} = (p'^m/m!)(p^n/[n!]_a), \quad p^K = 0,$$

$$m = 0, 1, ...; \quad n = 0, 1, ..., K-1.$$

This involves a new generator p', and the new relation $p^K = 0$. The expressions obtained for the Universal T-matrices are thus modified, when a is a root of unity, by the replacement of all twisted exponentials, according to the rule

$$e_a^{xp} \rightarrow \sum_m (x^K p')^m/m! \sum_{n=0}^{K-1} (xp)^n/[n!]_a,$$

with $p^K = 0$.

What is significant is the appearance of a new independent generator p' that replaces p^K. The structure of the dual algebra is drastically changed; it is no longer the Drinfeld quantum group $U_{<q>,a}(\mathcal{G})$. Of course, this latter still exists at roots of unity, but its dual is not the pseudogroup $A(<q>, a)$.

The structure of the dual of $A(<q>, a)$ at $a^K = 1$ can be found by the same methods as in the generic case, but much more quickly by the prescription

$$p' := \lim_{a^K \rightarrow 1} p^K/[K]_a.$$

Thus one obtains for the dual of $A(<q>, a)$ the additional relations

$$[P_i, P'_j] = 0, \quad [P'_i, P'_j] = 0,$$

$$[H_k, P'^j_i] = K(\delta_{ki} - \delta_{kj})P'^j_i,$$

$$[P_i, Q'_i] = (a-1)(q^{12})^{1-H_1-H_2}\left(Q_1^{K-1}a^{-H_2} - a^{-H_1}Q_1^{K-1}\right).$$

The last relation is noteworthy; it shows that the generators P'_i **do not** generate an ideal; consequently, the irreducible modules **do not** become nondecomposable at roots of unity.

The dual of $U_{<q>,a}(gl_N)$ can be obtained in the same way. The result is that $(X^j_i)^K$ and $(Y^j_i)^K$ vanish and new independent generators X'^j_i and Y'^j_i appear. In this case X'^j_i and Y'^j_i do generate ideals.

In the simplest case of $U_{q,q'}(gl_2)$ the dual pseudogroup is generated by X, X', ρ_1, ρ_2, Y and Y', with relations

$$X^K = Y^K = 0, \quad q = e^h, \; q' = e^{h'},$$

$$[\rho_1, X] = hX, \; [\rho_1, X'] = KhX',$$

$$[\rho_2, X] = h'X, \; [\rho_2, X'] = Kh'X', etc.$$

The associated Woronowicz matrix; that is, UT evaluated in the 2-dimensional representation of $gl(2)$, is

$$\begin{pmatrix} 1 & 0 \\ X & 1 \end{pmatrix}\begin{pmatrix} e^{\rho_1} & 0 \\ 0 & e^{\rho_2} \end{pmatrix}\begin{pmatrix} 1 & Y \\ 0 & 1 \end{pmatrix},$$

in which X' and Y' do not appear. To obtain the full dual one has to take UT in a representation of quantum $gl(N)$ in which $P^K \neq 0$.

Duality at roots of unity has already been described by Frölich and Kerler [20], for the case of standard quantum $sl(2)$.

Classical q-functions are always investigated in the domain $| q |< 1$. Indeed, many of these functions, including the twisted exponential, cease to exist as q tends to a root of unity. We have seen that the twisted exponential, within the algebraic contest, does have a natural generalization to roots of unity: but that this generalization is highly nontrivial and even mysterious in a purely analytical context. It is proposed to take up the study of q-functions on a broader basis, with a view to finding natural extensions to roots of unity. Perhaps even the Rogers-Ramanujan identies can be generalized to $q^K = 1$.

7. Quantum $gl(N)$ and Quantum $sl(N)$.

The (Drinfeld) quantum group $U_{<q>,a}(sl_N)$ is the subquotient of $U_{<q>,a}(gl_N)$ defined by the ideal generated by the element

$$\sum_{i+1}^{N} H_i \equiv \mathcal{H}.$$

The universal T- and R-matrices of quantum $gl(N)$ reduce to those of quantum $sl(N)$ under the projection that anulls $\mathcal{H}$. We shall calculate this reduced R-matrix, in the fundamental representation of $sl(N)$.

Restricting the Universal R-matrix (5.3) to any faithful N-dimensional representation gives the formula

$$R = \left[1 + \sum_{i>j} P_i^j \Phi(X_i^j)\right] \sum_{i,j} (\tilde{q}^{ij})^{H_i \oplus H_j}, \tag{7.1}$$

with $\tilde{q}^{ij} = aq^{ij}$ for $i < j$ and $\tilde{q}^{ij} = q^{ij}$ for $i \geq j$. With $\mathcal{H} = 0$, in the N-dimensional representation of $sl(N)$ we have $H_i = M_i^i - 1/N$. Substituting this into (7.1) we obtain

$$R = (1-a) \sum_{i<j} (\kappa_i/\kappa_j M_j^i \otimes M_i^j + \sum_{i,j} \hat{q}^{ij} M_i^i \otimes M_j^j, \tag{7.2}$$

with

$$\hat{q}^{ij} = (\kappa_i/\kappa_j)q^{ij}, \quad \kappa_i = \left(a^i \prod_k q^{ki}\right)^{1/N} . \tag{7.3}$$

This R-matrix for $sl(N)$ differs from that of $gl(N)$ in two particulars. (1) q^{ij} is replaced by $\hat{q}^{ij}$. These new parameters are not independent:

$$\prod_i \hat{q}^{ij} a^j = a^{(N+1)/2}. \tag{7.4}$$

If the original parameters satisfy

$$\prod_i q^{ij} a^j = a^{(N+1)/2}, \tag{7.5}$$

then $\hat{q}^{ij} = q^{ij}$. Therefore, reducing from $gl(N)$ to $sl(N)$ as we have done, from arbitrary initial parameters, gives the same result as restricting the parameters to satisfy (7.5). (2) The factor κ_i/κ_j can be removed by the isomorphism $M_i^j \to (\kappa_i/\kappa_j)M_i^j$.

The restriction (7.5) was found by Schirrmacher [12] from the requirement that the quantum determinant be unity.

The deformations of $A(<q>, a)$ have recently been calculated [15]. This algebra is rigid for essential, first order deformations except for very special values of the parameters. There are several series of deformations of twisted quantum $gl(N)$, here we illustrate the simplest one, in the case of $gl(3)$.

The parameters are q^{12}, q^{23}, q^{13} and a. The deformation that consists of adding the following piece

$$\delta R = \epsilon \left(q^{13} M_1^2 \otimes M_3^2 - M_3^2 \otimes M_1^2\right)$$

to (1.18), is exact and essential. It exists if and only if the parameters satisfy

$$q^{12} = q^{23} \quad \text{and} \quad q^{13} = (q^{12})^2.$$

The projection to $sl(3)$ fixes the value of a:

$$q^{12} = q^{23} := q, \quad q^{13} = q^2. \quad a = q^{-3}.$$

That is why, for $sl(3)$, unlike $gl(3)$, the term "roots of unity" applies to the values of the q's.

This esoteric form of quantum $sl(3)$ is not included in the list produced by Jimbo [2], but the classical limit is in the classification of Belavin and Drinfeld [21].

Appendix

The R-matrix for twisted, quantum $gl(N)$, in the fundamental representation, was given in (1.18),

$$R = \sum_i M_i^i \otimes M_i^i + \sum_{i<j} (q^{ji} M_j^j \otimes M_i^i + a q^{ij} M_i^i \otimes M_j^j + (1-a) M_j^i \otimes M_i^j),$$

The inverse matrix is given by the same formula, with the parameters q^{kl} and a replaced by their inverses.

The relations (2.4) of $A(<q>, a)$, written out in full detail, are

$$\begin{aligned} z_i^a z_i^b &= q^{ab} z_i^b z_i^a, \\ z_i^a z_j^a &= (a q^{ij})^{-1} z_j^a z_i^a, \ i < j, \\ z_i^a z_j^b &= (a q^{ab}/q^{ij}) z_j^b z_i^a, \ i > j, \ a < b \\ q^{ij} z_i^a z_j^b - q^{ab} z_j^b z_i^a &= (a-1) z_j^a z_i^b, \ i > j, \ a > b. \end{aligned} \tag{A.1}$$

The quotient algebras A_+, A_- satisfy relations that one gets from these by annulling the generators in I_+, I_-,

$$A_- : z_i^j \to \begin{cases} \tilde{X}_i^j, & j \leq i \\ 0, & i < j \end{cases} ; \quad A_+ : z_i^j \to \begin{cases} \tilde{Y}_i^j, & i \leq j \\ 0, & j < i \end{cases} \tag{A.2}$$

Finally, one gets from (2.12) the following relations for A_-,

$$x_i x_j = x_j x_i,$$

$$x_k X_i^j = \begin{cases} q^{ik} q^{kj} X_i^j x_k, & k < j < i \text{ or } j < i \leq k, \\ (1/a) q^{ik} q^{kj} X_i^j x_k, & j \leq k < i \end{cases} \tag{A.3}$$

These become relations of A after substitution of z_i for x_i. Likewise, the following relations for A^+ become relations of A if y_i is replaced by z_i,

$$y_i y_j = y_j y_i,$$

$$y_k Y_i^j = \begin{cases} q^{ik} q^{kj} Y_i^j y_k, & k \leq i < j \text{ or } i < j < k, \\ a q^{ik} q^{kj} Y_i^j y_k, & i < k \leq j. \end{cases} \tag{A.3'}$$

The coefficients in (2.15-19) are

$$k_{ij} = q^{i+1,j} q^{ji} / q^{i+1,j+1} q^{j+1,i},$$

$$k_i = q^{i+1,i-1} / q^{i+1,i} q^{i,i-1}. \tag{A.4}$$

We need some formulas for the coproduct of A_-. From (2.12) and

$$\Delta(\tilde{X}_i^j) = \sum_k \tilde{X}_i^k \otimes \tilde{X}_k^j \tag{A.5}$$

one gets

$$\Delta(x_i) = x_i \otimes x_i, \tag{A.6}$$

$$\Delta(X_i^j) = \sum_k X_i^k (x_k / x_j) \otimes X_k^j = X_i^j \otimes 1 + (x_i / x_j) \otimes X_i^j + \dots. \tag{A.7}$$

Only the first two terms are relevant for our calculation of T^-; neglecting the rest we have

$$\Delta(X) = A + B, \quad A = X_i^j \otimes 1, \quad B = (x_i / x_j) \otimes X_i^j. \tag{A.8}$$

One has $BA = aBA$ from (A.3) and thus $(A+B)^n = A^n + A^{n-1} B (1 + a + \dots + a^{n-1})$ and finally the result

$$\Delta(X_i^j)^n = (X_i^j)^n \otimes 1 + [n]_a (X_i^j)^{n-1}(x_i/x_j) \otimes X_i^j + \dots,$$

that was used in (3.4).

For the evaluation of $[P_i, Q_i]$ we need to know some terms in $\Delta(z_i)$,

$$\Delta(z_k) = (z_k z_k')^{H_k}\Big(1 + (q^{k,k+1}/a) Y_k X_k' - q^{k-1,k} X_{k-1}' Y_{k-1} \dots\Big)$$

$$\big(\Delta(z_k)\big)^{H_k} = (zz')^{H_k}(1 + [H_k]_a (q^{k,k+1}/a) Y_k X_k' - [H_k]_{1/a} q^{k-1,k} Y_{k-1} X_{k,k-1}' + \dots).$$

The relation (4.6) now follows easily in the same way.

The representations π and π' are given by (2.10) and (1.18), and more explicitly by vskip-3mm

$$\pi(z_j^i) = \begin{cases} (1-a)M_i^j, & i < j, \\ 0 \quad, & i > j, \end{cases} \quad \pi(z_k^k)_i^j = \delta_i^j q^{ki} a^{(i>k)},$$

$$\pi'(z_i^j) = \begin{cases} (1-1/a)M_j^i, & i < j, \\ 0, & i > j, \end{cases} \quad \pi'(z_k^k)_i^j = \delta_i^j q^{ki} a^{-(i<k)}, \qquad (A.9)$$

where $(i < k) = 1$ if $i < k$ and zero otherwise.

Acknowledgments

This work was supported in part by the National Science Foundation.

References

1. V.G. Drinfeld, Int.Congr.Math., Berkeley 1986 798-820.
2. M.Jimbo, Commun.Math.Phys. **102** (1986) 537-547.
3. R.J. Baxter, Ann.Phys. **70** (1971) 193-228.
4. E.K. Sklyanin and L.D. Faddeev, Sov.Phys.Dokl. **23** (1978) 902-904.
5. L. Woronowicz, Publ.Res.Ins.Math.Sci., Kyoto Univ., **23** (1987) 117-181.
6. Y. Manin, *Topics in noncommutive geometry*, Princeton Univ. Press, 1991.

7. Y.N. Reshetikhin, L. Takhtajan and L.D. Faddeev, Leningrad Math.J. **1** (1998) 193-225; A. Sudbery, J.Phys.A **20** (1990), L697.

8. P. Bonneau, M. Flato and G. Pinczon, Lett.Math.Phys. bf 25 (1992) 75-84.

9. S.P. Vokos, B. Zumino and J. Wess, Z.Phys.C **48** (1990) 65-74; C. Fronsdal, Lett.Math.Phys. **24** (1992) 73-78.

10. C. Fronsdal and A. Galindo, The dual of a Quantum Group, Lett.Math.Phys. **27** (1993) 59-71; The Universal T-Matrix, Proc. Joint Summer Res.Conf. (AMS-IMS-SIAM) on Conformal Field Theory, Topological Field Theory and

11. N.Y. Reshetikhin, Lett.Math.Phys. **20** (1990) 331-336.

12. A. Schirrmacher, Z.Phys.C **50** (1991) 321.

13. A. Sudbery, J.Phys.A **23** (1990) L697.

14. J.C. Baez, Lett.Math.Phys. **23** (1991) 133.

15. C.Fronsdal and A.Galindo, Deformations of Multiparameter quantum $gl(N)$. Preprint UCLA/92/TEP/52.

16. P. Truini and V.S. Varadarajan, Lett.Math.Phys. **21** (1991) 287.

17. M.V. Pimsner, A Class of Markov Traces, preprint.

18. J.F. Cornwell, J.Math.Phys. **33** (1992) 3963-3977).

19. Kirillov and N.Y. Reshetikhin, Adv.Series in Math.Phys. **7** (1988) 285-339.

20. J. Frölich and T.Kerler, *Quantum Groups, Quantum Categories and Quantum Field Theory*, to be published.

21 A.A. Belavin and V.G. Drinfeld, Sov.Sci.Rev.Math. **4** (1984) 93-165.

C. Fronsdal

Department of Physics

University of California, Los Angeles, CA 90024-1547

UNITARY REPRESENTATIONS OF QUANTUM LORENTZ GROUP

W. PUSZ and S.L.WORONOWICZ

ABSTRACT.Recent results concerning representation theory of quantum Lorentz group are presented.

1 Introduction

In this talk we describe recent results of the representation theory of quantum Lorentz group. This review is based on [6] and [5]. It was inspired by the classsical results for the Lorentz group as presented by M.A.Najmark [3] and I.M.Gelfand, M.I.Graev, N.J.Vilenkin [2]

A quantum Lorentz group considered here means the quantum deformation of the Lorentz group described in [4] corresponding to a fixed value of the deformation parameter $\mu = q \in]0,1[$. The group will be denoted by *QLG*.

In the first part we describe all irreducible unitary representations of *QLG*. They split into principal and two complementary series. Beside that we have two 1-dimensional representations including the trivial one.

In the second part we investigate a large family of (not necessarily unitary) representations of *QLG* induced by 1-dimensional representations of the parabolic subgroup $P \subset QLG$ consisting of all upper-triangular matrices. The representations act on the space of smooth sections of (quantum) line boundles over the homogeneous space $P \setminus QLG$. This spaces denoted by D_χ (where χ runs over the set of 1-dimensional representations of P) play the fundamental role in [2]. We call them Gelfand spaces.

A deeper investigation (with the technique of invariant bilinear forms) of D_χ shows that in principle all results concerning the classical Lorentz group contained in Chapter 3 of [2] remains in power in the quantum case. In particular the conditions distinguishing unitary representations are of the same form and lead in a natural way to principal and complementary series.

The difference between the classical and quantum case consists in a slightly different topological structure of the set of 1-dimensional representations of the group P. In the classical case this representations are labeled by pairs (n_1, n_2) where $n_1, n_2 \in \mathbf{C}$ with $n_1 - n_2 \in \mathbf{Z}$ and the different pairs correspond to the different representations.

In the quantum case the correspondence between the pairs (n_1, n_2) and the representations of P is no longer one-to-one: the pairs (n_1, n_2) and $(n_1 + \frac{2\pi i}{\log q}, n_2 + \frac{2\pi i}{\log q})$ (where

E. A. Tanner and R. Wilson (eds.), Noncompact Lie Groups and Some of Their Applications, 453–472.

q is the deformation parameter) gives rise to the same representations of P.

As we shall see the above difference between the classical and the quantum case explains in a simple way all the suprising features of the theory of unitary representations of quantum Lorentz group such as the new topological structure of the principal series, the existence of two (instead of one) complementary series and the existence of non-trivial 1-dimensional representation.

2 Quantum groups and their representations

To introduce the basic notions of the representation theory of quantum groups we consider at first the classical case.

Let G be a compact group, $A = C(G)$ be the C^*-algebra of continuous functions on G. The group structure of G is encoded in the comultiplication $\Delta \in \mathrm{Mor}(A, A \otimes A)$ introduced by

$$(\Delta a)(g, g') = a(gg')$$

where a is a continuous function on G and g, g' runs over G.

In the compact case it is sufficient to consider finite dimensional representations. Let H be a finite dimensional Hilbert space. We have natural bijections:

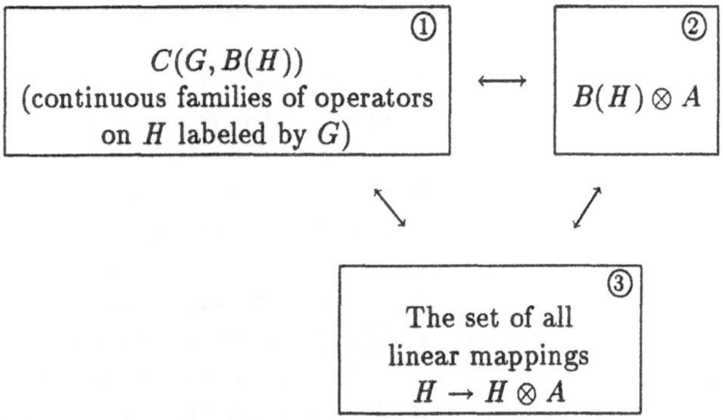

In what follows for any linear $v: H \to H \otimes A$ we shall use the same letter to denote the corresponding elements of $B(H) \otimes A$ and $C(G, B(H))$.

One can easily verify that

① $v \in C(G, B(H))$, $v(gg') = v(g)v(g')$ $\Longleftrightarrow$ ② $v \in B(H) \otimes A$, $(\mathrm{id} \otimes \Delta)v = v_{12}v_{13}$

⇘ ⇗

③ The diagram

$$\begin{array}{ccc} H & \xrightarrow{\ v\ } & H \otimes A \\ {\scriptstyle v}\downarrow & & \downarrow{\scriptstyle \mathrm{id} \otimes \Delta} \\ H \otimes A & \xrightarrow{\ v \otimes \mathrm{id}\ } & H \otimes A \otimes A \end{array}$$

is commutative

In ② we used leg numbering notation [4]: if $v = \sum m_i \otimes a_i$ then

$$v_{12} = \sum m_i \otimes a_i \otimes I$$

$$v_{13} = \sum m_i \otimes I \otimes a_i$$

$$v_{23} = \sum I \otimes m_i \otimes a_i$$

We refer to ② saying that v is a representation of G in H and to ③ saying that v is an action of G on H. For example one may consider Δ as an action of G on A. More generally if D is a finite dimensional vector subspace of A invariant under the right shifts, then:

$$\Delta : D \longrightarrow D \otimes A$$

and $v = \Delta|_D$ is an action of G on D. This kind of action we meet in the theory of induced representations.

The non-compact case is more complicated. Let G be a non-compact group. In this case we can associate with G different algebras which are the same in the compact case: $C_\infty(G)$ - the (non-unital) C^* - algebra of continuous functions on G tending to 0 at infinity, $C_{\text{bounded}}(G)$ - the (unital) C^* - algebra of bounded continuous functions on G and $C(G)$ - the *-algebra of continuous functions on G.

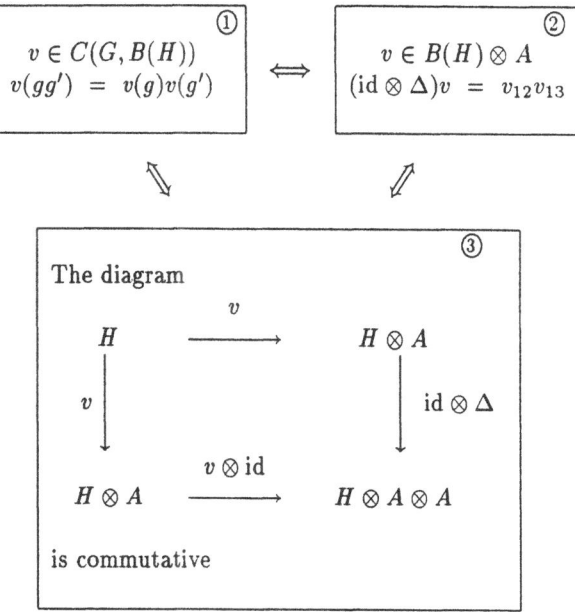

As the basic algebra related to G we take $A = C_\infty(G)$. Due to the famous Gelfand-Naimark theorem A contains the full information of G as a topological locally compact space. In particular the other algebras such as $C_{\text{bounded}}(G)$ and $C(G)$ can be reconstructed in a purely algebraic way once $C_\infty(G)$ is given. We have:

$$C_{\text{bounded}}(G) = M(C_\infty(G))$$

$$C(G) = C_\infty(G)^\eta$$

where for any C^* - algebra A, $M(A)$ is the multiplier algebra and A^η is the set of all elements affiliated with A.

Let us recall that for any (non-unital) C^* - algebra A

$$M(A) = \begin{array}{c}\text{The largest } C^*\text{-algebra that}\\ \text{contains } A \text{ as a separating ideal}\end{array}$$

It is always the unital C^* - algebra. If $A \subset B(H)$ (non-degenerated embeding) then

$$M(A) = \{b \in B(H) : ba \in A, ab \in A \text{ for all } a \in A\}$$

It is clear that for the unital C^* - algebra A : $M(A) = A$ and if $A = CB(H)$ is the C^* - algebra of compact operators on H then $M(CB(H)) = B(H)$. The reader should notice that $M(C_\infty(G)) = C_{\text{bounded}}(G) = C(\overline{G})$ where $\overline{G}$ is the Čech-Stone compactification of G. The multiplier functor M is an algebraic counterpart of the Čech-Stone compactification of locally compact spaces.

We have not enough time to explain the notion of affiliated elements. The affiliation relation is denoted by η :

$$T\eta A \Longleftrightarrow T \in A^\eta$$

In any case $A^\eta \supset M(A)$. Elements of A^η may be regarded as unbounded multipliers acting on A [7]. It turns out that $C_\infty(G)^\eta = C(G)$ and

$$CB(H)^\eta = \begin{array}{c}\text{The set of all closed}\\ \text{operators on } H\end{array}$$

The later example shows that in general A^η is not even a vector space. If A is unital then $A^\eta = A$.

In many cases we reconstruct an algebra A from a given sets of affiliated elements. One says that $A = C^*(\alpha, \beta, \gamma, \delta, \ldots)$ is generated by $\alpha, \beta, \gamma, \delta, \ldots$ if A is in a sense the smallest C^* - algebra such that $\alpha, \beta, \gamma, \delta, \ldots \, \eta A$.

For example if $H = L^2(R)$, $\hat{x}$ = multiplication by x and $\hat{p} = \frac{1}{i}\frac{d}{dx}$ then the C^* - algebra generated by $\hat{x}$ and $\hat{p}$ coincides with $A = CB(H)$.

For two C^* - algebras A, B we shall denote by $\mathrm{Mor}(A, B)$ the space of morphisms from A to B : $\Phi \in \mathrm{Mor}(A, B)$ means that Φ is a non-degenerated *-algebra homomorphism of A into $M(B)$. The nondegeneracy means that $\Phi(A)B$ is dense in B. Clearly $\mathrm{Mor}(A, CB(H))$ $= \mathrm{Rep}(A, H)$ where $\mathrm{Rep}(A, H)$ is the set of all non-degenerate representations of A in H.

This notion of a morphism is a generalization of a morphism in the category of commutative C^* - algebras and corresponds to a continuous map between locally compact spaces. Any $\Phi \in \mathrm{Mor}(A, B)$ can be extended in the canonical way to $M(A)$ and A^η :

$$\begin{array}{rcl}\Phi : M(A) & \longrightarrow & M(B)\\ \Phi : A^\eta & \longrightarrow & B^\eta\end{array}$$

In the case of non-compact groups one has to consider representations in infinite dimensional Hilbert spaces. Let H be a seperable Hilbert space. Then the correspondence ② $\Longleftrightarrow$ ③ becomes more complicated and reflects different continuity properties of infinite-dimensional representations:

v belongs to $\Longleftrightarrow$	$v : G \longrightarrow B(H)$ is
$B(H) \otimes A$	norm continuous and $v(g) \to 0$ for $g \to \infty$
$B(H) \otimes M(A)$	norm continuous, bounded and $v(G)$ is almost finite-dimensional
$M(B(H) \otimes A)$	norm continuous and bounded
$M(CB(H) \otimes A)$	*-strong continuous and bounded

In the quantum group case the algebra $A = C_\infty(G)$ is no longer commutative. To introduce a quantum group G one has to fix a C^* - algebra A and a coassociative morphism $\Delta \in \mathrm{Mor}(A, A \otimes A)$. Elements of A may be considered as "continuous functions vanishing at infinity" on non-compact quantum space G whereas Δ encodes the group structure of G. In brief we write $G = (A, \Delta)$. This point of view is sufficient for the purposes of the present paper. We shall deal only with a few concrete quantum groups not entering the general theory so there is no necessity to present a formal definition of a quantum group.

Considerations presented above lead to the following notion of unitary strongly continuous representation of a quantum grup G.

Definition 1 *Let $G = (A, \Delta)$ be a quantum group and H be a separable Hilbert space. We say that v is a unitary representation of G in H if v is an unitary element of $M(CB(H) \otimes A)$ and*

$$(\mathrm{id} \otimes \Delta)v = v_{12}v_{13}$$

3 Quantum SU(2) group and its representation theory

The quantum Lorentz group contains the quantum $SU(2)$ group and its quantum Pontryagin dual group as subgroups. This fact gives a deeper insight into the strucure of the quantum Lorentz group so we consider these subgroups first.

Let q be a fixed number of the interval]0,1[and A_c be the C^* -algebra generated by two elements α_c, and γ_c satisfying the well known commutation relations

$$\left.\begin{array}{c} \alpha_c^*\alpha_c + \gamma_c^*\gamma_c = I, \qquad \alpha_c\alpha_c^* + q^2\gamma_c^*\gamma_c = I, \\ \alpha_c\gamma_c = q\gamma_c\alpha_c, \qquad \alpha_c\gamma_c^* = q\gamma_c^*\alpha_c, \qquad \gamma_c\gamma_c^* = \gamma_c^*\gamma_c \end{array}\right\} \tag{3.1}$$

The quantum $SU(2)$ group is by definition

$$S_qU(2) = (A_c, \Delta_c)$$

where $\Delta_c \in \mathrm{Mor}(A_c, A_c \otimes A_c)$ is uniquely defined by its values on the generators

$$\begin{aligned} \Delta_c(\alpha_c) &= \alpha_c \otimes \alpha_c - q\gamma_c^* \otimes \gamma_c \\ \Delta_c(\gamma_c) &= \gamma_c \otimes \alpha_c + \alpha_c^* \otimes \gamma_c \end{aligned}$$

According to (3.1) the elements α_c and γ_c are bounded. Therefore A_c is unital and $S_qU(2)$ is compact.

Let

$$\mathcal{A}_c = \mathrm{Pol}(\alpha_c, \gamma_c, \alpha_c^*, \gamma_c^*)$$

be the smallest *-subalgebra of A_c containing α_c and γ_c. Then $\mathcal{A}_c$ is dense in A_c and the set $\{\alpha_{ck}\gamma_c^m\gamma_c^{*n} \; : \; k \in \mathbf{Z}, m, n = 0, 1, 2 \ldots\}$ where

$$\alpha_{ck} = \begin{cases} \alpha_c^k & \text{for } k \geq 0 \\ (\alpha_c^*)^{-k} & \text{for } k \leq 0 \end{cases}$$

is a linear basis in $\mathcal{A}_c$.

We consider linear functionals

$$f_0, \; f_+, \; f_- : \mathcal{A}_c \longrightarrow \mathbf{C}$$

defined on the elemets of the basis in the following way:

$$f_0(\alpha_{ck}\gamma_c^m\gamma_c^{*n}) = \begin{cases} q^{-\frac{k}{2}} & \text{for } m = n = 0 \\ 0 & \text{otherwise} \end{cases}$$

$$f_+(\alpha_{ck}\gamma_c^m\gamma_c^{*n}) = \begin{cases} q^{\frac{k}{2}} & \text{for } m = 1, n = 0 \\ 0 & \text{otherwise} \end{cases}$$

$$f_-(\alpha_{ck}\gamma_c^m\gamma_c^{*n}) = \begin{cases} -q^{\frac{k-2}{2}} & \text{for } m = 0, n = 1 \\ 0 & \text{otherwise} \end{cases}$$

Let v_c be an unitary representation of $S_qU(2)$ acting on a finite-dimensional Hilbert space H. Then $v_c \in B(H) \otimes \mathcal{A}_c$ and setting

$$\left.\begin{array}{rcl} q^{J_3} &=& (\mathrm{id} \otimes f_0)v_c \\ J_+ &=& (\mathrm{id} \otimes f_+)v_c \\ J_- &=& (\mathrm{id} \otimes f_-)v_c \end{array}\right\} \tag{3.2}$$

we introduce three operators q^{J_3}, J_+, J_- acting on H. They satisfy the commutation relations:

$$\left.\begin{array}{ll} q^{J_3}J_+ = qJ_+q^{J_3}, & q^{J_3}J_- = q^{-1}J_-q^{J_3}, \\ [J_+, J_-] = \dfrac{q^{-2J_3} - q^{2J_3}}{q^{-1} - q}, & \\ (J_+)^* = J_-, & q^{J_3} > 0 \end{array}\right\} \tag{3.3}$$

Any strongly continuous unitary representation v_c of $S_qU(2)$ acting on a (infinite - dimensional) Hilbert space H is a direct sum of irreducible finite dimensional representations. In this case (3.2) are (in general unbounded) closed operators acting on H. They have a common invariant dense essential domain (a core) consisting of vectors belonging to finite

- dimensional v_c - invariant subspaces of H.

The set of irreducible representations of $S_qU(2)$ is labeled by spin parameter $s = 0, 1/2, 1, 3/2, \ldots$. Let s be one of this number. The corresponding unitary representation denoted by u^s acts on (2s+1)-dimensional Hilbert space H^s : $u^s \in B(H^s) \otimes \mathcal{A}_c$. In this case the operators (3.2) are denoted by $q^{J_3^s}, J_+^s, J_-^s \in B(H^s)$. For example:

$$u^{1/2} = \begin{pmatrix} \alpha_c & -q\gamma_c^* \\ \gamma_c & \alpha_c^* \end{pmatrix}$$

and

$$q^{J_3^{1/2}} = \begin{pmatrix} q^{-1/2} & 0 \\ 0 & q^{1/2} \end{pmatrix} \quad J_+^{1/2} = \begin{pmatrix} 0 & 0 \\ 1 & 0 \end{pmatrix} \quad J_-^{1/2} = \begin{pmatrix} 0 & 1 \\ 0 & 0 \end{pmatrix}$$

To introduce the Pontryagin dual of $S_qU(2)$ we consider the C^* - algebra

$$A_d = \sum_s{}^{\oplus} B(H^s).$$

Let π^s be the canonical projection $\pi^s \in \mathrm{Mor}(A_d, B(H^s))$. Any element $a \,\eta\, A_d$ is uniquely determined by sequence $(\pi^s(a))_{s=0,1/2,1,\ldots}$. Any sequence $(a_s)_{s=0,1/2,1,\ldots}$ where $a_s \in B(H^s)$ can be obtained in this way. An element a belongs to A_d ($M(A_d)$ respectively) if $\|\pi^s(a)\|$ goes to 0 for $s \to \infty$ (is bounded respectively). The reader should notice that in this case A_d^η carries a natural *-algebra structure.

Let

$$u = \sum_s{}^{\oplus} u^s$$

and

$$q^{J_{d3}} = \sum_s{}^{\oplus} q^{J_3^s}, \qquad J_{d+} = \sum_s{}^{\oplus} J_+^s, \qquad J_{d-} = \sum_s{}^{\oplus} J_-^s$$

Then $u \in M(A_d \otimes A_c)$ and $q^{J_{d3}}, J_{d+}, J_{d-}$ are unbounded elements affiliated with A_d satisfying relations (3.3). One can show that

$$A_d = C^*(q^{J_{d3}}, J_{d+}, J_{d-}).$$

Moreover there exists one and only one $\Delta \in \mathrm{Mor}(A_d, A_d \otimes A_d)$ such that

$$\Delta_d(J_{d\pm}) = q^{J_{d3}} \otimes J_{d\pm} + J_{d\pm} \otimes q^{-J_{d3}}$$

$$\Delta_d(q^{J_{d3}}) = q^{J_{d3}} \otimes q^{J_{d3}}$$

Δ_d is coassociative and the quantum Pontryagin dual $S_qU(2)$ group is:

$$\widehat{S_qU(2)} = (A_d, \Delta_d)$$

To explain why we refer to the Pontryagin duality let us notice that

$$\begin{aligned}(\mathrm{id} \otimes \Delta_c)u &= u_{12}u_{13} \\ (\Delta_d \otimes \mathrm{id})u &= u_{23}u_{13}\end{aligned}$$

This relation expresses the bicharacter property of u : u is a representation of $S_qU(2)$ and u^{-1} is a representation of $\widehat{S_qU(2)}$. The bicharacter u plays the same role in representation theories of $S_qU(2)$ and $\widehat{S_qU(2)}$ as a bicharacter e^{ipx} in the representation theories of $\mathbf{R}$ and its Pontryagin dual group $\widehat{\mathbf{R}} = \mathbf{R}$.

Using this property one can prove a duality theorem [5]:

Theorem 2

$$\left(\begin{array}{c} v_c \in M(CB(H)\otimes A_c) \\ \text{is a unitary representation} \\ \text{of } S_qU(2) \end{array}\right) \iff \left(\begin{array}{c} v_c \;=\; (\psi_d \otimes \mathrm{id})u \\ \text{where} \\ \psi_d \in \mathrm{Rep}(A_d, H) \end{array}\right)$$

$$\left(\begin{array}{c} v_d \in M(CB(H)\otimes A_d) \\ \text{is a unitary representation} \\ \text{of } \widehat{S_qU(2)} \end{array}\right) \iff \left(\begin{array}{c} v_d \;=\; (\psi_c \otimes \mathrm{id})\tau(u^{-1}) \\ \text{where } \psi_c \in \mathrm{Rep}(A_c, H) \\ \text{and } \tau \text{ is a flip :} \\ A_d \otimes A_c \rightarrow A_c \otimes A_d \end{array}\right)$$

It means that $S_qU(2)$ and $\widehat{S_qU(2)}$ are mutually Pontryagin dual groups to each other and there is one-to-one correspondence between strongly continuous unitary representations of the group and nondegenerate representations of "the algebra of functions" on the dual group. This dual approach in the representation theory of quantum groups is quite natural and corresponds to the Lie algebra approach in the Lie group representation theory. A description of a representation of an algebra is often simpler and more convenient then a description of a group action. For example to introduce $\psi_c \in \mathrm{Rep}(A_c, H)$ it is enough to fix two operators $\alpha_c, \gamma_c \in B(H)$ satisfying (3.1) Similarly to given $\psi_d \in \mathrm{Rep}(A_d, H)$ there correspond operators q^{J_3}, J_+, J_- acting in H and satisfying (3.3).

4 QLG and its irreducible unitary representations

To consider the quantum Lorentz group

$$QLG \;=\; (A, \Delta)$$

we have to describe C^* - algebra A and a comultiplication Δ.

We fix $q \in]0,1[$ and let A be a (non-unital) C^* - algebra generated by four unbounded elements α, β, γ and δ satisfying the following 17 relations proposed by Podleś:

$$
\begin{aligned}
&\alpha\beta = q\beta\alpha, && \alpha\gamma = q\gamma\alpha, && \alpha\delta - q\beta\gamma = I,\\
&\beta\delta = q\delta\beta, && \beta\gamma = \gamma\beta, && \delta\alpha - q^{-1}\beta\gamma = I,\\
&\gamma\delta = q\delta\gamma, && && \\
&\alpha\alpha^* = \alpha^*\alpha + (1-q^2)\gamma^*\gamma, && && \gamma\alpha^* = q\alpha^*\gamma,\\
&\beta\alpha^* = q^{-1}\alpha^*\beta + q^{-1}(1-q^2)\gamma^*\delta, && && \delta\alpha^* = \alpha^*\delta,\\
&\beta\beta^* = \beta^*\beta + (1-q^2)[\delta^*\delta - \alpha^*\alpha] - (1-q^2)^2\gamma^*\gamma, && && \\
&\delta\beta^* = q\beta^*\delta - q(1-q^2)\alpha^*\gamma, && && \gamma\beta^* = \beta^*\gamma,\\
&\delta\delta^* = \delta^*\delta - (1-q^2)\gamma^*\gamma, && && \gamma\gamma^* = \gamma^*\gamma,\\
&\delta\gamma^* = q^{-1}\gamma^*\delta. && &&
\end{aligned}
\tag{4.1}
$$

One can prove that there is the unique morphism $\Delta \in \mathrm{Mor}(A, A\otimes A)$ such that

$$
\begin{aligned}
\Delta(\alpha) &= \alpha\otimes\alpha + \beta\otimes\gamma, & \Delta(\beta) &= \alpha\otimes\beta + \beta\otimes\delta,\\
\Delta(\gamma) &= \gamma\otimes\alpha + \delta\otimes\gamma, & \Delta(\delta) &= \gamma\otimes\beta + \delta\otimes\delta.
\end{aligned}
$$

This morphism is coassociative and encodes a group structure in *QLG*.

The above commutation relations are complicated. Fortunately it was realized that any matrix $\begin{pmatrix} \alpha, & \beta \\ \gamma, & \delta \end{pmatrix}$ where $\alpha, \beta, \gamma, \delta$ are operators in a Hilbert space satisfying this relations is of the form

$$
\begin{pmatrix} \alpha, & \beta \\ \gamma, & \delta \end{pmatrix} = \begin{pmatrix} \alpha_c, & -q\gamma_c^* \\ \gamma_c, & \alpha_c^* \end{pmatrix} \begin{pmatrix} q^{J_3}, & (1-q^2)q^{-1/2}J_+ \\ 0, & q^{-J_3} \end{pmatrix}
\tag{4.2}
$$

(and a similar formula for adjoints) where operators α_c, γ_c satisfy (3.1) and q^{J_3}, J_+, satisfy (3.3). Moreover any operator from the set $\{\alpha_c, \gamma_c\}$ commutes with any operator from the set $\{q^{J_3}, J_+, J_-\}$. Above formula is a quantum version of the Iwasawa decomposition of the classical Lorentz group. It shows that

$$
A = A_c \otimes A_d
$$

Let

$$
p_c = \mathrm{id}\otimes e_d, \qquad p_d = e_c\otimes \mathrm{id}
$$

where $e_c \in \mathrm{Mor}(A_c, \mathbf{C})$, $e_d \in \mathrm{Mor}(A_d, \mathbf{C})$ are counits of $S_qU(2)$ and $S_q\hat{U}(2)$ respectively (e_c, e_d are the unique morphisms such that $e_c(\alpha) = 1$, $e_c(\gamma) = 0$; $e_d(q^{J_3}) = 1$, $e_d(J_\pm) = 0$). Then

$$
p_c \in \mathrm{Mor}(A, A_c) \qquad p_d \in \mathrm{Mor}(A, A_d)
$$

and they correspond to embeddings

$$
S_qU(2) \longrightarrow QLG, \qquad S_q\widehat{U(2)} \longrightarrow QLG
$$

One can check that

$$
\Delta_c p_c = (p_c\otimes p_c)\Delta \qquad \Delta_d p_d = (p_d\otimes p_d)\Delta
$$

This means that $S_qU(2)$ and $S_q\widehat{U(2)}$ are subgroups of *QLG* : $\Delta\,|_{S_qU(2)} = \Delta_c$ and $\Delta\,|_{S_q\widehat{U(2)}} = \Delta_d$.

The group structure of *QLG* can be reproduced from that of $S_qU(2)$ and $S_q\widehat{U(2)}$. Let $\sigma \in \mathrm{Mor}(A_c \otimes A_d, A_d \otimes A_c)$ be given by

$$\sigma(a \otimes x) = u(x \otimes a)u^{-1}$$

where as before u is a bicharacter $u = \sum_s^{\oplus} u^s$. Then

$$\Delta = (\mathrm{id} \otimes \sigma \otimes \mathrm{id})(\Delta_c \otimes \Delta_d)$$

Summarizing: the quantum Lorentz group

$$QLG = (A_c \otimes A_d, \ (\mathrm{id} \otimes \sigma \otimes \mathrm{id})(\Delta_c \otimes \Delta_d)).$$

This fact is of great importance and simplifies the study of the representation theory for *QLG*. Any representation v of *QLG* can be described as the pair (v_c, v_d) of representations $S_qU(2)$ and $S_q\widehat{U(2)}$ respectively acting in the same space and satisfying a compatibility condition.

Let

$$v \in \mathrm{M}(CB(H) \otimes A)$$

be an unitary representation of the quantum Lorentz group *QLG* acting in a Hilbert space H and let

$$v_c = v\,|_{S_qU(2)} := (\mathrm{id} \otimes p_c)v$$

$$v_d = v\,|_{S_q\widehat{U(2)}} := (\mathrm{id} \otimes p_d)v$$

Then $v_c \in \mathrm{M}(CB(H) \otimes A_c)$ and $v_d \in \mathrm{M}(CB(H) \otimes A_d)$ are unitary representations of $S_qU(2)$ and $S_q\widehat{U(2)}$ respectively

$$v = (v_c)_{12}(v_d)_{13}$$

and

$$(v_d)_{12}(v_c)_{13} = (\mathrm{id} \otimes \sigma)(v_c)_{12}(v_d)_{13}$$

Conversely, if unitary representations v_c, v_d acting in the same Hilbert space satisfy the last condition then $v := (v_c)_{12}(v_d)_{13}$ is an unitary representation of *QLG*. We shall refer to this as the compatibility condition.

Now we can associate with v two sets of operators $\{J_+, J_-, q^{J_3}\}$, $\{\alpha_c, \gamma_c\}$ acting in H via the correspondence:

$v_c = v\,\vert_{S_qU(2)}$ representation of $S_qU(2)$	$\Longleftrightarrow$	$\psi_d \in \mathrm{Rep}(A_d, H)$	$\Longleftrightarrow$	operators J_+, J_-, q^{J_3} satisfying (3.3)	
					(4.3)
$v_d = v\,\vert_{S_q\widehat{U(2)}}$ representation of $S_q\widehat{U(2)}$	$\Longleftrightarrow$	$\psi_c \in \mathrm{Rep}(A_c, H)$	$\Longleftrightarrow$	operators $\alpha_c, \gamma_c, \alpha_c^*, \gamma_c^*$ satisfying (3.1)	

The compatibility condition in terms of this operators means:

$$\left.\begin{array}{ll} q^{J_3}\alpha_c = \alpha_c q^{J_3}, & q^{J_3}\gamma_c = q^{-1}\gamma_c q^{J_3}, \\ J_+\alpha_c = q\alpha_c J_+ - q^{\frac{3}{2}}\gamma_c^* q^{J_3}, & J_+\gamma_c = \gamma_c J_+ + q^{-\frac{1}{2}}(\alpha_c^* q^{J_3} - \alpha_c q^{-J_3}), \\ J_+\alpha_c^* = q^{-1}\alpha_c^* J_+ + q^{-\frac{1}{2}}\gamma_c^* q^{-J_3}, & J_+\gamma_c^* = \gamma_c^* J_+ \end{array}\right\} \quad (4.4)$$

These relations (as well as the relations (3.3)) have to be supplemented by regularity conditions (like the famous integrability condition of Nelson for Lie algebra relations) stating the existence of sufficiently large invariant domain on which the relations hold. The regularity conditions give the precise meaning to the commutation relations involving unbounded operators. In our case the regularity conditions (as well as the relations themselves) follow from the fact that the considered operators $(J_+, J_-, q^{J_3}, \alpha_c, \gamma_c)$ are related to a unitary representation of the quantum Lorentz group. We have no time to formulate these conditions explicitely. It should be stressed however that they play the essential role in our analysis. This analysis leads to the complete classification of all irreducible representations of *QLG*. In what follows we briefly present the results.

As we know representation v restricted to $S_qU(2)$ is a representation v_c. Any such a representation is a direct sum of irreducible ones. Let Sp v be the spin spectrum of v : a (half-) integer $s \in \mathrm{Sp}\, v$ if and only if u^s is contained in v_c, and p be the minimal element of Sp v. Assume that v is irreducible. Then for any $s \in \mathrm{Sp}\, v$, the multiplicity of u^s in v_c is 1:

$$H = \sum_{s \in \mathrm{Sp} v} \oplus H^s$$

Any H^s is a v_c - invariant subspace in H. Therefore

$$J_+, J_-, q^{J_3} \ : \ H^s \longrightarrow H^s$$

and the action of this operators is well known. It turns out that operators

$$\begin{array}{l} \alpha_c, \gamma_c \\ \alpha_c^*, \gamma_c^* \end{array} \ : H^s \longrightarrow H^{s-1} \oplus H^s \oplus H^{s+1}$$

(H^{s-1} does not exist for $p = 0$).
Using commutation relations (3.1) , (3.3) and (4.4) one can show that the Casimir operator:

$$C(v) = q^{-1/2}(1 - q^2)\gamma_c J_+ - \alpha_c^* q^{J_3+1} - \alpha_c q^{-J_3-1} \quad (4.5)$$

commutes with all operators related to v. In the irreducible case

$$C(v) = c(v)I$$

where $c(v)$ is a (complex) eigenvalue of $C(v)$.

The eigenvalue of $C(v)$ together with Sp v completely determines (up to a unitary equivalence) an irreducible representation v. The table below presents all the possibilities:

Table 1:

Sp v	The eigenvalue of $C(v)$	Remarks
$\{0\}$	$c = \pm(q+q^{-1})$	1- dimensional representations. Sign "-" corresponds to the trivial one
$\{p, p+1, p+2, \ldots\}$ p - positive (half-)integer	$\mid c-2 \mid + \mid c+2 \mid = 2(q^p + q^{-p})$	Principal series
$\{0, 1, 2, \ldots\}$	$-r < c < r$ where $r = q + q^{-1}$	There are three cases: 1. $c \in [-2, 2]$ Principal series with $p = 0$ 2. $c \in]-(q+q^{-1}), -2[$ 3. $c \in]2, (q+q^{-1})[$ Two complementary series

In Fig.1 we showed the admissible values of $c(v)$ described in Table 1. We see that the values corresponding to the principal series belong to ellipses with common focuses located at points -2 and 2. The size of the ellipses depends on initial spin p. For $p = 0$ the ellipse degenerates to the interval [-2,2]. Besides the principal series we have two complementary

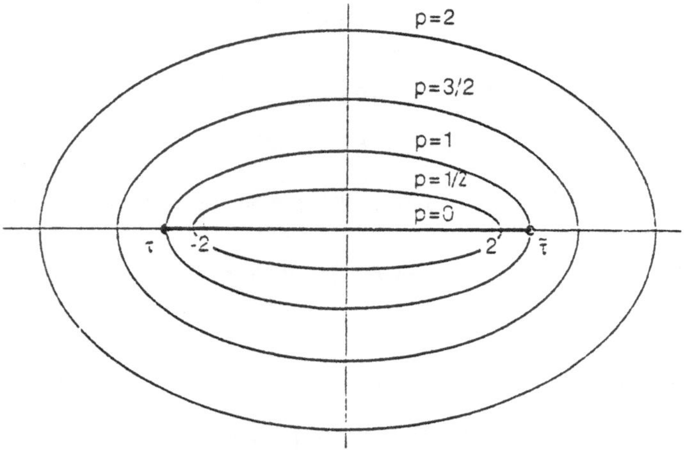

Figure 1.

series corresponding to intervals $]-(q+q^{-1}), -2[$ and $]2, q+q^{-1}[$. For these series $p = 0$. The two 1-dimensional representations τ and $\tilde{\tau}$ corresponds to points $\mp(q+q^{-1})$.

To compare this result with the representation theory of classical Lorentz group we use selfadjoint Casimir operators Δ and Δ' considered in [3, p.167 and statement on p.187]. The eigenvalues of Δ and Δ' together with the spin spectrum completely determine an irreducible unitary representation v of the classical Lorentz group. Table 2 shows all the possibilities.

Table 2:

Sp v	The eigenvalue d of $\Delta + i\Delta'$	Remarks
$\{0\}$	$d = 0$	1- dimensional trivial representation.
$\{p, p+1, p+2, \ldots\}$ p - positive (half-)integer	$\mid d-2 \mid - \mathrm{Re}\, d = 2(2p^2 - 1)$	Principal series
$\{0, 1, 2, \ldots\}$	$0 < d$	There are two cases: 1. $d \in [2, \infty[$ Principal series with $p = 0$ 2. $d \in]0, 2[$ Complementary series

The admissible values of eigenvalues of $\Delta + i\Delta'$ are presented on Fig.2. In this case the values corresponding to principal series belong to parabolas with common focus at the point 2 and directrices depending on the minimal spin p. For $p = 0$ the parabola degenerates to a half-line $[2, \infty[$.

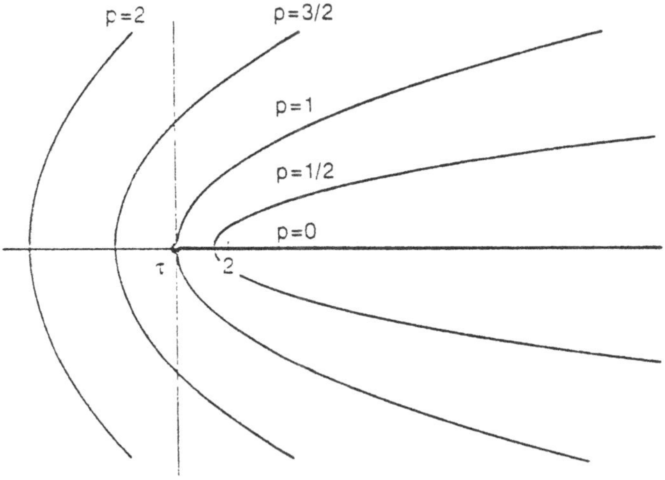

Figure 2.

There is only one complementary series. It corresponds to the interval $]0, 2[$ and as before $p = 0$ is the associated minimal spin. There is also only one 1-dimensional representation - the trivial one τ. It corresponds to the point 0.

To analyse the classical limit $q \to 1$ one should consider the rescaled quantum Casimir operator

$$C' = 2\frac{C + q + q^{-1}}{q - 2 + q^{-1}}$$

The reader easily verify that with this transformation the ellipses of Fig.1 tend (as $q \to 1$) to parabolas of Fig.2. In particular the degenerated ellipse transforms onto the degenerated parabola. The interval $]-(q+q^{-1}),-2[$ corresponding to the first complementary series transforms onto the interval $]0,2[$. The interval corresponding to the second complementary series is moved onto $]2+t_q, 4+t_q[$ where $t_q = 8(q+q^{-1}-2)^{-1}$. For $q \to 1$, the interval is shifted to infinity and the corresponding complementary series disappears. The same holds for the additional 1-dimensional representation.

5 Gelfand spaces and induced action of QLG

The methods used in the previous sections permit to describe all irreducible representations of the quantum Lorentz group *QLG* in an implicit way. Within this approach one can only derive formulae which show how the generators $\alpha_c, \gamma_c, q^{J_3}, J_+, J_-$ of the representation act on some basic vectors of the carrier Hilbert space. We have no explicit expressions descibing the action of *QLG* itself on that space. In the classical case such expressions can be obtained by realization of the carrier space as a space of functions (satisfying some conditions as e.g. a homogeneity conditions) on some G-manifolds. In many cases it is convenient to impose also some regularity conditions on considered functions: working with nuclear spaces of smooth functions one may use very powerful methods of distribution theory. This technique for the classical Lorentz group was proposed in [2]. The deeper analysis shows that in effect it consists in inducing representations of the Lorentz group from 1-dimensional representations of its parabolic subgroup.

In this section we try to mimic this approach in the quantum case. It turns out that in this context one has to consider also non-unitary representations of *QLG* with no hilbertian structure on the underlying vector space. To this aim it is necessary to generalize the framework presented before.

First of all let us notice that the notion of representation as introduced in Definition 1 is formulated in the C^* -algebra language and is not applicable if H is not endowed with a Hilbert space structure. Instead we shall use the concept of action of *QLG* on a vector space which better suits. For the purpose of our paper it is sufficient to consider right invariant subspaces of $\mathcal{A}$ on which *QLG* acts by right shifts. In this sentence $\mathcal{A}$ denotes the space of all "smooth elements" affiliated with A (the precise definition is given later). $\mathcal{A}$ is a *-subalgebra of A^η. It turns out that $\Delta(\mathcal{A}) \subset \mathcal{A}\hat{\otimes}\mathcal{A}$, where $\hat{\otimes}$ denotes the algebraic tensor product followed by a suitable completion. $D \subset \mathcal{A}$ is right invariant if $\Delta(D) \subset D\hat{\otimes}\mathcal{A}$.

Working with a non-unitary representations of the quantum Lorentz group one may also use the results contained in Sections 2 and 3. One should notice however that in non-unitary case the representations ψ_c and ψ_d are not *-homomorphisms (after all the representation space is not endowed with an invariant scalar product and there is no natural *-involution in the space of operators). Therefore in the sequence of operators $\alpha_c, \alpha_c^*, \gamma_c, \gamma_c^*, J_+, J_-, q^{J_3}$ constructed out of considered representation of *QLG* in the way described in (4.3), $\alpha_c^*(\gamma_c^*, J_-, q^{J_3}$ respectively) is no longer related by a hermitian conjugation to $\alpha_c(\gamma_c, J_+, q^{J_3}$ respectively). $\alpha_c, \alpha_c^*, \gamma_c, \gamma_c^*, J_+, J_-, q^{J_3}$ should be treated as independent variables subjected to the relations (3.1), (3.3) (except the last row) and (4.4) suplemented by their formal hermitian conjugation: For example the relation $\alpha_c\gamma_c = q\gamma_c\alpha_c$ should be suplemented by $\gamma_c^*\alpha_c^* = q\alpha_c^*\gamma_c^*$.

The formula (4.5) expressing the Casimir operator remains in power.

We briefly describe the construction of induced representation in quantum case.

To introduce the parabolic subgroup $P = (A_P, \Delta_P)$ one has to complete the set of Podleś relations (4.1) adding the relation "$\gamma = 0$" (we remind that in the classical case P consists of all upper-triangular matrices belonging to $SL(2,\mathbf{C})$). The corresponding C^* - algebra will be denoted by A_P. By definition

$$A_P = A/I_\gamma$$

where I_γ is the closed two sided ideal of A generated by γ.

Let $\pi \in \mathrm{Mor}(A, A_P)$ be the canonical epimorphism and $\dot\alpha, \dot\beta, \dot\gamma, \dot\delta\ \eta\ A_P$ be the elements corresponding to $\alpha, \beta, \gamma, \delta\ \eta\ A$: $\dot\alpha = \pi(\alpha)$ and so on. Then $\dot\gamma = 0$ and

$$\begin{array}{ll}
\dot\alpha, \dot\delta \text{ - normal} & \\
\dot\alpha\dot\beta = q\dot\beta\dot\alpha, & \dot\beta\dot\alpha^* = q^{-1}\dot\alpha^*\dot\beta, \\
\dot\beta\dot\delta = q\dot\delta\dot\beta, & \dot\delta\dot\alpha^* = \dot\alpha^*\dot\delta, \\
\dot\alpha\dot\delta = I = \dot\delta\dot\alpha, & \dot\delta\dot\beta^* = q\dot\beta^*\dot\delta, \\
\dot\beta\dot\beta^* = \dot\beta^*\dot\beta + (1-q^2)(\dot\delta^*\dot\delta - \dot\alpha^*\dot\alpha). &
\end{array}$$

Applying π to the matrix elements of (4.2) we see that $\pi(\gamma_c) = 0$, $\pi(\alpha_c^*)$ is unitary and $\dot\delta = \pi(\delta) = \pi(\alpha_c^*)\pi(q^{-J_3})$. Therefore $\dot\delta^*\dot\delta = \pi(q^{-2J_3})$, $\dot\delta^*\dot\delta$ is an invertible element affiliated with A_P and

$$\mathrm{Sp}\,\dot\delta^*\dot\delta \subset q^{\mathbf{Z}} \cup 0 \tag{5.1}$$

The group structure of P is the one induced by that of QLG. The comultiplication Δ_P is the unique element of $\mathrm{Mor}(A_P, A_P \otimes A_P)$ such that

$$\Delta_P \circ \pi = (\pi \otimes \pi)\Delta.$$

In particular

$$\Delta_P(\dot\alpha) = \dot\alpha \otimes \dot\alpha, \qquad \Delta_P(\dot\beta) = \dot\alpha \otimes \dot\beta + \dot\beta \otimes \dot\delta,$$
$$\Delta_P(\dot\delta) = \dot\delta \otimes \dot\delta.$$

Now we shall consider characters of P i.e. 1-dimensional representations of P. We do not assume neither unitarity nor even boundedness. More precisely χ is a character if χ is an invertible element affiliated with A_P and $\Delta_P\chi = \chi \otimes \chi$. It turns out that any character of P is of the form

$$\chi = \dot\delta^{n_1-1}(\dot\delta^*)^{n_2-1} = (\mathrm{Phase}\,\dot\delta)^{n_1-n_2} \mid \dot\delta \mid^{n_1+n_2-2} \tag{5.2}$$

where $n_1, n_2 \in \mathbf{C}$, $n_1 - n_2 \in \mathbf{Z}$.

Remark: We have inserted -1 in the exponents to have better correspondence with the Gelfand notation [2] .

Due to the spectral condition (5.1) two pairs $(n_1, n_2), (n_1', n_2')$ give rise to the same character if and only if

$$n_1 - n_1' = n_2 - n_2' = \frac{2k\pi i}{\log q} \quad \text{for some } k \in \mathbf{Z}.$$

In such a case we write $(n_1, n_2) \equiv (n_1', n_2')$.

The induced representations considered in this paper act on spaces of "smooth functions" on *QLG*. We say that an element a, affiliated with $A = A_c \otimes A_d$ is smooth if for any $s = 0, 1/2, 1, \ldots$:

$$(\mathrm{id} \otimes \pi^s)a \in \mathcal{A}_c \otimes B(H^s).$$

The set of smooth elements will be denoted by $\mathcal{A}$. It is clear that $\mathcal{A}$ is a *-subalgebra of A^η. One may also consider smooth elements affiliated with $A \otimes A$. By definition $a \, \eta \, A \otimes A$ is smooth if

$$(\mathrm{id} \otimes \pi^s \otimes \mathrm{id} \otimes \pi^{s'})a \in (\mathcal{A}_c \otimes B(H^s)) \otimes_{\mathrm{alg}} (\mathcal{A}_c \otimes B(H^{s'})$$

for any $s, s' = 0, 1/2, 1, \ldots$. The set of smooth elements affiliated with $A \otimes A$ may be denoted by $\mathcal{A} \hat{\otimes} \mathcal{A}$ where $\hat{\otimes}$ is the algebraic tensor product followed by a suitable completion. It turns out that $\Delta(\mathcal{A}) \subset \mathcal{A} \hat{\otimes} \mathcal{A}$.

Let χ be a character of P. The representation of *QLG* induced by χ acts by right shifts on the space D_χ of smooth elements which transform under the left action of P according to the representation χ :

$$D_\chi = \{a \in \mathcal{A} : (\pi \otimes \mathrm{id})\Delta a = \chi \otimes a \}. \tag{5.3}$$

The reader should notice that the transformation low

$$(\pi \otimes \mathrm{id})\Delta a = \chi \otimes a \tag{5.4}$$

coincides in the classical case with $a(pg) = \chi(p)a(g)$ for all $p \in P$ and $g \in G$ (cf.[1, p.473, formula (1)]).

Since the left and the right shifts commute, D_χ is invariant under the right shifts :

$$\Delta(D_\chi) \subset D_\chi \hat{\otimes} \mathcal{A}.$$

Therefore $v_\chi := \Delta|_{D_\chi}$ is a smooth action of *QLG* on D_χ. In other words D_χ carries a representation of *QLG*. This is the representation induced by χ.

To make our notation close to the one used in [2] we write $D_{n_1 n_2}$ (where $n_1, n_2 \in \mathbf{C}$ and $n_1 - n_2 \in \mathbf{Z}$) instead of D_χ for χ given by (5.2). The relation (5.4) can be solved explicitly:

$$D_{n_1 n_2} = \{\sigma^{-1}(q^{-(n_1+n_2-2)J_3} \otimes \alpha_{ck} \gamma_c^m \gamma_c^{*n}) : m - n - k = n_1 - n_2\}^{\substack{\text{linear} \\ \text{span}}} \tag{5.5}$$

The space $D_\chi = D_{n_1 n_2}$ in the classical setting appeared for the first time in the beautiful monograph [2] by Gelfand and collaborators. To commemorate this fact D_χ will be called the Gelfand spaces. We have

Theorem 3

Let $n_1, n_2 \in \mathbf{C}$, $n_1 - n_2 \in \mathbf{Z}$ and $p = \frac{1}{2} \mid n_1 - n_2 \mid$.

Then the Casimir operator (4.5) *and the spin spectrum of the representation $v_{n_1 n_2}$ of G induced by the character* (5.2) *is given by*

$$\begin{aligned} \mathrm{Sp}\, v_{n_1 n_2} &= \{p, p+1, p+2, \ldots\} \\ C(v_{n_1 n_2}) &= -(q^{n_1} + q^{-n_1})I \end{aligned} \tag{5.6}$$

Moreover the spin spectrum is simple: each u^s enters to $v_{n_1 n_2}|_{S_qU(2)}$ at most once.

The technique of generalized functions (distributions) developed in [2] works in our case as well. It gives the full description of :

- Invariant bilinear and sesquilinear functionals on $D_\chi \times D_{\chi'}$
- Intertwining operators $D_\chi \longrightarrow D_{\chi'}$
- The set of all χ such that on D_χ there exists a positive invariant sesquilinear form.

Let χ be the character of P related to the pair (n_1, n_2) via the formula (5.2). Then χ^* is related to $(\bar{n}_2, \bar{n}_1)$ and (cf.(5.3))

$$(D_{n_1 n_2})^* = D_{\bar{n}_2 \bar{n}_1}.$$

The same relation follows from (5.5). Due to this fact the invariant bilinear functionals on $D_{n_1 n_2} \times D_{n'_1 n'_2}$ are in one-to-one correspondence with invariant sesquilinear functionals on $D_{\bar{n}_2 \bar{n}_1} \times D_{n'_1 n'_2}$.

Let

$$\begin{array}{rccc} S: & D_\chi \times D_{\chi'} & \longrightarrow & \mathbf{C} \\ & (x, y) & \longmapsto & (x \mid y)_S \end{array}$$

be a sesquilinear form on $D_\chi \times D_{\chi'}$. Then S gives rise to an $\mathcal{A}$-valued sesquilinear form on $(D_\chi \hat{\otimes} \mathcal{A}) \times (D_{\chi'} \hat{\otimes} \mathcal{A})$:

$$(x \otimes a \mid y \otimes b)_S := (x \mid y)_S \, a^* b.$$

We say that S is invariant if

$$(\Delta\, x \mid \Delta\, y)_S = (x \mid y)_S \, I_{\mathcal{A}}$$

for any $x \in D_\chi$, $y \in D_{\chi'}$.

Theorem 4

Let $n_1, n_2, n'_1, n'_2 \in \mathbf{C}$, $n_1 - n_2, n'_1 - n'_2 \in \mathbf{Z}$. Assume that there exists a non-zero invariant sesquilinear form on $D_{n_1 n_2} \times D_{n'_1 n'_2}$. Then we have the following four possibilities:

1.

$$(n'_1, n'_2) \equiv (-\bar{n}_2, -\bar{n}_1) \tag{5.7}$$

2.

$$(n'_1, n'_2) \equiv (\bar{n}_2, \bar{n}_1) \tag{5.8}$$

3.

$$(n'_1, n'_2) \equiv (-\mathrm{Re}\, n_2, \mathrm{Re}\, n_1) \tag{5.9}$$

where $\mathrm{Re}\, n_1 = 1, 2, \ldots$ *and* $\mathrm{Im}\, n_1 \equiv 0 \bmod (2\pi / \log q)$

4.

$$(n'_1, n'_2) \equiv (\mathrm{Re}\, n_2, -\mathrm{Re}\, n_1) \tag{5.10}$$

where $\mathrm{Re}\, n_2 = 1, 2, \ldots$ *and* $\mathrm{Im}\, n_2 \equiv 0 \bmod (2\pi / \log q)$

In all these cases the invariant sesquilinear form is unique (up to a scalar factor).

The above theorem leads in a standard way to the following description of non-trivial intertwining operators acting between Gelfand spaces. An intertwiner $T : D_\chi \longrightarrow D_{\chi'}$ is trivial if $T = 0$ or $\chi = \chi'$ and $T = \lambda I$.

Theorem 5

1. Let n_1, n_2 be positive integers, $\varepsilon = 0$ or $i\pi/\log q$ and $D^\varepsilon_{n_1 n_2} := D_{n_1+\varepsilon, n_2+\varepsilon}$. Then $D^\varepsilon_{n_1 n_2}$ contains the only one nontrivial invariant subspace $E^\varepsilon_{n_1 n_2}$ and $\dim E^\varepsilon_{n_1 n_2} = n_1 n_2$, $D^\varepsilon_{-n_1,-n_2}$ contains the only one nontrivial invariant subspace $F^\varepsilon_{-n_1,-n_2}$ and $\operatorname{codim} F^\varepsilon_{n_1 n_2} = n_1 n_2$, $D^\varepsilon_{-n_1,n_2}$, $D^\varepsilon_{n_1,-n_2}$ have no nontrivial invariant subspace. Moreover we have the following diagram of nontrivial intertwiners (except the ones that starts or ends at 0; these are obviously trivial):

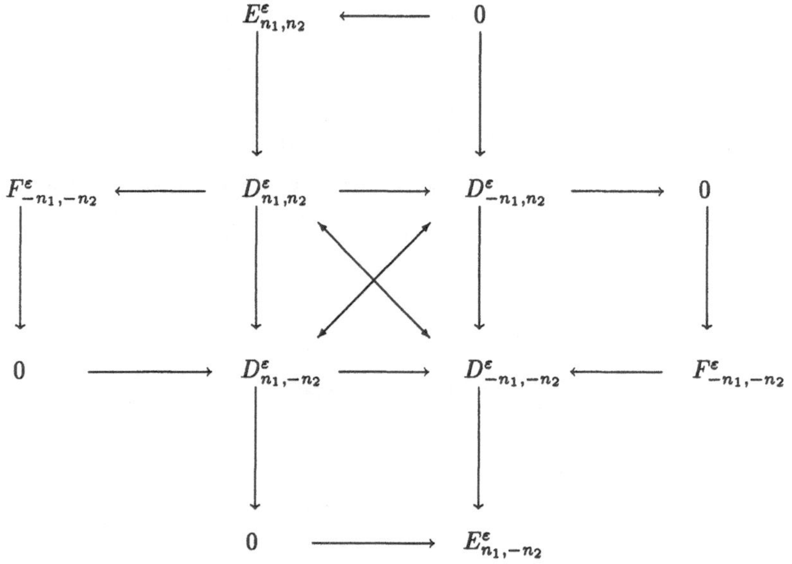

(5.11)

All intertwiners are unique up to a complex factor and any subsequence containing exactly two spaces $D^\varepsilon_{\pm n_1,\pm n_2}$ in succession is exact.

2. Let χ be a character of P such that the space D_χ has not appeared in the diagram (5.11) and $\chi' := (\dot{\delta}\dot{\delta}^*)^{-2}\chi^{-1}$. (The reader should notice that χ' corresponds to the pair $(-n_1, -n_2)$, where (n_1, n_2) is related to χ via (5.2)). Then there exists unique (up to a scalar factor) bijective intertwiner

$$D_\chi \overset{T}{\longleftrightarrow} D_{\chi'}.$$

Spaces D_χ and $D_{\chi'}$ contain no non-trivial invariant subspace.

3. The intertwiners listed in the above two points are the only non-trivial intertwiners acting between the Gelfand spaces.

Remark: The finite-dimensional representations acting on $E^{\varepsilon}_{n_1 n_2}$ of point 1 were studied in [4] . They all are non-unitary excepting the the cases of two 1-dimensional representations E^{ε}_{11}. Let us note also that by virtue of the diagram (5.11) the representations acting on $F^{\varepsilon}_{-n_1,-n_2}$, $D^{\varepsilon}_{-n_1,n_2}$ and $D^{\varepsilon}_{n_1,-n_2}$ are equivalent.

6 Gelfand spaces with unitary actions of QLG

Using Theorem 4 one can easily select all Gelfand spaces $D_{n_1 n_2}$ endowed with an invariant sesquilinear form $S : D_{n_1 n_2} \times D_{n_1 n_2} \to \mathbf{C}$. Due to the uniqueness of S it is automatically hermitian (after a suitable choise of the phase of the numerical factor). If S is positive then by the standard procedure $D_{n_1 n_2}$ can be completed to a Hilbert space $H_{n_1 n_2}$ and the action of *QLG* on $D_{n_1 n_2}$ extends in a natural way to a unitary representation (denoted again by $v_{n_1 n_2}$) of *QLG* on $H_{n_1 n_2}$. We shall show that in this way we can obtain all infinite-dimensional irreducible unitary representations listed in Table 1.

Let in Theorem 4, $(n'_1, n'_2) = (n_1, n_2)$. One can easily check that this relation is incompatible with (5.9) and with (5.10). Therefore only the first two possibilities remain.

Solving (5.7) we get

$$(n_1, n_2) = (p + \frac{i\rho}{2}, -p + \frac{i\rho}{2}) \tag{6.1}$$

where $p \in \mathbf{Z}/2$ and $\rho \in \mathbf{R}$. In this case S is automaticaly positive. Clearly ρ is defined mod $(4\pi/\log q)$ so we may assume that $\rho \in]2\pi/\log q, -2\pi/\log q]$. Moreover we may assume that $p \geq 0$ (and that $\rho \geq 0$ for $p = 0$): according to Theorem 5 simultaneous change of sign of p and ρ leads to an equivalent representations. The same theorem shows that $v_{n_1 n_2}$ is irreducible. Using Theorem 3 we get

$$\begin{aligned} Sp\, v_{n_1 n_2} &= \{p, p+1, p+2, \ldots\} \\ c(v_{n_1 n_2}) &= a\cos\varphi + ib\sin\varphi \end{aligned} \tag{6.2}$$

where $a = 2\cosh(p\log q)$, $b = 2\sinh(p\log q)$ and $\varphi = \pi + (\rho\log q)/2 \in [0, 2\pi[$ ($\varphi \in [0,\pi]$ for $p = 0$). For fixed p the values of (6.2) runs over the whole (degenerated for $p = 0$) ellipse $| c - 2 | + | c + 2 | = 2(q^p + q^{-p})$. Therefore (cf.Table 1) the representations $v_{n_1 n_2}$ (where n_1, n_2 are given by (6.1)) exhaust all the representations of the principal series.

Solving (5.8) we get

$$(n_1, n_2) = (\rho + \varepsilon, \rho + \varepsilon)$$

where $\rho \in \mathbf{R}$ and $\varepsilon = 0, i\pi/\log q$. In this case S is strictly positive if and only if $| \rho |< 1$. As before we may assume that $\rho \geq 0$. The case $\rho = 0$ was covered by (6.1). Therefore it is sufficient to consider $\rho \in]0, 1[$. Using Theorem 3 we get

$$\begin{aligned} Sp\, v_{n_1 n_2} &= \{0, 1, 2, \ldots\} \\ c(v_{n_1 n_2}) &= \mp 2\cosh(\rho\log q) \end{aligned} \tag{6.3}$$

where the upper (lower) sign corresponds to $\varepsilon = 0$ ($\varepsilon = i\pi/\log q$). The values of (6.3) covers the two intervals $] - (q + q^{-1}), -2[$ and $]2, q + q^{-1}[$. Therefore (cf. Table 1) in this case $v_{n_1 n_2}$ runs over all representations of the two complementary series. For the classical case the solution with $\varepsilon = i\pi/\log q$ do not exist and we have only one complementary series.

References

[1] Barut,A.O., Rączka,R.: Theory of Group Representations and Applications. PWN-Polish Scientific Publishers, Warszawa 1977

[2] Gelfand,I.M.,Graev,M.I.,Vilenkin,N.J.:Generalized Functions vol.5:Integral Geometry and Representation Theory. Academic Press, New York-London 1966

[3] Naimark,M.A.: Linear Representations of the Lorentz Group. Inernational Series on Monographs in Pure and Applied Mathematics vol.63, The Macmillan Company New York 1964

[4] Podleś,P.,Woronowicz,S.L.: Quantum deformation of Lorentz group. Commun.Math.-Phys. **130**, 381-431 (1990)

[5] Pusz,W.: Irreducible unitary representations of quantum Lorentz group. Commun. Math.Phys.**152**,591-626(1993)

[6] Pusz,W.,Woronowicz,S.L.: Representations of quantum Lorentz group on Gelfand spaces. in preparation

[7] Woronowicz,S.L.: Unbounded elements affiliated with C^* - algebras and non-compact quantum groups. Commun.Math. Phys. **136**, 399-432 (1991)

Department of Mathematical Methods in Physics
Faculty of Physics, University of Warsaw
Hoza 74, PL 00-682 Warsaw
Poland

CONTRACTION OF QUANTUM GROUPS AND LATTICE PHYSICS.

Riccardo Giachetti.

ABSTRACT. Inhomogeneous quantum groups obtained by contraction of semisimple ones are shown to be a natural symmetry for dynamical systems on a lattice. The pseudoeuclidean group in dimensions $1+1$ is applied to harmonic excitations of a crystal and the one dimensional Galilei group is used to study spin chains. The Hopf algebra structure permits to determine the spectral properties both of the single particle and of composite systems. A general picture of the possible deformations of the two dimensional Euclidean group is finally presented using an R-matrix approach.

1. Introduction.

A certain number of inhomogeneous quantum algebras of large physical interest has recently been determined by extending the contraction procedure from semisimple Lie algebras to their quantized version (6; 7). In order to preserve the Hopf algebra structure, it may occurs that sometimes the quantum parameter itself must undergo a rescaling, so that, after the contraction, the parameter can acquire a physical dimension. It emerges then naturally a picture in which the quantum algebra represents the symmetry of a dynamical system with a fundamental length or time scale (8) and in which the coalgebra establishes the rules for combining the elementary excitations (4; 5). Here we shall firstly review two possible contractions of $SU_q(2)$ giving different quantizations of the two dimensional pseudoeuclidean (or 1+1 – Poincaré) group, $E(1,1)$ (7): the associated physical system to one of them is that of *phonons*, namely the excitations of a harmonic crystal (1; 4). We then are going to exhibit two quantum analogues of the Galilei group, one of which represents the symmetry of linear magnetic chains and describes the properties of the quantization of their linear modes, the *magnons* (9; 5). Finally we shall consider the problems involved with the determination of the R-matrices for the contracted structures and we shall discuss some new results for the quantization of $E(2)$, (2), which, on the one hand, allow to recover at a Hopf algebra level the results recently obtained by contracting the C^*-algebra of the "non-commuting representative functions" of $SU_q(2)$ (14) and, on the other, provide a unified picture for the possible deformations of the two dimensional Euclidean group.

E. A. Tanner and R. Wilson (eds.),
Noncompact Lie Groups and Some of Their Applications, 473–485.

2. Two dimensional pseudoeuclidean quantum group and phonons.

Let us consider the defining relations for $SU_q(1,1)$:

$$[J_3, J_1] = -iJ_2, \qquad [J_3, J_2] = -iJ_1, \qquad [J_1, J_2] = (i/2)\,[2J_3]_z \equiv (i/2)\,\frac{\sin(zJ_3)}{\sin(z/2)}.$$

The usual (Euclidean) $SU_q(2)$ is obtained by taking a positive sign in the right hand side of the first commutator and changing z into iz. The coalgebra is

$$\Delta J_3 = \mathbf{1}\otimes J_3 + J_3\otimes \mathbf{1}, \qquad \Delta J_\ell = e^{-izJ_3/2}\otimes J_\ell + J_\ell\otimes e^{izJ_3/2}$$

with $\ell = 1,2$, while counit and antipode read

$$\gamma(J_3) = -J_3, \qquad \gamma(J_\ell) = -e^{izJ_3/2}\,J_\ell\,e^{-izJ_3/2}: \qquad \epsilon(J_\ell) = \epsilon(J_3) = 0.$$

Transforming the generators according to

$${}^t(P_1,\ P_2,\ J,\ z) = \text{diag}\ \{\varepsilon,\ \varepsilon,\ 1,\ 1\}\ {}^t(J_1,\ J_2,\ J_3,\ z)$$

and taking the limit $\varepsilon \to 0$, we find

$$[J, P_1] = -iP_2, \qquad [J, P_2] = -iP_1, \qquad [P_1, P_2] = 0.$$

Analogously we get

$$\Delta J = \mathbf{1}\otimes J + J\otimes \mathbf{1}, \qquad \Delta P_\ell = e^{-izJ/2}\otimes P_\ell + P_\ell\otimes e^{izJ/2},$$

for the coproduct,

$$\gamma(J) = -J, \qquad \gamma(P_\ell) = -e^{izJ/2}\,P_\ell\,e^{-izJ/2},$$

for the antipode. The counit is vanishing on the generators.

We have thus determined a Hopf algebra whose commutators exactly reproduce the corresponding Lie algebra, while the coalgebra maintains the deformation. We shall refer to this algebra as to $E_q(1,1)$ - or $E_q(2)$ for its Euclidean version. The dual of the latter has been investigated in (13) in a Von Neumann algebras context and the q-analog of Bessel functions has been introduced; a C^*-algebra approach has then been discussed in (14) and the Hopf algebra of the noncommutative representative functions has recently been obtained by a contraction procedure in (15). We shall comment on this later on.

A different pseudoeuclidean quantum algebra is found by means of the following rescaling:

$${}^t(J,\ P_0,\ P,\ a) = \text{diag}\ \{1,\ \varepsilon,\ \varepsilon,\ \varepsilon^{-1}\}\ {}^t(J_1,\ J_2,\ J_3,\ z).$$

Taking the limit $\varepsilon \to 0$ a nontrivial quantum behaviour is maintained. Indeed:

$$[J, P] = iP_0, \qquad [J, P_0] = (i/a)\,\sin(aP), \qquad [P, P_0] = 0.$$

Introducing the generators $k^{\pm 1} = e^{\pm iaP}$ in order to remove the ambiguity of the determination of P up to an additive term $2\pi n/a$, $n \in \mathbf{Z}$, we find the following relations:

$$kP_0k^{-1} = P_0\,, \qquad kJk^{-1} = J + aP_0\,, \qquad JP_0 - P_0J = (k - k^{-1})/(2a)\,,$$
$$kk^{-1} = 1\,.$$

We refer to this quantum algebra as to $E_\ell(1,1)$ (subscript ℓ stands for "lattice"). The coproducts, antipodes and counits are

$$\Delta(J) = k^{-1/2} \otimes J + J \otimes k^{1/2}\,, \qquad \Delta(P_0) = k^{-1/2} \otimes P_0 + P_0 \otimes k^{1/2}\,,$$
$$\Delta(k) = k \otimes k\,,$$
$$\gamma(J) = -J - (a/2)\,P_0\,, \qquad \gamma(k) = k^{-1}\,, \qquad \gamma(P_0) = -P_0\,,$$
$$\epsilon(J) = \epsilon(P_0) = 0\,, \qquad \epsilon(k) = 1\,.$$

The Casimir of $E_\ell(1,1)$ is

$$C = P_0^2 - (2/a)^2 \sin^2(aP/2)\,.$$

The Euclidean analogue of the previous algebra will be denoted by $E_\ell(2)$.

Let us now show how $E_q(1,1)$ can be brought to bear to some aspects of phonon physics. Let us consider the equations of motion of a linear chain of equal masses lying at a distance a from one another with nearest neighbour interaction and external restoring force:

$$\ddot{z}_j(t) = \omega^2\,(z_{j-1}(t) + z_{j+1}(t) - 2z_j(t)) - c^2 z_j(t)\,.$$

Periodic boundary conditions are assumed and initial data $z_j(0)$, $\dot{z}_j(0)$ assigned. Letting $v = \omega/a$, $m = c/v^2$, we embed the system into the single PDE

$$\left(\partial_t^2 + (2v/a)^2 \sin^2(-ia\partial_x/2) + m^2v^4\right)\, z(x,t) = 0\,,$$

with periodic conditions $z(0,t) = z(L,t)$ and Cauchy data $z(x,0)$, $\partial_t z(x,0)$. If $z(ja,0) = z_j(0)$, $\partial_t z(ja,0) = \dot{z}_j(0)$ for all j, the solutions of the system are obtained by $z_j(t) = z(ja,t)$ irrespectively of the behaviour at $x \neq ja$. If we take the following differential realization

$$P_0 = (i/v)\,\partial_t\,, \qquad k = \exp(a\partial_x)\,, \qquad J = i(x/v)\partial_t - (vt/a)\sin(-ia\partial_x)\,,$$

we see that the above PDE is precisely the eigenvalue equation

$$C\; z(x,t) = m^2v^2\; z(x,t)$$

for the Casimir of $E_\ell(1,1)$. The case $m = 0$ describes the phonons. In the momentum representation, where $P = p$ is diagonal, some mechanical features of the system are well described. Introducing the position operator $X = i\partial_p$, we also have

$$J = (1/a)\,\{\sin(ap/2), X\}_+\,, \qquad P_0 = (2/a)\,\sin(ap/2)$$

and the first Brillouin zone $0 \leq p < 2\pi/a$, (1), P_0 is positive. The position operator can be directly determined as

$$X = (1/2)\left\{P_0^{-1}, J\right\}_+$$

and its time derivative

$$v_g = \dot{X} = iv\,[P_0, X] = v\ \cos(aP/2)$$

reproduces the well known expression for the group velocity of the phonons.

Let us now show how the coproduct gives the rules for the fusion of the phonons. Recall that when a Lie algebra acts on a system, the generators of the global symmetry are obtained by summing the generators of the elementary constituents. Equivalently, since for any generator G of the Lie algebra we have

$$\Delta G = G \otimes \mathbf{1} + \mathbf{1} \otimes G \equiv G^{(1)} + G^{(2)}$$

and Δ is a algebra homomorphism, $G^{(1)} + G^{(2)}$ generates the symmetry of the composite system, $G^{(i)}$ acting on the i-th elementary component. This property holds also if the system presents a quantum group symmetry, although the generators may well be non primitive. In the case at hand, the global variables obtained by the coproduct are

$$P_0 = e^{-iaP^{(1)}/2}\,P_0^{(2)} + P_0^{(1)}\,e^{iaP^{(2)}/2} \quad , \qquad J = e^{-iaP^{(1)}/2}\,J^{(2)} + J^{(1)}\,e^{iaP^{(2)}/2}\,;$$

$$k \;=\; k^{(1)}k^{(2)}\,.$$

From the last $P = P^{(1)} + P^{(2)} + 2\pi n/a$, with n chosen so that P is kept in the fixed Brillouin zone: quantum symmetry implies Umklapp process.

Explicitly, for two possibly differently polarized phonons with velocity parameters $v^{(1)}$, $v^{(2)}$ and dispersion relations

$$\Omega^{(r)} = v^{(r)}P_0^{(r)} = (2v^{(r)}/a)\,\sin(aP^{(r)}/2)\,, \qquad (r = 1, 2)$$

the coproduct P_0 reads

$$\begin{aligned} P_0 \;&=\; e^{-iaP^{(1)}/2}\,(2/a)\sin(aP^{(2)}/2) + (2/a)\sin(aP^{(1)}/2)\,e^{iaP^{(2)}/2} \\ &=\; (2/a)\sin(a(P^{(1)} + P^{(2)})/2)\,. \end{aligned}$$

Therefore, by the energy conservation, the physical process actually occurs if there exists a velocity v such that

$$\Omega = \Omega^{(1)} + \Omega^{(2)}$$

where $\Omega = (2v/a)\sin(aP/2)$ is the dispersion relation of the composite system.

The position global operator is

$$X = \frac{1}{2}(X^{(1)} + X^{(2)}) + \frac{1}{2}\left\{\frac{\sin(a(P^{(1)} - P^{(2)})/2)}{\sin(a(P^{(1)} + P^{(2)})/2)}\frac{1}{2}(X^{(1)} - X^{(2)})\right\}_+$$

and reproduces the Heisenberg algebra $[X, P] = i$. The procedure can be applied to any number of phonons. Energy and momentum come out in a directly symmetric form. Not so the global boost: however its global symmetrized form still closes an $E_\ell(1,1)$ algebra.

3. One dimensional Galilei quantum group and linear magnetic chains.

A deformation $G_\ell(1)$ of the one-dimensional Galilei algebra is obtained, in analogy with the classical case, by a further contraction of $E_\ell(1,1)$. Letting

$$^t(B,\ M,\ K,\ \ell) = \text{diag}\ \{\varepsilon,\ 1,\ \varepsilon^{-1},\ \varepsilon/2\}\ ^t(J,\ P_0,\ P,\ a)\,,$$

where B represents the Galileian boost, M the mass and K the momentum, for $\varepsilon \to 0$ we have

$$[B,K] = iM\,, \qquad [B,M] = [K,M] = 0\,.$$

The coproducts and antipodes are

$$\Delta B = e^{-i\ell K} \otimes B \;+\; B \otimes e^{i\ell K}\,, \qquad \Delta M = e^{-i\ell K} \otimes M + M \otimes e^{i\ell K}\,,$$
$$\Delta K = \mathbf{1} \otimes K + K \otimes \mathbf{1}\,,$$
$$\gamma(B) = -B \;-\; \ell M\,, \qquad \gamma(M) = -M\,, \qquad \gamma(K) = -K\,.$$

For the physical applications that will be presently shown, we need an extension $\Gamma_\ell(1)$ of $G_\ell(1)$ by an additional primitive generator T, the "kinetic energy" such that

$$[B,T] = (i/\ell)\,\sin(\ell k)\,, \qquad [T,K] = [T,M] = 0$$

and $\gamma(t) = -T$. The Casimir of $\Gamma_\ell(1)$ reads

$$C = MT - (1/\ell^2)\Big(1 - \cos(\ell K)\Big)$$

and this algebra admits the differential realization

$$B \;=\; mx\,, \qquad M = m\,, \qquad K = -i\partial_x\,,$$
$$T \;=\; (m\ell^2)^{-1}\Big(1 - \cos(-i\ell\partial_x)\Big) + c/m\,,$$

where c is the constant value of the Casimir.

The physical system that can be studied by means of the quantum symmetry $\Gamma_\ell(1)$ is a spin 1/2 system referred to as XXZ model and known to be integrable by the Bethe Ansatz method (3; 10). Its Hamiltonian is

$$\mathcal{H} = 2J \sum_{i=1}^{N}\Big((1-\alpha)\,(S_i^x S_{i+1}^x + S_i^y S_{i+1}^y) + S_i^z S_{i+1}^z\Big)\,,$$

with $S_{N+1}^a = S_1^a$. Following the standard method for analyzing such models, we see that the "vacuum state" $|0\rangle$, *i.e.* the state with all spins directed downwards, has energy $\epsilon_0 = JN/2$ and that, in terms of the states $\psi = \sum_i f_i S_i^+ |0\rangle$ with a single spin deviate, the eigenvalue equation for $\mathcal{H}$ becomes the algebraic system

$$2Js\Big((1-\alpha)\,(f_{i-1} + f_{i+1}) - 2f_i\Big) = (\epsilon - \epsilon_0) f_i\ .$$

With the same procedure of the previous section, this system is embedded into the PDE

$$-4Js\Big(1 - (1-\alpha)\cos(-i\ell\partial_x)\Big) f(x) = (\epsilon - \epsilon_0) f(x)\ ,$$

whose left hand side operator coincides with the differential realization of T, once we make the identifications

$$(m\ell^2)^{-1} = -2J(1-\alpha), \qquad c/m = -2J\alpha.$$

Considering as well the Galileian position operator $X = B/M$, we find the magnon velocity

$$\dot{X} = i\,[T, X] = J\ell \sin(\ell K)$$

in agreement with well known results.

Let us discuss the two magnon states $\psi = \sum_{ij} f_{ij} S_i^+ S_j^+ |0\rangle$, where $f_{ij} = f_{ji}$, $i \neq j$, while f_{ii} are physically meaningless and have no part in the theory. The system for the coefficients f_{ij} results in

$$\Big(\epsilon - \epsilon_0 + 8Js\Big) f_{ij} - 2s(1-\alpha)\sum_n \Big(J_{nj}\, f_{in} + J_{in}\, f_{nj}\Big) = J_{ij}\Big((1-\alpha)(f_{ii} + f_{jj}) - f_{ij} - f_{ji}\Big)$$

where the bonds J_{ij} are equal to J when the label (ij) are nearest neighbor pairs and vanish otherwise. For $s = 1/2$ the amplitudes f_{ii} cancel in pairs. The Bethe ansatz method imposes the separate vanishing of the two sides of the previous relation, taking the vanishing of the right part as a boundary condition for the homogeneous free equation that is obtained by the vanishing of the left part.

We now use the $\Gamma_\ell(1)$ symmetry. From ΔT we find the two magnon energy

$$T_{12} = T_1 + T_2 = (M_1\ell^2)^{-1}\Big(1 - \cos(\ell K_1)\Big) + (M_2\ell^2)^{-1}\Big(1 - \cos(\ell K_2)\Big) + (c_1/M_1) + (c_2/M_2)\,.$$

For $M_1 = M_2 = m$, using the previous identifications and the differential realization $K_1 = -i\partial_{x_1}$, $K_2 = -i\partial_{x_2}$, the eigenvalue equation $T_{12} f(x_1, x_2) = (\epsilon - \epsilon_0) f(x_1, x_2)$ for the two magnon amplitude $f(x_1, x_2)$ reads

$$T_{12}\, f(x_1, x_2) = -8Js\, f(x_1, x_2) + 2Js(1-\alpha)\Big(f(x_1, x_2 + \ell) + f(x_1, x_2 - \ell) + f(x_1 + \ell, x_2) + f(x_1 - \ell, x_2)\Big)$$

and is equivalent to the free system of the Bethe ansatz. Recalling that from ΔM and ΔK we have

$$M_{12} = M_1\, e^{i\ell K_2} + M_2\, e^{i\ell K_1}, \qquad K_{12} = K_1 + K_2,$$

we can write for ΔT and ΔC

$$T_{12} = 1/(M_{12}\ell^2)\Big(1 - \cos(\ell K_{12})\Big) + U_{12}, \qquad C_{12} = M_{12}\, U_{12},$$

where

$$U_{12} = (c_1/M_1) + (c_1/M_1) - \frac{(M_{12} - M_1 - M_2)^2}{2\,\ell^2\, M_{12}\, M_1\, M_2}\,.$$

The two magnon bound states are obtained by requiring that the energy has a homogeneous dependence of degree -1 upon the total mass, exactly as in the single magnon cases. This gives the equation

$$U_{12} + M_{12}\,(\partial U_{12}/\partial M_{12}) = (\partial C_{12}/\partial M_{12}) = 0$$

and corresponds therefore to the critical behaviour of C_{12} with respect to the total mass M_{12}. We obtain:

$$M_{12} = M_1 + M_2 + \ell^2 M_1 M_2\,(c_1/M_1 + c_1/M_1)\,.$$

For $M_1 = M_2 = M$ we find $M_{12} = 2M/(1-\alpha)$ and the energy of bound states is

$$T_{12} = (\ell^2\,M_{12})^{-1}\Big(1 - \cos(\ell K_{12})\Big) + U_{12} = -2J\Big(1 - (1-\alpha)^2\,\cos^2(\ell K/2)\Big)\,.$$

The procedure can be extended to any number of magnons. The relations are

$$T_{12\ldots n} = \sum_{k=1}^{n} T_k = (\ell^2 M_{12\ldots n})^{-1}\Big(1 - \cos(\ell K_{12\ldots n})\Big) + U_{12\ldots n}$$

where $K_{12\ldots n} = \sum_{k=1}^{n} K_k$ and

$$U_{12\ldots n} = \sum_{k=1}^{n} U_k - \frac{1}{2\ell^2}\sum_{k=2}^{n}\frac{(M_{12\ldots k} - M_{12\ldots(k-1)} - M_k)^2}{M_{12\ldots k}M_{12\ldots(k-1)}M_k}$$

and where $M_{12\ldots k}$ are defined by iterating the coproduct and using the coassociativity:

$$M_{l\ldots k} = M_{l\ldots(h-1)}\;e^{i\ell(K_h+\ldots+K_k)} + M_{h\ldots k}\;e^{-i\ell(K_l+\ldots+K_{h-1})}\,, \qquad l < h \le k\,.$$

The coproducts of the Casimir $C_{12\ldots k}$ are then found to be

$$C_{12\ldots k} = M_{12\ldots k}\,U_{12\ldots k}$$

and their critical behaviour with respect to $M_{12\ldots k}$ determines the bound states. The vanishing of the derivatives of $C_{12\ldots k}$ with respect to $M_{12\ldots k}$ for $k = 2, \ldots, n$, gives

$$M_{12\ldots k} = M_{12\ldots(k-1)} + M_k + \ell^2\,M_{12\ldots(k-1)}\,M_k\,(U_{12\ldots(k-1)} + U_k)\,, \quad k = 2, \ldots n\,.$$

The solution of these recurrence relations is

$$M_{12\ldots k} = -\Big(2J(1-\alpha)\,\ell^2\Big)^{-1}\,\mathcal{U}_{k-1}(1/(1-\alpha))\,, \quad k = 2, \ldots n\,,$$

$$T_{12\ldots n} = \frac{-2J(1-\alpha)}{\mathcal{U}_{n-1}(1/(1-\alpha))}\Big(\mathcal{T}_n(1/(1-\alpha)) - \cos(\ell K_{12\ldots n})\Big)\,,$$

so that the bound state energy of the n magnon bound states has a closed form in terms of the Tchebischeff polynomials $\mathcal{U}_k$ and $\mathcal{T}_k$.

4. Contractions of R-matrices and deformations of E(2).

In studying the physical applications of quantum groups that we have shown in the previous sections, a natural question is whether there is more than one deformed structure for a same Lie symmetry. We shall answer this question in the case of the Euclidean group $E(2)$, whose pseudoeuclidean version has been applied to lattice systems. Useful tool for investigating this problem are provided by R-matrices of the contracted structures, when they exist. Indeed quasitriangularity of a Hopf algebra can fail to be preserved by the contraction, since singularities can develop in the procedure. In order to be concrete, we shall here briefly review three cases: (*i*) the three dimensional Euclidean quantum group, technically more complicated, but conceptually simpler, where the divergences in the R-matrix contraction cancel; (*ii*) the one dimensional Heisenberg quantum group, where the contracted R-matrix exists, provided we renormalize it by a divergent central factor; (*iii*) two different versions of the two dimensional Euclidean quantum group where the contraction of R-matrices are badly divergent and do not exist in algebraic form: we shall however show that R-matrices can be found in two dimensional representations and used to construct the dual algebra of the "noncommutative representative functions". For later use we recall the expression of the R-matrix of $SU_q(2)$, namely (11)

$$R = R_a\, R_b$$

with

$$R_a = e^{zJ_3 \otimes J_3}\,,$$
$$R_b = \sum_{k\geq 0} \frac{(1-e^{-z})^k}{[k]!}\, e^{-zk(k-1)/4} \left(e^{zkJ_3/2}(J_+)^k \otimes e^{-zkJ_3/2}(J_-)^k\right),$$

and $q = e^z$.

(*i*) Consider the product of two copies of $SU_q(2)$:

$$[J_3^k, J_\pm^\ell] = \pm\delta_{k\ell}\, J_p^k m\,, \qquad [J_+^k, J_-^\ell] = \delta_{k\ell}\, \frac{\sinh(zJ_3^k)}{\sinh(z/2)}\,,$$

where the first copy is deformed with $q = e^z$ and the second with $q^{-1} = e^{-z}$. Let us define

$$J_s = J_s^1 + J_s^2\,, \qquad N_s = J_s^1 - J_s^2\,, \qquad (s = \pm, 3)$$

and rescale the generators according to

$${}^t(P_s,\ J_s,\ w) = \mathrm{diag}\ \{\varepsilon,\ 1,\ \varepsilon^{-1}\}\ {}^t(N_s,\ J_s,\ z)\,.$$

We then get the $E_q(3)$ Hopf algebra defined by

$$[J_3, J_\pm] = \pm J_\pm\,, \qquad [J_3, P_\pm] = [P_3, J_\pm] = \pm P_\pm\,,$$
$$[J_s, J_s] = [P_\ell, P_s] = 0\,,$$
$$[J_+, J_-] = 2J_3\,, \cosh(wP_3)\,, \qquad [J_\pm, J_\mp] = \pm(2/w)\,, \sinh(wP_3)\,,$$

$$\Delta P_\pm = e^{-wP_3/2} \otimes P_\pm + P_\pm \otimes e^{wP_3/2} \,,$$
$$\Delta J_\pm = e^{-wP_3/2} \otimes P_\pm + P_\pm \otimes e^{wP_3/2} - (w/2)\Big(e^{-wP_3/2} J_3 \otimes P_\pm - P_\pm \otimes e^{wP_3/2} J_3 \Big) \,,$$
$$\Delta J_3 = J_3 \otimes \mathbf{1} + \mathbf{1} \otimes J_3 \,, \qquad \Delta P_3 = P_3 \otimes \mathbf{1} + \mathbf{1} \otimes P_3 \,,$$

$$\gamma(J_3) = -J_3 \,, \qquad \gamma(J_\pm) = -(J_\pm \pm wP_\pm) \,, \qquad \gamma(P_s) = -P_s \,.$$

The R-matrix of $SO_q(4)$ is given by the product

$$R = R_a^1 \, R_a^2 \, R_b^1 \, R_b^2 \,.$$

In the contraction, the divergences of $R_a^1$and R_a^2 cancel in pairs, giving the finite result

$$R_a^1 \, R_a^2 \;\rightarrow\; e^{w(J_3 \otimes P_3 + P_3 \otimes J_3)} \,.$$

The situation is more difficult for the other two factors, (see (7) for details). However here again the result, found by expansion in series, is finite and can be written in terms of hypergeometric functions, namely

$$R_b^1 \, R_b^2 \;\rightarrow\; {}_2F_1(a, 1-a; 1/2; -w^2A^2) + wA(1-2a)\; {}_2F_1(a+1/2, 3/2-a; 3/2; -w^2A^2)$$

where

$$\begin{aligned} a &= -B/(wA) + 1/2 \,, \\ A &= (w/2)\, Q_+ \otimes Q_- \,, \\ B &= (w/2)(L_+ \otimes Q_- + Q_+ \otimes L_-) - \\ &\qquad (w^2/2)(Q_+ \otimes Q_- - (1/2)(J_3Q_+ \otimes Q_- + Q_+ \otimes J_3Q_-)) \,, \end{aligned}$$

and

$$Q_\pm = e^{\pm wP_3/2} P_\pm \,, \qquad L_\pm = e^{\pm wP_3/2} J_\pm \,.$$

Taking into account the particular arguments of the hypergeometric series, we have the following result

$$R = \exp\{w(J_3 \otimes P_3 + P_3 \otimes J_3)\} \; \exp\{2B \operatorname{arcsinh}(wA)/(wA)\} \; (1 + w^2A^2)^{-1/2} \,.$$

This expression is an R-matrix for $E_q(3)$, *i.e.* it satisfies the relation

$$R \, \Delta \, R^{-1} = \sigma \circ \Delta$$

(σ being the twist operator for the tensor product) and solves the Yang-Baxter equation

$$R_{12} \, R_{13} \, R_{23} \;=\; R_{23} \, R_{13} \, R_{12} \,.$$

(ii) Consider now the product $SU_q(2) \otimes U(1)$ and rescale the four generators and the quantum parameter $z = \log q$ as follows

$$\begin{pmatrix} A \\ A^\dagger \\ N \\ H \\ w \end{pmatrix} = \begin{pmatrix} \varepsilon & & & & \\ & \varepsilon & & & \\ & & -1 & \varepsilon^{-2} & \\ & & & 2 & \\ & & & & \varepsilon^{-2} \end{pmatrix} \begin{pmatrix} J_+ \\ J_- \\ J_3 \\ G \\ z \end{pmatrix},$$

where G is the new $U(1)$ generator. In the limit $\varepsilon \to 0$ we get

$$[A, A^\dagger] = \frac{\sinh(wH/2)}{w/2}, \qquad [N, A] = -A, \qquad [N, A^\dagger] = A^\dagger, \qquad [H, \cdot\,] = 0,$$

as well as

$$\Delta A = e^{-wH/4} \otimes A + A \otimes e^{wH/4}, \qquad \Delta A^\dagger = e^{-wH/4} \otimes A^\dagger + A^\dagger \otimes e^{wH/4},$$

while for H and N we have obviously

$$\Delta N = \mathbf{1} \otimes N + N \otimes \mathbf{1}, \qquad \Delta H = \mathbf{1} \otimes H + H \otimes \mathbf{1}.$$

We thus determine a Hopf algebra which is a central extension of the Heisenberg algebra and that we call $H_q(1)$. For the R-matrix, neglecting higher orders in ε, we find the divergent expression

$$R \rightarrow e^{(w/4\varepsilon^2)H\otimes H} \sum_{k\geq 0} 1/(k!) \left(w^{1/2} e^{wH/4}\, A\right)^k \otimes \left(w^{1/2} e^{-wH/4}\, A^\dagger\right)^k.$$

However the divergent factor is central and therefore inessential for the R-matrix. The finite contribution of $e^{zJ_3 \otimes J_3}$ in the contraction limit comes from the next order of the expansion in ε, which is $e^{-(w/2)(H\otimes N + N \otimes H)}$ since $J_3 = H/(2\varepsilon^2) - N$. We then have:

$$R = \exp\Big\{-(w/2)(H \otimes N + N \otimes H)\Big\} \exp\Big\{w\, e^{H\otimes \mathbf{1} - \mathbf{1}\otimes H}\, (A + A^\dagger)\Big\}.$$

In order to quantize the group we use 3×3 matrices of the form

$$T = \begin{pmatrix} 1 & \alpha & \beta \\ & 1+\gamma & \delta \\ & & 1 \end{pmatrix}.$$

If accordingly we represent the generators $A, A^\dagger, H, N$ and the R-matrix, the equation

$$R\, T_1\, T_2 = T_2\, T_1\, R$$

with $T_1 = T \otimes \mathbf{1}$ and $T_2 = \mathbf{1} \otimes T$, determines the relations for the algebra of the noncommutative representative functions $Fun(E_q(3))$, namely

$$\alpha\beta - \beta\alpha = (w/2)\,\alpha, \qquad \alpha\delta = \delta\alpha, \qquad \delta\beta - \beta\delta = (w/2)\,\delta,$$

γ being central. Coproducts, antipodes and counits are then found in the standard way (11).

(*iii*) Let us finally consider the deformations $E_q(2)$, $E_\ell(2)$ of the two dimensional Euclidean group introduced in section 2 and obtained by contracting $SU_q(2)$ with two different rescalings. Using the same rescalings to contract the R-matrix of $SU_q(2)$, bad divergences arise in the limit $\varepsilon \to 0$, so that no algebraic R-matrix is available. In spite of this difficulty, however, we shall determine the Hopf algebras of the noncommutative functions $Fun(E_q(2))$ and $Fun(E_\ell(2))$ as well as a picture of the possible deformations of the Euclidean group. This is done using a two dimensional representation by matrices of the form

$$T = \begin{pmatrix} v & n \\ 0 & 1 \end{pmatrix},$$

with v unitary and n complex. The coproducts of the matrix elements are

$$\Delta v = v \otimes v, \qquad \Delta n = v \otimes n + n \otimes 1, \qquad \Delta 1 = 1 \otimes 1 .$$

We then extend the algebra by the conjugate $\bar{v}$ of the group-like unitary generator v, $\bar{v}v = v\bar{v} = 1$, and by the conjugate $\bar{n}$ of n; we have

$$\Delta \bar{v} = \bar{v} \otimes \bar{v}, \qquad \Delta \bar{n} = \bar{v} \otimes \bar{n} + \bar{n} \otimes 1,$$

as well as

$$\gamma(v) = \bar{v}, \qquad \gamma(\bar{v}) = v, \qquad \gamma(1) = 1,$$

$$\gamma(n) = -\bar{v}n, \qquad \gamma(\bar{n}) = -v\bar{n},$$

$$\epsilon(v) = \epsilon(\bar{v}) = \epsilon(1) = 1, \qquad \epsilon(n) = \epsilon(\bar{n}) = 0 .$$

We finally observe that, in the chosen representation, the infinitesimal generators have the following form

$$P_x = \begin{pmatrix} 0 & 1 \\ 0 & 0 \end{pmatrix} \qquad P_y = \begin{pmatrix} 0 & i \\ 0 & 0 \end{pmatrix} \qquad J = \begin{pmatrix} -1 & 0 \\ 0 & 0 \end{pmatrix} .$$

We now look for a 4×4 matrix which transposes the coproduct and satisfies the Yang-Baxter equation. For $E_q(2)$, besides the trivial solution $R_q = \sigma$, up to an equivalence we find the unique solution

$$R_q = \begin{pmatrix} 1 & & & \\ & & 1 & \\ & 1 - e^{-z} & e^{-z} & \\ & & & 1 \end{pmatrix} .$$

The usual prescription $R(T \otimes 1)(1 \otimes T) = (1 \otimes T)(T \otimes 1)R$ of the R-matrix approach, gives the following relations between the generators of $Fun(E_q(2))$:

$$vn = e^z\, nv, \qquad n\bar{v} = e^z\, \bar{v}n, \qquad v\bar{v} = 1 .$$

For real z we also have

$$\bar{n}\bar{v} = e^{z}\,\bar{v}\bar{n}\,, \qquad v\bar{n} = e^{z}\,\bar{n}v\,.$$

No relation between n and $\bar{n}$ is found from the R-matrix; it can be obtained by requiring the self-consistency of the whole set of commutation rules and the compatibility with homomorphism property of the coproduct. The result is

$$n\bar{n} \;=\; e^{z}\bar{n}n\,,$$

and a straightforward change of variables shows that $Fun(E_q(2))$ is exactly the algebra given in (14; 15; 12).

The very same procedure applied to $E_\ell(2)$ yields the R matrix:

$$R_\ell = \begin{pmatrix} 1 & 1 & -1 & \\ & 1 & & \\ & & 1 & \\ & & & 1 \end{pmatrix},$$

which gives the equation

$$vn - nv = v^2 - v\,,$$

the conjugate one and those deduced by using of $v\bar{v} = \bar{v}v = 1$. For the relation involving n and $\bar{n}$ we proceed as previously, getting

$$[n, \bar{n}] = -\,(n + \bar{n})\,,$$

which completes the structure of the Hopf algebra $Fun(E_\ell(2))$. In (2) it has been shown that the dual of this last algebra, $Fun(E_\ell(2))'$, reproduces our initial algebra $E_\ell(2)$.

To conclude with, we shall consider the relationship between the two deformations of $E(2)$. Starting again with the two dimensional representation we have previously introduced, we can look for general expressions of the products vn and $n\bar{n}$ which are polynomial in the generators, selfconsistent and compatible with the coproduct. Using a straightforward recurrence procedure we find

$$vn = e^{z}\,nv + w\,(v^2 - v)\,, \qquad n\bar{n} = e^{z}\,\bar{n}n - w\,(n + \bar{n})\,,$$

where z and w are arbitrary parameters. The results for $Fun(E_q(2))$ and $Fun(E_\ell(2))$ are respectively recovered for $w = 0$ and $z = 0$. Moreover a nonvanishing w can be eliminated by rescaling the generators n and $\bar{n}$ and it can be directly verified that the whole structure is a genuine Hopf algebra which deforms the algebra of the representative functions of $E(2)$. However, for $z \neq 0$, we can define (16)

$$m = n - \frac{w}{1 - e^{z}}\,(v - 1)$$

and we see that v, m, $\bar{m}$ generate $Fun(E_q(2))$: the only singularity of the transformation is the value $z = 0$, where the deformation $Fun(E_\ell(2))$ arises. This completes the picture of the possible deformations of the two dimensional Euclidean group.

References

[1] N.W.Ashcroft, N.D.Mermin, *Solid State Physics*, (HRS International Editions, Philadelphia, PA, 1987).

[2] A. Ballesteros, E. Celeghini, R. Giachetti, E. Sorace and M. Tarlini, *An R-matrix Approach to the Quantization of the Euclidean Group* $E(2)$, preprint DFF-182/1/93, Firenze (1993).

[3] H.Bethe, Z.Phys. **71**, 205 (1931).

[4] F. Bonechi, E. Celeghini, R. Giachetti, E. Sorace and M. Tarlini, Phys. Rev. Lett. **68**, 3718 (1992).

[5] F. Bonechi, E. Celeghini, R. Giachetti, E. Sorace and M. Tarlini, J. Phys. A 25 (1992) L939, Phys. Rev. B in press.

[6] E. Celeghini, R. Giachetti, E. Sorace and M. Tarlini, J. Math. Phys. **31**, 2548 (1990); J. Math. Phys. **32**, 1155 (1991); J. Math. Phys. **32**, 1159 (1991).

[7] E. Celeghini, R. Giachetti, E. Sorace and M. Tarlini, "*Contractions of quantum groups*", in *Quantum Groups*, Lecture Notes in Mathematics n. 1510, 221, (Springer-Verlag, 1992).

[8] E. Celeghini, R. Giachetti, E. Sorace and M. Tarlini, Phys. Lett. B 280 (1992) 180.

[9] D.C. Mattis, Rev. Mod. Phys. **58** 361, (1986).

[10] R. Orbach, Phys. Rev. **112**, 309 (1958).

[11] N.Yu. Reshetikhin, L.A. Takhtadzhyan and L.D. Faddeev, Leningrad Math. J. **1**, 193 (1990).

[12] P. Schupp, P. Watts and B. Zumino, Lett. Math. Phys. **24**, 141 (1992).

[13] L.L. Vaksman and L.I. Korogodski, Sov. Math. Dokl. **39**, 173 (1989).

[14] S.L. Woronowicz, Lett. Math. Phys. **23**,251 (1991), and Comm. Math. Phys. **144**, 417 (1992).

[15] S.L. Woronowicz, Comm. Math. Phys. **149**, 637 (1992).

[16] S.L. Woronowicz, private communication.

Dipartimento di Matematica, Università di Bologna
Piazza di Porta S. Donato 5,
40127 Bologna, Italy.

A QUANTUM POINCARÉ GROUP AND THE DIRAC-COULOMB PROBLEM

L.C. Biedenharn

Marco Tarlini

ABSTRACT. The construction of quantum group realizations and the contraction procedure is discussed, briefly, and the κ-Poincaré quantum group construction is reviewed. Gauging the κ-Poincaré-Dirac equation yields the κ-Dirac-Coulomb problem which is confronted with recent quantum optics experimental data on the hydrogen atom to estimate the quantum parameter κ.

1 Introduction

One of the most interesting recent developments in quantum physics, and mathematics, is that of a *quantum group*. This structure was developed, nearly simultaneously, in several different ways: (a) in the statistical mechanics of Ising type models[1] (b) in inverse scattering theory[2]; and (c) by mathematicians[3] seeking to define non-commutative differential geometry. For a physicist, there are two essential new ideas in quantum groups (a) *deformation*: the Lie algebra structure constants are functions of the quantum parameter q such that for $q \to 1$, the standard Lie algebra is obtained and (b) *non-commutative co-multiplication*. To understand intuitively the latter concept it is helpful to recognize that the addition of angular momentum defines a *commutative* co-multiplication. In the q-group generalization, this addition depends on order and becomes non-commutative. In consequence of the non-commutative co-multiplication the coordinates of the carrier space become non-commutative.

For *semi-simple* Lie algebras, the most important properties of the representation theory have been generalized to their quantum analogs[1–6] and the theory and applications are by now well-developed.[7–9] The situation is very different for *non-semisimple* groups where the quantum deformation structure is not unique.[10] Some of the most fundamental, and

E. A. Tanner and R. Wilson (eds.), Noncompact Lie Groups and Some of Their Applications, 487–495.

interesting, symmetries in physics - most importantly the Poincaré symmetries of the present conference - fall into this category.

There is accordingly no unique quantum group extension of the Poincaré group and there are two distinct approaches in the literature, one based on inducing a quantum Poincaré group from the symmetries of non-commutative Minkowski space[11] and the second, based on contraction of the (unique) Drinfeld-Jimbo realization of the $SO_q(3,2)$ quantum group[12,13] We will survey the latter procedure here. As an application we shall extend this (latter) quantum Poincaré group to the problem of the hydrogen atom – the Dirac-Coulomb problem – in order to confront this new symmetry structure with the highly precise recent measurements of this physical system.

2 The Construction of Quantum Group Realizations

Professor Mackey in his lecture at this conference has emphasized the extraordinary utility of quadratic boson algebras in developing (classical) symmetry group structures. It would appear useful to indicate here just how this occurs in the standard (Lie group) case and how this situation generalizes to quantum groups via the generalization to "quantum bosons" (q-bosons).

Let us consider the $n\times n$ defining matrix realization of the Lie algebra su_n. Let m_α, ($\alpha = 1...n^2-1$) denote the $n \times n$ matrices over $\mathbf{C}$ corresponding to the abstract generators g_α. Define the n commuting boson operators a_i and $\bar{a}_i$ (the adjoint boson operator), $i = 1...n$ by the commutation relations: $[\bar{a}_i, a_j] = \delta_{ij}$ with all other commutators vanishing.

Then we have the *Jordan-Schwinger* mapping: $\mathcal{J} : g_\alpha \rightarrow \mathcal{J}(g_\alpha)$, where $\mathcal{J}(g_\alpha) \equiv \sum_{i,j}(m_\alpha)_{ij}\bar{a}_j a_i$.

Theorem : The Jordan–Schwinger mapping is a homomorphism of Lie algebras, that is,

$$\mathcal{J}([g_\alpha, g_\beta]) = [\mathcal{J}(g_\alpha, \mathcal{J}(g_\beta)].$$

For the quantal angular momentum group ($SU(2)$) this construction yields *every finite dimensional unitary irrep.*[14] This result extends immediately to all $SU(n)$ if we extend the map $\mathcal{J}$ to n commuting boson algebra copies of the original map (formally speaking this extension uses (commuting) co-multiplication of the (commuting) independent boson copies). This construction accordingly yields *all* unitary irreps of *all* $SU(n)$, and moreover of its subgroups. It is indeed of remarkable utility as Professor Mackey emphasized.

As one might guess, there is a quantum group extension of these results using q-analog-boson operators[15–17]

Consider the *q–creation operator* a^q, and its Hermitian conjugate the *q–destruction operator* $\bar{a}^q$. The *q – boson vacuum ket* $|0>_q$ is defined by

$$\bar{a}^q|0>_q \equiv 0. \tag{2.1}$$

We postulate the algebraic relation

$$\bar{a}^q a^q - q^{\frac{1}{2}} a^q \bar{a}^q = q^{-\frac{N_q}{2}}, \tag{2.2}$$

where N_q is the (Hermitian) *number operator* with:

$$[N_q, a^q] = a^q, \tag{2.3}$$

$$[N_q, \bar{a}^q] = -\bar{a}, \text{ with } N_q|0> \equiv 0. \tag{2.4}$$

This algebra is a q-analog of the algebra of boson operators which is recovered in the limit $q \to 1$.

Using these q-analog bosons one can now define (for the $SU_q(n)$ quantum group) the q-analog of the Jordan-Schwinger map (based on the $n \times n$ matrix realization of the $su_q(n)$ quantum-group algebra). Moreover, by using the (non-commutative) co-multiplication for independent q-analog boson realizations, one obtains the most general realization *(for generic q)* and accordingly all finite dimensional unitary irreps of the quantum group $SU_q(n)$.

We remark that there is a generalization of this construction applicable to a non-compact realization of the symplectic group, $Sp(n)$, for both the standard Lie group and quantum group versions. However, one does not now obtain all unitary irreps. (For the standard Lie group case this construction is called the Dirac map[18].)

3 The Contraction Procedure for Obtaining Some Non-Compact Quantum Groups

Let us consider the quantum group $SU_q(2)$, which has the commutation relations:

$$[J_3, J_\pm] = \pm J_\pm, \text{ and } [J_+, J_-] = \frac{q^{J_3} - q^{-J_3}}{q^{\frac{1}{2}} - q^{-\frac{1}{2}}} = \frac{\sinh(zJ_3)}{\sinh(z/2)}, \tag{3.1}$$

where $z = \ln q$, with $q \in \mathbf{R}^+$. We define the contraction process for this quantum group to be the substitution:

$$\epsilon^{\frac{1}{2}} J_+ \to A, \;\; \epsilon^{\frac{1}{2}} J_- \to A^+, \;\; 2\epsilon J_3 \to H, \;\; \epsilon^{-1} z \to \omega, \tag{3.2}$$

followed by the limit: $\epsilon \to 0$, $z \to \infty$ such that $\epsilon^{-1} z = \omega$ is finite.

This limit yields a new quantum group whose generators obey the commutation relations:

$$[H, A] = [H, A^+] = 0, [A, A^+] = \frac{\sinh(\frac{\omega H}{2})}{\omega/2}. \tag{3.3}$$

The corresponding co-multiplication rules are found to be:

$$\begin{aligned} \Delta(H) &= \mathbf{1} \otimes H + H \otimes \mathbf{1}, \\ \Delta(A) &= e^{-\frac{\omega H}{4}} \otimes A + A \otimes e^{\frac{\omega H}{4}}, \\ \Delta(A^+) &= e^{-\frac{\omega H}{4}} \otimes A^+ + A^+ \otimes e^{\frac{\omega H}{4}}. \end{aligned} \tag{3.4}$$

This is a quantum group analog of the Heisenberg group.[19]

The quantum group $SU_q(2)$ admits a second, distinct, contraction procedure. Using the substitution, $J_1 = \frac{1}{2}(J_+ + J_-)$; $J_2 = \frac{1}{2}(J_+ - J_-)$, the commutation relations of $SU_q(2)$ become:

$$[J_1, J_2] = \frac{i}{2} \frac{\sinh(zJ_3)}{\sinh(z/2)}, [J_2, J_3] = iJ_1, [J_3, J_1] = iJ_2. \tag{3.5}$$

Now let us use the contraction: $\epsilon J_1 \to P_y$, $J_2 \to J$, $\epsilon J_3 \to P_x$ and $\epsilon^{-1} z \to \omega$. Taking the limit: $\epsilon \to 0, z \to \infty$ with $\epsilon^{-1}z = \omega$=finite, one obtains the new quantum algebra: $[P_x, P_y] = 0, [J, P_x] = iP_y, [J, P_y] = -\frac{i}{\omega}\sinh(\omega P_x)$. The co-product structure is found to be:

$$\begin{aligned} \Delta(J) &= e^{-\frac{\omega P_x}{2}} \otimes J + J \otimes e^{\frac{\omega P_x}{2}}, \\ \Delta(P_y) &= e^{-\frac{\omega P_x}{2}} \otimes P_y + P_y \otimes e^{\frac{\omega P_x}{2}}, \\ \Delta(P_x) &= \mathbf{1} \otimes P_x + P_x \otimes \mathbf{1}. \end{aligned} \tag{3.6}$$

This defines the quantum Euclidean group, $E_q(2)$, in two dimensions.[19]

4 The κ-Poincaré Quantum Group

The approach we shall discuss for obtaining a deformed Poincaré algebra is to apply the contraction process to the standard q-deformation of the (anti) de Sitter algebra $so(3,2)$. Taking the limit of the de Sitter radius $R \to \infty$ with an accompanying limit of the deformation parameter q, such that $\lim(R \ \ln q) = \kappa^{-1}$, one obtains the *$\kappa$-Poincaré quantum group.*[12]

[Confusingly, Dirac's quantum number in the Dirac-Coulomb problem is also called kappa, and to avoid confusion, we shall henceforth denote the *inverse* of the quantum group kappa parameter by ϵ and Dirac's quantum number, as customary, by κ.]

The κ-Poincaré quantum group has been developed by Lukierski, Nowicki and Ruegg[12] and extended by the Lodz group,[20] and by Nowicki, Sorace and Tarlini.[13] We follow ref.13 in summarizing this structure. The algebra structure is:

$$\begin{aligned} &[P_i, P_j] = 0 \,, && [P_i, P_0] = 0 \,, \\ &[M_i, P_j] = i\epsilon_{ijk}\, P_k \,, && [M_i, P_0] = 0 \,, \\ &[L_i, P_0] = iP_i \,, && [L_i, P_j] = i\epsilon^{-1}\delta_{ij} \sinh(\epsilon P_0) \,, \\ &[M_i, M_j] = i\epsilon_{ijk}\, M_\kappa \,, && [M_i, L_j] = i\epsilon_{ijk}\, L_\kappa \,, \\ &[L_i, L_j] = -i\epsilon_{ijk}(M_k \cosh(\epsilon P_0) - \frac{1}{4}\epsilon\, P_\kappa\, P_l M_l) \,. \end{aligned} \tag{4.1}$$

Here $P_\mu \equiv \{P_0, P_i\}$ are the deformed generators for energy and momenta, the M_i are the spatial rotation generators (they close on an undeformed Hopf co-commutative subalgebra), and the L_i are the deformed boost generators.

The coalgebra (Δ is the co-multiplication operation) and the antipode (S is the quantum group analog of an inverse) are:

$$\begin{aligned} &\Delta M_i = M_i \otimes I + I \otimes M_i, && \Delta P_0 = P_0 \otimes I + I \otimes P_0, \\ &\Delta P_i = P_i \otimes \exp(\frac{\epsilon P_0}{2}) + \exp(-\frac{\epsilon P_0}{2}) \otimes P_i, && \Delta L_i = L_i \otimes \exp(\frac{\epsilon P_0}{2}) + \\ &\exp(-\frac{\epsilon P_0}{2}) \otimes L_i + \frac{\epsilon}{2}\epsilon_{ijk}(P_i \otimes M_\kappa \exp(\frac{\epsilon P_0}{2}) + \exp(-\frac{\epsilon P_0}{2}) M_j \otimes P_\kappa), \end{aligned} \tag{4.2}$$

$$S(P_\mu) = -P_\mu,\ S(M_i) = -M_i,\ S(L_i) = -L_i + \epsilon\frac{3i}{2}P_i. \tag{4.3}$$

The two deformed invariant operators of the κ-Poincaré group are:

$$C_1 = \left(\frac{2}{\epsilon}\sinh(\frac{\epsilon P_0}{2})\right)^2 - P_iP_i, \quad C_2 = \left(\cosh(\epsilon P_0) - \frac{\epsilon^2 P_iP_i}{4}\right) W_0^2 - W_iW_i \tag{4.4}$$

where $W_0 = P_iM_i$ and $W_i = \frac{1}{\epsilon}\sinh(\epsilon P_0)M_i + \epsilon_{ijk}P_jL_k$.

The Lorentz Lie algebra representation using the Dirac γ-matrices is also a representation of the Poincaré Lie algebra with the four-momentum generators represented by zero. This representation also fulfills the commutation relations (4.1) of the κ-Poincaré quantum group. (This is not surprising because in general the lowest dimensional representation of a Lie algebra is a representation of the corresponding quantum deformation.) We can then "add" a spinless representation and the Dirac γ-representation by using the co-multiplication of the κ-Poincaré quantum group to obtain the desired spinorial representation for $s = \frac{1}{2}$. We obtain:

$$\mathcal{P}_\mu = P_\mu, \qquad \mathcal{M}_i = M_i + m_i, \qquad \mathcal{L}_i = L_i + \exp\left(-\epsilon\frac{P_0}{2}\right) l_i - \frac{\epsilon}{2}\epsilon_{ijk}m_jP_k, \tag{4.5}$$

where $m_i = \frac{i}{4}\epsilon_{ijk}\gamma_i\gamma_k$ and $l_i = -\frac{i}{2}\gamma_0\gamma_i$.

The κ-deformed Dirac operator – call it $\mathcal{D}$ – is required to be invariant under the generators $\{\mathcal{P}_\mu, \mathcal{M}_i, \mathcal{L}_i\}$ in (4.5). From the commutation relations (and properties of the Dirac matrices) one can verify that the operator $\mathcal{D}$:

$$\mathcal{D} \equiv -\exp(-\frac{\epsilon P_0}{2})\gamma_iP_i + \gamma_4\frac{1}{\epsilon}\sinh(\epsilon P_0) - \frac{i\epsilon}{2}\gamma_4P_iP_i, \tag{4.6}$$

obeys the required relations: $[\mathcal{D}, \mathcal{L}_i] = [\mathcal{D}, \mathcal{M}_i] = [\mathcal{D}, \mathcal{P}_\mu] = 0$.

Moreover, the square of $\mathcal{D}$ can be found to be:

$$\mathcal{D}^2 = C_1(1 + \frac{\epsilon^2}{4}C_1) = -\frac{4}{3}C_2. \tag{4.7}$$

Thus the κ-Dirac equation may be written in the explicit form:

$$\left(-\exp(-\frac{\epsilon P_0}{2})\gamma_iP_i + \gamma_4\frac{1}{\epsilon}\sinh(\epsilon P_0) - \frac{i\epsilon}{2}\gamma_4P_iP_i\right)\psi = m(1 + \frac{\epsilon^2m^2}{4})^{\frac{1}{2}}\psi, \tag{4.8}$$

where $m^2 = C_1$. (Note that in the limit $\epsilon \to 0$, one recovers the usual Dirac equation.)

Let us expand the κ-Dirac equation, (4.8) (multiplied for convenience by $\exp\left(\frac{\epsilon P_0}{2}\right)$ on both sides), in powers of ϵ. The resulting equation, to an error $\approx \epsilon^2$, is:

$$[(\gamma_4P_0 - \gamma_iP_i) + \frac{\epsilon}{2}(\gamma_4(P_0^2 - P_iP_i) - mP_0)]\psi = m\psi. \tag{4.9}$$

5 The κ-Dirac-Coulomb Equation

In order to obtain the κ-Dirac-Coulomb equation, it is necessary now to gauge eq.(4.9), thereby introducing the Coulomb potential. In general, gauging of quantum groups is a current research problem,[21] but for the $U(1)$ group of electromagnetism and a one-body equation (no co-multiplication) the straightforward gauging $P_\mu \to P_\mu - eA_\mu$ is quite satisfactory.[22] (Note that in eq.(4.9), unlike eq.(4.8), there are no operator ordering problems after the gauging that would require symmetrization.)

Accordingly, we gauge eq.(4.9), $P_0 \to H + \dfrac{\alpha Z}{r}$, introducing the (attractive) Coulomb potential. We obtain the first order κ-Dirac Coulomb equation:

$$H\psi = \{(\gamma_4\gamma_i P_i + m\gamma_4 - \frac{\alpha Z}{r}) + \frac{\epsilon}{2}((H + \frac{\alpha Z}{r})^2 - P_i P_i - \gamma_4 m(H + \frac{\alpha Z}{r}))\}\psi. \tag{5.1}$$

To an error of order ϵ^2, we can identify H in the perturbation terms with the Dirac Hamiltonian ($H_{\mathcal{D}}$, eq.(5.1) with $\epsilon = 0$). The perturbation conserves angular momentum and parity.

Since the unperturbed (bound-state) Dirac-Coulomb problem has solutions ($H_{\mathcal{D}} \to E_{\mathcal{D}}$, $N =$ principal quantum number, $\kappa =$ Dirac's quantum number) degenerate in the sign of κ (except for $\kappa = -N$), one would normally apply 2×2 degenerate perturbation theory for an order ϵ calculation of the shifted energy levels. However, parity is conserved by the perturbation, so that the 2×2 matrix becomes diagonal. The diagonal matrix elements of the perturbation in eq.(5.1) are to be taken between eigenstates of the unperturbed Dirac-Coulomb Hamiltonian. Using the quadratic Dirac Hamiltonian for $P_i P_i$ we can put the perturbation in a more convenient form. Denoting the perturbation in eq.(5.1) (the term multiplied by $\frac{\epsilon}{2}$) by H_{pert} we find: (Dirac's notation)

$$H_{pert} = m^2 - \frac{i\alpha Z}{r^2}\rho_1 \boldsymbol{\sigma}\cdot\hat{\boldsymbol{r}} - E_{\mathcal{D}} m\rho_3 - m\rho_3\frac{\alpha Z}{r}. \tag{5.2}$$

To our initial surprise, matrix elements of eq.(5.2) were found to be *zero*. In fact, the perturbation is *identically zero*. It is easy to show this: with (5.2) in operator form,

$$H_{pert} = m^2 - \frac{1}{2}\{H_{\mathcal{D}}, m\rho_3\} - m\rho_3\frac{\alpha Z}{r}. \tag{5.3}$$

Using the Dirac Hamiltonian eq.(5.3) is seen to vanish. *The first order correction to the κ-Poincaré Dirac-Coulomb problem is identically zero.*[23] This is quite remarkable and indicates that the κ-Poincaré quantum group tends to make minimal changes in standard structures.

6 Further Remarks and Conclusion

In an attempt to determine some experimental limit on the κ-Poincaré length scale, it would be natural now to proceed to the second-order corrections in the κ-Poincaré Dirac-Coulomb problem. It is easily seen from eq.(4.8) that this procedure must fail, *since the second-order perturbation is singular.* This singularity comes from the term $\frac{1}{6}\left(\frac{\alpha Z}{r}\right)^3$ introduced via $\frac{1}{\epsilon}\sinh(\epsilon P_0)$ in (4.8), and cannot be eliminated. (The full second-order correction is complicated and need not be given here.)

Thus we are at an impasse in this attempt to bound the κ-Poincaré length scale experimentally. We remark, however, that there is a rather speculative way to proceed further.

One of the motivations underlying the quantum group approach is that the quantum parameter may possibly serve as a convergence factor. The singular terms in the κ-Poincaré-Dirac-Coulomb problem enter through exponentials of the P_0 operator, which suggests that finite (time) displacements occur. Thus it is not unreasonable to hope that the κ-Poincaré length scale may somehow cut-off the singular terms in the Coulomb problem. For the singular second order perturbation term the cut-off enters logarithmically, and is thus relatively insensitive to the cut-off. Using the ground state second-order (cut-off) energy shift determines, in fact, a length-scale that is self-consistent (the cut-off scale in the singular integral is of the same order of magnitude as that scale determined by the cut-off energy shifts).

The ground-state second-order (cut-off) energy shift is found to have the order of magnitude:

$$\Delta E(1s_{\frac{1}{2}}) \approx (m\epsilon)^2 \cdot (m\alpha^6). \tag{6.1}$$

Let us now compare this estimated shift, $\Delta E(1s_{\frac{1}{2}})$, with the observed experimental limits[24] on the accuracy with which the $1s_{\frac{1}{2}}$ eigen-energy is known. A reasonable estimate from the results in ref.24 is that any deviation from the eigen-energy of the $1s_{\frac{1}{2}}$ state is less than approximately $10^{-3} MHz$. Since the Rydberg, $\frac{1}{2}m\alpha^2$, is 3289 THz this implies:

$$(m\epsilon)^2 \left(m\alpha^6\right) \approx \left(m\alpha^2\right) \times 10^{-12}, \quad \text{or} \quad m\epsilon \approx 10^{-2}. \tag{6.2}$$

This implies that the κ-Poincaré length scale is: $\epsilon \approx 10^{-13} cm$ or smaller. This self-consistent estimate of the κ-Poincaré length scale is both speculative and not very accurate, despite the high experimental precision, in direct consequence of the remarkable properties found for the κ-Poincaré Dirac-Coulomb problem which has no first-order perturbation and a singular second-order perturbative effect.

7 Acknowledgements

We would like to thank Professor Mark Raizen, University of Texas (Austin) for discussions on precision measurements in atomic hydrogen and Professor Berndt Mueller, Duke University, for discussion on the κ-Dirac-Coulomb problem.

REFERENCES AND FOOTNOTES

1. M. Jimbo, *Lett.Math.Phys.* **10**, (1985) 63; Commun. Math. Phys. **102**, (1986) 537.

2. E.K. Sklyanin, Funct.Anal.Appl. **16**, (1982) 262.

3. A. Connes, *Publ.Math.IHES* **62**, (1985) 257.

4. V.G. Drinfeld, *Proceedings of the ICM*, Berkeley, CA, edited by A.M. Gleason (American Mathematical Society, Providence, RI, 1986), p. 798.

5. S. Woronowicz, Publ. RIMS (Kyoto Univ.) **23**, 117 (1987); Commun. Math. Phys. **122**, 125 (1989).

6. L. Faddeev, N. Yu. Reshetikhin, and L. Takhtajan, in *Braid Group, Knot Theory and Statistical Mechanics*, edited by C.N. Yang and M.L. Ge (World Scientific, Singapore, 1989).

7. A.N. Kirillov and N. Yu. Reshetikhin, LOMI preprint E-9-88 (1988).

8. L.C. Biedenharn and M. Tarlini, Lett. Math. Phys. **20**, 271 (1990).

9. L.A. Takhtajan, Adv. Stud. Pure Math. **19**, 435 (1989).

10. E. Celeghini, R. Giachetti, E. Sorace and M. Tarlini, J. Math. Phys. **31**, 2548 (1990); **32**, 1155, 1159 (1991); "Contraction of quantum groups" in Lect. Notes in Math. n. 1510, pg. 221 (Springer-Verlag, Berlin 1992).

11. O. Ogievetsky, W.B. Schmidke, J. Wess and B. Zumino, "q-Deformed Poincaré Algebra" MPI-Ph/91-98 November, 1991 preprint.

12. J. Lukierski, A. Nowicki and H. Ruegg, Phys. Lett. **B 293** (1992) 344.

13. A. Nowicki, E. Sorace and M. Tarlini, "The Quantum Deformed Dirac Equation from the κ-Poincaré Algebra", preprint DFF 177/12/92 (Firenze, Italy), Phys. Lett. **B** (in press).

14. L.C. Biedenharn and J.D. Louck, *Angular Momentum in Quantum Physics Theory and Application*, Vol. 8, *Ency. of Math. and Appl.* (G.-C. Rota, Ed.) Cambridge University Press (Cambridge, U.K.) 1989.

15. L.C. Biedenharn, *J. Phys. A: Math. Gen.* **22**, (1989) L873.

16. A.J. Macfarlane, *J. Phys. A: Math. Gen.* **22**, (1989) 4581.

17. C.P. Sun and H.C. Fu, *J. Phys. A: Math. Gen.* **22**, (1989) L983.

18. N. Mukunda, H. van Dam and L.C. Biedenharn, *Relativistic Models of Extended Hadrons Obeying a Mass-Spin Trajectory Constraint*, Lecture Notes in Physics #165, Springer-Verlag Berlin (1982) cf. p.16.

19. E. Celeghini, R. Giachetti, E. Sorace and M. Tarlini, *J.Math.Phys.* **31** 2548.

20. S. Giller, J. Kunz, P. Kosinski, M. Majewski and P. Maslanka, "*On q-covariant Wave Functions*", Lodz University Preprint, August 1992.

21. L. Castellani, "Gauge Theories of Quantum Groups", preprint DFTT-19/92 (Torino, Italy).

22. We wish to thank Leonardo Castellani for discussion on this point.

23. In a preprint just received, J. Lukierski, H. Ruegg and W. Rühl, "From κ-Poincaré Algebra to κ-Lorentz Quasigroup: A Deformation of Relativistic Symmetry", preprint KL-TH-92/22 find that the first order correction vanishes in the *non-relativistic* approximation.

24. M. Weitz, F. Schmidt-Kaler, and T.W. Hänsch, Phys.Rev.Lett. **68**, 1120 (1992).

Center for Particle Physics,
Department of Physics,
University of Texas at Austin,
Austin, TX 78712

INFN, Sezione di Firenze
Dipartimento di Fisica
Universitá degli Studi di Firenze
Largo E. Fermi 2, 50125
Firenze, Italy